Chicago Public Schools

Wonderful opportunities in industry lie ahead for these two trainees who are working on a numerical control system for the automated drill press. As craftsmen in today's industry they will be in position to take advantage of the new developments in automatic controls that provide greater speed and accuracy in many manufacturing processes. As craftsmen who know the fundamentals of their trade and the basic mathematics that underlie numerical control systems they can become key men. They will be in positions to make the best use of machines and materials and to improve the cost picture by increasing efficiency and eliminating needless expense. But to be able to achieve they must master the skills of their trade and gain an understanding of related fields, especially mathematics.

Mathematics
for Industry

by
S. E. RUSINOFF

 American Technical Publishers, Inc.
Alsip, Illinois 60658

The tables of "Natural Trigonometric Functions" and "Logarithms of Numbers" appearing in the back of this book are reprinted from *Machinery's Handbook*, copyright by The Industrial Press, 148 Lafayette Street, New York.

COPYRIGHT, © 1949, 1958, 1968

BY AMERICAN TECHNICAL PUBLISHERS, INC.

ALL RIGHTS RESERVED

3456789-68-9

LIBRARY OF CONGRESS CATALOG CARD NO. 68-14731
ISBN: 0-8269-2200-7

NO PORTION OF THIS BOOK MAY BE REPRODUCED BY THE MIMEOGRAPH PROCESS OR BY ANY OTHER PROCESS WITHOUT PERMISSION OF THE PUBLISHERS

PRINTED IN THE UNITED STATES OF AMERICA

PREFACE

This text presents in simplified form the mathematics most often needed in the field of practical engineering and machine shop practice. It concentrates on actual problems of engineering, shop, and drafting room. It applies arithmetic, algebra, geometry, trigonometry, and their various phases to jobs encountered in everyday industry.

A conspicuous quality of the book, and one that strongly recommends it, is its simplicity and clarity. The emphasis being on actual practice, the use of simple terms and expressions to describe methods and procedures, and numerous illustrations throughout the text to assist the student and instructor.

The text itself is a direct outgrowth of many years of not only classroom teaching of students engaged in industry, but also of constant contact with shopmen for consultation. The material used in this text and the methods outlined in it rely heavily upon up-to-date practices in industry. As a result of experience in being called upon to assist in solving problems growing out of a particular job at hand, the author has had the opportunity of converting his classroom into a clearing center. Through it there has flowed an exchange of valuable information covering views and experiences of experts in the various branches of industry. The instruction proper in the text has been implemented with the findings of these meetings.

The first two chapters present a review of shop arithmetic before proceeding with the balance of the text material. Simplified algebra is treated in the following two chapters. Then follows plane geometry in the fifth and sixth chapters, which present a review of the principles of plane geometry and a brief treatment of solid geometry. Shop trigonometry and its application in the solution of numerous problems chosen from actual practice are treated extensively in the seventh chapter. Chapters VIII and IX treat the application of shop mathematics to screw threads and gears. The usefulness of logarithms in the solution of shop problems is described in Chapter X. The operation of the slide rule is illustrated in Chapter XI. Chapter XII demonstrates the application of simplified mathematics in problems dealing with engineering computations such as strength of materials. Chapter XIII explains the use of graphs as a convenient method for showing relationships between facts. The mechanisms required for computing deviations of a process or product from prescribed conditions is discussed in Chapter XIV.

Chapter XV deals with inspection and quality control in manufacturing. Problems that confront supervisory personnel in selecting tools and equipment are covered in a new chapter, Chapter XVI, Tool Engineering Mathematics. The considerations of cost, efficiency, output, depreciation, service life, and other factors that influence industrial decisions are covered by practical examples in this chapter. The chapter on graphs has been rewritten to include new types.

A self-examination has been added at the front of the book. This enables the reader to detect the areas where he needs review and to refresh his knowledge. Shop problems have been used to bring the material to a practical level and to arouse interest.

Useful tables, including natural trigonometric functions and common logarithms, are given in the Appendix. These tables will furnish users of the book with the data they need in solving many different problems.

S. E. RUSINOFF

Chicago, Illinois

CONTENTS

CHAPTER	PAGE
To the Student—A Self-Test	IX

A test on practical problems to let the student evaluate his proficiency in basic mathematics.

I. Review of Arithmetic 1

Common fractions and a method for checking fundamental arithmetic operations. Practice problems applicable to the shop and toolroom on page 20.

II. Review of Arithmetic—*Continued* 29

Working with decimal fractions, square root, and cube root. Practice problems applicable to the shop and toolroom on page 48.

III. Algebra 58

The fundamentals of algebra and algebraic expressions and the solution of equations with one unknown. Practice problems applicable to the shop and toolroom on page 84.

IV. Algebra—*Continued* 92

Ratios, proportions, percentages, and the solution of equations having more than one unknown. Algebraic principles and problems applied to the shop and toolroom on page 114. Practice problems applicable to the shop and toolroom on page 128.

V. Geometry 142

The principles, important theorems, and figures of plane geometry. Practice problems applicable to the shop and toolroom on page 172.

VI. Geometry—*Continued* 183

The measurement of plane figures and problems in construction. Practice problems applicable to the shop and toolroom on page 219.

VII. Trigonometry 232

The principles and functions of trigonometry and its practical shop application. Practice problems applicable to the shop and toolroom on page 282.

VIII. Screw Threads 292

Definitions, types and measurement of standard screw threads. Also, checking of thread gages and cutting screw threads on a lathe. Practice problems applicable to the shop and toolroom on page 315.

CHAPTER	PAGE
IX. GEARS	321

An explanation of the uses, types, and principles of gears in the shop, and the mathematics involved in computing gear problems. Practice problems applicable to the shop and toolroom on page 353.

X. LOGARITHMS 359

The use of this rapid means of solving calculations employing multiplication, division, raising numbers to various powers, and extracting roots. Practice problems applicable to the shop and toolroom on page 368.

XI. THE SLIDE RULE 372

The employment of the slide rule in rapidly calculating mathematical processes, such as proportions, trigonometry, multiplication, division, logarithms, etc. Practice problems applicable to the shop and toolroom on page 394.

XII. ENGINEERING COMPUTATIONS 402

Methods followed in solving engineering problems frequently encountered in the shop, such as strength of materials, efficiency of machines, estimating time and power required for cutting operations, etc. Practice problems applicable to the shop and toolroom on page 425.

XIII. USE OF GRAPHS IN SOLUTION OF ENGINEERING PROBLEMS . 432

Employment of the graph as a convenient method for showing relationships between facts. Practice problems applicable to the shop and toolroom on page 436.

XIV. COMPUTATION OF AUTOMATIC CONTROLS FOR AUTOMATION . 438

The mechanisms and mathematics required for detecting and measuring deviations of a process or product from prescribed conditions. Practice problems applicable to the shop and toolroom on page 456.

XV. PROBLEMS IN INSPECTION AND QUALITY CONTROL 457

The operations and mechanisms involved in inspecting and controlling the quality of a product.

XVI. TOOL ENGINEERING MATHEMATICS 479

The mathematics of operating costs, output, depreciation, efficiency, service life, and other factors as they affect selection of machine tools and equipment.

ANSWERS TO SELF-TEST PROBLEMS 493
APPENDIX OF USEFUL TABLES 495
INDEX 569

A SELF-EVALUATING TEST

TO THE STUDENT

The crying need of industry today is for people with a good background in basic arithmetic, algebraic equations, and intuitive geometry but before you say "I've already had arithmetic, algebra, and geometry" test yourself on the following self-evaluating examination covering the basic principles in the first few chapters of this book.

Grade yourself by counting off two points for each problem omitted or answered incorrectly. Subtract this total from 100 and if your grade is below 80 you will know that you need a review of the four chapters of this book covered in the tests. (Answers on page 493.)

SELF-TESTING PROBLEMS PERTAINING TO CHAPTER I. (ARITHMETIC)

1. Determine the overall length of the link in Fig. 1. (All linear dimensions in the following problems are in inches unless otherwise specified.)

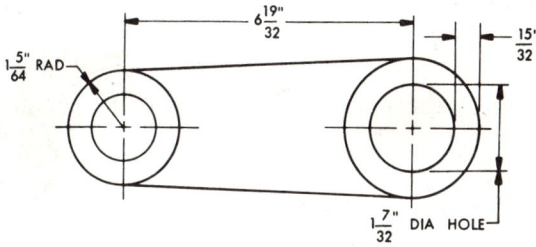

Fig. 1.

2. If the overall length of the link is $9\frac{19}{64}$ what is the center distance between the two end holes? See Fig. 1.
3. What is the radius at the right end of Fig. 1?
4. Determine the dimensions A and B shown in Fig. 2.
5. If A in Fig. 2 is $7\frac{9}{32}$ what would be dimension B?

ix

6. If B in Fig. 2 is $\frac{13}{16}$ what would A become?

7. If A in Fig. 2 is $8\frac{5}{8}$ what would the dimension $4\frac{37}{64}$ refer to?

8. A in Fig. 2 is $8\frac{1}{8}$ and B is $\frac{15}{16}$. What is the diameter of the center circle if all other dimensions in the figure are the same?

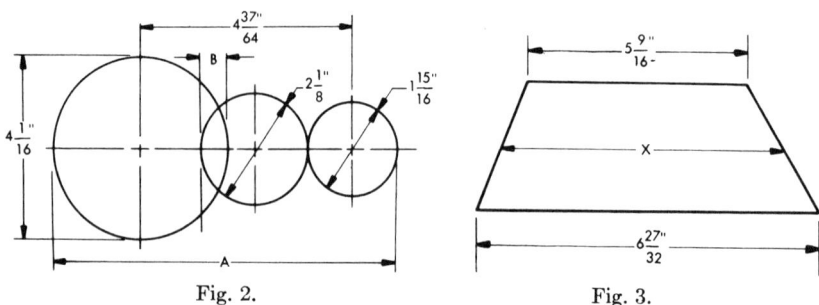

Fig. 2. Fig. 3.

9. What is the length of a line drawn parallel and equidistant from the top and bottom lines of the trapezoid shown in Fig. 3?

10. If dimension at the top is $6\frac{1}{8}$ instead of $5\frac{9}{16}$ what would be dimension X in Fig. 3?

11. If the top dimension were $5\frac{29}{32}$ what would dimension X become in Fig. 3?

12. If the dimension at the bottom is $6\frac{5}{32}$ instead of $6\frac{27}{32}$ in Fig. 3 what would be dimension X?

13. In Fig. 3. If the dimension at the bottom is $6\frac{11}{16}$ what is dimension X?

14. Find the taper per foot of the cone-shaped part shown in Fig. 4. (Note: After determining the correct value of the taper per inch multiply this value by 12 for the correct value of the taper per foot. The taper per inch is obtained by subtracting the small diameter of the part from the large diameter and dividing the result by the length of the part.)

5. If the large diameter is $4\frac{11}{32}$ instead of $4\frac{9}{32}$ what would be the taper per foot of the part in Fig. 4?

16. If the large diameter in Fig. 4 becomes $5\frac{15}{32}$ what does the taper per foot become?

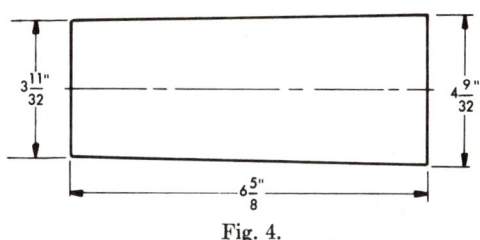

Fig. 4.

17. The small diameter in Fig. 4 is $3\frac{9}{32}$. What is the taper per foot?

18. If the small diameter is $3\frac{1}{32}$ instead of $3\frac{11}{32}$ what is the taper per foot of the cone in Fig. 4?

19. The length of the cone-shaped part in Fig. 4 is changed to $8\frac{3}{4}$. The other dimensions are the same. What is the correct taper per foot?

20. An aluminum-silicon alloy used for casting is about $\frac{45}{50}$ aluminum, $\frac{1}{15}$ silicon and the balance iron.
Find the weight of each metal in a casting if 317 lbs of the alloy were required to make 7 castings.

21. Assume a row of 19 rivets spaced $2\frac{29''}{32}$ apart are used to join two strips of metal. What is the distance between centers of the first and last rivets?

22. $24\frac{7}{16} \div 5\frac{7}{64} = ?$

23. $22\frac{11}{64} \div 4\frac{7}{32} = ?$

24. $17\frac{11}{64} \times 9\frac{5}{21} \times 7\frac{3}{5} = ?$

25. $21\frac{21}{64} \times \frac{37}{39} \times 7\frac{18}{23} \times 5\frac{9}{17} = ?$

26. An aluminum-copper alloy used for casting certain components consists of $\frac{45}{50}$ aluminum, $\frac{3}{50}$ copper and the balance of other elements, such as silicon, iron, etc.
Find the weight of aluminum, copper, and other elements combined in each component if 436 lbs of the alloy were required to make 9 castings.

27. A beryllium-copper alloy used for casting military aircraft components consists of $\frac{47}{50}$ copper, $\frac{2}{50}$ beryllium, and a small balance of other elements combined.
Find the weight of copper, beryllium, and other elements if 562 lbs of the alloy were required to make 17 castings.

SELF-TESTING PROBLEMS PERTAINING TO CHAPTER II
(ARITHMETIC)

28. Find dimensions A and B of the contour gage shown in Fig. 5.

29. If the radius at the top of Fig. 5 is changed from .215 to .227, what will the dimensions A and B become due to the change?

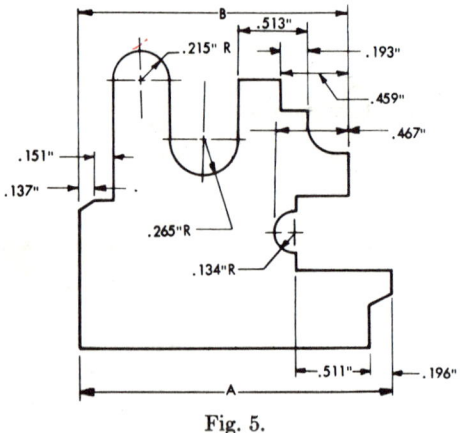

Fig. 5.

30. In Fig. 5. If radius .265 is changed to .247 what will be the dimensions A and B as a result of the change? All other dimensions remain the same.

31. If dimension B in Fig. 5 is equal to 2.017 what should radius .265 be changed to? All other dimensions remain the same. What will dimension A be under these conditions?

32. Calculate the unknown elements of a $\frac{5''}{8}$ square headed bolt and nut in accordance with the proportions shown in Fig. 6.

d = diameter of bolt = $\frac{5''}{8}$.
$A = 1.5000 \times d = ?$
$B = 1.4142 \times A = ?$
$C = .6666 \times d = ?$
$D = .875 \times d = ?$

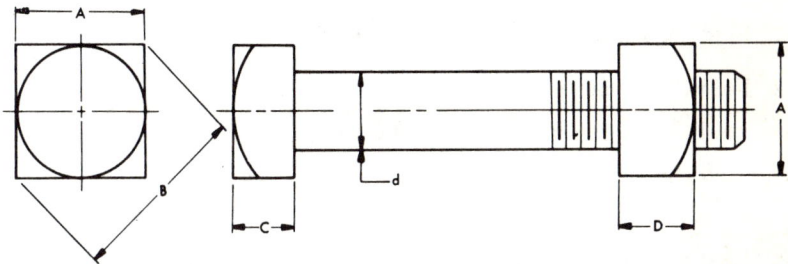

Fig. 6.

33. Calculate the dimensions A, B, C, and D if diameter of bolt d in Fig. 6 is equal to:

$\frac{1''}{2}$

$\frac{9}{16}$

$\frac{3}{4}$

(Note: Use same formula as indicated in Problem 32.)

34. Retaining three decimal places, what dimension would you recommend for the cylinder bore diameter if the calculated piston diameter is 3.485514″?

35. If the calculated piston diameter in Problem 34 has six decimal places, what would you recommend for the diameter if only three decimal places were wanted?

36. Convert the following decimal fractions to common fractions:
.890625
.90625
37. $637.0029 \times 34.073 = ?$
38. $.827503 \div 75.00408 = ?$
39. Convert the following common fractions to decimal fractions:
$$\frac{57}{64}$$
$$\frac{39}{64}$$
40. Find in the decimal equivalent chart the common fraction that is closest to .85084.
(Note: See Table 3 in the Appendix of Useful Tables in this book.)
41. Find the square root of $\frac{144}{625}$ by inspection.
42. Find the square root of 521,284.
43. Find the square root of 432 by factoring.
44. Find the cube root of 3249.6839 to three decimal places.
45. Find the cube root of $\dfrac{87\frac{29}{32}}{3\frac{5}{8}}$ to the nearest one decimal place.

SELF-TEST PROBLEMS PERTAINING TO CHAPTER III (ELEMENTARY ALGEBRA)

46. $(48\,ax^2y - 37a^3 \times Xy^2 + 62a^2x^3y^3) \times (14a^2x^2y^2 - 12x^3y + 25a^3y - 8a^3y) = ?$

47. $$\frac{276mn\ x^3y^4 + 368n^2\ x^4y^2 - 414mn^2x^5y^5 - 384m^2x}{23\ nx^2y - 32\ mxy^3}$$
$$\frac{{}^2y^6 - 512mnx^3y^4 + 576m^2n\ x^4y^7}{23nx^2y - 32mxy^3} = ?$$

48. Find the factors of the following expression: $a - 32a + 192 = ?$
49. Simplify or solve the following: $72.06 \times 10^8 = ?$
50. Simplify or solve: $.00653x10^9 = ?$
51. Evaluate $\dfrac{24x^3 - 6y^2}{4XZ} + \dfrac{(53)^2}{X + YZ} - \dfrac{X + Y^2 + Z^2}{7Y^3}$
if $X = 4$, $Y = 3$, $Z = 2$.
52. The area A of the cross sectioned portion shown in Fig. 7 can be

expressed by the formula $A = \dfrac{\pi d^2}{4} - bh$. Solve for h and d.

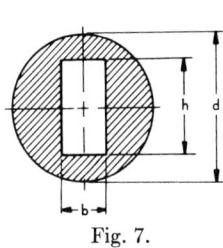

Fig. 7.

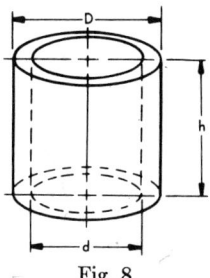

Fig. 8.

53. Calculate the cross-sectioned area A in Fig. 7 if d is equal to 1-3/4″, b is equal to 1/2″, and h is equal to 7/8″.

Use the same formula $A = \dfrac{\pi d^2}{4} - bh$

54. The volume V of the solid material of the hollow cylinder shown in Fig. 8 can be expressed by the formula $V = \dfrac{\pi}{4}(D^2 - d^2)h$. Solve for h, D, and d.

55. Calculate the volume V in Fig. 8 if $D = 3\text{-}1/2''$, $d = 1\text{-}5/8''$, and $h = 3/8''$.

Use the same formula as indicated in Problem 54.

SELF-TESTING PROBLEMS PERTAINING TO CHAPTER V (ELEMENTARY GEOMETRY)

56. Determine the following angles in Fig. 9:
ACB
ACD
DCE
CAF

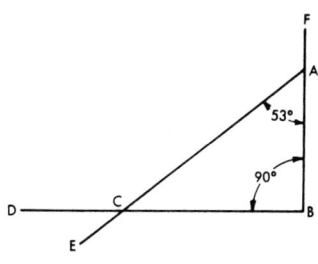

Fig. 9.

57. Find graphically the points of tangency of circular arcs *A*, *B*, *C*, and *D* as shown in Fig. 10.

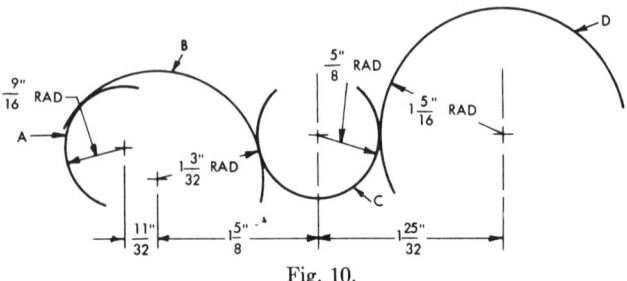

Fig. 10.

58. Determine the following in Fig. 11:
X
Y
Z
U
V

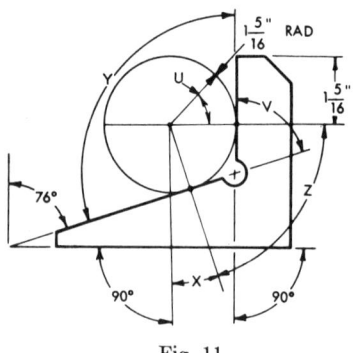

Fig 11.

59. Calculate the radii *R* and *r* in Fig. 12.

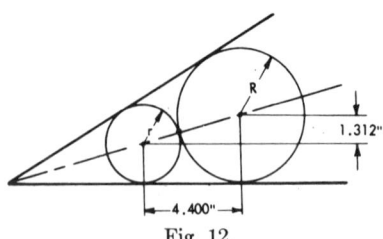

Fig. 12.

CHAPTER I
REVIEW OF ARITHMETIC

COMMON FRACTIONS

The importance of correctly using fractions is obvious to everyone and, above all, to men in industry. In acknowledgment of this fact, it seems proper to begin this text with a brief review of fractions.

DEFINITIONS

If an object or a unit is divided into a definite number of equal parts, we speak of one or more of these parts as *fractions* of the object, meaning parts of the whole.

Let us examine a 6-inch ruler. An inch usually is divided into a number of equal parts, say 16. A distance equal to seven of these parts is expressed as a fraction of an inch (see Fig. 1-1). It would be written, $\frac{7''}{16}$.

Another example of a fraction is a portion of the area of an octagon. Assume that three of the eight equal parts of an octagon are cross-sectioned, forming a solid area. This area is referred to as $\frac{3}{8}$ of the octagon, Fig. 1-2.

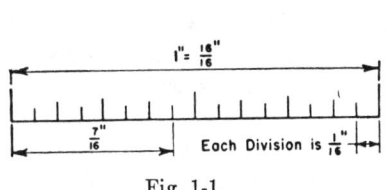
Fig. 1-1.

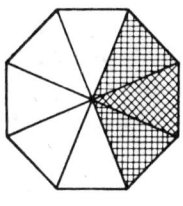
Fig. 1-2.

The number of equal parts into which a unit is divided is known as the *denominator*. It indicates the denomination or value of each of the parts with respect to the whole. The denominator of the 6-inch ruler, Fig. 1-1, is a sixteenth; the denominator of the octagon, Fig. 1-2, is an eighth.

The number of equal parts constituting a fraction of the whole is known as the *numerator*. Since seven parts of the whole are chosen in

Fig. 1-1, the numerator is 7, and the division is indicated by seven parts, each representing a sixteenth part of the whole. In Fig. 1-2, the numerator is 3, and the fraction is expressed as $\frac{3}{8}$.

Fractions expressed in the foregoing manner are known as *common fractions*. There is another method of expressing fractions of a whole, and that is the decimal method, which will be explained later.

A fraction may be regarded as an indicated division, in which the numerator is the dividend (the number to be divided) and the denominator the divisor (the number by which you divide). If the numerator in a common fraction is smaller than the denominator, the fraction is known as a *proper fraction;* if the numerator is equal to or greater than the denominator, the fraction then becomes an *improper fraction*.

Examples of proper fractions are $\frac{3}{8}, \frac{17}{32}, \frac{11}{16}$; examples of improper fractions are $\frac{16}{16}, \frac{11}{8}, \frac{39}{32}$.

A *mixed number* is composed of a whole number (often called an integer) and a proper fraction. Examples of mixed numbers are $1\frac{3}{4}, 7\frac{1}{8}, 10\frac{1}{16}$.

REDUCTION OF FRACTIONS

Reduction as used in mathematics means simply to change from one form to another form without altering the value. Thus a whole or mixed number may be changed to a fraction and an improper fraction may be changed to whole or mixed numbers without altering their value in any way.

Reduction of Whole and Mixed Numbers. A whole number is reduced to a fraction of a given denominator by first changing the unit 1 to a fraction of a given denominator and then multiplying the whole number by the numerator of the fraction. For example, let us reduce a whole number (7) to eighths.

$$1 = \frac{8}{8}, \frac{7 \times 8}{8} = \frac{56}{8}$$

Mixed numbers may be reduced to improper fractions by the following method: The whole number is multiplied by the denominator of the fraction and to this is added the numerator of the fraction, giving the new numerator of the resulting fraction over the original denominator. For example, we will reduce a mixed number $\left(5\frac{7}{16}\right)$ to sixteenths.

$$\frac{(5 \times 16) + 7}{16} = \frac{87}{16}$$

Reduction of Improper Fractions. Improper fractions can be reduced to whole or mixed numbers as follows: The numerator of the fraction is divided by the denominator, and the resulting quotient (or answer) is the whole number. If there is a remainder, it is shown as a fraction with the remainder becoming the numerator and the original denominator of the improper fraction becoming the denominator.

Example 1: Reduction of the improper fraction $\frac{49}{7}$ to a whole number.

$$49 \div 7 = 7$$

Example 2: Reduction of the improper fraction $\frac{53}{16}$ to a mixed number.

$$16 \overline{\smash{\big)}\,53} = 3\frac{5}{16}$$
$$\underline{48}$$
$$5$$

Example 3: Reduction of the improper fraction $\frac{71}{32}$ to a mixed number by long division.

$$32 \overline{\smash{\big)}\,71}2\frac{7}{32}$$
$$\underline{64}$$
$$7$$

Reduction of Fractions to Their Lowest Terms. When the numerator and the denominator of a fraction are not divisible by a whole common number or factor, they are prime to each other, and the fraction is at its lowest terms. Examples of such fractions are $\frac{3}{7}, \frac{4}{9}, \frac{6}{13}$.

Examples of fractions that can be reduced to their lowest terms are $\frac{12}{18}, \frac{15}{25}, \frac{21}{49}$.

The reduction of a fraction to its lowest terms consists of dividing the numerator and the denominator by the largest number common to both. This changes the appearance but not the value of the fraction. In reverse, both elements of the fraction can be multiplied by the same number without changing the actual value of the fraction. Such an operation is often desirable for reasons that will be explained later.

The three fractions, given previously as examples, can be reduced to their lowest terms in this manner:

$$1.\ \frac{12}{18} = \frac{12 \div 6}{18 \div 6} = \frac{2}{3}$$

$$2.\ \frac{15}{25} = \frac{15 \div 5}{25 \div 5} = \frac{3}{5}$$

$$3.\ \frac{21}{49} = \frac{21 \div 7}{49 \div 7} = \frac{3}{7}$$

FINDING FACTORS COMMON TO NUMERATOR AND DENOMINATOR

To facilitate the finding of a factor common to both the numerator and the denominator, the following suggestions are made.

1. All even numbers, such as 2, 4, 6, 8, etc., are divisible by 2.

   Examples: 18, 32, 60, 74.

2. All numbers ending with 5 or 0 are divisible by 5.

   Examples: 135, 1045, 6340.

3. If the last two digits are divisible by 4, or are zeros, the number itself is divisible by 4.

   Examples: 2328, 6300, 7384.

4. If the last three digits of a number are divisible by 8, or are zeros, the number itself is divisible by 8.

   Examples: 73608, 26512, 23000.

5. If the sum of all the digits of a number is divisible by 3, the number itself is divisible by 3.

   Examples: 585, 6342, 4857.

6. If the given number is even and the sum of its digits is divisible by 3, the number itself is divisible by 6.

Examples: 354, 5712, 8514.

7. If the sum of all the digits of a number is divisible by 9, the number itself is divisible by 9.

Examples: 2358, 7461, 435816.

The sum of the digits 435816 is equal to 27, which is divisible by 9; the given number itself, therefore, is divisible by 9. For the same reason, the given numbers, 2358 and 7461, are divisible by 9.

The Least Common Denominator. In many mathematical operations there are instances where two or more fractions appear with various denominators. It is necessary, therefore, to change these fractions to a form where all the denominators are the same without affecting their value. The procedure consists of finding a common denominator which is divisible by every denominator of the given fractions. It is also desired to find the smallest possible number which is divisible by all the given denominators. Such a number is called the *least common denominator*.

To maintain the real values of the given fractions, it is necessary to divide the found denominator by the denominator of each given fraction; then each resulting quotient is multiplied by each numerator to determine the numerators of the new fractions.

Numerous methods have been devised for the purpose of obtaining the least common denominator of a group of given fractions. One of these methods is illustrated in the paragraphs which follow. Let us take a few fractions, such as $\frac{3}{8}, \frac{7}{12}, \frac{17}{30}, \frac{21}{42}$, and convert them into equivalent fractions having a least common denominator.

PROCEDURE FOR FINDING THE LEAST COMMON DENOMINATOR

Each denominator is expressed as a product of its prime factors. (Prime factors consist of prime numbers that, when multiplied together, give a given product. A prime number is divisible only by one and by itself, e.g., 2, 3, 5, 7, 11, 13, 17, 19 and so on.) The procedure of finding the least common denominator consists of seeking the prime factors in an ascending order.

Step 1. The denominators of the given fractions, $\frac{3}{8}, \frac{7}{12}, \frac{17}{30}, \frac{21}{42}$, are rewritten in a column.

Step 2. The prime factors of each denominator are now determined. It is a valuable aid to recall suggestion 1, 2, and 5, previously given under the heading "Finding Factors Common to Numerator and Denominator." Taking the first denominator 8, we notice that it is an even number and for this reason it is divisible by 2. Thus 2 is our first prime factor. Dividing 2 into 8 we have 4. Since 4 is an even number it is also divisible by 2, and 2 becomes our second prime factor. Our third prime factor is also 2. ($2 \div 2 = 1$). This gives us $8 = 2 \times 2 \times 2$.

The second denominator, 12, is an even number and divisible by 2. Dividing 12 by 2 gives 6, which is also divisible by 2. Dividing 6 by 2 we have 3, which is a prime number. Thus $12 = 2 \times 2 \times 3$.

In following this procedure for the remaining denominators also, we now have:

$$8 = 2 \times 2 \times 2$$
$$12 = 2 \times 2 \times 3$$
$$30 = 2 \times 3 \times 5$$
$$42 = 2 \times 3 \times 7$$

Step 3. The least common denominator must contain as many prime factors of each denomination listed as may be necessary to make it divisible by each given denominator.

Prime factor 2 must be taken three times in order to satisfy the given denominator 8 and all the rest of the denominators.

Prime factor 3 may be taken once, as there is no given denominator which includes this prime factor more than once.

Prime factor 5 is only needed once for given denominator 30.

Finally, prime factor 7 may be included once to satisfy given denominator 42. Collecting the minimum number of chosen prime factors: $2 \times 2 \times 2 \times 3 \times 5 \times 7 = 840$. The least common denominator is 840.

Step 4. The next step is dividing 840 by each given denominator to obtain quotients, which in turn are multiplied by each given numerator, thus transforming given fractions into their equivalents, using a common denominator.

$$(840 \div 8) \times 3 = 315$$
$$(840 \div 12) \times 7 = 490$$
$$(840 \div 30) \times 17 = 476$$
$$(840 \div 42) \times 21 = 420$$

REVIEW OF ARITHMETIC

Step 5. The final step is the tabulation of the result.

The given fractions are $\frac{3}{8}, \frac{7}{12}, \frac{17}{30}, \frac{21}{42}$.

The equivalent fractions are $\frac{315}{840}, \frac{490}{840}, \frac{476}{840}, \frac{420}{840}$.

Example of Finding the Least Common Denominator: For the fractions $\frac{5}{12}, \frac{7}{30}, \frac{9}{63}, \frac{11}{84}$, find their equivalent fractions with the same denominator.

Steps 1 and 2.

$$12 = 2 \times 2 \times 3$$
$$30 = 2 \times 3 \times 5$$
$$63 = 3 \times 3 \times 7$$
$$84 = 2 \times 2 \times 3 \times 7$$

Step 3. The least common denominator is $2 \times 2 \times 3 \times 3 \times 5 \times 7 = 1260$.

Step 4. The new numerators are:

$$(1260 \div 12) \times 5 = 525$$
$$(1260 \div 30) \times 7 = 294$$
$$(1260 \div 63) \times 9 = 180$$
$$(1260 \div 84) \times 11 = 165$$

Step 5. The given fractions are $\frac{5}{12}, \frac{7}{30}, \frac{9}{63}, \frac{11}{84}$.

The equivalent fractions are $\frac{525}{1260}, \frac{294}{1260}, \frac{180}{1260}, \frac{165}{1260}$.

ADDITION OF FRACTIONS

When adding two or more common fractions, it is first necessary to examine their denominators. Fractions with unlike denominators must be treated differently than fractions with like denominators.

Adding Fractions with Like Denominators. When the denominators are similar, the numerators of the given fractions are added and the result becomes the final numerator, the denominator remaining unchanged. If the result is an improper fraction, the fraction must be converted into a whole or mixed number.

Example:
$$\frac{3}{8} + \frac{5}{8} + \frac{7}{8} + \frac{1}{8} = \frac{16}{8} = 2$$

Adding Fractions with Unlike Denominators. In the case of varied denominators, the least common denominator is found and the equivalent fractions are determined. The sum of the equivalent fractions is obtained by adding the numerators and using the least common denominator.

Example 1: Add the fractions $\frac{5}{8}, \frac{17}{32}, \frac{9}{16}, \frac{23}{64}$.

On examination, the least common denominator is found to be 64. Most fractions occurring in shop problems have denominators that are multiples of 2, and the least common denominator is found by inspection.

$$\frac{5}{8} + \frac{17}{32} + \frac{9}{16} + \frac{23}{64} = \frac{40 + 34 + 36 + 23}{64} = \frac{133}{64} = 2\frac{5}{64}$$

Example 2: Add the fractions $\frac{7}{24}, \frac{5}{21}, \frac{9}{35}, \frac{13}{49}$.

The least common denominator of these fractions is not readily found and the procedure previously explained must be used.

$$24 = 2 \times 2 \times 2 \times 3$$
$$21 = 3 \times 7$$
$$35 = 5 \times 7$$
$$49 = 7 \times 7$$

The least common denominator is $2 \times 2 \times 2 \times 3 \times 7 \times 7 \times 5 = 5880$.

$$\frac{7}{24} + \frac{5}{21} + \frac{9}{35} + \frac{13}{49} = \frac{1715 + 1400 + 1512 + 1560}{5880} = \frac{6187}{5880} = 1\frac{307}{5880}$$

REVIEW OF ARITHMETIC

Adding Mixed Numbers. When mixed numbers are added, the whole numbers are added first and then the fractions are added in accordance with the outlined rules.

Example:

$$3\frac{7}{64} + \frac{5}{8} + 2\frac{17}{32} + 12\frac{9}{16} = 17\frac{7 + 40 + 34 + 36}{64} = 17\frac{117}{64} = 18\frac{53}{64}$$

It should be noted that after adding the whole numbers, care must be exercised to carry their sum over to the final step, in which a possible improper fraction may be converted into a mixed number.

Another though longer method that can be used when adding mixed numbers is first to reduce all the given mixed numbers to improper fractions, and, when their sum is found, to convert the resulting improper fraction into a mixed number.

SUBTRACTION OF FRACTIONS

The subtraction of common fractions, as is the case with the subtraction of whole numbers, involves the use of the subtrahend and the minuend. The fraction from which you subtract is the *minuend*. The fraction to be subtracted is the *subtrahend*. The result or answer is the *difference* or *remainder*.

Subtracting Fractions with Like Denominators. Where the denominators are similar the operation is simply subtracting the numerator of the subtrahend from the numerator of the minuend.

Example:

$$\frac{12}{16} - \frac{7}{16} = \frac{12 - 7}{16} = \frac{5}{16}$$

Subtracting Fractions with Unlike Denominators. The method used to subtract fractions with varied denominators is similar to that used in adding fractions with varied denominators. The given fractions must first be reduced to their least common denominator and then the numerator of the subtrahend is subtracted from the numerator of the minuend.

Example 1: Subtract $\frac{15}{32}$ from $\frac{39}{64}$. The least common denominator can be determined by inspection.

$$\frac{39}{64} - \frac{15}{32} = \frac{39 - 30}{64} = \frac{9}{64}$$

Example 2: Subtract $\frac{13}{21}$ from $\frac{41}{56}$. The least common denominator must be determined.

$$56 = 2 \times 2 \times 2 \times 7$$
$$21 = 3 \times 7$$
$$\text{L.C.D.} = 2 \times 2 \times 2 \times 7 \times 3 = 168$$
$$\frac{41}{56} - \frac{13}{21} = \frac{123 - 104}{168} = \frac{19}{168}$$

Subtracting Mixed Numbers. In subtracting mixed numbers the procedure is to first subtract the fractions and then the whole numbers. In the case of mixed numbers whose fraction in the subtrahend is greater than in the minuend, the difference of the whole numbers is reduced by 1. The fraction in the minuend is increased sufficiently by this value of 1 to facilitate the subtraction.

Example 1:

$$3\frac{15}{64} - 1\frac{9}{16} = 1\frac{79 - 36}{64} = 1\frac{43}{64}$$

In the given example, $\frac{15}{64}$ is smaller than $\frac{9}{16}$. One (1) is added to $\frac{15}{64}$, making it $1\frac{15}{64}$ or $\frac{79}{64}$, reducing the whole number in the minuend to 2.

Example 2:

$$4\frac{11}{32} - \frac{7}{8} = 3\frac{43 - 28}{32} = 3\frac{15}{32}$$

Example 3:

$$103\frac{7}{64} - 3\frac{5}{16} = 99\frac{71 - 20}{64} = 99\frac{51}{64}$$

ADDITION AND SUBTRACTION OF FRACTIONS COMBINED

Two methods are given for the combined operation of adding and subtracting fractions.

Method 1. The least common denominator of all the given fractions is first found and the numerators are then collected in accordance with their signs.

REVIEW OF ARITHMETIC

Example:

$$5\frac{13}{32} + 2\frac{5}{16} - 4\frac{3}{8} = 1\frac{13 + 10 - 12}{32} = 1\frac{11}{32}$$

Method 2. A method that offers less possibility of error is one in which all the mixed numbers are converted into improper fractions, their common denominator found, and the numerators collected according to their signs.

Example:

$$3\frac{13}{32} + 2\frac{5}{16} - 4\frac{3}{8} = \frac{109}{32} + \frac{37}{16} - \frac{35}{8} = \frac{109 + 74 - 140}{32} = \frac{43}{32} = 1\frac{11}{32}$$

Examples of Methods 1 and 2:

1. $6\frac{7}{64} + 3\frac{5}{32} - 4\frac{11}{16} = 5\frac{7 + 10 - 44}{64} = 4\frac{37}{64}$, or $6\frac{7}{64} + 3\frac{5}{32} - 4\frac{11}{16} =$

$\frac{391}{64} + \frac{101}{32} - \frac{75}{16} = \frac{391 + 202 - 300}{64} = \frac{293}{64} = 4\frac{37}{64}$

2. $5\frac{9}{16} - \frac{7}{8} + 3\frac{15}{64} - 2\frac{1}{4} + \frac{23}{32} = 6\frac{36 - 56 + 15 - 16 + 46}{64} = 6\frac{25}{64}$, or

$5\frac{9}{16} - \frac{7}{8} + 3\frac{15}{64} - 2\frac{1}{4} + \frac{23}{32} = \frac{89}{16} - \frac{7}{8} + \frac{207}{64} - \frac{9}{4} + \frac{23}{32} =$

$\frac{356 - 56 + 207 - 144 + 46}{64} = \frac{409}{64} = 6\frac{25}{64}$

3. $4\frac{9}{28} - 2\frac{11}{21} + \frac{13}{15} = 2\frac{135 - 220 + 364}{420} = 2\frac{279}{420} = 2\frac{93}{140}$, or

$4\frac{9}{28} - 2\frac{11}{21} + \frac{13}{15} = \frac{121}{28} + \frac{53}{21} - \frac{13}{15} = \frac{1815 - 1060 + 364}{420} = \frac{1119}{420} =$

$2\frac{279}{420} = 2\frac{93}{140}$

4. $7\frac{3}{4} - 3\frac{5}{8} - \frac{7}{16} - 1\frac{9}{32} - \frac{41}{64} = 3\frac{48 - 40 - 28 - 18 - 41}{64} = 1\frac{49}{64}$, or

$7\frac{3}{4} - 3\frac{5}{8} - \frac{7}{16} - 1\frac{9}{32} - \frac{41}{64} = \frac{31}{4} - \frac{29}{8} - \frac{7}{16} - \frac{41}{32} - \frac{41}{64} =$

$\frac{496 - 232 - 28 - 82 - 41}{64} = \frac{113}{64} = 1\frac{49}{64}$

MULTIPLICATION OF FRACTIONS

The multiplication of numbers in general is to be regarded as simplified addition. If we are required to multiply 13 × 4, we take 13 four times (13 + 13 + 13 + 13), which is equal to 13 × 4 = 52.

In multiplication three terms are used, the multiplicand, the multiplier, and the product. The number or fraction multiplied is called the *multiplicand*. The number or fraction by which you multiply is called the *multiplier*. The result of the multiplication is called the *product*.

Multiplying a Fraction and a Whole Number. Since multiplication is simplified addition, if it is required to multiply $\frac{7}{16}$ by 3, we take $\frac{7}{16}$, the multiplicand, three times.

$$\frac{7}{16} \times 3 = \frac{7}{16} + \frac{7}{16} + \frac{7}{16} = \frac{21}{16} = 1\frac{5}{16}$$

The operation is simplified by multiplying the numerator by 3, using the same denominator.

$$\frac{7}{16} \times 3 = \frac{7 \times 3}{16} = \frac{21}{16} = 1\frac{5}{16}$$

Multiplying Fractions by Fractions. If the problem involves multiplication of two or more fractions, a procedure similar to multiplication of a fraction and a whole number is followed. The example $\frac{7}{16} \times 3$ may be written:

$$\frac{7}{16} \times \frac{3}{1} = \frac{7 \times 3}{16 \times 1} = \frac{21}{16} = 1\frac{5}{16}$$

The result was obtained by multiplying the numerators of the given fraction to get the numerator of the final fraction. The final denominator is the result of multiplication of the given denominators.

Examples:

1. $\dfrac{5}{16} \times \dfrac{3}{4} = \dfrac{15}{64}$

2. $\dfrac{3}{8} \times \dfrac{1}{2} = \dfrac{3}{16}$

REVIEW OF ARITHMETIC 13

3. $\dfrac{13}{32} \times \dfrac{3}{2} = \dfrac{39}{64}$

4. $\dfrac{5}{8} \times \dfrac{3}{4} \times \dfrac{7}{2} = \dfrac{105}{64} = 1\dfrac{41}{64}$

Multiplying Mixed Numbers. If one or all of the given factors are mixed numbers, they should be reduced to improper fractions first. Afterwards, they should be multiplied, as explained.

Examples:

1. $3\dfrac{1}{2} \times \dfrac{7}{8} = \dfrac{7}{2} \times \dfrac{7}{8} = \dfrac{49}{16} = 3\dfrac{1}{16}$

2. $\dfrac{3}{4} \times 2\dfrac{3}{8} = \dfrac{3}{4} \times \dfrac{19}{8} = \dfrac{57}{32} = 1\dfrac{25}{32}$

3. $2\dfrac{1}{2} \times 4\dfrac{5}{8} \times 7\dfrac{3}{4} = \dfrac{5}{2} \times \dfrac{37}{8} \times \dfrac{31}{4} = \dfrac{5735}{64} = 89\dfrac{39}{64}$

Cancellation Method. The multiplication of fractions can be simplified in many cases by first reducing the given fractions to their lowest terms. This is accomplished by dividing all possible numerators and denominators by a common number. It is well to note that any numerator of the given fractions can be divided by a number common to any denominator of the fractions. This operation, preliminary to multiplication, is called cancellation.

Examples:

1. $\dfrac{\cancel{7}^{1}}{\cancel{32}_{4}} \times \dfrac{\cancel{40}^{5}}{\cancel{49}_{7}} = \dfrac{5}{28}$

2. $2\dfrac{15}{64} \times 3\dfrac{19}{39} = \dfrac{\cancel{143}^{11}}{\cancel{64}_{8}} \times \dfrac{\cancel{136}^{17}}{\cancel{39}_{3}} = \dfrac{187}{24} = 7\dfrac{19}{24}$

3. $2\dfrac{21}{22} \times 4\dfrac{29}{34} \times \dfrac{17}{25} = \dfrac{\cancel{65}^{13}}{\cancel{22}_{2}} \times \dfrac{\cancel{165}^{\cancel{15}^{3}}}{\cancel{34}_{\cancel{2}_{1}}} \times \dfrac{\cancel{17}^{1}}{\cancel{25}} = \dfrac{39}{4} = 9\dfrac{3}{4}$

Note: In the case of mixed numbers, conversion to improper fractions must be completed before cancellation is attempted. Do not cancel a fraction that is only part of a given quantity. Consider the entire quantity.

DIVISION OF FRACTIONS

Division is generally an operation determining how many times one quantity is contained in another. As an example, when 24 is divided by 4, the purpose is to find how many times the quantity 4 is contained in the quantity 24. It is stated: $24 \div 4 = 6$.

Conversely, if we take the quantity 4 six times, we have $4 \times 6 = 24$. Because of this fact it can be stated that the division of numbers is multiplication in reverse. One process proves the other. This is also true when dividing numbers by fractions, fractions by numbers, and fractions by fractions.

In division the number or fraction to be divided is called the *dividend*. The number or fraction by which you divide is called the *divisor*. The result of division is called the *quotient*.

The division of fractions is simplified by the use of reciprocals. A *reciprocal* of a number is 1 divided by that number. The reciprocal of a fraction is also 1 divided by the fraction. As division is the opposite of multiplication, the reciprocal of a fraction is the fraction inverted. Thus, in dividing common fractions, change the fraction in divisor to its reciprocal and multiply.

Example 1:

$$7 \div \frac{3}{5} = \frac{7 \times 5}{3} = \frac{35}{3} = 11\frac{2}{3}$$

Dividing 7 by $\frac{3}{5}$ is the same as multiplying 7 by the reciprocal of $\frac{3}{5}$, which is $\frac{5}{3}$.

Example 2:

$$12 \div \frac{4}{9} = \frac{\cancel{12}^{3} \times 9}{\cancel{4}_{1}} = 27$$

Example 3:

$$\frac{13}{32} \div 7 = \frac{13}{32} \times \frac{1}{7} = \frac{13}{224}$$

Dividing by 7 is the same as multiplying by $\frac{1}{7}$.

REVIEW OF ARITHMETIC

Example 4:

$$\frac{27}{64} \div \frac{9}{16} = \frac{\cancel{27}^3}{\cancel{64}_4} \times \frac{\cancel{16}^1}{\cancel{9}_1} = \frac{3}{4}$$

In dividing mixed numbers, the numbers must be converted to improper fractions, after which one fraction is multiplied by the reciprocal of the other.

Examples:

1. $7\frac{23}{32} \div 4 = \frac{247}{32} \times \frac{1}{4} = \frac{247}{128} = 1\frac{119}{128}$

2. $5\frac{7}{16} \div \frac{29}{32} = \frac{\cancel{87}^3}{\cancel{16}_1} \times \frac{\cancel{32}^2}{\cancel{29}_1} = 6$

3. $32\frac{9}{64} \div 7\frac{9}{16} = \frac{2057}{64} \div \frac{121}{16} = \frac{\cancel{2057}^{17}}{\cancel{64}_4} \times \frac{\cancel{16}^1}{\cancel{121}_1} = \frac{17}{4} = 4\frac{1}{4}$

4. $\frac{1}{16} \div \frac{1}{32} = \frac{1}{\cancel{16}_1} \times \frac{\cancel{32}^2}{1} = 2$

5. $24 \div \frac{1}{16} = 24 \times 16 = 384$

6. $\frac{1}{64} \div 8 = \frac{1}{64} \times \frac{1}{8} = \frac{1}{512}$

7. $1 \div \frac{1}{64} = 1 \times 64 = 64$

8. $3 \div \frac{1}{32} = 3 \times 32 = 96$

CHECKING FUNDAMENTAL OPERATIONS OF ARITHMETIC BY THE PROCESS OF ELIMINATION OF NINES

A rapid method of checking fundamental arithmetical operations is explained as follows:

Let us assume a number, say 13. When 13 is divided by 9, the quotient will be 1 and the remainder will be 4. The remainder represents a value called the *excess of nines* of the given number. For the number assumed (13), it is seen that when as many nines as possible have been taken from it, the amount remaining (4) constitutes the excess of nines.

Let us take another number, say 23. Dividing 23 by 9, the quotient is 2 and the remainder is 5, which is the excess of nines of the given number.

A number, such as 7, when divided by 9, will yield 0 in the quotient and a remainder of 7, the excess of nines being 7.

A number, such as 27, when divided by 9, yields 3 as the quotient and a remainder of 0, signifying that the excess of nines in 27 is 0.

A shorter method of determining the excess of nines of a given number is to add all the digits of the number in any order, and, if the sum of the digits is less than 9, the excess of nines is equal to the total of the digits. If the sum of the digits in the given number is over 9, all possible nines are canceled out, and the remainder is the excess of nines.

Examples of determining the excess of nines:

1. Given the number 13, the excess of nines may be found by adding the digits $1 + 3$ and obtaining 4 (the excess of nines) instead of dividing 13 by 9 and getting a remainder of 4 as the excess of nines.
2. In the given number 23, the excess of nines is the digit 2 plus the digit 3 which equals 5.
3. In the given number 7, the excess of nines equals the digit 7.
4. In the given number 27, the excess of nines equals $2 + 7 = 9$. Then 9 is canceled out, leaving 0.
5. In the given number 678, the excess of nines equals the digits $6 + 7 + 8 = 21$. Adding the digits $2 + 1$ we obtain 3, the net excess of nines in 678.
6. In the given number 7,652, the excess of nines equals $7 + 6 + 5 + 2 = 20 = 2 + 0 = 2$. A still shorter method is to eliminate the $7 + 2$ immediately, leaving $6 + 5 = 11 = 1 + 1 = 2$.
7. In the given number 2,475, the excess of nines equals 0. $2 + 7 = 9$ and $4 + 5 = 9$, cancelling out.
8. In the given number 378,546, the excess of nines equals $7 + 8 = 15$. The $3 + 6 = 9$ and the $5 + 4 = 9$, cancelling out, and leaving $1 + 5 = 6$.
9. In the given number 756,843, the excess of nines equals $7 + 8 = 15 = 1 + 5 = 6$. The $6 + 3 = 9$ and the $4 + 5 = 9$, cancelling out.
10. In the given number 6,020,703, the net excess of nines equals 0. The $2 + 7 = 9$ and the $6 + 3 = 9$, cancelling out.

REVIEW OF ARITHMETIC

CHECKING MULTIPLICATION OF NUMBERS

Multiplication is sometimes checked by using the multiplier as the multiplicand and the multiplicand as the multiplier and then multiplying. This, however, is a long process and there is possibility of error. A certain check for multiplication is determining the excess of nines. The process is as follows:

The excess of nines of the multiplicand and the excess of nines of the multiplier are found and multiplied one by the other. If the result is equal to the excess of nines of the product, the multiplication of the given numbers has been correctly performed. If when the excess of nines of each of the given factors is multiplied one by the other the result produces a number greater than 9, the cancellation of nines is continued until the next excess is either less than 9 or is 0.

Example 1:

```
   23    excess of nines = 2 + 3 =  5
  ×15    excess of nines = 1 + 5 =  6
  ―――                                ―
  115               excess of nines = 30 = 3 + 0 = 3
   23
  ―――
  345    excess of nines = 3, the 4 + 5 = 9, cancelling out.
```

The excess of nines in the product is equal to that of the two factors multiplied by each other, which proves the operation to be correct.

Example 2:

```
    678    excess of nines = 6 + 7 + 8 = 21 = 2 + 1 =  3
   ×384    excess of nines = 3 + 8 + 4 = 15 = 1 + 5 =  6
   ――――                                                ――
   2712                         excess of nines = 18 = 1 +
   5424                                   8 = 9 = 0
   2034
  ――――――
  260352   excess of nines = 0.  The operation is correct.
```

Example 3:

```
    4,383   excess of nines = 4 + 3 + 8 + 3 = 18 = 1 + 8 = 9 = 0
   ×7,356   (There is no need of finding the excess of nines of the
   ――――――    multiplier, because the excess of nines of the product
    26298    must be equal to zero since the excess of nines of the
    21915    multiplicand is equal to zero.)
    13149
    30681
  ――――――――
  32241348  excess of nines = 0. The operation is correct.
```

18   MATHEMATICS FOR INDUSTRY

Example 4: Check multiplication of 325 × 478 × 673.

```
     325    excess of nines = 1
   ×478    excess of nines = 1
   ─────
    2600
    2275
    1300
   ──────
   155350
    ×673    excess of nines = 7
   ──────
   466050
  1087450
   932100
  ────────
 104550550   excess of nines = 7
```

Multiplying the excess of nines of given factors by each other, we obtain $1 \times 1 \times 7 = 7$, which checks with the excess of nines of the product, hence the operation is correct.

CHECKING DIVISION OF NUMBERS

As division is the opposite of multiplication, division can be checked by multiplying the divisor by the quotient and adding the remainder. Since this is a long method, the checking of division can be carried out more quickly and simply by the excess of nines' method which is as follows:

The excess of nines of the divisor is multiplied by that of the quotient; and, if this product is equal to the excess of nines of the dividend, the operation of the division is correct. If there is a remainder in the division, the excess of nines of the remainder is added to the product obtained from multiplication of the excess of nines of the divisor by that of the quotient. If this final result is equal to the excess of nines of the dividend, the operation of the division is correct.

Example 1:

```
                    15     excess of nines of quotient = 6
                 23)345    excess of nines of dividend = 3
   excess      23
   of nines    ───
   of divi-    115
   sor = 5     115
               ───
               000
```

Multiplying the excess of nines of the divisor (5) by that of the quotient (6), the result is $5 \times 6 = 30 = 3 + 0 = 3$, which is equal to the excess of nines of the dividend (3), proving the operation has been correctly performed.

Example 2:

```
                        384    excess of nines of quotient = 6
                678|260352     excess of nines of dividend = 0
                    2034
 excess         ─────────
 of nines          5695
 of divi-          5424
 sor = 3         ─────────
                   2712
                   2712
                 ─────────
                   0000
```

The excess of the divisor is multiplied by that of the quotient, $3 \times 6 = 18 = 1 + 8 = 9 = 0$. The net result, which is zero, is equal to the excess of nines of the dividend, which is also zero, proving the operation has been correctly performed.

Example 3:

```
                       4381    excess of nines of quotient  = 7
              7356|32231675    excess of nines of dividend  = 2
                   29424
 excess          ─────────
 of nines          28076
 of divi-          22068
 sor = 3         ─────────
                   60087
                   58848
                 ─────────
                   12395
                    7356
                 ─────────
                    5039    excess of nines of remainder = 8
```

The excess of nines of the divisor $\times$ the excess of nines of the quotient $+$ the excess of nines of the remainder $=$ the excess of nines of the dividend: $(3 \times 7) + 8 = 21 + 8 = 29 = 2 + 0 = 2$. The operation of division has been correctly performed.

The rapid method of checking other fundamental operations by the excess of nines will be taken up in the chapter which follows, in which decimal fractions are reviewed, and in subsequent chapters dealing with squares, square roots, cubes, and cube roots.

PRACTICE PROBLEMS APPLICABLE IN THE SHOP AND TOOLROOM

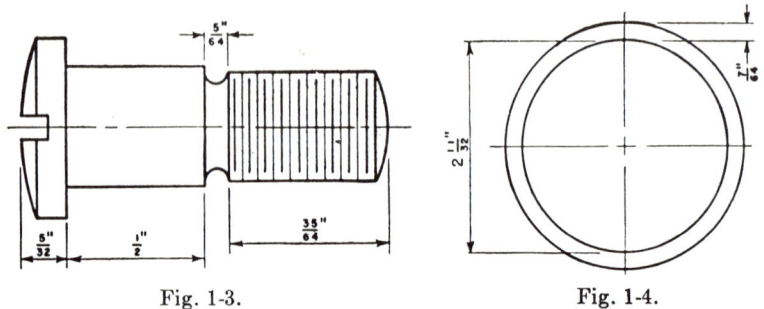

Fig. 1-3.   Fig. 1-4.

1. Find the total length of the shoulder pin in Fig. 1-3.
2. What is the outside diameter of the ring in Fig. 1-4?

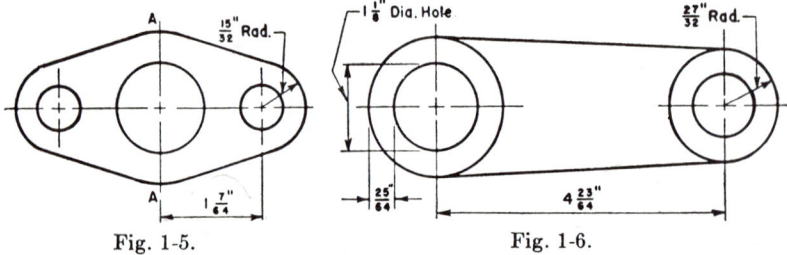

Fig. 1-5.   Fig. 1-6.

3. Calculate the over-all length of the gasket in Fig. 1-5.
   Note: The gasket is symmetrical about A-A.
4. Determine the over-all length of the crankhandle in Fig. 1-6.

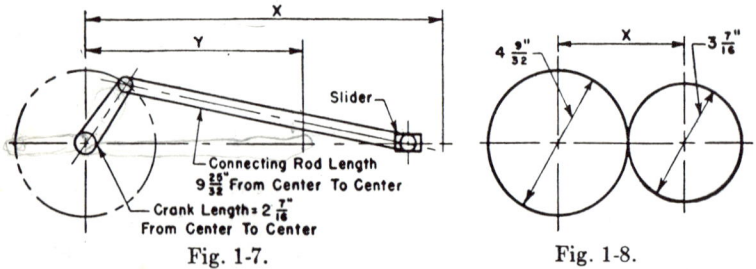

Fig. 1-7.   Fig. 1-8.

5. Determine distances X and Y, when the slider is at the extreme positions (dead centers) in **Fig. 1-7**.

REVIEW OF ARITHMETIC 21

6. Calculate the center distance X between the 2 disks in Fig. 1-8.

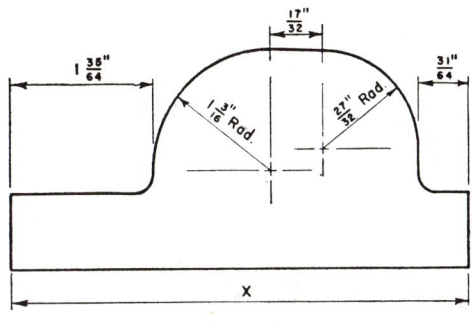

Fig. 1-9.

7. Determine the over-all length X of the part shown in Fig. 1-9.

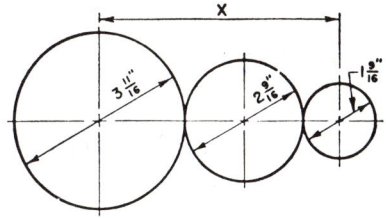

Fig. 1-10.

8. Calculate the center distance X between the end disks in Fig. 1-10.

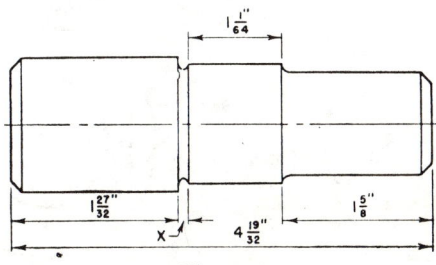

Fig. 1-11.

9. Determine dimension X in Fig. 1-11.

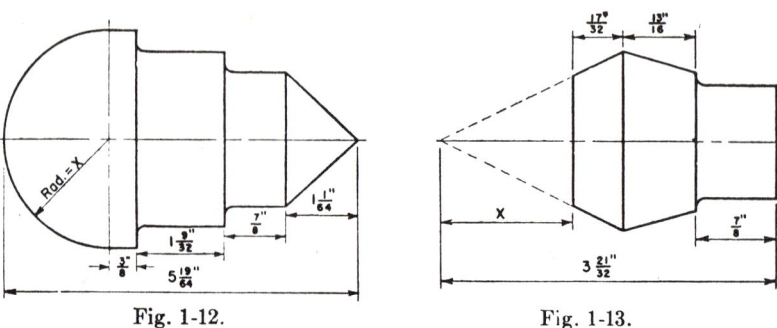

Fig. 1-12.   Fig. 1-13.

10. Calculate the radius X in Fig. 1-12.
11. Determine dimension X in Fig. 1-13.

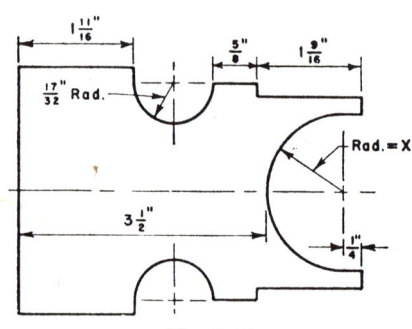

Fig. 1-14.

12. Calculate the radius X in Fig. 1-14.

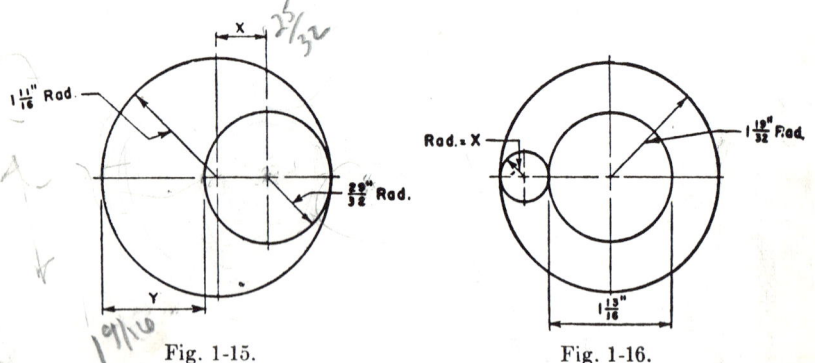

Fig. 1-15.   Fig. 1-16.

13. Calculate dimensions X and Y in Fig. 1-15.
14. Calculate the radius X in Fig. 1-16.

REVIEW OF ARITHMETIC

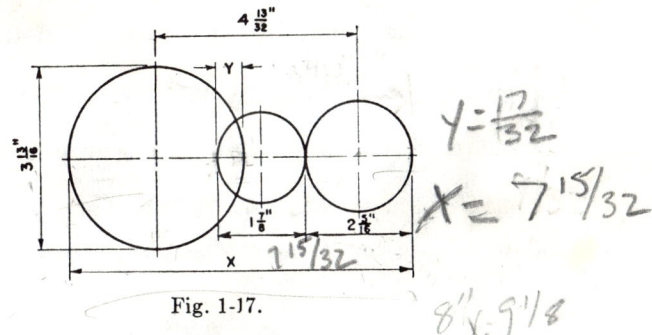

Fig. 1-17.

15. Determine dimensions X and Y in Fig. 1-17.

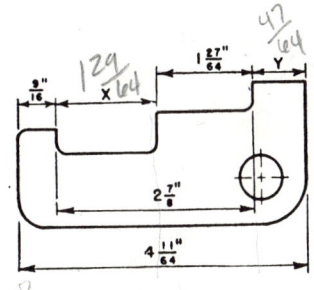

Fig. 1-18.

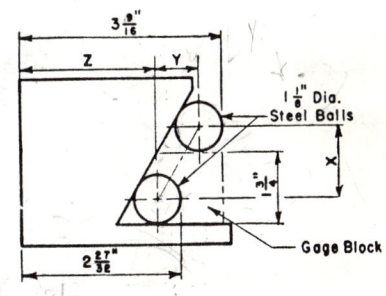

Fig. 1-19.

16. Determine dimensions X and Y in Fig. 1-18.
17. Calculate dimensions X, Y, and Z in Fig. 1-19.

18. $3\frac{7}{26} + 4\frac{11}{65} + 7\frac{9}{35} + 12\frac{15}{28} = ?$

19. $13\frac{19}{85} + \frac{23}{64} + 6\frac{7}{18} + 4\frac{8}{15} = ?$

20. $24\frac{7}{44} + 9\frac{11}{28} + \frac{17}{23} + 17\frac{13}{21} = ?$

21. $37\frac{9}{35} + 13\frac{17}{56} + \frac{15}{96} + \frac{37}{42} = ?$

22. $12\frac{19}{32} + 9\frac{11}{28} - 2\frac{17}{21} + 5\frac{19}{63} - 3\frac{23}{42} = ?$

23. $27\frac{35}{64} - 14\frac{18}{55} + 7\frac{15}{32} - 2\frac{31}{44} - \frac{41}{88} = ?$

24   MATHEMATICS FOR INDUSTRY

24. A brass rod is $23\tfrac{3''}{8}$ long. Three pieces are cut off, the pieces measuring $4\tfrac{11''}{32}$, $3\tfrac{9''}{16}$, and $5\tfrac{7''}{8}$ long, respectively. If the thickness of the sawblade is $\tfrac{1''}{16}$, how long is the remaining piece of the rod?

25. Assume $\tfrac{2}{7}$ of a machine is assembled the first day, $\tfrac{1}{3}$ the second day, and the balance the third day. What part of the machine is assembled on the third day?

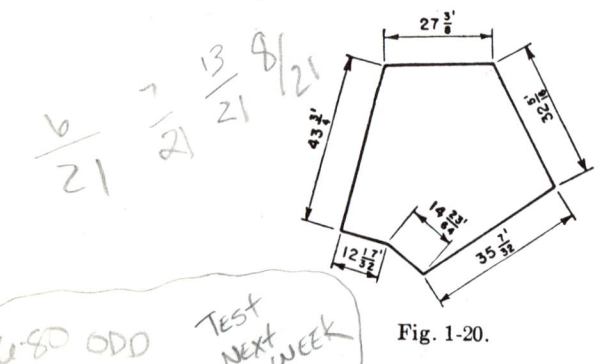

Fig. 1-20.

26. Find the perimeter or the boundary of a lot, shown in the plan view of Fig. 1-20.

27. How much music wire is required to wind three coil springs, if the developed lengths of the springs are $17\tfrac{9''}{64}$, $23\tfrac{5''}{8}$, and $16\tfrac{9''}{32}$, respectively? Assume a wastage in the cutting of $\tfrac{1''}{8}$ per spring.

28. $4 \times \tfrac{3}{7} = ?$

29. $6 \times \tfrac{13}{18} = ?$

30. $14 \times \tfrac{17}{21} = ?$

31. $\tfrac{7}{13} \times 6 = ?$

32. $\tfrac{5}{27} \times 12 = ?$

33. $\tfrac{9}{32} \times 24 = ?$

34. $\tfrac{7}{16} \times \tfrac{3}{4} = ?$

35. $\tfrac{11}{32} \times \tfrac{1}{2} = ?$

36. $\tfrac{39}{64} \times \tfrac{28}{45} = ?$

37. $\tfrac{27}{32} \times \tfrac{28}{45} = ?$

38. $\tfrac{7}{25} \times \tfrac{5}{28} \times \tfrac{4}{13} = ?$

39. $\tfrac{9}{65} \times \tfrac{26}{63} \times \tfrac{14}{23} = ?$

REVIEW OF ARITHMETIC

40. $\dfrac{7}{48} \times \dfrac{9}{35} \times \dfrac{12}{21} = ?$
41. $7\dfrac{3}{16} \times 4\dfrac{4}{9} \times 2\dfrac{4}{7} = ?$ 82 1/7

42. $14\dfrac{3}{8} \times 7\dfrac{7}{23} \times 8\dfrac{3}{5} = ?$
43. $24\dfrac{7}{16} \times 6\dfrac{2}{17} \times 4\dfrac{9}{65} = ?$ 618 1547/2210

44. $12\dfrac{3}{8} \times \dfrac{16}{33} \times 6\dfrac{5}{24} \times \dfrac{15}{32} = ?$
45. $2\dfrac{5}{32} \times 3\dfrac{3}{7} \times 4\dfrac{13}{23} \times 6\dfrac{2}{15} = ?$ 207

46. $12\dfrac{4}{16} \times 17\dfrac{6}{32} \times 7\dfrac{5}{8} \times 5\dfrac{44}{64} = ?$
47. $22\dfrac{55}{64} \times 13\dfrac{13}{35} \times 9\dfrac{18}{23} \times 4\dfrac{9}{38} = ?$

48. The circumference of a circle is close to $3\dfrac{1}{7}$ times its diameter. Calculate the circumference of a circle, the diameter of which is equal to $13\dfrac{15}{64}''$.

49. An aluminum-silicon alloy used for casting is about $\dfrac{23}{25}$ aluminum, $\dfrac{1}{14}$ silicon, and the balance iron. Find the weight of each metal in a casting, if 273 pounds of the alloy was required to make 8 castings. 34 1/8

50. Dowmetal-H, used for many commercial and aircraft sand castings, is composed of the following metals: $\dfrac{1}{17}$ of its weight is aluminum, $\dfrac{1}{34}$ is zinc, $\dfrac{1}{200}$ is silicon, $\dfrac{1}{500}$ is manganese, and the balance is magnesium. Assuming that a sand casting weighs $17\dfrac{1}{4}$ pounds, what is the weight of the magnesium in the casting?

51. A project in a machine shop requires the following operators and their time: three lathe hands, 4 days, $8\dfrac{1}{2}$ hours per day, at $\$1.12\dfrac{1}{2}$ per hour; two drill press operators, 3 days, $8\dfrac{1}{2}$ hours per day, at $87\dfrac{1}{2}$ cents per hour; three other men for miscellaneous operations, including material of the processed units, 3 days, $8\dfrac{1}{2}$ hours per day, at \$1.25 per hour. What is the selling price of the completed units, if it is $1\dfrac{3}{8}$ times the cost of the material and the labor combined?

Note: Do not convert the given data into decimal fractions. Use mixed numbers and common fractions, as they are presented in the problem.

52. If the spindle of a lathe makes 160 revolutions per minute, and it

takes 58 spindle revolutions to turn a shaft to a required diameter, how many shafts can be turned in $4\frac{17}{32}$ hours?

53. $6 \div \frac{3}{7} = ?$
54. $7 \div \frac{13}{18} = ?$
55. $15 \div \frac{21}{32} = ?$
56. $\frac{7}{18} \div 14 = ?$
57. $\frac{5}{27} \div 15 = ?$
58. $\frac{9}{32} \div 24 = ?$
59. $\frac{7}{16} \div \frac{3}{4} = ?$
60. $\frac{11}{32} \div \frac{1}{2} = ?$
61. $\frac{39}{64} \div \frac{21}{8} = ?$
62. $\frac{32}{27} \div \frac{28}{45} = ?$
63. $\frac{7}{25} \div \frac{28}{5} = ?$
64. $7\frac{3}{16} \div \frac{23}{32} = ?$
65. $\frac{7}{24} \div 14\frac{7}{8} = ?$
66. $\frac{16}{33} \div 8\frac{8}{35} = ?$
67. $24\frac{7}{16} \div 5\frac{7}{64} = ?$
68. $22\frac{11}{64} \div 4\frac{7}{32} = ?$
69. $5\frac{1}{28} \div 34\frac{4}{35} = ?$
70. $1 \div \frac{3}{64} = ?$

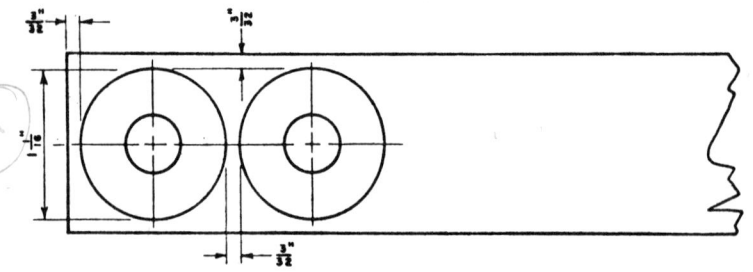

Fig. 1-21.

71. *a)* How wide should a steel strip be ordered to allow $\frac{3}{32}''$ on each side by punching out flat washers, shown in Fig. 1-21?

REVIEW OF ARITHMETIC

b) What will be the minimum length of the strip to punch out 750 washers?

72. a) What is the weight of the minimum length of a strip required in Problem 71b, if the thickness of the metal is $\frac{3}{32}''$, and a cubic inch of steel weighs about $\frac{7}{25}$ of a pound?

b) What is the cost of the strip on the basis of $8\frac{17}{32}$ cents per pound?

73. What is the length of a line drawn parallel and equidistant from the top and bottom lines of the trapezoid shown in Fig. 1-22?

Note: The length of such a line (X) is equal to ½ the sum of the parallel lines in the trapezoid.

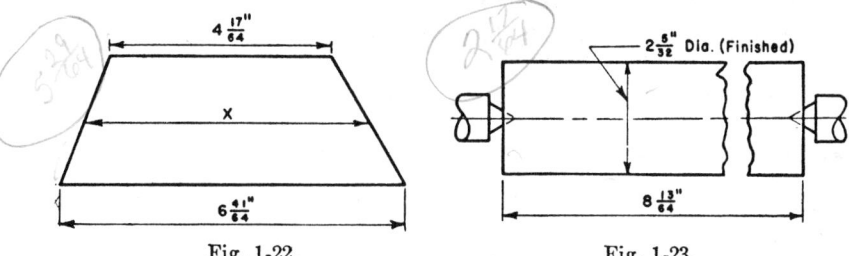

Fig. 1-22.   Fig. 1-23.

74. Assume a row of 17 rivets spaced $3\frac{9}{64}$ inches apart are used to join two sheets of metal. What is the distance between centers of the first and last rivets?

75. What should be the stock diameter of the cylindrical piece turned on a lathe, Fig. 1-23, if the roughing cut is $\frac{3}{32}''$ deep, and the finishing cut $\frac{1}{64}''$ deep?

76. How long will it take in Fig. 1-23 to turn a length of $8\frac{13}{64}''$ stock in one operation if the cutting tool moves $\frac{3}{128}''$ along the work for each revolution of the work, and the number of revolutions is 34 per minute?

77. Find the taper per foot of the cone-shaped part in Fig. 1-24.

Formula: The taper per foot of the conical part shown is equal to the difference between the end diameters divided by the length of the part and multiplied by 12.

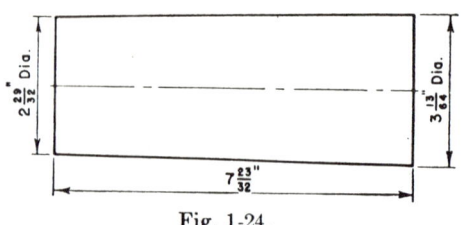

Fig. 1-24.

78. How many equally spaced holes can be punched through a sheet steel strip $41\frac{15''}{16}$ long, if the center to center distance is equal to $2\frac{27''}{64}$, and the end hole centers are $1\frac{19''}{32}$ from each end?

79. A machine operator new to the work completes a project in $18\frac{5}{32}$ hours, earning 72 cents per hour. A similar project, according to records, was completed by an experienced operator in $13\frac{7}{16}$ hours at an hourly wage of 96 cents an hour. Which of the two projects cost less in labor to make, and what was the difference?

80. A mechanic is able to assemble a machine in $17\frac{1}{4}$ hours, another mechanic completes the same task in $15\frac{3}{8}$ hours. Assume that the two men are working together at their normal capacities on the same machine, how long will it take them to assemble it? What part of the project will be done by each mechanic?

Chapter II

REVIEW OF ARITHMETIC—CONTINUED

DECIMAL FRACTIONS

Fractions that have 10, 100, 1000, etc., for denominators are called decimal fractions. In a decimal fraction, then, the denominator is always a multiple of 10. It can be determined what the denominator is from the position of the decimal point. To the right of the point there will be as many digits as there are zeros or ciphers in the assumed denominator. In a mixed number, the whole number is to the left of the decimal point.

Examples of Decimal Fractions:

1. $\dfrac{3}{10} = .3$ 2. $\dfrac{7}{100} = .07$

3. $\dfrac{374}{1000} = .374$ 4. $\dfrac{83}{10{,}000} = .0083$

5. $1\dfrac{7}{10} = 1.7$ 6. $23\dfrac{74}{100} = 23.74$

7. $372\dfrac{9}{1000} = 372.009$ 8. $2467\dfrac{34}{10{,}000} = 2467.0034$

Note: When there are less figures in the numerator than zeros in the denominator, the necessary number of decimal places are made by adding zeros on the left of the figures.

The correct reading of a decimal is very important. If a decimal is read incorrectly, the value of the quantity will be misinterpreted.

Take two of the given examples, say .0083 and 2467.0034. In the first decimal fraction, there are four digits to the right of the decimal point. The assumed denominator will have four ciphers, and the fraction is read .0083 = $\dfrac{83}{10{,}000}$ or eighty-three ten-thousandths. In the mixed decimal the whole number is read to the left of the decimal point; to the right of the point there are four digits, which means that the denominator will have four ciphers. The mixed decimal is read $2467.0034 = 2467\dfrac{34}{10,000}$ or two thousand four hundred sixty-seven and thirty-four ten-thousandths.

An understanding of this simple rule will be the guide in all operations in which decimals are involved.

ADDITION OF DECIMAL FRACTIONS

In adding decimal fractions, the given numbers are arranged in a vertical column in such manner that all of the decimal points are in a straight line. Thus units will be added to units, tens to tens, hundredths to hundredths, etc. The decimal point of the sum is placed below all the given decimal points. Then the figures are added the same way whole numbers are added.

Examples:

	1.	23.43	2.	126.4387	3.	6.4375
		.16		43.007		.5625
		467.072		.0709		7.03125
		490.662		4872.304		24.546875
				5041.8206		45.953125
						84.531250

SUBTRACTION OF DECIMAL FRACTIONS

In the subtraction of decimal fractions, as in the addition of decimal fractions, the given numbers are first arranged so that the decimal points are under each other. The decimal point of the result is placed directly below the other decimal points and subtraction of the numbers follows.

In the operation of adding decimal fractions, any number of decimals may be used; in subtraction, only two decimals are involved. These are the minuend, the number from which you subtract, and the subtrahend, the number subtracted. The result or answer is the difference or remainder.

Examples:

1.	38.6473 (minuend)	2.	27.3562	3.	472.07082
	−23.3785 (subtrahend)		−13.4937		−47.89435
	15.2688 (difference)		13.8625		424.17647
4.	35.0092	5.	26.42010	6.	123.040506
	−.8357		−.00728		−44.56789
	34.1735		26.41282		78.472616

MULTIPLICATION OF DECIMAL FRACTIONS

As in multiplication of whole numbers, the first number of the decimal fraction is the multiplicand, the second number is the multiplier, and the result is the product. The operation is the same as multiplication of whole numbers except for the placement of the decimal point. In multiplication of decimals, proper placement of the decimal point in the product is very important in obtaining the correct answer. The rule for the placing of the decimal point is: The number of decimal places in the product must be equal to those of the multiplicand and the multiplier together. Prefix ciphers when necessary. (The decimal places in the given number and in the product are to the right of the decimal point.)

Example:

```
         6.8472 (Multiplicand)
        ×7.3568 (Multiplier)
         547776
         410832
         342360
         205416
         479304
      50.37348096 (Product)
```

Since there are four decimal places in each of the given numbers, there must be eight decimal places in the product.

Checking Multiplication of Decimal Fractions. Multiplication of decimals can be checked by the excess of nines method. In the previous example, the excess of nines in the multiplicand is equal to zero, which makes the excess of nines in the product equal to zero. Zero of nines in the multiplicand multiplied by any excess of nines in the multiplier will result in zero for the excess of nines in the product. It should be noted that the checking of the result by the excess of nines does not determine the correct position of the decimal point.

Example 1:

$$23.0476 \text{ excess of nines} = 4$$
$$\times 746.805 \text{ excess of nines} = 3$$

```
    1152380              12 = 1 + 2 = 3
    1843808
    1382856
     921904
    1613332
  17212.0629180 excess of nines = 3
```

The number of decimal places in the product is equal to that of the given numbers, which is $4 + 3 = 7$. Checking the multiplication by the excess of nines, it is found to be correct.

When multiplying large numbers or decimal fractions of many digits, the multiplication may be checked step by step to avoid error. This will save time and further work.

Example 2:

```
      237.0246   excess of nines = 6
    ×652.7084   excess of nines = 5
      9480984   excess = 6 = 6 × 4 = 24 = 2 + 4 =  6     (correct)
     18961968   excess = 3 = 6 × 8 = 48 = 4 + 8 = 12 = 3 (correct)
     16591722   excess = 6 = 6 × 7 = 42 = 4 + 2 =  6     (correct)
      4740492   excess = 3 = 6 × 2 = 12 = 1 + 2 =  3     (correct)
     11851230   excess = 3 = 6 × 5 = 30 = 3 + 0 =  3     (correct)
     14221476   excess = 0 = 6 × 6 = 36 = 3 + 6 =  9 = 0 (correct)
  154707.94742664 excess = 3 = 6 × 5 = 30 = 3 + 0 =  3   (correct)
```

The excess of nines in the multiplicand is 6. The excess multiplied by the first digit of the multiplier, which is 4, will result in $6 \times 4 = 24 = 2 + 4 = 6$. The excess of nines in the first product, $9 + 4 + 8 + 0 + 9 + 8 + 4$, is also 6, proving that the first step is correct. The remaining partial products may be checked in the same way, multiplying the excess of nines in the multiplicand (a constant) by the successive digits of the multiplier. The final product may be checked by investigating its excess of nines, which is equal to 3. The excess of nines of the multiplicand is 6. When it is multiplied by the excess of nines in the multiplier, the result will be 3, which proves that the final product is correct.

Multiplying by 10, 100, 1000, etc. Since in a decimal fraction the position of the decimal point determines the value of the denominator, it follows that to multiply a decimal by 10, 100, 1000, etc., it is only necessary to move the decimal point as many places to the right as there are ciphers in the multiplier. Thus multiplying a decimal by 10, one needs only to move the decimal point one place to the right. Moving the decimal point two places to the right multiplies by 100; moving it three places to the right multiplies by 1000; etc. If there are not figures enough, annex ciphers to the right.

DIVISION OF DECIMAL FRACTIONS

Division of decimals is similar in operation to division of whole numbers, the only difference being that the decimal point must be correctly placed in the quotient. The quantities involved are the dividend, the divisor, the quotient, and, if anything is left over, the remainder.

Since division is the opposite of multiplication, the number of decimal places of the quotient is the difference between the number of decimal places of the dividend and the divisor.

The procedure in the division of decimals, especially when the decimal consists of a large number of digits, should be orderly.

PROCEDURE OF DIVISION OF DECIMAL FRACTIONS

Step 1. Given quantities are arranged as in long division. If there is a decimal point in the divisor, it should be moved to the extreme right, making the divisor a whole number.

Step 2. To insure a correct result, the decimal point in the dividend is moved an equal number of places to the right. Observe that in moving the decimal point to the right the decimal fraction is multiplied. If the dividend and the divisor are multiplied by the same factor the result will not be disturbed. If the dividend is either a whole number or does not contain sufficient decimal places to balance the number of places the decimal point has been moved in the divisor, ciphers are added to the dividend to make up the difference.

Step 3. The decimal point of the quotient is placed directly above that of the balanced dividend.

Step 4. The first digit of the quotient is placed directly above the last digit of the dividend immediately involved. The subsequent digits of the quotient are placed directly above the last digit of the dividend involved in the particular division. The operation is continued as far as is necessary to obtain the number of decimal places desired in the quotient.

34 MATHEMATICS FOR INDUSTRY

Example: Divide 374.2835 by 32.47.

```
                       11.52    (Quotient)
(Divisor) 32 47. | 374 28.35    (Dividend)
                   324 7
                    49 58
                    32 47
                    17 113
                    16 235
                       8785
                       6494
                       2291    (Remainder)
```

A final check confirms the fact that the number of decimal places in the quotient is equal to that of the dividend, less the number of decimal places in the divisor. In the given example, the number of decimal places in the dividend is four and in the divisor it is two, therefore the quotient has two decimal places.

If a more precise result is desired and the operation is continued, ciphers may be added to the right of the dividend and thus additional decimal places are obtained in the quotient.

Checking Division of Decimal Fractions. The division of decimals may be easily checked by the excess of nines method. If there is no remainder, the excess of nines in the quotient multiplied by the excess of nines in the divisor should equal the excess of nines in the dividend. If there is a remainder, its excess of nines is added to the excess of nines in the product obtained by multiplying the excess of nines in the quotient by the excess in the divisor and this should equal the excess of nines in the dividend. In the previous example, the excess of nines in the dividend equals 5, in the divisor 7, in the quotient 0, and in the remainder 5. $7 \times 0 + 5 = 5$, the same as in the dividend.

Example 1: Divide 64.04 by 28.0354.

```
                    2.28
28 0354. | 64 0400.00
           56 0708
            7 96920
            5 60708
            2 362120
            2 242832
              119288   (Remainder)
```

REVIEW OF ARITHMETIC

To check the number of decimal places in the quotient, the number of decimal places in the divisor, four, is subtracted from the number of decimal places in the dividend, six, and the result is the two decimal places in the quotient, which is correct.

In checking the division by the excess of nines, it is seen that the excess of nines in the dividend equals 5, in the divisor 4, in the quotient 3, and in the remainder 2. $4 \times 3 = 12 = 1 + 2 = 3$, which is the product of the excess of nines in the divisor multiplied by the excess in the quotient. The remainder excess, 2, is added to the 3, giving 5, which is equal to the excess of nines in the dividend and indicates that the operation has been correctly performed.

Example 2: Divide 127 by 34.207.

```
                    3.7126      excess of nines = 1
      34 207. | 127 000.0000    excess of nines = 1
                102 621
                 24 3790
  excess         23 9449
  of nines         43410        7 × 1 + 3 = 10 ⎫
  of divi-         34207            1 + 0 =  1 ⎬ The division is cor-
  sor = 7          92030                1 =  1 ⎭ rect.
                   68414
                  236160
                  205242
                   30918      excess of nines = 3
```

Example 3: Divide 73.256 by 284.

```
                   .2579    excess of nines = 5
           284 | 73.256     excess of nines = 5
                 56 8
  excess         16 45      5 × 5 + 7 = 32 ⎫
  of nines       14 20          3 + 2 =  5 ⎬ The division
  of divi-        2 256              5 =  5 ⎭ is correct.
  sor = 5         1 988
                  2680
                  2556
                   124      excess of nines = 7
```

Example 4: Divide 697 by .000475.

```
                  1 467368.     excess of nines = 8
       000475. ⟌ 697 000000.    excess of nines = 4
                  475
                  ―――
                  222 0
   excess         190 0
   of nines       ――――
   of divi-       32 00
   sor = 7        28 50          7 × 8 + 2 = 58 ⎫
                  ――――           5 + 8 = 13    ⎬ The division
                   3 500         1 + 3 =  4    ⎬ is correct.
                   3 325                 4 =  4⎭
                   ――――
                    1750
                    1425
                    ――――
                    3250
                    2850
                    ――――
                    4000
                    3800
                    ――――
                     200   excess of nines = 2
```

Dividing by 10, 100, 1000, etc. Decimal fractions can be divided by 10 by moving the decimal point one place to the left. Moving the decimal point two places to the left divides by 100; moving it three places to the left divides by 1000; etc. If necessary, zeros should be prefixed.

ROUNDING OFF NUMERICAL VALUES

In many mathematical operations in which decimals are involved, the number of decimal places retained in the result depends upon the degree of precision desired. For example, if we multiply decimal fractions by each other, the number of decimal places in the product is equal to the sum of decimal fractions in the given factors. This number of decimal places may be too many for the degree of accuracy desired; in that case only a sufficient number of places is retained in the product, and the unnecessary decimals are omitted. For the same reason, in dividing decimals, the operation may be endlessly or unduly prolonged. To speed up calculation, therefore, the operation of division is stopped as soon as a quotient of the desired degree of accuracy is obtained. The operation is never stopped abruptly, however, without considering beforehand the number of decimal places that are desired to be retained in the result. The last place in the result is rounded off, which means that the last decimal place is either unaffected or is increased by one.

According to the American Standards Association, the rule to be followed in rounding off decimals is to keep the last figure retained unchanged when the first figure discarded is less than 5, and to increase the last figure by 1 when the first figure discarded is more than 5.

When the first figure discarded is 5, some computers increase the last retained figure by 1 if the discarded figure following the 5 is greater than zero, or if the last retained figure is an odd number. If these conditions do not exist, the last retained figure remains unchanged.

When the first discarded figure is 5, other computers allow the last retained figure to remain unchanged or increase it by 1, depending upon manufacturing tolerances and fits of the mating parts.

Say that the diameter of a piston and the bore of a cylinder are dimensioned. If the dimension is applied to the bore of the cylinder, the last retained figure is increased by 1. In dimensioning the diameter of the piston, the last figure retained is unchanged. This is a process spoken of as "rounding off" of the parts. How to proceed is entirely a matter of judgment.

Examples:

1. 7.357432 = 7.35743 = 7.3574 = 7.357 = 7.36
2. 23.436789 = 23.43679 = 23.4368 = 23.437 = 23.44
3. 32.748502 = 32.748
4. 32.747502 = 32.748
5. 32.748532 = 32.749
6. Piston diameter = 1.249354 calculated = 1.2493 specified
7. Cylinder bore = 1.249754 calculated = 1.2498 specified

For purposes of factory production, it is logical to specify the piston diameter and the cylinder bore on the basis of the calculated dimensions.

When dealing with component dimensions, as those of shoulder pins, shafts, and so forth, a good practice is to carry an additional decimal place beyond the degree of accuracy required in every involved dimension. In that way, when all component dimensions are assembled, the result will approach a correct interpretation of the true state of conditions.

CONVERSION OF DECIMALS AND FRACTIONS

As previously stated, a decimal fraction has an assumed denominator, which is a multiple of 10. Therefore the decimal fraction can be written with the given decimal as the numerator and the denominator is determined by the position of the decimal point. This fraction can then be reduced to simplest forms by finding factors common to numerator and denominator in successive stages.

Examples:

1. $.1875 = \dfrac{1,875}{10,000} = \dfrac{75}{400} = \dfrac{3}{16}$

2. $.09375 = \dfrac{9,375}{100,000} = \dfrac{375}{4,000} = \dfrac{3}{32}$

3. $.5625 = \dfrac{5,625}{10,000} = \dfrac{225}{400} = \dfrac{9}{16}$

4. $.828125 = \dfrac{828,125}{1,000,000} = \dfrac{33,125}{40,000} = \dfrac{1,325}{1,600} = \dfrac{53}{64}$

Since a common fraction is considered an indicated division, it follows that it may be reduced to a decimal by carrying out the division. The procedure is to divide the numerator by the denominator and carry the division out until the required number of decimal places is obtained. Since most fractions do not have an exact decimal equivalent, the last figure in the retained number of decimal places is rounded off.

Examples:

1. $\dfrac{37}{64} = .578125$ 2. $\dfrac{17}{38} = .4473+$ 3. $\dfrac{19}{65} = .2923+$

```
       .578125              .4473              .2923
  64 )37.000000         38 )17.0000         65 )19.0000
     32 00                 15 20               13 00
     ─────                 ─────               ─────
      5 00                  1 80                6 00
      4 48                  1 52                5 85
      ─────                 ─────               ─────
        520                   280                 150
        512                   266                 130
        ───                   ───                 ───
         80                   140                 200
         64                   114                 195
         ───                   ───                 ───
         160          26 (Remainder)      5 (Remainder)
         128
         ───
         320
         320
         ───
         000
```

REVIEW OF ARITHMETIC 39

DECIMAL EQUIVALENT CHART

The unit of measure used in the shop is the inch. Its subdivisions or fractions are multiples of 2. Often it is desired to show these fractions in decimals, or vice versa. To facilitate the operation of conversion, a decimal equivalent chart, Table 3, is included in the Appendix. The smallest value in the chart is $\frac{1}{64}$ and the largest value is 1. Successive values vary by $\frac{1}{64}$.

Examples:

1. $\frac{7}{64} = .109375$
2. $\frac{21}{32} = .65625$
3. $\frac{13}{16} = .8125$

INCH-MILLIMETER CONVERSION

To facilitate this conversion, a chart based on the conversion factors has been devised. See Table **7** of the Appendix.

1 inch = 25.4 millimeters
1 millimeter = .03937 inch

Occasions arise in machine operations of the aircraft and automotive industries when it is found necessary or desirable to resort to conversion. In the manufacture of ball bearings and roller bearings, also, millimeter dimensioning is required, and the millimeter equivalent in decimals of an inch is sought in order that the correct cutting tools for manufacture of the bearings and the proper instruments and gages for checking the accuracy of the parts may be used.

As an example, a self-aligning ball-thrust bearing, Type 55BTSM, has an inside diameter, specified 55 mm., and an outside diameter, specified 88 mm. From the chart it can be determined that the equivalent inside diameter is equal to 2.1654 inches and the equivalent outside diameter to 3.4646 inches.

SQUARE ROOT

The square root of a number is a number that when multiplied by itself is equal to the number given. The operation of extracting the square root is signified by the radical sign $\sqrt{}$. The index of the root is the small number that appears in the "vee" of the radical sign, as $\sqrt[3]{}$, for example. For a square root, no index is required. A *perfect square* is a number of which the square root leaves no remainder. An *imperfect square* is a number that has no exact square root since it will always be found to leave a remainder.

Examples of perfect squares are 4, 9, 16, 25, 36, 49, etc.

Examples of imperfect squares are 2, 3, 5, 7, 10, etc.

EXTRACTING SQUARE ROOT OF A PERFECT SQUARE

The operation of extracting the square root of a perfect square will be explained in five steps. The square root of 121,104 will be found as an example and the completed example will first be presented, followed by the steps of the procedure.

Example:

$$\sqrt{12\ 11\ 04.}\ \overline{348.}$$

		9
Trial divisor	3 × 20 = 60	3 11
Completed divisor	60 + 4 = 64	2 56
Trial divisor	34 × 20 = 680	55 04
Completed divisor	680 + 8 = 688	55 04
		00 00

Step 1. The square of single digits, except 1, 2, and 3, is composed of two digits. The first step in the operation is to point off the given numbers into periods of two digits, beginning at the decimal point. If the given number consists of an odd number of digits, the period at the left consists of a single digit. The number of periods formed in this manner is equal to the number of figures in the root.

Step 2. Find the nearest square root of the first period at the left. This is 3 and will be the first digit of the root. It is written above the radical sign.

Step 3. Now square the first digit of the root (3) and subtract the result (9) from the first period of the given number (12), which will be the first remainder. In continuing the operation, the next period of the given number (11) is brought down to the first remainder, and a trial divisor is made by multiplying the first digit in the root by the *constant*

REVIEW OF ARITHMETIC 41

20 (3 × 20 = 60). This operation is carried out in the extreme left-hand space, as demonstrated in the example.

Step 4. A second digit in the root is found by dividing the trial divisor (60) in the number created by the first remainder and the second period of the given number (311). The figure 60 will go in 311 five times but, as will be explained, a figure one unit less is used (4). This second digit (4) is placed above the radical sign and is added to the trial divisor (60) giving the completed divisor (64). The second digit (4) is then multiplied by the completed divisor (64), and the product (256) is subtracted from the appropriate portion of the given number (311). If this product is larger than the given number (as would have been the case if the figure 5 was used as the second digit), it would indicate that the trial figure is too large and a figure one unit smaller must be used. To the second remainder (55) is added the next period of the given number (04).

Step 5. A new trial divisor is created by multiplying the figures of the root thus far obtained by 20 (34 × 20 = 680). The third digit in the root is obtained by dividing the new trial divisor (680) into the number derived in the previous step (5504). The second completed divisor is obtained by adding the third digit in the root (8) to the second trial divisor. This completed divisor is multiplied by the third digit in the root (8) and the product (5504) is subtracted from the appropriate portion of the given number (5504). Here again, if the product is larger than the given number, a trial figure one unit smaller must be used. Step 5 is repeated until there are no more periods of figures to move down from the given number. Since we are dealing now with perfect squares the last subtraction will leave no remainder.

EXTRACTING SQUARE ROOT OF AN IMPERFECT SQUARE

When working with an imperfect square, a remainder will be present after the last period of figures has been moved down. A decimal point is placed after the figures already obtained in the root. One must now decide how many decimal places he wishes in the answer and annex two ciphers (one period) for each decimal place desired. The procedure then continues in the usual manner.

CHECKING THE EXTRACTION OF THE SQUARE ROOT

The square root of a number may be checked by multiplying the root by itself to see if it equals the given number. If the square is imperfect the figures obtained by multiplying the root by itself will very nearly equal the given number. Square root is more often checked by the simple method of the excess of nines.

In checking the extraction of the square root by the excess of nines method, the excess of nines of the given number and of the root is determined. The process of checking consists of squaring the excess of nines of the root. If the result is equal to the excess of nines of the given number, the operation of extracting the square root has been correctly performed. When the squares are imperfect, the excess of nines of the remainder is added to the square of the excess of nines of the root. The sum should be equal to the excess of nines of the given number.

Example 1:

$$\sqrt{5\ 61\ 69.}\ \ 237.\quad \text{excess of nines} = 3$$
$$\text{excess of nines} = 0$$

```
              4
   2 × 20 = 40      | 1 61
   40 + 3 = 43      | 1 29
   23 × 20 = 460    |   32 69    3 × 3 = 9 = 0
   460 + 7 = 467    |   32 69
                    ───────
                        00 00
```

Example 2:

$$\sqrt{23\ 75\ 84.}\ \ 487.+\quad \text{excess of nines} = 1$$
$$\text{excess of nines} = 2$$

```
              16
   4 × 20 = 80      7 75
   80 + 8 = 88      7 04          1 × 1 + 1 = 2
   48 × 20 = 960      71 84
   960 + 7 = 967      67 69
                     ─────
                       4 15    (Remainder)
                               excess of nines = 1
```

EXTRACTING SQUARE ROOT OF A DECIMAL FRACTION

In extracting the square root of a decimal fraction, the procedure is to point off to the right of the decimal, in the given fraction, periods consisting of two places each. The method to be followed is then the same as for whole numbers. The important thing is that every pointed-off period in the given fraction will net one digit in the root.

If the given decimal fraction is not a perfect square, or if it is long,

REVIEW OF ARITHMETIC 43

the necessary number of periods is pointed off to give the required number of decimal places in the root, depending upon the degree of precision desired in the result. If the decimal fraction is short, any number of ciphers may be added to the right in pointing off the necessary number of periods.

Example 1:

$$\begin{array}{r} .3764 \\ \sqrt{.14\ 16\ 76\ 96} \end{array}$$
excess of nines = 2
excess of nines = 4

$$\begin{array}{r|l} & 9 \\ \hline 3 \times 20 = 60 & 5\ 16 \\ 60 + 7 = 67 & 4\ 69 \\ \hline 37 \times 20 = 740 & 47\ 76 \\ 740 + 6 = 746 & 44\ 76 \\ \hline 376 \times 20 = 7520 & 3\ 00\ 96 \\ 7520 + 4 = 7524 & 3\ 00\ 96 \\ \hline & 0\ 00\ 00 \end{array}$$

$2 \times 2 = 4$

Four periods, each consisting of two decimal places, are pointed off to the right of the decimal point in the given fraction. As a result, four decimal places are found in the root. The given fraction is a perfect square, and there is no remainder. The operation may be checked by the excess of nines.

Example 2: Extract the square root of the fraction .5727854 to five decimal places.

$$\begin{array}{r} .75682 \\ \sqrt{.57\ 27\ 85\ 40\ 00} \end{array}$$
excess of nines = 1
excess of nines = 2

$$\begin{array}{r|l} & 49 \\ \hline 7 \times 20 = 140 & 8\ 27 \\ 140 + 5 = 145 & 7\ 25 \\ \hline 75 \times 20 = 1500 & 1\ 02\ 85 \\ 1500 + 6 = 1506 & 90\ 36 \\ \hline 756 \times 20 = 15120 & 12\ 49\ 40 \\ 15120 + 8 = 15128 & 12\ 10\ 24 \\ \hline 7568 \times 20 = 151360 & 39\ 16\ 00 \\ 151360 + 2 = 151362 & 30\ 27\ 24 \\ \hline & 8\ 88\ 76 \end{array}$$

$1 \times 1 + 1 = 2$

(Remainder)
excess of nines = 1

To obtain the required number of decimal places in the root, the number of periods in the given decimal fraction must be five. As many ciphers are added at the right as are proper to the condition.

Example 3: Extract the square root of the fraction .0005673 to four decimal places.

$$\sqrt{.00\ 05\ 67\ 30} \quad \begin{array}{l} .0238 \\ \end{array}$$

	4
$2 \times 20 = 40$	1 67
$40 + 3 = 43$	1 29
$23 \times 20 = 460$	38 30
$460 + 8 = 468$	37 44
	86

excess of nines = 4
excess of nines = 3

$4 \times 4 = 16 = 1 + 6 = 7$
$7 + 5 = 12 = 1 + 2 = 3$

(Remainder)
excess of nines = 5

In this example the first period to the right of the decimal point is 00, which makes the first figure in the root 0. The operation then proceeds in the manner of the previous example. The first trial divisor is obtained after the first significant figure is derived in the root.

When mixed decimal fractions are given, the procedure of extracting the square root begins by pointing off the periods to the right and the left of the decimal point. The periods pointed off to the left are for the whole number in the root and the periods to the right are for the decimal fraction in the root.

Example 4: Extract the square root of 572.04358 to three decimal places.

$$\sqrt{5\ 72.\ 04\ 35\ 80} \quad \begin{array}{l} 23.917 \\ \end{array}$$

	4
$2 \times 20 = 40$	1 72
$40 + 3 = 43$	1 29
$23 \times 20 = 460$	43 04
$460 + 9 = 469$	42 21
$239 \times 20 = 4780$	83 35
$4780 + 1 = 4781$	47 81
$2391 \times 20 = 47820$	35 54 80
$47820 + 7 = 47827$	33 47 89
	2 06 91

excess of nines = 4
excess of nines = 7

$4 \times 4 = 16 = 1 + 6 = 7$
$7 + 0 = 7$

(Remainder)
excess of nines = 0

REVIEW OF ARITHMETIC 45

In this example there are two periods to the left of the decimal point, therefore there are two digits to the left of the decimal point in the root. The number of decimal places in the root is three, which corresponds to the three periods to the right of the decimal point in the given number.

EXTRACTING SQUARE ROOT OF COMMON FRACTIONS

The square root of common fractions and mixed numbers is obtained by converting the given quantities into decimal fractions or mixed decimals, except when in common fractions the numerator and the denominator are perfect squares.

Examples:

1. $\sqrt{\dfrac{25}{64}} = \dfrac{5}{8}$ 2. $\sqrt{\dfrac{9}{16}} = \dfrac{3}{4}$ 3. $\sqrt{\dfrac{16}{81}} = \dfrac{4}{9}$ 4. $\sqrt{\dfrac{9}{100}} = \dfrac{3}{10}$

5. $\sqrt{\dfrac{4}{25}} = \dfrac{2}{5}$ 6. $\sqrt{\dfrac{37}{64}} = \sqrt{.57\ 81\ 25\ 00} = .7603$

$$
\begin{array}{r|l}
 & 49 \\
\hline
7 \times 20 = 140 & 8\ 81 \\
140 + 6 = 146 & 8\ 76 \\
\hline
76 \times 20 = 1520 & 5\ 25\ 00 \\
15200 + 3 = 15203 & 4\ 56\ 09 \\
\hline
\text{(Remainder)} & 68\ 91
\end{array}
$$

7. $\sqrt{\dfrac{5}{64}} = \sqrt{.07\ 81\ 25} = .279$

$$
\begin{array}{r|l}
 & 04 \\
\hline
2 \times 20 = 40 & 3\ 81 \\
40 + 7 = 47 & 3\ 29 \\
\hline
27 \times 20 = 540 & 52\ 25 \\
540 + 9 = 549 & 49\ 41 \\
\hline
\text{(Remainder)} & 2\ 84
\end{array}
$$

8. $\sqrt{27\dfrac{13}{64}} = \sqrt{27.20\ 31\ 25} = 5.215$

$$
\begin{array}{r|l}
 & 25 \\
\hline
5 \times 20 = 100 & 2\ 20 \\
100 + 2 = 102 & 1\ 04 \\
\hline
52 \times 20 = 1040 & 16\ 31 \\
1040 + 1 = 1041 & 10\ 41 \\
\hline
521 \times 20 = 10420 & 5\ 90\ 25 \\
10420 + 5 = 10425 & 5\ 21\ 25 \\
\hline
\text{(Remainder)} & 69\ 00
\end{array}
$$

EXTRACTING CUBE ROOT

The cube root of a number is a number that, when taken three times as a factor, is equal to the number given. The extraction of the cube root of a number is signified by the radical sign and the small number 3 which is the index.

EXTRACTING CUBE ROOT OF A PERFECT CUBE

For the purposes of illustrating the procedure of extracting the cube root of a perfect cube, the cube root of 52,313,624 will be extracted. The completed problem will be first presented and then the process involving seven steps is given.

Example:

$$
\begin{array}{r}
374 \\
\sqrt[3]{52\ 313\ 624} \\
27 \\
\end{array}
$$

$$
\begin{array}{rr}
& 25\ 313 \\
3^2 \times 300 = 2700 & \\
(3^2 \times 300 \times 7) + (3 \times 30 \times 7^2) + 7^3 = & 23\ 653 \\
& 1\ 660\ 624 \\
(37)^2 \times 300 = 410700 & \\
[(37)^2 \times 300 \times 4] + (37 \times 30 \times 4^2) + 4^3 = & 1\ 660\ 624 \\
\hline
& 0\ 000\ 000 \\
\end{array}
$$

Step 1. The given number is pointed off into periods of *three* digits, starting with the unit digit.

Step 2. Find the nearest cube root of the first period. This is 3. It is cubed and subtracted from the first period (52) and the first remainder (25) is obtained.

Step 3. Bring the next period down to the remainder, and a trial divisor is obtained by squaring the first digit in the root and multiplying it by the *constant 300*, giving $3^2 \times 300 = 2700$.

Step 4. Find a trial figure for the second digit in the root by dividing 2700 into 25313. If it is too high, a lower trial figure is used. After trial the second digit is found to be 7.

Step 5. Square the first digit in the root, then multiply the result by the constant 300 and multiply this product by the second digit in the root ($3^2 \times 300 \times 7$). Another quantity, which consists of the first digit in the root, is multiplied by the *constant 30*, and again multiplied by the square of the second digit in the root. It is then added to the quantity just obtained and, finally, the cube of the second digit is added. The expression is $(3^2 \times 300 \times 7) + (3 \times 30 \times 7^2) + 7^3$, and is subtracted from 25313, which gives the second remainder, 1660.

Step 6. Bring down the last period. A trial divisor is found in the manner previously described, except that in this operation the first two digits of the root are squared and multiplied by the *constant 300*, $(37)^2 \times 300 = 410700$.

Step 7. This step is similar to Step 5. The sum of the quantities $[(37)^2 \times 300 \times 4] + (37 \times 30 \times 4^2) + 4^3$ is obtained. Note that the

REVIEW OF ARITHMETIC

first quantity $(37)^2 \times 300 \times 4$ is the square of the first two digits in the root times the constant 300 times the third digit in the root. Finally, the sum is subtracted from the partial given number and the result is zero, indicating that the given number is a perfect cube.

EXTRACTING CUBE ROOT OF AN IMPERFECT CUBE

In extracting the cube root of an imperfect cube it must be decided how many decimal places are desired in the cube root. For each decimal place desired three ciphers (one period) are added after the decimal point. The procedure is then the same as for perfect cubes.

CHECKING THE EXTRACTION OF THE CUBE ROOT

The extraction of the cube root may be proved by repeating the root as a factor three times; the product must equal the number given. When the cube is not perfect the product should be almost equal to the given number. Extraction of the cube root may also be easily checked by the excess of nines method.

The excess of nines in the root is cubed, and, if it is equal to the excess of nines of the given number, the operation has been correctly performed. In case of a remainder, the excess of nines of the remainder is added to the cubed excess of nines of the root and is checked against the excess of nines of the given number.

Example: Extract the cube root of 673829.

$$
\begin{array}{r}
87.6 \\
\sqrt[3]{673\ 829\ 000} \\
512 \\
\hline
\end{array}
$$

$$
\begin{array}{rr}
8^2 \times 300 = 19200 & 161\ 829 \\
(8^2 \times 300 \times 7) + (8 \times 30 \times 7^2) + 7^3 = & 146\ 503 \\
\hline
(87)^2 \times 300 = 2370700 & 15\ 326\ 000 \\
\lbrack(87)^2 \times 300 \times 6\rbrack + (87 \times 30 \times 6^2) + 6^3 = & 13\ 718\ 376 \\
\hline
& 1\ 607\ 624
\end{array}
$$

excess of nines $= 3$
excess of nines $= 8$
$3^3 = 27 = 9 = 0$
$0 + 8 = 8$
(Remainder)
excess of nines $= 8$

Note: If it is desired to obtain in the root a result that offers a larger number of decimal places than does the given example, the operation may be extended.

EXTRACTING CUBE ROOT OF A DECIMAL FRACTION

When extracting the cube root of a decimal fraction, point off periods of three figures each to the right of the decimal point. Every pointed off period will result in one digit of the root. The procedure is then the same as for whole numbers.

When working with mixed decimal fractions, begin pointing off the periods to the right and the left of the decimal point and proceed to extract the root.

EXTRACTING CUBE ROOT OF COMMON FRACTIONS

If the given quantity from which the cube root is to be extracted is a common fraction, or is a mixed number containing a common fraction, the common fraction is converted into a decimal unless the numerator and the denominator are perfect cubes. If such is the case the cube root is extracted from the numerator and the denominator separately.

PRACTICE PROBLEMS APPLICABLE IN THE SHOP AND TOOLROOM

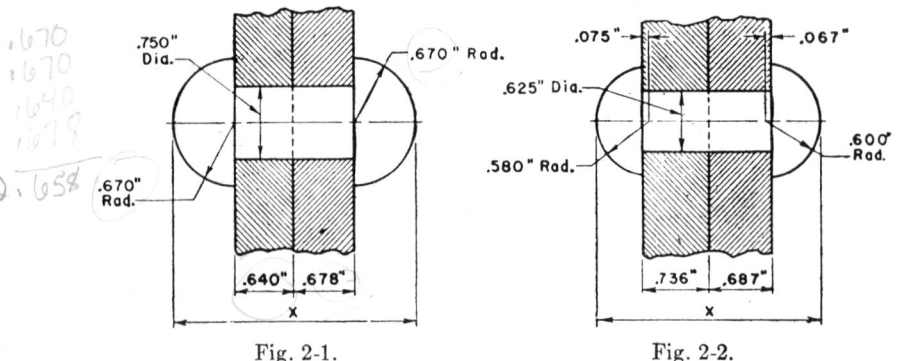

Fig. 2-1.   Fig. 2-2.

1. Find the total length X of the round-head assembled rivet shown in Fig. 2-1.
2. Find the total length X of the buttonhead assembled rivet shown in Fig. 2-2.

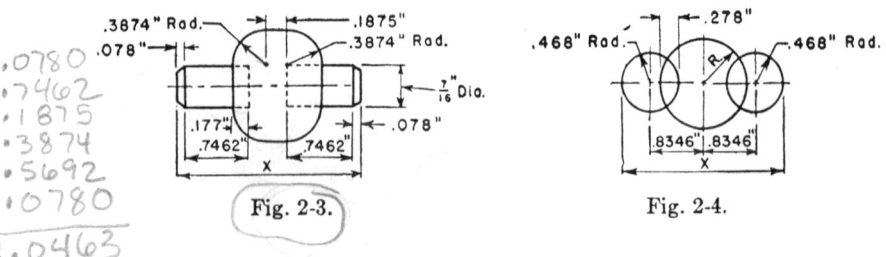

Fig. 2-3.   Fig. 2-4.

3. Find the total length X of the assembled unit shown in Fig. 2-3.
4. Find the radius R of the center disk and the over-all dimension X shown in Fig. 2-4.

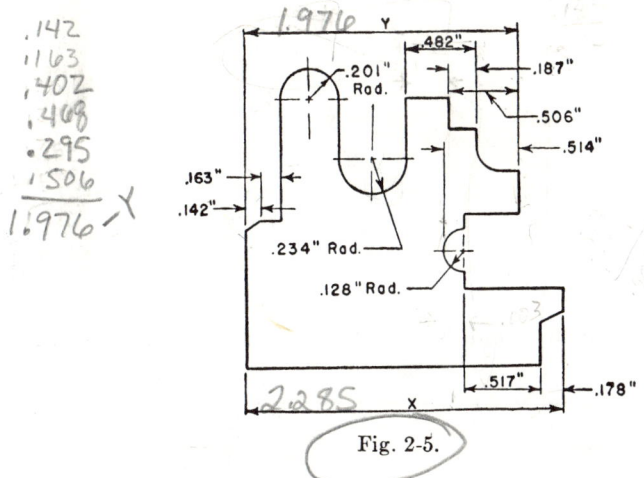

Fig. 2-5.

5. Find dimensions X and Y of the profile gage shown in Fig. 2-5.

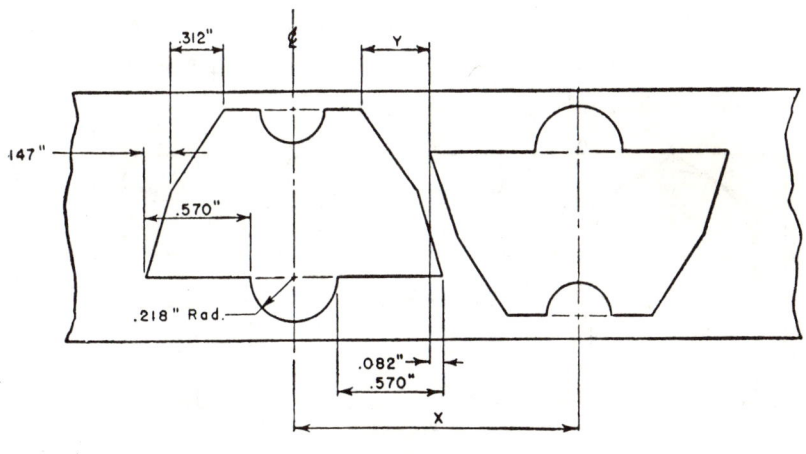

Fig. 2-6.

6. Find dimensions X and Y between identical blanks reversed to each other, shown in Fig. 2-6. Note that the blanks are symmetrical about their vertical center lines.

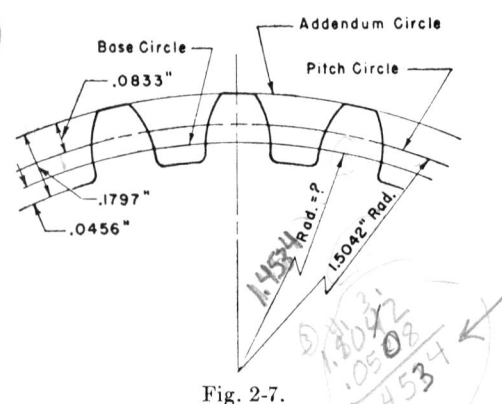

Fig. 2-7.

7. Find the base circle radius of an involute spur gear shown in Fig. 2-7.

Fig. 2-8.

8. A cylinder rod $1''$ in diameter is turned on a lathe to a final diameter $.840''$. A rough cut $\dfrac{1}{16}''$ deep is taken in the first operation. How deep is the finishing cut? See Fig. 2-8.

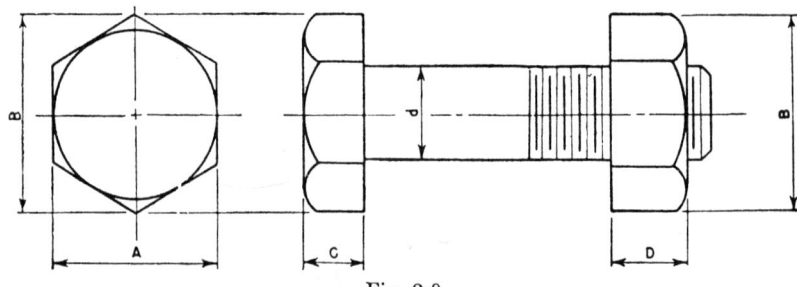

Fig. 2-9.

9. Dimension the unknown elements of a $\dfrac{7''}{8}$ hexagon-head bolt and nut in accordance with the proportions shown in Fig. 2-9.

d = diameter of bolt = $\dfrac{7''}{8}$

REVIEW OF ARITHMETIC

$A = 1.500 \times d = ?$ 1.3125
$B = 1.1547 \times A = ?$ 1.5155438
$C = .6666 \times d = ?$ $.583275$
$D = .875 \times d = ?$ $.765625$

10. Dimension the unknown elements of an $\frac{11''}{16}$ hexagon-head bolt and nut in accordance with the proportions shown in Fig. 2-9.

    $d =$ diameter of bolt $= \frac{11''}{16}$

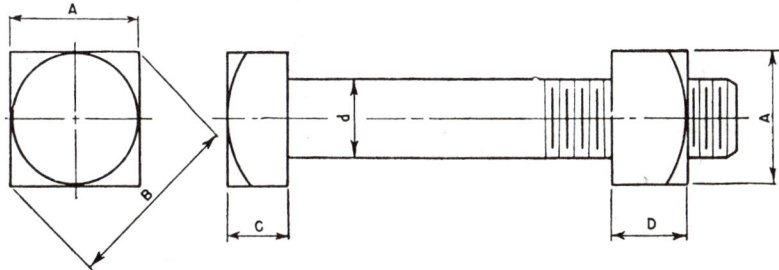

Fig. 2-10.

11. Dimension the unknown elements of a $\frac{3''}{4}$ square-head bolt and nut in accordance with the proportions shown in Fig. 2-10.

    $d =$ diameter of a bolt $= \frac{3''}{4} = .75$

    $A = 1.500 \times d = ?$ 1.125
    $B = 1.4142 \times A = ?$ $2.124995(68)$
    $C = .6666 \times d = ?$ $.49995$
    $D = .875 \times d = ?$ $.655625$

12. Dimension the unknown elements of a $\frac{13''}{16}$ square-head bolt and nut in accordance with the proportions shown in Fig. 2-10.

    $d =$ diameter of bolt $= \frac{13''}{16}$

    $\frac{19}{16}$

13. Dimension the unknown elements of a $1\frac{3''}{16}$ square-head bolt and nut in accordance with the proportions shown in Fig. 2-10.

    $d =$ diameter of bolt $= 1\frac{3''}{16} = 1.1875$

    $A = 1.78125$
    $B = 2.5190438$
    $C = .7915875$
    $D = 1.0390625$

52   MATHEMATICS FOR INDUSTRY

14. Dimension the unknown elements of a $\frac{3''}{8}$ round-head machine screw in accordance with the proportions shown in Fig. 2-11.

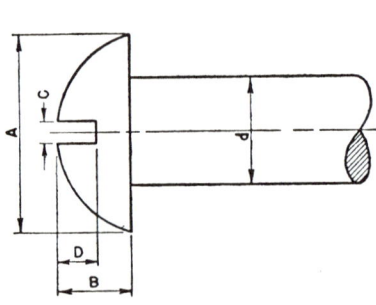
Fig. 2-11.

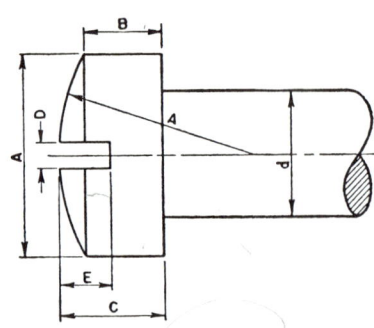
Fig. 2-12.

d = diameter of screw = $\frac{3''}{8}$
$A = (1.850 \times d) - .005 = ?$
$B = .700 \times d = ?$
$C = (.1730 \times d) + .015 = ?$
$D = (.500 \times B) + .010 = ?$

15. Dimension the unknown elements of a $\frac{7''}{16}$ oval fillister-head machine screw in accordance with the proportions shown in Fig. 2-12.

d = diameter of screw = $\frac{7''}{16}$ = .4375
$A = (1.640 \times d) - .009 = ?$ .7085
$B = (.660 \times d) - .002 = ?$ .28675
$C = (.134 \times A) + B = ?$ .381689
$D = (.173 \times d) + .015 = ?$ .0906375
$E = \frac{1}{2} \times C = ?$ .1908445

16. Dimension the unknown elements of a $\frac{9''}{16}$ flat-head machine screw in accordance with the proportions shown in Fig. 2-13.

d = diameter of screw = $\dfrac{9''}{16}$

$A = (2 \times d) - .008 = ?$

$B = \dfrac{d - .008}{1.739} = ?$

$C = (.173 \times d) + .015 = ?$

$D = \dfrac{1}{3} \times B = ?$

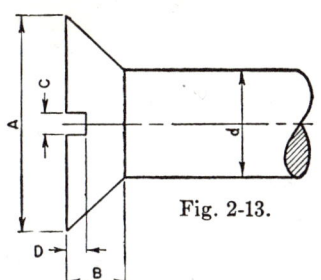

Fig. 2-13.

Calculate the number of threads per inch of the machine screws #3, #6, #12, #24, #28, and #30 in Problems 17 to 22, inclusive. Use the formula:

$N = \dfrac{6.5}{d + .020}$ = Nearest even whole number.

N = the number of threads per inch.
d = diameter of screw.

17. # 3 screw = .099″ diameter screw.
18. # 6 screw = .138″ diameter screw.
19. # 12 screw = .216″ diameter screw.
20. # 24 screw = .372″ diameter screw.
21. # 28 screw = .424″ diameter screw.
22. # 30 screw = .450″ diameter screw.

Round off the numerical values in Problems 23 to 28, inclusive, and retain the number of decimal places as indicated in each case:

23. 27.64132. Retain three decimal places.
24. 9.75478214. Retain four decimal places.
25. 18.438457. Retain three decimal places.
26. 372.637584. Retain three decimal places.
27. 78.7921503. Retain four decimal places.
28. 34.0399599. Retain four decimal places.
29. Retaining three decimal places, what dimension would you recommend for the cylinder bore diameter if the calculated diameter is equal to 4.598513″?
30. If the calculated diameter in Problem 29 is intended for the piston diameter, what dimension retaining three decimal places would you recommend for the piston diameter?
31. 374.0724 × 82.392 = ? Check result by excess of nines.
32. .8372 × .07468 = ? Check result by excess of nines.
33. 17.3068 × 8.0037 = ? Check result by excess of nines.

34. 637.0029 × 34.073 = ? Check result by excess of nines.
35. 84.30702 × 72.0093 = ? Check result by excess of nines.
36. 73.0465 × .4305 = ? Check result by excess of nines.
37. .003756 × .0009327 = ? Check result by excess of nines.
38. .3076 ÷ .825 = ? Check result by excess of nines.
39. .03458 ÷ .726 = ? Check result by excess of nines.
40. .008575 ÷ .653 = ? Check result by excess of nines.
41. .82567 ÷ .000758 = ? Check result by excess of nines.
42. 4.00364 ÷ .00754 = ? Check result by excess of nines.
43. 72.50302 ÷ 6.0437 = ? Check result by excess of nines.
44. .827503 ÷ 75.00408 = ? Check result by excess of nines.
45. .000578 ÷ 375.00823 = ? Check result by excess of nines.

Convert the decimal fractions in Problems 46 to 50 inclusive to common fractions:
46. .203125
47. .65625
48. .703125
49. .890625
50. .90625

Convert the common fractions in Problems 51 to 53 inclusive to decimal fractions:

51. $\dfrac{35}{64}$

52. $\dfrac{51}{64}$

53. $\dfrac{59}{64}$

54. Convert the common fraction $\dfrac{17}{45}$ to a decimal fraction consisting of five decimal places.

55. Convert the common fraction $\dfrac{23}{37}$ to a decimal fraction consisting of six decimal places.

56. Convert the common fraction $\dfrac{27}{53}$ to a decimal fraction consisting of four decimal places.

57. Convert the common fraction $\dfrac{25}{49}$ to a decimal fraction consisting of three decimal places.

58. Find in the decimal equivalent chart the common fraction that is closest to .58379.

59. Find in the decimal equivalent chart the common fraction that is closest to .85084.

REVIEW OF ARITHMETIC 55

60. Find in the decimal equivalent chart the common fraction that is closest to $\frac{47}{83}$.

    Hint: Convert to a decimal fraction and check with the chart.

61. Find in the decimal equivalent chart the common fraction that is closest to $\frac{59}{89}$.

62. No. 20 U.S. Standard gage sheet steel is .0359″ thick. Find its thickness in millimeters (four decimal places).

63. No. 14 B. & S. gage brass wire is .0641″ thick. Find its thickness in millimeters (four decimal places).

64. An aluminum sheet is 3.4 millimeters thick. Find its thickness in inches (three decimal places).

65. What is the equivalent of an 18 millimeter spark plug in inches (four decimal places)?

66. The pitch diameter of a $\frac{7''}{8}$—18 threads spark plug is .8384″ maximum.

    What is the equivalent of this decimal dimension in millimeters (three decimal places)?

67. Find the square root of 144 by inspection.
68. Find the square root of 625 by inspection.
69. Find the cube root of 1728 by inspection.
70. Find the cube root of 125 by inspection.
71. Find the square root of $\frac{144}{625}$ by inspection.

72. Find the square root of 521,284. Check result by excess of nines.
73. Find the square root of 6,775,609. Check result by excess of nines.
74. Find the square root of 664.6084. Check result by excess of nines.
75. Find the square root of 27. Carry the operation to four decimal places. Check by excess of nines.
76. Find by factoring the square root of 72, that is by treating the given number as the product of two factors, one of which is a perfect square and the other a single digit.

    Hint: $72 = 36 \times 2$. $\sqrt{72} = \sqrt{36} \times \sqrt{2}$.

77. Find the square root of 432 by factoring.

    Hint: $432 = 144 \times 3$.

78. Find the square root of 726 by factoring.

MATHEMATICS FOR INDUSTRY

Note: The following table will be useful for finding the square root of numbers by factoring.

$\sqrt{2} = 1.4142 \qquad \sqrt{3} = 1.7320 \qquad \sqrt{5} = 2.2360$

$\sqrt{6} = 2.4494 \qquad \sqrt{7} = 2.6457$

Find the square root of the numbers in Problems 79 to 91, inclusive. Check result by excess of nines:

79. 729
80. 45,796 — 214.
81. .075076
82. .555025
83. 4.301476
84. 1028.51055616
85. 1277.3476
86. .0725 (Four decimal places in the root.)
87. .32456 (Five decimal places in the root.)
88. 7 (Five decimal places in the root.)
89. 3 (Five decimal places in the root.)
90. $\dfrac{2}{3}$ (Three decimal places in the root.)
91. $7\dfrac{5}{8}$ (Three decimal places in the root.)

92. Find the square root of $\dfrac{4\frac{11}{16}}{7\frac{7}{64}}$ to three decimal places.

93. Find the square root of $\dfrac{23\frac{9}{32}}{4\frac{13}{64}}$ to four decimal places.

94. Find the square root of $\dfrac{43\frac{7}{16}}{7\frac{19}{32}}$ to four decimal places.

95. Find the square root of $\dfrac{49.0307}{6.3046}$ to four decimal places.

Find the cube root of the numbers in Problems 96 to 113, inclusive. Check results by the excess of nines.

96. 1
97. 2
98. 3
99. 4
100. 5
101. 6
102. 7
103. 9
104. 79,507
105. 373.248

REVIEW OF ARITHMETIC

106. 4,913
107. 110,592
108. 185,193
109. 1,124,864
110. 8,120,601
111. 731.432701
112. 121.4404
113. 1031.219362727
114. Find the cube root of 37.0082 to three decimal places. Check result by excess of nines.
115. Find the cube root of 3274.0706 to three decimal places. Check result by excess of nines.
116. Find the cube root of 48765.37 to three decimal places. Check result by excess of nines.
117. Find the cube root of 362975.108 to three decimal places. Check result by excess of nines.

118. Find the cube root of $\dfrac{7\frac{11}{64}}{23\frac{9}{16}}$ to three decimal places.

119. Find the cube root of $\dfrac{276\frac{11}{16}}{31\frac{5}{32}}$ to three decimal places.

120. Find the cube root of $\dfrac{87\frac{53}{64}}{3\frac{5}{8}}$ to three decimal places.

CHAPTER III

ALGEBRA

In the first two chapters of this text, most of the instruction pertained to definite numbers. Methods by which fundamental operations may be performed and various ways by which results can be checked were shown. The fundamental operations of which we speak are the processes of *arithmetic*, which is the science of positive real numbers. Two examples follow:

Example 1: The arithmetical statement $13^2 - 6^2 = (13 + 6) \times (13 - 6) = 19 \times 7 = 133$, demonstrates a method by which the difference of squares of two given numbers may be computed. If the given numbers are replaced by two other numbers, the method of computation is the same.

Example 2: The statement $(16 - 5)^2 = 16^2 - (2 \times 16 \times 5) + 5^2 = 121$, demonstrates a method by which the square of the difference of the two given numbers may be computed. The square of the difference of any two numbers would be computed in the same manner.

In order to make a rule that can be applied in examples of the kind just given, and in other kinds of examples as well, another branch of mathematics, called *algebra*, is employed. This useful branch of the science of mathematics is discussed in this and the succeeding chapter.

Algebra is an extension of arithmetic. Like arithmetic, it employs operations basic to all processes of computation; but for stating general rules, algebra makes use of symbols (usually letters) instead of definite numbers—though, frequently, numbers are used in combination with letters. The various signs, among which are $+$, $-$, $\times$, $\div$, etc., are used in algebra in the same way as in arithmetic.

ALGEBRAIC SYMBOLS

A symbol is a letter by which it is agreed to represent some object or thing. When working a problem in algebra the first thing necessary is to choose the symbols to be used, that is assign certain letters to each of the different objects concerned with the problem. In the Example 1, preceding, the 13 is to be represented by the symbol a, and the 6 by the symbol b. Then the example will be expressed algebraically in this manner: $a^2 - b^2 = (a + b) \times (a - b)$.

Similarly, the values of Example 2 can be expressed algebraically as

ALGEBRA

$(a - b)^2 = a^2 - 2ab + b^2$, where the number 16 is replaced by the symbol a, and the number 5 by the symbol b.

In form of expression these rules are *equations* or *formulas*, as the need occasions.

By use of algebra, we group simplified rules, such as formulas, systematically in order that they may be applied when operations involving definite numbers are required. The use of symbols for statement of general rules is the distinguishing feature of algebra. All of the rules laid down for the processes of arithmetic, as presented in the preceding chapters, are equally applicable to the processes of algebra, the difference being the substitution in algebra of letters and symbols for numbers.

ALGEBRAIC EXPRESSIONS

Algebraic expressions are mathematical statements representing certain relationships or conditions. They are divided into general classifications: *numerical* and *literal*. Examples 1 and 2, as first given, express definite numbers and the signs combining them and are of the numerical group. Examples 1 and 2 as given, where letters have been substituted for numbers, are of the literal group. An algebraic expression of the numerical group is $7, -4, 17 + 4 - (8 - 3)$; one of the literal group is $m + 7, n - 4, 17 + 4x - (8x - 3)$. The letters used in algebraic expressions represent numerical values. When we speak of a value of an algebraic expression, we mean the number it represents.

As stated before, the signs used in algebraic operations are similar to those used in arithmetic, except that in multiplication the sign may be expressed in different ways. For an example, m multiplied by n can be expressed, $m \times n$, $m \cdot n$, mn.

At times, it is necessary to group definite portions of an algebraic expression, and we may employ parenthesis ( ), brackets [ ], and braces { }.

Examples:

1. $\{[43a - (7b - 5a)] \times 23\} - 4b$

2. $[(43a - 7b + 5a) \times 23] - 4b$

These two algebraic expressions have the same value.

Any portion of an algebraic expression separated by a plus or a minus sign is a *term*. The sign prefixing the term is a part of the term. Algebraic expressions containing a single term are called *monomials*.

Examples of Monomials:

1. $3a$ 2. $7ab$

3. $11x^2y$ 4. $\dfrac{13a^2y}{4x}$

5. $\sqrt{169a^4x^2}$

Algebraic expressions containing two terms are called *binomials*.

Examples of Binomials:

1. $7ab - 3a$ 2. $11x^2y - \dfrac{13a^2y}{4x}$

Note: In the first example, $7ab$ and $-3a$ are the terms of the expression; similarly, in the second example, $11x^2y$ and $-\dfrac{13a^2y}{4x}$ are the terms of the expression.

Algebraic expressions containing two or more terms are called *polynomials*.

In the expression $11x^2y$ the factors of the expression are 11, x^2, and y. (When two or more quantities are multiplied to yield a product, each quantity is a factor.) Any one of these three factors, or the product of any number of them, is the *coefficient* of the remaining factors. Consequently, $11x^2$ is the coefficient of y and $11y$ is the coefficient of x^2. In the expression $\dfrac{13a^2y}{4x}$, the coefficient of $\dfrac{y}{x}$ is $\dfrac{13a^2}{4}$.

When speaking of coefficients, we usually are speaking about the numerical part of the expression. We state this as the numerical coefficient. The numerical coefficient of the term $11x^2y$ is 11 and the numerical coefficient of the term $\dfrac{13a^2y}{4x}$ is $\dfrac{13}{4}$.

If a literal term does not contain a numerical value, the numerical coefficient of the term is 1. In the term mn, no numerical part is expressed but 1 is understood and the term may be expressed as $1mn$.

If a group of factors constituting a product are equal to each other, the product of the factors is called the *power* of one of the factors. Thus $b \times b \times b$ is a power and it may be expressed in the form b^3. The factor b is called the *base*, and the small number 3 above and to the right of b is called the *exponent* of the power. In the expression b^3, 3 is the exponent and signifies that b is taken three times as a factor. The exponent may be a numerical value or it may be expressed as a letter; it may be a positive or a negative number or a whole or a fractional number.

ALGEBRA

In the expression $\sqrt{169a^4x^2}$, 4 is the exponent of a and 2 is the exponent of x. In the same expression, the symbol $\sqrt{}$ indicates the square root. Other symbols, such as $\sqrt[3]{}$ and $\sqrt[4]{}$, indicate, respectively, the cube root, the fourth root, and so on. The small number written above and to the left of the symbol $\sqrt{}$ indicates the root to be taken. This number is called the *index* of the root. In a square root, the index is not written.

DETERMINING VALUES OF NUMERICAL ALGEBRAIC EXPRESSIONS

A numerical algebraic expression consists of numbers and appropriate signs. The value of such an expression may be determined by carrying out the operations indicated by the signs.

Examples: The numerical value of:

1. $6 - 3 + 7 - 4 = 6$
2. $12 - 5 - 2 + 7 = 12$
3. $13 + \sqrt{64} = 21$
4. $4^2 + 3^3 - 5^2 = 18$
5. $\sqrt[3]{1728} + 5^3 - \sqrt{169} = 124$

DETERMINING VALUES OF LITERAL ALGEBRAIC EXPRESSIONS

The value of a literal algebraic expression is the number that it represents. In order to find the value, numbers are used in place of letters and symbols. The mathematical operations indicated by the signs are carried out as in numerical algebraic expressions.

Examples:

1. Find the value of $\frac{1}{4}pr$, if $p = 3.5$ and $r = 5$. Replacing the given letters by the appropriate numbers, we obtain $\frac{1}{4}pr = \frac{1}{4} \times 3.5 \times 5 = 4.375$.

2. Find the value of $\frac{\pi}{4}d^2$, if $\pi = 3.1416$ and $d = 3$. (π (pi) is equal to ratio of circumference of circle to its diameter.) Replacing the given letters by the corresponding numbers, we obtain $\frac{\pi}{4}d^2 = \frac{3.1416}{4} \times 3^2 = .7854 \times 9 = 7.0686$.

3. Find the value of $(a - b)^2 = a^2 - 2ab + b^2$, if $a = 9$ and $b = 4$. Replacing the given letters by the corresponding numbers, we obtain $(a - b)^2 = (9 - 4)^2 = 25$, or $a^2 - 2ab + b^2 = 81 - 72 + 16 = 25$.

POSITIVE AND NEGATIVE ALGEBRAIC NUMBERS AND EXPRESSIONS

The purpose of numbers in general is to provide a means by which the magnitude of quantities may be evaluated. For example, the net profit of a certain tool shop for the month of January is equal to $1,375.00. This value is a *positive* number. Other examples of positive numbers are: temperature, 84 F.; altitude, 1,200 feet above sea level.

Where values diminish below zero we deal with minus quantities. If the tool shop just mentioned incurred a loss of $625.00 for the month of February, the number 625 represents a *negative* number and is so indicated by placing a minus sign before it, viz., -625. Other examples of negative numbers are: $-23°$ F., which means 23 degrees below zero; -180 ft., which means 180 feet below sea level. In general, negative numbers are used to evaluate magnitudes below zero—or, as one might put it, negative numbers express magnitude of quantity in reverse. Positive and negative numbers, including 0, are called algebraic numbers. The value of a number without reference to its sign is called the absolute or numerical value or magnitude. Positive numbers may or may not have the positive sign prefixed to them but negative numbers always have the negative or minus sign $(-)$ prefixed.

Addition of Positive and Negative Algebraic Numbers. When adding two or more positive numbers or two or more negative numbers, their absolute magnitudes are added and the sum has their common sign.

Examples:

1. $(+14) + (+8) = 14 + 8 = 22$
2. $(-23) + (-12) = -(23 + 12) = -35$

When adding a group of positive and negative numbers, the positive numbers are added as a unit and the negative numbers are added as another unit, then the difference of the two sums is obtained and the sign of the greater sum is prefixed to the result.

Examples:

1. $17 + 4 - 5 + 8 - 12 - 6 = (17 + 4 + 8) + (-5 - 12 - 6) = 29 - 23 = 6$

ALGEBRA 63

2. $-34 - 18 + 27 + 4 - 26 = (27 + 4) + (-34 - 18 - 26) = 31 + (-78) = -(78 - 31) = -47$

The following rules, then, are suggested when two or more numbers are added algebraically:

1. When adding two or more numbers which have like signs, their algebraic sum is the sum of their absolute values, with the common sign prefixed.
2. When adding two or more numbers which have unlike signs, their algebraic sum is the difference between the sum of the positive values and the sum of the negative values, with the sign of the greater sum prefixed to the result.

Subtraction of Positive and Negative Algebraic Numbers. Subtraction of algebraic numbers is the opposite of addition of algebraic numbers. When a sum of two or more numbers is known, and one or more of the numbers is also known, the unknown number is found by subtraction, just as in arithmetic, except that in arithmetic the minuend is always greater than the subtrahend, while in algebra it may be greater or smaller, and, for another thing, one or all of the given numbers may be either positive or negative.

The following rules are suggested when two or more numbers are subtracted algebraically:

1. Subtracting a positive number or adding a numerically equal negative number produces the same result.

Example: $23 - (+16)$ is the same as $23 + (-16)$, the result in each case is equal to 7.

2. Subtracting a negative number or adding a numerically equal positive number produces the same result.

Example: $42 - (-17)$ is the same as $42 + 17$, the result in each case being equal to 59.

The following rule concerning disposition of the signs in subtraction may be suggested: The sign of the subtrahend is considered to be changed and the subsequent operation is similar to addition of algebraic numbers.

Examples:

1. $63 - (+27) = 63 - 27 = 36$ (The sign of the subtrahend 27 is changed from $+$ to $-$, and the operation performed is similar to addition.)

2. $84 - (+38) = 84 + 38 = 122$ (The sign of the subtrahend 38 is changed from $-$ to $+$, and the operation performed is similar to addition.

Addition and Subtraction of Literal Algebraic Expressions. In arithmetic it is known that one cannot add or subtract unlike things. For example, quarts and pounds cannot be added or subtracted. This is also true in algebra. One cannot add $4x$ to $3y$. The addition can only be indicated like $4x + 3y$.

When adding or subtracting literal algebraic expressions, similar terms, which are generally monomials, are grouped together and are added or subtracted from each other by adding or subtracting only the numerical coefficients, leaving the letters and symbols unchanged. If the monomials are not similar, the operations of addition or subtraction are merely indicated.

Examples of Addition of Monomials:

1. $-13mn$
    $+6mn$
    $-3mn$
    $+24mn$
    ───────
    $+14mn$

2. $+7ab$
    $-3ab$
    $+5ab$
    $-4ab$
    ───────
    $+5ab$

3. $+27xy$
    $-14xy$
    $+6xy$
    $-33xy$
    ───────
    $-14xy$

4. $+18x^2y^3$
    $-23x^2y^3$
    $+42x^2y^3$
    $-16x^2y^3$
    ───────
    $+21x^2y^3$

5. $+46mn$
    $-25x^2y$
    $+14abx$
    $-21mn$
    ───────
    $+25mn - 25x^2y + 14abx$

Examples of Subtraction of Monomials:

1. $17ab^2 - 12ab^2 = 5ab^2$
2. $28ab - (-13xy) = 28ab + 13xy$
3. $124abx - 46mny = 124abx - 46mny$
4. $-23x^2y^3 - 18x^2y^3 = -41x^2y^3$
5. $-42mn - (+26mn) = -16mn$
6. $93x^2y - (-mn) = 93x^2y + mn$

In adding or subtracting polynomials, the procedure is similar to that

ALGEBRA

indicated for monomials. Care must be taken to group similar terms in the same column so they can be combined as are monomials.

Example of Addition of Polynomials:

$$\begin{array}{r} -7ax - 12by^2 + 3mn \\ +12ax + 18by^2 - 14mn \\ -24ax - 17by^2 + 23mn \\ -41ax + 32by^2 - 47mn \\ +29ax + 76by^2 + 38mn \\ \hline -31ax + 97by^2 + 3mn \end{array}$$

Example of Subtraction of Polynomials: Subtract $(24xy^3 - 12mn + 18pr)$ from $(43xy^3 + 24mn - 32ab)$.

$$\begin{array}{r} 43xy^3 + 24mn - 32ab \\ 24xy^3 - 12mn \qquad\quad + 18pr \\ \hline 19xy^3 + 36mn - 32ab - 18pr \end{array}$$

Checking Addition and Subtraction of Algebraic Expressions. A simple operation may be performed if proof of an algebraic result is desired. In the preceding examples dealing with addition and subtraction of polynomials, arbitrarily chosen numerical values may be substituted for the letters given. Usually a value of 1 for all letters is chosen for the reason that the operation is simplified thereby.

Example of Proof of Addition of Polynomials: The value of 1 is used. $a = 1, b = 1, m = 1, n = 1, x = 1, y = 1.$

Example:	Proof:
$-7ax - 12by^2 + 3mn$	$-7 - 12 + 3 = -16$
$+12ax + 18by^2 - 14mn$	$+12 + 18 - 14 = +16$
$-24ax - 17by^2 + 23mn$	$-24 - 17 + 23 = -18$
$-41ax + 32by^2 - 47mn$	$-41 + 32 - 47 = -56$
$+29ax + 76by^2 + 38mn$	$+29 + 76 + 38 = +143$
$-31ax + 97by^2 + 3mn$	$-31 + 97 + 3 = +69$

Example of Proof of Subtraction of Polynomials: The value of 1 is used. $a = 1, b = 1, m = 1, n = 1, p = 1, r = 1, x = 1.$

Example:	Proof:
$43xy^3 + 24mn - 32ab$	$+43 + 24 - 32 \qquad = +35$
$24xy^3 - 12mn \qquad + 18pr$	$+24 - 12 \qquad + 18 = +30$
$19xy^3 + 36mn - 32ab - 18pr$	$+19 + 36 - 32 - 18 = +5$

66   MATHEMATICS FOR INDUSTRY

Algebraic results generally may be tested in this manner, and arbitrarily chosen numerical values may be substituted for the letters and symbols of the given and resulting algebraic expressions provided these values are used consistently throughout the entire operation. For rapid proof, the value of 1 is suggested for the reason previously stated.

Multiplication of Positive and Negative Algebraic Numbers and Expressions. When multiplying algebraic numbers and expressions, two fundamental rules must be followed, namely, the Rule of Signs and the Rule of Exponents.

1. *Rule of Signs.* When two algebraic quantities, both positive or both negative, are multiplied, the result is positive in both cases. When one of the quantities is positive and the other is negative, the result is negative.

Examples:

1. $a \times b = ab$ 2. $m \times n = mn$ 3. $(-x) \times (-y) = xy$

4. $(+c) \times (-d) = -cd$ 5. $-x^2 \times y = -x^2 y$ 6. $-m \times x^3 = -mx^3$

2. *Rule of Exponents.* When any single letter appears in both factors, the same letter will appear in the product with an exponent equal to the sum of the exponents of both factors. When no exponent is written, the exponent 1 is understood.

Examples:

1. $a^4 \times a^3 = a^{4+3} = a^7$ 2. $m^3 \times m^2 = m^5$

3. $ab^2c^3 \times a^2b^3c^4 = a^3b^5c^7$ 4. $ax^2y \times b^2x^3y^2 = ab^2x^5y^3$

The general rule applied when two algebraic quantities are multiplied is: The numerical coefficients are multiplied and the product is preceded by a sign, in accordance with the Rule of Signs. The next step consists of annexing letters that appear in the given factors, determining the proper exponents in accordance with the Rule of Exponents.

Examples:

1. $-13ax^2$
   $\times$
   $\underline{4a^2x^3}$
   $-52a^3x^5$

2. $17m^2y^3$
   $\times$
   $\underline{24my^2}$
   $408m^3y^5$

3. $-26n^2x^2y^3$
   $\times$
   $\underline{-12nx^2y^2}$
   $312n^3x^4y^5$

4. $27m^2n^2x^3y^4$ 5. $-6\frac{3}{5}x^2y^3z^4$

$\times -8mn^3x^2y^2$ $\times -10xy^2z^3$

$\overline{-216m^3n^5x^5y^6}$ $\overline{66x^3y^5z^7}$

The above algebraic expressions are monomials. When multiplying a polynomial by a monomial, the same procedure is followed, except that each term of the polynomial is multiplied by the monomial, the product having the same number of terms as the polynomial.

Examples of Multiplying a Polynomial by a Monomial:

1. $7a^2x^3+3ay^2$ 2. $-14m^2x^2-6m^3y^4$

$\times 4a^3x^2$ $\times -8my^3$

$\overline{28a^5x^5+12a^4x^2y^2}$ $\overline{112m^3x^2y^3+48m^4y^7}$

3. $7mnx+9nx^2-mnp$ 4. $-14prx^2-8ry^3+6r^2x^3$

$\times mx$ $\times 7rx^3$

$\overline{7m^2nx^2+9mnx^3-m^2npx}$ $\overline{-98pr^2x^5-56r^2x^3y^3+42r^3x^6}$

5. $5m^{2n} \times (7m^{3n}-3m^{2n}x) = 35m^{5n}-15m^{4n}x$

6. $(-6\frac{1}{2}p^{2n}+4p^{3n}x-7p^{4n}y)\times(-8p^{3n}z) = 52p^{5n}z-32p^{6n}xz+56p^{7n}yz$

When multiplying two polynomials, all terms of the multiplicand are multiplied by each term of the multiplier, thus as many partial products will occur as there are terms in the multiplier. These partial products are combined algebraically in a final product. For convenience, the partial products are arranged in a sequence that unites similar terms in a convenient form.

Examples of Multiplying a Polynomial by a Polynomial:

1. $4a^2y^3-7a^3y^4+12ay^2$

$\times$

$5ay+8a^2y^2$

$\overline{}$

$20a^3y^4-35a^4y^5+60a^2y^3$

$96a^3y^4+32a^4y^5-56a^5y^6$

$\overline{116a^3y^4-3a^4y^5+60a^2y^3-56a^5y^6}$

2. $-13xy^2-7x^2y^3+8x^3y^4$
$\times$
$\underline{\quad 12xy+9x^2y^2-14x^3y^3\quad}$
$-156x^2y^3-84x^3y^4+96x^4y^5$
$\quad\quad -117x^3y^4-63x^4y^5+72x^5y^6$
$\underline{\quad\quad\quad\quad 182x^4y^5+98x^5y^6-112x^6y^7}$
$-156x^2y^3-201x^3y^4+215x^4y^5+170x^5y^6-112x^6y^7$

Checking Multiplication of Positive and Negative Algebraic Numbers and Expressions. If proof of the algebraic result is desired when multiplying algebraic expressions, the procedure suggested for operations involving addition and subtraction may be followed. Numerical values of arbitrary choice may be substituted for the letters in the given and resulting expressions, but a value of 1 is recommended for simplicity and speed of operation.

Example of Proof of Multiplication of Polynomials:

   Example: *Proof:*
$\quad 4a^2y^3-7a^3y^4+12ay^2 \quad\quad 4\ -7\ +12\ = \quad\quad\quad\quad 9$
$\times$
$\underline{\quad 5ay+8a^2y^2 \quad\quad\quad\quad\quad\quad 5\ +8\quad\quad =\quad\quad\quad\quad 13\quad}$
$20a^3y^4-35a^4y^5+60a^2y^3 \quad 20\ -35\ +60\ +32\ -56\ +96\ =\ 117$
$\underline{96a^3y^4+32a^4y^5\quad\quad\quad\quad\quad\quad -56a^5y^6\quad\quad\quad\quad\quad\quad\quad\quad\quad}$
$116a^3y^4-3a^4y^5+60a^2y^3-56a^5y^6$

It should be noted that this rapid proof will test the coefficients of the result, but it will not discover an error made in the addition of exponents. For the latter test, any value except 1 may be substituted for the letters. Let us choose a value of 2 and substitute for all letters in the example. The operation is as follows:

          Proof:
$\quad (4\times4\times8)-(7\times8\times16)+(12\times2\times4)=128-896+96=\quad -672$
$\times$
$\underline{\quad (5\times2\times2)+(8\times4\times4)\quad\quad\quad\quad\quad\quad =20+128\quad\quad =\quad\quad 148\quad}$
$2560-17920+1920+16384-114688+12288\quad\quad\quad\quad=-99456$

Division of Positive and Negative Algebraic Numbers and Expressions. When dividing algebraic numbers and expressions, the fundamental rules regarding signs and exponents are:
1. *Rule of Signs.* Like signs in the dividend and the divisor will result

in a positive quotient. Unlike signs in the dividend and the divisor will result in a negative quotient.

Examples:

1. $ab \div a = b$ 2. $-ab \div -a = b$ 3. $ab \div -a = -b$
4. $mn \div n = m$ 5. $-mn \div -n = m$ 6. $-mn \div n = -m$

2. *Rule of Exponents.* Since division is the opposite of multiplication, the Rule of Exponents for division is the opposite of that given for multiplication. The rule states: The exponent of the quotient is equal to the difference between the exponents of the dividend and the divisor. When the exponents of the dividend and divisor are equal then the quotient is 1.

Examples:

1. $a^5 \div a^2 = a^{5-2} = a^3$

2. $a^3 \div a^5 = a^{3-5} = a^{-2}$ *Note:* $a^{-2} = \dfrac{1}{a^2}$

3. $x^5 y^3 \div x^2 y = x^{5-2} y^{3-1} = x^3 y^2$

4. $mx^4 y^3 \div mx^2 y = \dfrac{m}{m} x^{4-2} y^{3-1} = x^2 y^2$

The general rule applied when two algebraic quantities are divided is: The numerical coefficients are divided and the quotient is preceded by a sign, in accordance with the Rule of Signs. The letters are then annexed that appear with the given numbers and the proper exponents are determined in accordance with the Rule of Exponents.

Examples of Dividing Monomials by Monomials:

1. $\dfrac{24 m^3 n^4}{6 m^2 n} = 4 m n^3$ 2. $\dfrac{-36 p^4 r^3}{9 p^2 r} = -4 p^2 r^2$

3. $\dfrac{-42 x^3 y^2}{-6 x^2 y} = 7xy$ 4. $\dfrac{52 a^3 x^4 y^3}{-13 a x^2 y} = -4 a^2 x^2 y^2$

When dividing a polynomial by a monomial, the process is similar to that of long division of numbers. The first term of the dividend is divided by the divisor to obtain the first term of the quotient; then the second term of the dividend is divided by the divisor to obtain the second term of the quotient, the process being continued until all the terms of the dividend are divided by the divisor.

Examples of Dividing a Polynomial by a Monomial:

1. $\dfrac{36m^4n^2 + 28m^6n^4}{4m^2n} = 9m^2n + 7m^4n^3$

2. $\dfrac{84x^3y^2 - 64x^4y^3 + 44x^5y^4}{16xy^2} = 5.25x^2 - 4x^3y + 2.75x^5y^2$

3. $\dfrac{-17a^3b^2x^4 + 85ab^4x^3 - 51a^2b^2x^2}{-34abx} = .5a^2bx^3 - 2.5b^3x^2 + 1.5abx$

When dividing a polynomial by a polynomial, the process is similar to long division of numbers. The procedure is illustrated step by step in the following example: Divide $6a^2 - ab - 2b^2$ by $3a - 2b$.

$$\begin{array}{r}2a + b = \text{Quotient}\\\text{Divisor} = 3a - 2b \;\overline{\smash{\big)}\;6a^2 - ab - 2b^2} = \text{Dividend}\\6a^2 - 4ab\\\hline 3ab - 2b^2\\3ab - 2b^2\end{array}$$

Step 1. If possible, arrange the terms in the descending order of powers of some common letter.

Step 2. The first term of the dividend is divided by the first term of the divisor, and the first term of the quotient is obtained.

Step 3. The first term of the quotient is multiplied by all the terms of the divisor and the product is subtracted from the dividend.

Step 4. Annex the last term of the dividend to the remainder and divide the first term of the remainder by the first term of the divisor, and the second term of the quotient is obtained.

Step 5. The second term of the quotient is multiplied by all the terms of the divisor and the product is subtracted from the newly obtained dividend. There is no remainder, which proves that the division is complete. Had the dividend contained more terms than the given expression, the division would have been carried further along in the foregoing procedure until all the terms of the dividend were involved in the operation.

In the event of a remainder being no longer divisible by the first term of the divisor, the remainder can be expressed as a numerator of a common fraction, the denominator of which is the given divisor.

ALGEBRA

(71)

Examples of Dividing a Polynomial by a Polynomial:

1.

Divisor = $3ax^2 \ -4bxy^3$ = Quotient
$8a^2x^3 + 5\,b^3x^2y$ $\overline{\left) 24a^3x^5 \ -32a^2bx^4y^3 \ +15ab^3x^4y \ -20b^4x^3y^4 \right.}$ = Dividend
$\qquad\qquad\qquad\;\, 24a^3x^5 \qquad\qquad\quad\; +15ab^3x^4y$
$\qquad\qquad\qquad\;\, \overline{\qquad\qquad -32a^2bx^4y^3 \qquad\qquad\; -20b^4x^3y^4}$
$\qquad\qquad\qquad\;\, \qquad\qquad -32a^2bx^4y^3 \qquad\qquad\; -20b^4x^3y^4$

2.

Divisor = $8bx^2y\ -12axy^3 +16bx^3y^2$

$\qquad 4axy^2 \ +6bx^2$
$\overline{\left) 32abx^3y^3 \ -48a^2x^2y^5 \ +64abx^4y^4 \ +48b^2x^4y \ -72abx^3y^3 \ +96b^2x^5y^2\right.}$
$\;\;\; 32abx^3y^3 \ -48a^2x^2y^5 \ +64abx^4y^4$
$\qquad\qquad\qquad\qquad\qquad\qquad\qquad\;\, \overline{48b^2x^4y \ -72abx^3y^3 \ +96b^2x^5y^2}$
$\qquad\qquad\qquad\qquad\qquad\qquad\qquad\;\, 48b^2x^4y \ -72abx^3y^3 \ +96b^2x^5y^2$

Checking Division of Positive and Negative Algebraic Numbers and Expressions. Division of algebraic expressions may be checked by multiplying the divisor by the quotient and adding the remainder. Division of algebraic expressions can also be tested by methods suggested for testing addition, subtraction, and multiplication of the same expressions. If a check of coefficients is desired, a value of 1 may be substituted for all the letters in the given and the final expressions. For checking the exponents, a value such as 2 may be substituted for all the letters in the given and final expressions. The procedure has been fully explained in the foregoing operations, and the reader is urged to apply the principles he has learned when testing the result obtained in division of algebraic expressions.

FACTORING

When multiplying two or more quantities to yield a product, each quantity is a factor of the product. Factoring is the inverse of this process. In factoring one is given a product and must discover the factors that have been multiplied together to form the product. Thus it may be said that a factor of an algebraic expression is an expression that will exactly divide the given expression (product). A *prime* expression is an expression that has no other factors than itself and one.

Special Products. As an aid to factoring, special products that occur frequently as a result of the multiplication of certain expressions should be known. This enables one to quickly classify problems according to one or more of these types and then follow the method used for factoring that type. It is the knowledge of these various types that enable the factoring of complex problems since many algebraic expressions contain portions which are in the form of special products. The products of the following types occur frequently:

Type 1. $a(x + y + z) = ax + ay + az$

Example: $4ax(3a + x + 4ax) = 12a^2x + 4ax^2 + 16a^2x^2$

Type 2. $(a + b)(a - b) = a^2 - b^2$

Example: $(4m + 5n)(4m - 5n) = (4m)^2 - (5n)^2 = 16m^2 - 25n^2$

Type 3. $(a + b)^2 = (a + b)(a + b) = a^2 + 2ab + b^2$

Example: $(3m + n^2)^2 = (3m)^2 + (2 \times 3m \times n^2) + (n^2)^2 = 9m^2 + 6mn^2 + n^4$

Type 4. $(a - b)^2 = (a - b)(a - b) = a^2 - 2ab + b^2$

Example: $(4x^2 - 7y^3)^2 = (4x^2)^2 - (2 \times 4x^2 \times 7y^3) + (7y^3)^2 = 16x^4 - 56x^2y^3 + 49y^6$

Type 5. $(x + a)(x + b) = x^2 + (a + b)x + ab$

Example: $(6r + 7m)(6r + 5n) = (6r)^2 + (7m + 5n)6r + (7m \times 5n) = 36r^2 + 42mr + 30nr + 35mn$

Type 6. $(a + b)(x + y) = x(a + b) + y(a + b) = ax + bx + ay + by$

Example: $(6m + 3n)(5a + 2b) = (6m \times 5a) + (3n \times 5a) + (6m \times 2b) + (3n \times 2b) = 30am + 15an + 12bm + 6bn$

Factoring by Inspection. Complicated algebraic expressions can be simplified by means of factoring. Factoring is also useful in the solution of difficult equations. It should always be remembered that the process of factoring is simply determining the factors (unknown) of the product which is known.

When factoring the first thing one should look for is a common factor. Then, after finding the common factor by inspection, each term of the expression should be divided by this factor. The type formulas in the previous article "Special Products" become formulas for factoring by reading them from right to left.

Type 1. $ax + ay + az = a(x + y + z)$. In looking first for a common factor we see that a is common to all three terms. The common factor a is written down and each term is divided by a giving $ax + ay + az = a(x + y + z)$. This can be proved by multiplying a by $x + y + z$.

Example: Factor $63m^4x^3y^2 - 42m^2x^2y^3 + 21mx^4y - 126m^3x^2y^2$

The common factor is $21mx^2y$.

The equivalent expression is $21mx^2y(3m^3xy - 2my^2 + x^2 - 6m^2y)$.

In many cases there is not a factor common to the terms of the expression. One must then determine if the expression to be factored fits any of the other type formulas.

Type 2. $a^2 - b^2 = (a + b)(a - b)$. Factor $9a^2 - 4b^2$ according to this formula. $9a^2 - 4b^2 = (3a)^2 - (2b)^2 = (3a + 2b)(3a - 2b)$. We speak of this operation as the difference of the squares. From the formula we can see that the difference of the squares of two numbers is equal to the sum of the two numbers multiplied by the difference of the two numbers.

Type 3. $a^2 + 2ab + b^2 = (a + b)(a + b) = (a + b)^2$. Factor $25m^2 + 20m + 4$. In accordance with the formula we have $25m^2 + 20m + 4 = (5m)^2 + 2 \times 5m \times 2 + 2^2 = (5m + 2)^2$.

Type 4. $a^2 - 2ab + b^2 = (a - b)(a - b) = (a - b)^2$. In accordance with this formula, $36x^4 - 36x^2y + 9y^2 = (6x^2)^2 - 2 \times 6x^2 \times 3y + (3y)^2 = (6x^2 - 3y)^2$.

These two type formulas represent trinomials which are perfect squares. It should be noticed that when two terms are perfect squares the third term is plus or minus twice the product of the square roots of the other terms.

Type 5. Trinomials of the form $x^2 + px + q$ are factored by the formula in Type 5, $x^2 + (a + b)x + ab = (x + a)(x + b)$. To factor trinomials of this form the a and b must be found so that their product is q and their sum is p. When factoring $x^2 + 7x + 12$ then, we have $x^2 + 7x + 12 = (x + 4)(x + 3)$. Analyzing this expression it is seen that the coefficient of the middle term (7) of the trinomial is the same as the sum of the last terms $(4 + 3)$ of the factors, and the last term (12) of the trinomial is the product of the last terms (4×3). This is the pattern by which many algebraic expressions can be recognized and factored.

Examples:

1. $y^2 + 6y - 16 = (y + 8)(y - 2)$
2. $x^2 - 12x + 32 = (x - 8)(x - 4)$
3. $x^2 + 9x + 14 = (x + 7)(x + 2)$
4. $x^2 + 7xy + 10y^2 = (x + 2y)(x + 5y)$

Factoring by Grouping. Some expressions upon first examination do not reveal a factor common to all terms of the expression or do not exactly fit any of the first five type formulas. In that case, grouping may be necessary where factors are discovered that are common to each group, resulting in an expression that can be factored further or grouping can be used to put expressions into the common type forms. Type formula 6 is an example of grouping.

Example 1:

$$6x^2 + 18mx + 48xy + 8nx + 24mn + 64ny = 6x(x + 3m + 8y) + 8n(x + 3m + 8y) = (6x + 8n)(x + 3m + 8y)$$

The expression factored has no factor common to all terms, but, upon close scrutiny, the first three terms of the expression are discovered to have a common factor, $6x$, and the last three terms a common factor, $8n$. The regrouped expression is then factored further.

Example 2:

$$x^3 - 3x^2 + 6x - 18 = (x^3 - 3x^2) + (6x - 18) = x^2(x - 3) + 6(x - 3) = (x - 3)(x^2 + 6)$$

The first two terms of the expression and the last two terms of the expression are factored separately, and another factoring is performed to obtain the simplified expression.

POWERS, EXPONENTS, AND ROOTS

Rules similar to the rules of signs and exponents for multiplication and division of algebraic expressions may be applied when raising an expression to a desired power or when extracting a desired root, since raising an expression to a certain power is equivalent to multiplying the expression by itself a number of times, and extracting its root is equivalent to performing division.

ALGEBRA

Powers and Exponents. The *Rules of Signs* as applied to the raising of an expression to a desired power are:
1. When a positive quantity is raised to any power, the result is positive.
2. When a negative quantity is raised to an even power, the result is positive.
3. When a negative quantity is raised to an odd power, the result is negative.

Examples:

1. $(+x)^2 = (+x)(+x) = x^2$
2. $(+x)^3 = (+x)(+x)(+x) = x^3$
3. $(-x)^2 = (-x)(-x) = x^2$
4. $(-x)^3 = (-x)(-x)(-x) = -x^3$

The *Rules of Exponents* as applied to the raising of an expression to a desired power are:

1. When raising an expression to a desired power, the exponents of the expression and the power are multiplied by each other.

Examples:

1. $(x^3)^2 = x^3 x^3 = x^{3 \times 2} = x^6$
2. $(y^4)^3 = y^{4 \times 3} = y^{12}$

2. When multiplying powers by each other, the exponents are added.

Examples:

1. $x^3 \times x^2 = x^{3+2} = x^5$
2. $x^4 y^3 \times x^3 y^2 = x^{4+3} y^{3+2} = x^7 y^5$
3. $7x^3 y^2 \times (-4x^2 y) = -28 x^5 y^3$

3. When dividing one power by another power, the exponents are subtracted from each other. A quantity raised to the zero power is equal to 1.

Examples:

1. $x^7 \div x^5 = x^{7-5} = x^2$
2. $3x^5 y^3 \div 2x^3 y = 1.5 x^2 y^2$

3. $x^5 \div x^5 = x^{5-5} = x^0$. But $\dfrac{x^5}{x^5} = 1$, therefore $x^0 = 1$

4. $\dfrac{x^5}{x^7} = x^{5-7} = x^{-2}$. But $\dfrac{x^5}{x^7} = \dfrac{1}{x^2}$, therefore $x^{-2} = \dfrac{1}{x^2}$

5. $\dfrac{3x^3}{7x^6} = \dfrac{3}{7}x^{3-6} = \dfrac{3}{7}x^{-3} = \dfrac{3}{7}\cdot\dfrac{1}{x^3} = \dfrac{3}{7x^3}$

6. $\dfrac{4x^5y^3}{5x^2y^5} = .8x^3y^{-2} = \dfrac{.8x^3}{y^2}$

7. $\dfrac{28x^3y^2}{5x^5y^6} = 5.6x^{-2}y^{-4} = \dfrac{5.6}{x^2y^4}$

Roots. The operation of extracting the root of an expression is contrary to that of raising the expression to a power. Therefore, the Rules of Exponents are likewise contrary to that used in raising an expression to a power. The *Rules of Exponents* are:

1. When extracting the root of an expression, the index of the root is divided into the exponent of the expression.

Examples:

1. $\sqrt{x^4} = x^{\frac{4}{2}} = x^2$

2. $\sqrt[3]{x^6} = x^{\frac{6}{3}} = x^2$

3. $\sqrt[3]{y^8} = y^{\frac{8}{3}} = y^{2\frac{2}{3}}$

4. $\sqrt[3]{64x^6y^9} = 4x^{\frac{6}{3}}y^{\frac{9}{3}} = 4x^2y^3$

5. $13\sqrt[4]{16x^8y^{12}} = 13\cdot 2x^{\frac{8}{4}}y^{\frac{12}{4}} = 26x^2y^3$

6. $\sqrt[4]{256x^2y^6} = 4x^{\frac{2}{4}}y^{\frac{6}{4}} = 4x^{\frac{1}{2}}y^{\frac{3}{2}}$

2. When extracting the root of a product, the roots of the individual factors are multiplied by each other.

Examples:

1. $\sqrt{x^4y^6} = \sqrt{x^4}\cdot\sqrt{y^6} = x^{\frac{4}{2}}y^{\frac{6}{2}} = x^2y^3$

2. $\sqrt[3]{x^6y^9z^{12}} = \sqrt[3]{x^6}\cdot\sqrt[3]{y^9}\cdot\sqrt[3]{z^{12}} = x^{\frac{6}{3}}\cdot y^{\frac{9}{3}}\cdot z^{\frac{12}{3}} = x^2y^3z^4$

3. $\sqrt[3]{8x^8\cdot y^6\cdot 27z^9} = 2x^{\frac{8}{3}}y^2 3z^3 = 6x^{\frac{8}{3}}y^2z^3$

ALGEBRA

Note that if an expression is in the form of a polynomial, its root is not the sum of the roots of the individual terms.

Example:

$$\sqrt[3]{x + y + z} \text{ is not equal to } \sqrt[3]{x} + \sqrt[3]{y} + \sqrt[3]{z}$$

The expression under the radical sign must be treated in its entirety and not broken up into its components.

Examples of Extracting the Root of an Expression:

1. $\sqrt{a} = a^{\frac{1}{2}}$, and since $a^{\frac{1}{2}} \times a^{\frac{1}{2}} = a^{\frac{1}{2}+\frac{1}{2}} = a^1 = a$, the following can be stated: $\sqrt{a} \cdot \sqrt{a} = a$.

2. $\sqrt[3]{a} = a^{\frac{1}{3}}$, and $a^{\frac{1}{3}} \times a^{\frac{1}{3}} \times a^{\frac{1}{3}} = a$; therefore, $\sqrt[3]{a} \cdot \sqrt[3]{a} \cdot \sqrt[3]{a} = a$.

3. $a^{\frac{1}{3}} \cdot a^{\frac{1}{3}} = a^{\frac{2}{3}} = \sqrt[3]{a^2}$

4. $a^{\frac{3}{2}} = \sqrt{a^3}$

5. $a^{\frac{3}{4}} = \sqrt[4]{a^3}$

It should be noted that the square of a positive or a negative quantity is always positive. Therefore, the square root of a positive quantity may be positive or negative.

Examples:

1. $\sqrt{a^2} = \pm a$ 2. $\sqrt{36} = \pm 6$

3. $\sqrt{9b^2} = \pm 3b$ 4. $\sqrt{144a^4b^2} = \pm 12a^2b$

The square root or, for that matter, any even root of a negative number does not really exist and is an imaginary expression. Imaginary expressions may be simplified as shown in the example.

Example:

$$\sqrt{-36} = \sqrt{36 \times (-1)} = \sqrt{36} \times \sqrt{-1} = 6\sqrt{-1} = 6i, \text{ where } i = \sqrt{-1}.$$

Frequently in engineering computations long expressions can be written in a shortened form, such as 3.75×10^7, or 437×10^{-9}, etc. These expressions are easier to handle than their equivalents of longer form, namely, $3.75 \times 10^7 = 3.75 \times 10,000,000 = 37,500,000$, and 437

$$\times 10^{-9} = 437 \times \frac{1}{10^9} = 437 \times \frac{1}{1,000,000,000} = \frac{437}{1,000,000,000} = .000000437.$$

SYMBOLS OF GROUPING

If an algebraic expression consists of a number of terms connected by the conventional signs of mathematical operations, multiplication and division are performed first and are followed by addition and subtraction.

Examples:

1. $6 + 4 \times 3 - 5 + 8 \div 2 - 7 + 9 = 6 + (4 \times 3) - 5 + (8 \div 2) - 7 + 9 = 6 + 12 - 5 + 4 - 7 + 9 = 31 - 12 = 19$

2. $3x - 2x \cdot y + 4a - a \div x + m \cdot n = 3x - 2xy + 4a - \frac{a}{x} + mn$

In the above operations, symbols such as ( ) parenthesis, [ ] brackets { } braces, etc., are used for grouping various terms of algebraic expressions in proper sequence. When removing or inserting parentheses, or other symbols of grouping, preceded by a plus sign, rewrite the terms within the symbols without altering their signs. When removing or inserting parentheses, or other symbols of grouping, preceded by a minus sign, rewrite the terms involved with their signs changed.

The symbols indicate the order in which the operations are performed. It is common to remove the innermost pair of symbols first.

Examples:

1. $\{16 + [9 + (36 \div 4)] - 5 + (4.5 \times 2) + 8\} \times (12 - 5) = [16 + (9 + 9) - 5 + 9 + 8] \times 7 = (16 + 18 - 5 + 9 + 8) \times 7 = 46 \times 7 = 322$

2. $[12 + 4(x - 3y)][4 - 7(5 - 2)] - 5x + 3y = (12 + 4x - 12y) \times [4 - 7(3)] - 5x + 3y = (12 + 4x - 12y)(4 - 21) - 5x + 3y = 48 + 16x - 48y - 252 - 84x + 252y - 5x + 3y = -73x + 207y - 204$

3. $447 - \{x^2y + [63 - (7x^2y + 13) + 12x^2y]\} = 447 - [x^2y + (63 - 7x^2y - 13 + 12x^2y)] = 447 - (x^2y + 63 - 7x^2y - 13 + 12x^2y) = 447 - x^2y - 63 + 7x^2y + 13 - 12x^2y = -6x^2y + 397$

ALGEBRA

Since the minus sign in the last example precedes the braces and the parentheses, all of the signs within the groupings change from $+$ to $-$, and from $-$ to $+$ upon removal of the braces and the parentheses.

FORMULAS

A mathematical rule that can be expressed by the proper signs and symbols is called a formula. The use of letters and symbols to represent numerical values makes it possible to solve various kinds of problems with comparative ease. Instead of applying certain mathematical principles to the solution of one particular problem, many problems of a particular category may be solved by the use of one properly selected formula. The construction of a formula by the use of letters and symbols, instead of numbers, condenses all necessary information into a limited space. When a specific problem is to be solved, its given values are inserted in the properly chosen formula, and the subsequent operations become that of simple arithmetic.

Examples of Formulas:

1. The circumference of a circle is equal to the constant π multiplied by the diameter of the circle. This statement can be condensed in a formula: Circumference $= \pi d$, where $\pi = 3.1416$ and $d =$ the diameter of the circle. The circumference of any circle may be found by the use of the formula πd. All that is necessary is to substitute the given diameter instead of symbol d.

2. The area of any circle can be expressed by the formula $\frac{\pi}{4}d^2$, where $\frac{\pi}{4} = \frac{3.1416}{4} = .7854$ is a constant and $d =$ the diameter of the circle.

3. The distance traveled by a train can be expressed by the formula $d = rt$, where $r =$ the rate of travel of train in miles or other units, and $t =$ time in hours, minutes, or seconds.

4. The formula for the indicated horsepower of a steam engine is:

$H = \dfrac{PLAN}{33,000}$, where $H =$ indicated horsepower of the engine.

$P =$ mean effective pressure on piston in pounds per square inch.
$L =$ length of piston stroke in feet.
$A =$ area of piston in square inches.
$N =$ number of strokes of piston per minute.

The indicated horsepower of a steam engine may be found from the preceding formula by substituting the known values of the engine in question and proceeding to work the problem by arithmetic.

SIMPLE EQUATIONS

A formula may also be called an equation, since it is a statement of equality between two algebraic expressions. An equation consists of two parts combined by the sign of equality ($=$).

Axioms. The following modifications of equations may be made without changing their real values. These statements are called *axioms*, since their truth is evident and no further derivation or proof is necessary.

1. The same quantity may be added or subtracted from both sides of the equation.

Examples:

1. If $3x^2y + 10 = 7xy^2 - 4$, then $3x^2y + 10 + 12 = 7xy^2 - 4 + 12$. (12 is added to both sides of the equation.)

2. If $\frac{6}{13}xy^2 - 8 = 1.4x^2y + 2$, then $\frac{6}{13}xy^2 - 8 - 7 = 1.4x^2y + 2 - 7$. (7 is subtracted from both sides of the equation.)

2. Each side of the equation may be multiplied or divided by the same quantity.

Examples:

1. If $12ax^2y = 7bxy^2$, then $(12ax^2y)x = (7bxy^2)x$, $12ax^3y = 7bx^2y^2$. (x is multiplied by each side of the equation.)

2. $\frac{12ax^2y}{x} = \frac{7bxy^2}{x}$, then $12axy = 7by^2$. (Each side of the equation is divided by x.)

3. Each side of the equation may be raised to the same power, and the same root may be extracted from each side of the equation.

Examples:

1. If $3ax^2y + 7m = 5axy^2 - 3n$, then $(3ax^2y + 7m)^3 = (5axy^2 + 3n)^3$. (Each side of the equation is raised to the third power.)

2. $\sqrt{3ax^2y + 7m} = \sqrt{5axy^2 - 3n}$. (The square root is extracted from each side of the equation.)

Solution of Simple Equations. When the numerical value of an unknown symbol in an equation can be determined, the equation can be solved.

The solution of equations may often be simplified and hastened by applying the foregoing axioms or by rearranging or transposing the equations, thereby definite values represented by various letters can be determined.

By transposition of a formula we mean that any independent term may be transposed from one side of the equation to the other, providing its sign is changed from $+$ to $-$ or from $-$ to $+$.

Also, a term which appears as a factor of every term in one side of the equation may be transposed to be a quotient of every term in the other side of the equation, or vice versa.

Examples:

1. If $14x + 8y + 3 = 47$, then $14x + 8y = 47 - 3$.
2. If $23x + 12a - 7 = 63$, then $23x + 12a = 63 + 7$.
3. If $17x^3y^2 + 27x^4y^3 - 12x^2y^4 = 28x^2y^3 - 13x^3y^2 + 19x^4y^3$, then $x^2y^2(17x + 27x^2y - 12y^2) = 28x^2y^3 - 13x^3y^2 + 19x^4y^3$, and $17x + 27x^2y - 12y^2 = \dfrac{28x^2y^3 - 13x^3y^2 + 19x^4y^3}{x^2y^2} = 28y - 13x + 19x^2y$.

Simple equations containing one unknown quantity can be solved by the following operations:
1. All terms containing the unknown quantity are transposed to one side of the equation, and all the other terms are transposed to the other side of the equation.
2. All similar terms are combined.
3. Both sides of the equation are divided by the coefficient of the unknown.

Examples:

1. Given equation: $7x + 16 = 14 + 8 - 5x$
   Operation 1: $7x + 5x = 14 + 8 - 16$
   Operation 2: $12x = 6$
   Operation 3: $x = \dfrac{6}{12} = .5$

2. Given equation: $7y + 12 = 11y - 27$
Operation 1: $12 + 27 = 11y - 7y$
Operation 2: $39 = 4y$
Operation 3: $y = \dfrac{39}{4} = 9\dfrac{3}{4}$

3. Given equation: $12.75x - 22 = 15 + 7.25x$
Operation 1: $12.75x - 7.25x = 15 + 22$
Operation 2: $5.50x = 37$
Operation 3: $x = \dfrac{37}{5.5} = 6.727$

Finding the Unknown Quantity by the Process of Elimination. Many simple equations can be solved by the process of elimination. The process of elimination consists of transposing the unknown quantity to one side of the equation and isolating it from the rest of the terms by operations that follow the principles of the foregoing axioms. The process will be illustrated by examples.

Examples:

1. Let us take the formula representing the total surface area of a cylinder: $S = 2\pi R^2 + 2\pi Rh$, where $R =$ the radius of the cylinder, and $h =$ the height of the cylinder.
Assume all values of the equation except h are known. It is required to solve for h. Then by applying the principles in the foregoing axioms, the given equation can be changed as follows:

Given equation: $S = 2\pi R^2 + 2\pi Rh$

Transposition: $S - 2\pi R^2 = 2\pi Rh$

Dividing both sides of the equation by $2\pi R$:

$$\dfrac{S - 2\pi R^2}{2\pi R} = \dfrac{2\pi Rh}{2\pi R}$$

Cancellation of $2\pi R$ in right side of the equation:

$$\dfrac{S - 2\pi R^2}{2\pi R} = h \text{ or}$$

$$h = \dfrac{S - 2\pi R^2}{2\pi R}$$

ALGEBRA

2. The formula representing the volume of a sphere is: $V = \frac{4}{3}\pi R^3$, where R = the radius of the sphere. If the volume of the sphere is known, its radius can be found by the process of elimination, thus:

$$V = \frac{4}{3}\pi R^3, \quad \frac{V}{\frac{4}{3}\pi} = \frac{\frac{4}{3}\pi R^3}{\frac{4}{3}\pi}, \quad \frac{V}{\frac{4}{3}\pi} = R^3, \quad \frac{3V}{4\pi} = R^3, \quad R = \sqrt[3]{\frac{3V}{4\pi}}.$$

If, on the other hand, the radius of the sphere is known, its volume can be found thus:

$R = \sqrt[3]{\frac{3V}{4\pi}} = \left(\frac{3V}{4\pi}\right)^{\frac{1}{3}}$. Note from the foregoing: $\sqrt[3]{\text{quantity}} = (\text{Quantity})^{\frac{1}{3}}$

$$R^3 = \left[\left(\frac{3V}{4\pi}\right)^{\frac{1}{3}}\right]^3 = \frac{3V}{4\pi}, \quad 4\pi \times R^3 = \frac{3V \times 4\pi}{4\pi}, \quad 4\pi R^3 = 3V, \quad \frac{4\pi R^3}{3} = \frac{3V}{3}, \quad \frac{4\pi R^3}{3} = V \text{ or } V = \frac{4}{3}\pi R^3.$$

3. The formula for finding the pitch diameter of a National form thread serew is: $E = M + \frac{.86602}{N} - 3G$,

where E = pitch diameter
M = measurement over the wires
N = number of threads per inch
G = wire diameter

Assume all values are given except N. To determine N in the given formula by the process of elimination, we proceed as follows:

$$E = M + \frac{.86602}{N} - 3G, \quad E - M + 3G = \frac{.86602}{N}, \quad (E - M + 3G) \times$$

$$N = \frac{.86602}{N} \times N = .86602, \quad \frac{(E - M + 3G) \times N}{E - M + 3G} = \frac{.86602}{E - M + 3G},$$

$$N = \frac{.86602}{E - M + 3G}.$$

4. The formula for the outside diameter of a spur gear is: $D = \frac{(N + 2)P}{\pi}$, where D = outside diameter, N = number of teeth in gear, and $P =$ circular pitch. Assume all values are given except N. It is required to find N.

$$D = \frac{(N+2)P}{\pi}, D\pi = \frac{(N+2)P \times \pi}{\pi} = (N+2)P, \frac{D\pi}{P} = \frac{(N+2)P}{P} = N+2, \frac{D\pi}{P} - 2 = N+2-2, \frac{D\pi}{P} - 2 = N.$$

Formulas which are statements of principles or rules represented by equations will be used extensively throughout this text for the purpose of solving shop problems. The ability to handle various formulas for rapid determination of unknown quantities is essential. Very often these formulas may be obtained from standard handbooks and, after substituting the known values for the letters and symbols, the unknown quantities can be determined. If it so happens that the unknown quantity is involved with other symbols and letters and cannot be detached readily for simple solution, the previously illustrated methods of isolating the unknown value are applied.

Frequently, the reader may be confronted with problems that demand exercises of sound judgment and it will be necessary for him to develop independently his own formulas or equations for the purpose of finding an unknown quantity or quantities. The formula, remember, is to be looked upon as a means to an end, and sufficient skill should be acquired to apply the chosen formula correctly in actual practice.

PRACTICE PROBLEMS APPLICABLE IN THE SHOP AND TOOLROOM

1. $14 \times 7 + 8 - 3 \times 9 = ?$
2. $224 \div 7 + 8 \times 3 - 23 = ?$
3. $-27 + 16 \times 7 + \frac{84}{12} - 8 = ?$
4. $-17 \times 8 + 42 - 17 + \frac{63}{-9} = ?$
5. $-23 + 14 - 8 \times 6 + 72 + \frac{-78}{-6} = ?$
6. $32 - 17 - 13 \times 7 + 61 - \frac{54}{-6} = ?$
7. $\sqrt{127 + 42} + 3^3 - \sqrt[3]{93 - 29} = ?$
8. $-13 + \sqrt{172 + \frac{72}{3}} - \frac{\sqrt[3]{27}}{3} = ?$

ALGEBRA

9. Find the value of $\frac{3}{8}ab$, if $a = 7$ and $b = 8$.

10. Find the value of $2\pi r$, if $\pi = 3.1416$ and $r = 1.25$.
11. Find the value of πd, if $d = 3.25$.

12. Find the value of $\frac{\pi}{4}d^2$, if $d = 2$.

13. Find the value of $\frac{\pi}{4}d^2$, if $d = 3.2$.

14. Find the value of $(a - b)^2$, if $a = 7$ and $b = 3$ and check result with $a^2 - 2ab + b^2$.
15. $(-14) - (-27) + (-7) = ?$
16. $28 + (-13) - (+6) = ?$
17. $37 - 3 + 16 - 12 - 23 = ?$
18. $-46 - (+35) + (-18) - (-23) + 18 = ?$

Add the monomials in Problems 19 to 22 inclusive:

19. $6ab, -13ab, -22ab, +32ab, -17ab.$ $-14ab$
20. $-27mn, -18mn, 46mn, 37mn, -8mn.$
21. $+28ax^2y^3, +34ax^2y^3, 46ax^2y^3, -57ax^2y^3, 23ax^2y^3.$ -50
22. $-64mn, -53x^2y, 43abx, -12mn, 46xy^2.$

Subtract the values in Problems 23 to 28, inclusive:

23. $26ab^2 - 17ab^2 = ?$ $-64mn$
24. $39a^2b - (-17ab^2) = ?$ $-12mn$
25. $-32axy^2 - 17axy^2 = ?$
26. $-27mny^2 - (-15mny^2) = ?$ $-76mn$
27. $74x^2y - (-mnx^2) = ?$
28. $\sqrt[3]{-125} \cdot x^2y - (-5x^2y) = ?$ -5
29. Find the algebraic sum of the following values: $3m^2nx, -17ax^2y, 23m^2nx, -11m^2nx, 16ax^2y, -23ax^2y, -26ax^2y.$

Add the polynomials in Problems 30 and 31:

30. $\quad 23ax + 21by^2 - 7mn$ 31. $-32mx^2y - 12nxy^2 + 6abx$
    $\quad -17ax - 81by^2 + 24mn$ $\quad -71mx^2y + 18nxy^2 - 42abx$
    $\quad -42ax + 71by^2 - 32mn$ $\quad 24mx^2y - 17nxy^2 + 23abx$
    $\quad 14ax - 23by^2 + 74mn$ $\quad -41mx^2y + 32nxy^2 - 47abx$
    $\quad -92ax + 67by^2 - 38mn$ $\quad 29mx^2y - 27nxy^2 + 34abx$

32. Subtract $42xy^2 - 21abx^2 + 17prx^3$ from $-73xy^2 - 41abx^2 - 62prx^3$.
33. Subtract $-24abx + 12my^2 - 74pz^3$ from $37abx - 12my^2 + 35pz^3 - 21xyz$.

Note: Test Problems 30, 31, 32, and 33 by substituting *1* wherever a letter appears in the given expressions.

34. $(-a) \times (-b) = ?$ $+ab$
35. $(m) \times (-n) = ?$ $-mn$
36. $x \times y \times z = ?$ xyz
37. $-x \times y^3 = ?$ $-xy^3$
38. $x^3 \times (-y^2) = ?$ $-x^3y^2$
39. $-x \times 3x^2 \times (-4x^3) = ?$ $+4x^4 - 12x^5$
40. $a^2bc^3 \times ab^2c^4 = ?$ $a^3b^3c^7$
41. $axy^2 \times a^2x^2y = ?$ $a^3x^3y^3$
42. $ax^2y^3 \times bc^2xy = ?$ $ac^2x^3y^4$
43. $(-23a^2x) \times (12ax^2) = ?$
44. $(26my^2) \times (-43my^2) = ?$
45. $\left(-12\frac{3}{8}ax^2y\right) \times (-8a^2xy^2) = ?$
46. $\left(-8\frac{3}{4}mx^3y^2\right) \times \left(-7\frac{1}{2}m^2xy^3\right) = ?$
47. $(24a^2x^3 + 8ay^3) \times 7a^2x^3 = ?$
48. $(-17mx^3 - 12m^2y^3) \times (-23m^3y^2) = ?$
49. $(16mn^2x^3 - 8n^3x^2 + mnp^2) \times m^2x = ?$
50. $(-23p^2rx^3 - 16r^2y^2 + 13r^3x^2 - 9x^2y^3) \times 13r^2x^2 = ?$
51. $13m^{3n} \times (18m^{5n} + 12m^{4n}x - 7m^n) = ?$
52. $\left(-13\frac{2}{3}p^{3n} - 6p^{4n}x + 8p^{2n}y\right) \times (-12p^{5n}z) = ?$
53. $(26ay^2 + 9a^2y^3 - 14a^3y) \times (13a^2y^2 - 7ay) = ?$
54. $(-32mx^2y - 16m^2xy^2 + 22m^3x^3y^3) \times (-9m^2xy^3 - 15mx^2y^2 + 5m^2x^4y^3 + m^4x^2y^5) = ?$
55. $(48ax^2y - 37a^3xy^2 + 62a^2x^3y^3) \times (14a^2x^2y^2 - 12x^3y + 25a^3y - 8a^3y) = ?$

Note: Test Problems 53, 54, and 55 by substituting *2* where a letter appears in the given expressions.

56. $pr \div r = ?$
57. $-mn \div n = ?$
58. $-xy \div -x = ?$
59. $a^7 \div a^3 = ?$
60. $x^4 \div x^6 = ?$
61. $x^7y^4 \div x^3y^2 = ?$
62. $ax^5y^3 \div ax^4y^2 = ?$
63. $72m^4n^2 \div 9m^2n = ?$
64. $-81a^4x^7y^4 \div -3a^2x^4y^2 = ?$
65. $-96m^2x^3y^2 \div 12m^2xy^2 = ?$
66. $63n^4y^2z \div -9n^2yz = ?$
67. $\dfrac{-123ax^3y^2}{3 \times 41ax^2xy^5} = ?$
68. $\dfrac{129m^3n^2 + 86m^5n^3}{43m^2n} = ?$
69. $\dfrac{-136a^2x^3y^2 - 85abx^4y^4}{-34ax^3y} = ?$
70. $\dfrac{12x^2 + 33xy + 18y^2}{4x + 3y} = ?$
71. $\dfrac{21x^3y^3 - 35x^2y^4 + 12x^4y^2 - 20x^3y^3}{5xy^2 - 3x^2y} = ?$
72. $\dfrac{84ab^2x^3y^3 - 112a^2b^2x^2y^4 - 72abx^4y^4 + 96a^2bx^3y^5}{14b^2xy^2 - 12bx^2y^3} = ?$

ALGEBRA

73. $\dfrac{276mnx^3y^4 + 368n^2x^4y^2 - 414mn^2x^5y^5 - 384m^2x^2y^6 - 512mnx^3y^4 + 576m^2nx^4y^7}{23nx^2y - 32mxy^3} = ?$

Note: Test Problems 70, 71, 72, and 73 by substituting 2 wherever a letter appears in the given expressions.

Solve the Problems 74 to 84 inclusive by considering the given expressions as special products.

74. $(5a + 3b)^2 = ?$
75. $(7p^2 + 4r)^2 = ?$
76. $(6m^2 + 8n^2)^2 = ?$
77. $(7x^3 - 4y^2)^2 = ?$
78. $(-12m^3 - 3n^2)^2 = ?$
79. $(6p + 3r)(6p - 3r) = ?$
80. $(4x^3 + 2y^3)(4x^2 - 2y^3) = ?$
81. $(5m + 3n)(16x + 4y) = ?$
82. $(7a + 2b)(8x + 3y) = ?$
83. $(8x + m)(8x + n) = ?$
84. $(12x + 7a)(12x + 4b) = ?$

Solve the Problems 85 to 98 inclusive with the aid of the special products:

85. $(x + 7)(x + 4) = ?$
86. $(ab + 3)(ab + 2) = ?$
87. $(x + 6)(x - 3) = ?$
88. $(x - 8)(x - 4) = ?$
89. $(2m + 3n)^2 = ?$
90. $(3a + r)^2 = ?$
91. $(x - 2p)^2 = ?$
92. $(2y - .5n)^2 = ?$
93. $(5m + 3n)(5m - 3n) = ?$
94. $(7p - 2.5r)(7p + 2.5r) = ?$
95. $(3x^2y + 2ab)(pr + 7) = ?$
96. $\left(\dfrac{1}{2}ax + \dfrac{1}{3}by\right)\left(1\dfrac{1}{4}my + 2\dfrac{1}{5}nx\right) = ?$
97. $(12xy^2 + 3mn)(12xy^2 + 5pr) = ?$
98. $\left(\sqrt{x + \dfrac{p}{r}}\right)\left(\sqrt{x + pr}\right) = ?$

Find the factors of the expressions in the Problems 99 to 109 inclusive:

99. $224a^3xy^2 + 42ax^3y - 98a^2x^2y^3 + 84ax^4y^2 = ?$
100. $12a^2 + 36ab + 96ax + 16ay + 48by + 128xy = ?$
101. $16x^2 - y^2 = ?$
102. $36z^2 - 72z + 36 = ?$
103. $x^3 - 3x^2 + 6x - 18 = ?$
104. $x^4 + 8x^2 - 48 = ?$
105. $y^2 - 24y + 128 = ?$
106. $x^4y^2 + 18x^2y - 144 = ?$
107. $a^6 + 30a^3 + 125 = ?$
108. $x^2 + 13xy + 42y^2 = ?$
109. $a^2 - 28a + 192 = ?$

Simplify or solve the expressions in Problems 110 to 143 inclusive by any convenient method:

110. $x^2 + 11x + 20 = ?$
111. $y^2 + 13y + 27 = ?$
112. $(x^3y^4)^3 = ?$
113. $(3x^2y^3)^4 = ?$

114. $(-7m^3n^4)^5 = ?$
115. $(-5m^2nx^2)^3 = ?$
116. $(-y^2z^4)^2 = ?$
117. $(6^m n^p)^2 = ?$
118. $(6m^a y^3)^b = ?$
119. $\left(\dfrac{3}{7}\dfrac{a^3}{x^2}\right)^4 = ?$
120. $\left(\dfrac{3}{8}\dfrac{m^2}{b^3}\dfrac{n}{x^3}\right)^4 = ?$
121. $\left[\left(\dfrac{x}{y}\right)^3\right]^4 = ?$
122. $(13a^x b^{2y} z^m)^{3n} = ?$
123. $\sqrt[3]{m^6 n^9 x^3} = ?$
124. $\sqrt[5]{\dfrac{3^5 m^{10} n^5}{4^{10} a^{15} b^{10}}} = ?$
125. $x^0 = ?$
126. $.6^0 = ?$
127. $4000^0 = ?$
128. $.7a^0 = ?$
129. $.007x^0 = ?$
130. $4^{-3} = ?$
131. $7^{-2} = ?$
132. $24^{\frac{1}{2}} = ?$
133. $27^{\frac{1}{3}} = ?$
134. $3^{-\frac{1}{2}} = ?$
135. $27^{\frac{1}{3}} = ?$
136. $m^{-4}x^{-\frac{1}{2}} \div m^{-5} = ?$
137. $5m^{\frac{1}{2}}n^{\frac{1}{4}} \times 3m^{\frac{1}{2}}n^{\frac{1}{4}} = ?$
138. $4.63 \times 10^6 = ?$
139. $34.07 \times 10^7 = ?$
140. $.0372 \times 10^7 = ?$
141. $72.06 \times 10^{-8} = ?$
142. $.724 \times 10^{-7} = ?$
143. $.00653 \times 10^9 = ?$

Express the expressions in Problems 144 to 147 inclusive in shorter forms:

144. $73{,}000{,}000 = ?$
145. $\dfrac{437}{10{,}000{,}000} = ?$
146. $.0000006734 = ?$
147. $\dfrac{37.829}{100{,}000{,}000} = ?$
148. $87 - [16 + (8 - 3)] = ?$
149. $34 + [73 - (26 + 8) + 9] = ?$
150. $36 + \{53 - [27 - (6 - 18) - 12]\} = ?$
151. $92 - \{46 - [48 - (36 - 14)] - (18 - 7)\} + 24 = ?$
152. $192 + \{68 - [24 + (21 - 6) - (26 - 13)]\} = ?$
153. $641 - \{24 + [28 - (28 - 12) + (8 - 3)]\} = ?$
154. Find the value of $\pi r^2 h$, if $\pi = 3.1416$, $r = 4.5$ and $h = 7$.
155. Find the value of $x^3 + 3x^2 y + 3xy^2 + y^3$, if $x = 5$ and $y = 4$.
156. Find the value of $\sqrt{m(m-a)(m-b)(m-c)}$, if $m = \dfrac{a+b+c}{2}$, $a = 42$, $b = 34$, and $c = 23$.
157. Evaluate $x^2 + 3y^3 + 4z^4$, if $x = 4$, $y = 3$, and $z = 2$.
158. Evaluate $m^3 + 3m^2 n + 6mn^2 - 4mn^3 + n^3$, if $m = 2$ and $n = 3$.
159. Evaluate $\dfrac{24x^3 - 6y^2}{4x^2} + \dfrac{5z^2}{x+y^2} - \dfrac{x+y^2+z^3}{7y^3}$, if $x = 4$, $y = 3$, $z = 2$.

ALGEBRA

160. Evaluate $(m - 2n + 3p)^2 - (n - 2p + 3r)^2 + (p - 2r + 3s)^2$, if $m = 1, n = 2, p = 3, r = 4$, and $s = 5$.

161. Evaluate $\sqrt{3a^2 + 4b^3 + c}$, if $a = 2, b = 3$, and $c = 4$.

162. Evaluate $(xy - mn)\sqrt{x^2ym + y^2mn - 4}$, if $x = 2, y = 3, m = 4$, and $n = 0$.

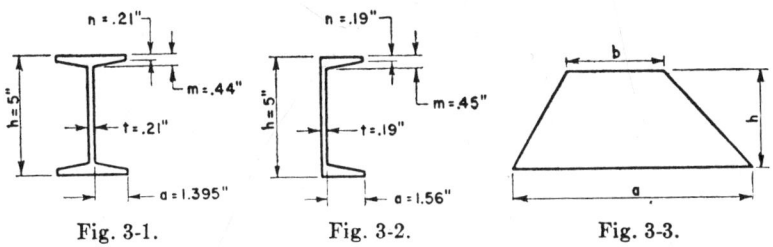

Fig. 3-1.   Fig. 3-2.   Fig. 3-3.

163. Find the area A of the cross section of the I beam shown in Fig. 3-1 in accordance with formula $A = ht + 2a(m + n)$.

164. Find the area A of the cross section of the channel iron shown in Fig. 3-2, in accordance with formula $A = ht + a(m + n)$.

165. The area A of a trapezoid shown in Fig. 3-3 can be expressed by the formula $A = \dfrac{h}{2}(a + b)$. It is required to solve for h. Use the principle of elimination of all terms except the unknown as previously explained.

166. Assuming the formula in Problem 165, solve for a.

167. Assuming the formula in Problem 165, solve for b.

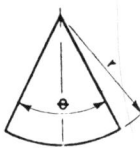 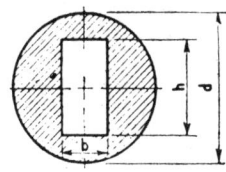 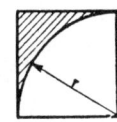

Fig. 3-4.   Fig. 3-5.   Fig. 3-6.

168. The area A of a sector shown in Fig. 3-4 can be expressed by the formula $A = \dfrac{\pi r^2 \theta}{360}$. Solve for r.

169. The area A of the cross-sectioned portion shown in Fig. 3-5 can be expressed by the formula $A = \dfrac{\pi d^2}{4} - bh$. Solve for h.

170. Assuming the formula in Problem 169, solve for d.

171. The area A of the cross-sectioned portion shown in Fig. 3-6 can be expressed by the formula $A = r^2 - \frac{\pi}{4}r^2$. Simplify this formula and solve for r.

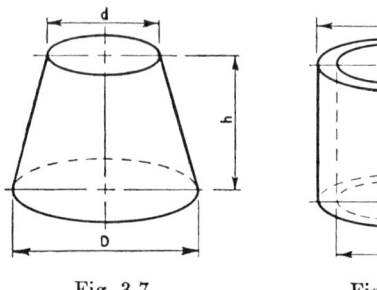

Fig. 3-7.   Fig. 3-8.

172. The volume V of the frustum of the cone shown in Fig. 3-7 can be expressed by the formula $V = \frac{\pi}{12}h\,(D^2 + d^2 + Dd)$. Solve for h.
173. The volume V of the solid material of the hollow cylinder shown in Fig. 3-8 can be expressed by the formula $V = \frac{\pi}{4}(D^2 - d^2)h$. Solve for h, solve for D and solve for d.
174. The weight W of the steel rivet shown in Fig. 3-9 can be expressed by the formula $W = .28 \times \left(\frac{2}{3}\pi R^3 + \frac{\pi}{4}d^2h\right)$, where W = weight in pounds, R, d, and h are in inches. Solve for R, solve for d, and solve for h.

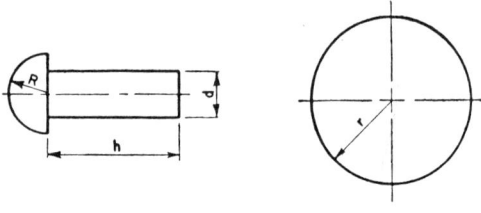

Fig. 3-9.   Fig. 3-10.

175. The weight, W, of the aluminum ball shown in Fig. 3-10 can be expressed by the formula, $W = .1 \times \left(\frac{4}{3}\pi r^3\right)$, where W = weight in pounds and r = radius in inches. Solve for r.
176. The formula for finding the pitch diameter of an Acme screw thread

is expressed by the following: $E = M + \dfrac{1.93334}{N} - 4.9939G$, where E = pitch diameter, M = measurement over the wires, N = number of threads per inch and G = wire diameter. Find M, find N, and find G.

177. The sum of two numbers is 37 and their difference is 3. Find the numbers.

Hint: Assume any one of the unknown numbers = x, and express the other unknown in terms of x. Develop an equation which shows the relationship between the unknown numbers and solve for x.

178. A tank can be filled by three inlet pipes operating separately in 9, 15, and 27 minutes respectively. How long will it take to fill the tank if all the pipes operate simultaneously?

Hint: Assume the volume of the tank = x, and express the filling capacity of each pipe per minute.

179. If 13 is added to four times the number, the result exceeds three times the number by 23. What is the number?

180. A square surface is changed to a rectangular surface of equal area by increasing one side of the square by 18 feet and decreasing the other side by 14 feet. What is the side of the original square?

181. A boy is $\dfrac{1}{3}$ as old as his father and $\dfrac{1}{5}$ as old as his grandfather. The sum of all three is 126. How old is each?

182. A boy is $\dfrac{1}{3}$ as old as his father, but in 13 years he will be only $\dfrac{1}{2}$ as old as his father. How old are father and son at present?

CHAPTER IV
ALGEBRA—CONTINUED

RATIOS

The relation of one quantity to another can be expressed mathematically in several different ways. One method frequently employed in comparison of like quantities is to state their relation in form of a ratio, which consists of dividing one quantity by another.

For example, if it is desired to compare 65 with 13, it may be done by expressing the relationship of the numbers as a ratio. Ratio is written as an indicated division or fraction in any of the following forms: $65:13$; $65 \div 13$; $\frac{65}{13}$. The first term (65) is called the *antecedent* and is the dividend, while the second term is called the *consequent* and is the divisor.

A ratio expresses the relationship of like quantities. For instance, the wealth of two people reckoned in dollars may be compared as a ratio. However, if the wealth of one person consists in ownership of a certain number of horses and the wealth of another in ownership of a given number of bushels of wheat, a ratio cannot be shown unless the wealth of both is measured by a single standard, as by sums of money or some other medium of exchange.

Examples:

1. The ratio of 14 to 2 is $\frac{14}{2} = \frac{7}{1}$ or $7:1$.

2. The ratio of m to $n = \frac{m}{n}$ or $m:n$.

3. The ratio of $18a$ to $3b = \frac{18a}{3b} = \frac{6a}{b}$ or $6a:b$.

4. The ratio of 12 to 18 $= \frac{12}{18} = \frac{2}{3}$ or $2:3$.

Inverse Ratios. An inverse ratio is a given ratio inverted. It is the reciprocal of the given ratio.

Examples:

1. If the given ratio of two numbers is $\frac{12}{5}$, the inverse ratio is $\frac{5}{12}$.

ALGEBRA

2. If the given ratio is $\dfrac{m}{n}$, the inverse ratio is $\dfrac{n}{m}$.

3. If the given ratio is $\dfrac{3a}{4b}$, the inverse ratio is $\dfrac{4b}{3a}$.

Reducing Ratios to Lowest Terms. Ratios may be reduced to their lowest terms by dividing both terms by a common factor. As ratios may be written in form of common fractions, the reduction is similar to that of fractions.

Examples:

1. The given ratio $\dfrac{20}{6}$ may be reduced to $\dfrac{10}{3}$.

2. The ratio $\dfrac{26}{4} = \dfrac{13}{2}$.

3. The ratio $\dfrac{12a^2b}{8ab^2} = \dfrac{3a}{2b}$.

PROPORTIONS

A proportion is a statement of equality between two ratios. The statement $16:4 = 20:5$ is a proportion and is read: 16 is to 4 as 20 is to 5. Each ratio constituting a proportion usually represents a comparison of like quantities as stated in the preceding article. For example, in the proportion $16 \div 4 = 20 \div 5$, the first ratio may indicate a comparison of hours used by a machine operator to complete two projects and the second ratio may indicate a comparison of costs of the two projects. The proportion may be read: 16 hours is to 4 hours as $20 is to $5. This proportion may be indicated in the following forms: $16 \div 4 = 20 \div 5$; $16:4::20:5$; $\dfrac{16}{4} = \dfrac{20}{5}$.

A proportion consists of four terms. The first and last terms are the *extremes* and the second and third terms are the *means* of the proportion. In the above proportion the terms 16 and 5 are the extremes and 4 and 20 are the means.

If the proportion is in the form $18 \div 12 = 12 \div 8$ (the second term of the proportion is equal to the third term), the term 12 is designated as the *mean proportional*.

Principles of Proportion. An important principle of proportion is that the product of the mean terms is equal to the product of the extreme terms. In the foregoing example $16 \div 4 = 20 \div 5$, the product of 16×5

is equal to the product 4 × 20. Problems dealing with proportions are solved by applying this principle. Usually three terms involved in the proportion are known, and the unknown term is found by equating one product to the other.

If, in the foregoing example, any one term is not known, say the third term, then $16 \times 5 = 4 \times x$, and $x = \dfrac{16 \times 5}{4} = 20$.

When, in a proportion, the mean proportional is an unknown quantity, it may be found by extracting the square root of the product of the extreme terms. If in the proportion $18 \div 12 = 12 \div 8$ the mean proportional is unknown, then it is equal to $\sqrt{18 \times 8} = \sqrt{144} = 12$.

PRINCIPLES USED IN ALTERING PROPORTIONS

Proportions may be altered in accordance with the following principles and still maintain the equality of both ratios:

1. Both sides of a proportion may be multiplied or divided by the same quantity.

   Examples:

   1. $\dfrac{36}{9} = \dfrac{56}{14}, \quad \left(\dfrac{36}{9}\right) \times 3 = \left(\dfrac{56}{14}\right) \times 3, \quad \dfrac{36}{9} \div 2 = \dfrac{56}{14} \div 2.$

   2. $4 = 4, 4 \times 3 = 4 \times 3, 4 \div 2 = 4 \div 2.$

2. In any proportion the first term is to the third as the second term is to the fourth, which in reality is alternating the mean terms.

   Example:

   If $36 \div 9 = 56 \div 14$, then $36 \div 56 = 9 \div 14$.

   Proof:

   $36 \times 14 = 9 \times 56, 504 = 504; 36 \times 14 = 56 \times 9, 504 = 504.$

3. In any proportion, the second term is to the first as the fourth term is to the third, which in reality is inversion of the proportion, making the mean terms take the place of the extreme terms, and vice versa.

   Example:

   If $36 \div 9 = 56 \div 14$, then $9 \div 36 = 14 \div 56$.

   Proof:

   $36 \times 14 = 9 \times 56, 504 = 504; 9 \times 56 = 36 \times 14, 504 = 504.$

4. In any proportion, the sum or the difference of the first two terms is to the second term as the sum or the difference of the last two terms is to the last term.

Example:

If $36 \div 9 = 56 \div 14$, then $\dfrac{36 + 9}{9} = \dfrac{56 + 14}{14}$ (quotient = 5) and $\dfrac{36 - 9}{9} = \dfrac{56 - 14}{14}$ (quotient = 3).

5. In any proportion, the sum or the difference of the first two terms is to the first term as the sum or the difference of the last two terms is to the third term.

Example:

If $36 \div 9 = 56 \div 14$, then $\dfrac{36 + 9}{36} = \dfrac{56 + 14}{56}$ (quotient = 1.25) and $\dfrac{36 - 9}{36} = \dfrac{56 - 14}{56} \left(\text{quotient} = \dfrac{3}{4} \right).$

6. Whenever a series of equal ratios appears in an expression, a proportion may be set up whereby one ratio consists of all numerators over all denominators, and the other ratio can be taken of any one numerator to its denominator.

Example:

If $\dfrac{36}{9} = \dfrac{56}{14} = \dfrac{80}{20} = \dfrac{72}{18}$, then $\dfrac{36 + 56 + 80 + 72}{9 + 14 + 20 + 18} = \dfrac{244}{61} = \dfrac{72}{18}$

(quotient = 4).

Direct and Inverse Proportions. Proportions may be direct or inverse, depending upon the conditions of the problem. Proportions are direct if an increase or reduction in one denomination will result in a corresponding increase or reduction in the other denomination. For example, if an operator grinds 43 pins in 3 hours, he will grind proportionately more pins in 7 hours. The greater number of pins can be taken as equal to x. Then the proportion may be set up in the following manner: $x \div 43 = 7 \div 3$. By the foregoing rule (product of the mean terms is equal to the product of the extreme terms), $3x = 43 \times 7$, and $x = \dfrac{43 \times 7}{3} = 100\dfrac{1}{3}$ pins. Usually a whole number of pins is stated as an answer to this example, namely, 100.

In an inverse proportion, an increase in one denomination will result in the corresponding reduction in the other denomination, and vice versa. For example, if 7 bricklayers complete a project in 24 days, a similar project will be completed by 12 bricklayers in a fewer number of days—more men at work but less time to complete the given task. The fewer number of days is taken as equal to x. Then the proportion may be set up as follows:

$$24 \div x = 12 \div 7, \quad 12x = 24 \times 7, \text{ and } x = \dfrac{24 \times 7}{12} = 14 \text{ days.}$$

The proportion can also be set up in this manner:

$$x \div 24 = 7 \div 12, \quad 12x = 24 \times 7, \text{ and } x = \dfrac{24 \times 7}{12} = 14 \text{ days.}$$

A suggestion is made to set up any proportion, direct or inverse, in a form wherein the first and third terms are always the greater denominations. If both ratios are written as fractions, the greater denominations will appear as the numerators. It is natural, it seems, to think of the greater values first, then why not take advantage of this fact in constructing the required proportion and show that one greater denomination is to its related smaller denomination as the contrasted greater denomination is to its related smaller denomination.

Examples Wherein Proportions Are Involved:

1. If a coil spring is elongated $\dfrac{3}{64}''$ when supporting a load of 4 pounds, how much will the spring stretch if a load of 12 pounds is suspended from it?

    This is a direct proportion, because an increase in the weight causes a proportional increase in the length of spring.

$$\dfrac{\text{Greater weight}}{\text{Smaller weight}} = \dfrac{\text{Greater elongation of spring}}{\text{Smaller elongation of spring}}$$

$$\dfrac{12}{4} = \dfrac{x}{\dfrac{3}{64}}, \quad 4x = 12 \times \dfrac{3}{64}, \quad x = \dfrac{12 \times \dfrac{3}{64}}{4} = \dfrac{9}{64}.$$

ALGEBRA

2. A pulley 8 inches in diameter is keyed to a motor shaft revolving 280 r.p.m. Another pulley 14 inches in diameter is driven by a belt connecting both pulleys. How fast does the larger pulley revolve? This is an inverse proportion, because a larger pulley runs slower. The speeds are inversely proportional to the pulley diameters.

$$\frac{\text{Greater r.p.m.}}{\text{Smaller r.p.m.}} = \frac{\text{Greater diameter}}{\text{Smaller diameter}}$$

$$\frac{280}{x} = \frac{14}{8}, \qquad 14x = 280 \times 8, \qquad x = \frac{\overset{20}{\cancel{280}} \times 8}{\underset{1}{\cancel{14}}} = 160.$$

The 14″ diameter pulley revolves 160 r.p.m.

3. A 32-tooth pinion drives a larger gear. The pinion speed is 450 r.p.m. and the gear speed is 180 r.p.m. How many teeth are in the gear?

This is an inverse proportion, because the number of teeth of the gear is inversely proportional to the gear speeds.

$$\frac{\text{Greater r.p.m.}}{\text{Smaller r.p.m.}} = \frac{\text{Greater number of teeth}}{\text{Smaller number of teeth}}$$

$$\frac{450}{180} = \frac{x}{32}, \qquad 180x = 450 \times 32, \qquad x = \frac{\overset{5}{\cancel{450}} \times \overset{16}{\cancel{32}}}{\underset{\underset{1}{2}}{\cancel{180}}} = 80$$

The larger gear has 80 teeth.

COMPOUND RATIOS AND PROPORTIONS

In some cases the relationship of the quantities in question must be expressed in terms of more than one ratio. In cases of this kind the ratios are compounded, and the corresponding proportions become compound proportions.

For example, if 3 men working 7 hours can unload 24 tons of coal, how many men working 10½ hours will be needed to unload 60 tons of coal, assuming that the efficiency of the men remains constant throughout the task?

Method 1. This example can be worked by analyzing independently each possible ratio and the corresponding proportions. Assume the same

98   MATHEMATICS FOR INDUSTRY

task for both groups of men, then the proportion is inverted and is shown as follows:

$$\frac{\text{More men}}{\text{Less men}} = \frac{\text{More hours}}{\text{Less hours}}, \frac{3}{x} = \frac{10.5}{7}, 10.5x = 3 \times 7, x = \frac{3 \times 7}{10.5} = 2 \text{ men}.$$

Now the second phase of the problem may be taken, and another proportion is shown as follows:

$$\frac{\text{More men}}{\text{Less men}} = \frac{\text{More tons}}{\text{Less tons}}, \frac{x}{2} = \frac{60}{24}, 24x = 60 \times 2, x = \frac{60 \times 2}{24} = 5 \text{ men}.$$

Method 2. A shorter and simpler method can be used in solving the problem just stated. The method consists of multiplying the group of ratios, provided the ratios are expressed correctly. Whether a quantity is increased or decreased depends, of course, upon the conditions of the problem. When multiplying the individual ratios of a problem, bear in mind that if a quantity is multiplied by a ratio smaller than 1, the quantity is decreased, and when it is multiplied by a ratio greater than 1, the quantity is increased.

In the foregoing problem, the ratios can be stated in the manner here shown. The initial operation consists of stating the given problem in order.

Men	Hours	Tons
3	7	24
x	$10\frac{1}{2}$	60

The first ratio is between x and 3, and is stated as $\frac{x}{3}$. Fewer men will be required for a period of $10\frac{1}{2}$ hours, and the second ratio may be stated as $\frac{7}{10.5}$. More men are needed for more tons, therefore the third ratio is expressed as $\frac{60}{24}$.

The final step consists of constructing the resulting proportion, which is expressed in the form:

$$\frac{x}{3} = \frac{7}{10.5} \times \frac{60}{24}, \text{ and } x = 3 \times \frac{7}{10.5} \times \frac{60}{24} = \frac{3 \times 7 \times \overset{5}{\cancel{60}}}{10.5 \times \underset{2}{\cancel{24}}} = \frac{105}{21} = 5 \text{ men}.$$

All ratios are constructed on the basis of quantities involving the unknown. In this case the number of men is the quantity unknown.

Example Involving a Greater Number of Ratios:

If 4 assemblers in a truck manufacturing plant working 7 hours a day assemble 32 trucks in 3 days, how many days will it take 7 assemblers working 9 hours a day to assemble 62 trucks?

The problem is stated in order.

Men	Hours per Day	Trucks	Days
4	7	32	3
7	9	62	x

All the ratios are based on the number (quantity) of days.

First ratio is $\frac{x}{3}$, the unknown number of days over the known.

Second ratio is $\frac{4}{7}$, more men will require less days.

Third ratio is $\frac{7}{9}$, more hours per day will require less days.

Fourth ratio is $\frac{62}{32}$, more trucks are assembled in more days.

The final expression is in the form:

$$\frac{x}{3} = \frac{4}{7} \times \frac{7}{9} \times \frac{62}{32}, \quad \text{and} \quad x = 3 \times \frac{4}{7} \times \frac{7}{9} \times \frac{62}{32} =$$

$$\frac{\cancel{3} \times \cancel{4} \times \cancel{7} \times \cancel{62}}{\cancel{7} \times \cancel{9} \times \cancel{32}} = \frac{31}{12} = 2\frac{7}{12} \text{ days.}$$

Compound ratios and proportions are used extensively in power transmission computations where the diameters and speeds of pulleys and gears are involved, or where the number of teeth of gears and their speeds are involved.

Example Involving Pulleys, Fig. 4-1:

Assume the speed of pulley A is given as 1200 r.p.m. Determine the speed of pulley B which is equal to x. (See Fig. 4-1.)

R.p.m.	Dia.	Dia.	Dia.	
1200	8	6	7	(Drivers)
x	12	14	16	(Driven)

In the case of pulleys and gears, the drivers and driven are designated as shown.

$\dfrac{x}{1200}$, the ratio of speed of pulley B to that of pulley A.

$\dfrac{8}{12}$, a larger diameter pulley will rotate at a slower speed.

$\dfrac{6}{14}$, a larger diameter pulley will rotate at a slower speed.

$\dfrac{7}{16}$, a larger diameter pulley will rotate at a slower speed.

$$\dfrac{x}{1200} = \dfrac{8}{12} \times \dfrac{6}{14} \times \dfrac{7}{16}, \qquad x = \cancel{1200}^{\,100\,\,25} \times \dfrac{\cancel{8}^{\,1}}{\cancel{12}_{\,1}} \times \dfrac{\cancel{6}^{\,1}}{\cancel{14}_{\,2}} \times \dfrac{\cancel{7}^{\,1}}{\cancel{16}_{\,2}} = 150 \text{ r.p.m.}$$

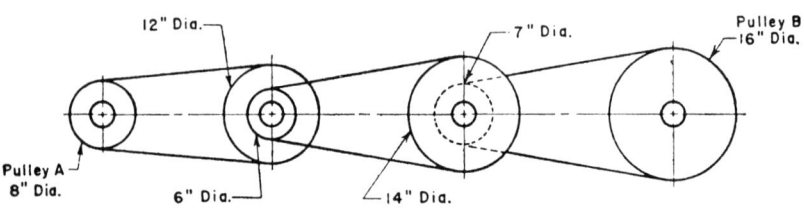

Fig. 4-1.

PERCENTAGES

Values or quantities may be compared by expressing them in terms of per cent. A part of a whole thing can be expressed as a fraction and, according to the preceding discussion, a fraction represents the ratio of the part to the whole. For example, if a coil spring 12″ long is stretched 3″, the stretch can be expressed as a fraction of the original length of the spring, namely, $\dfrac{3}{12} = \dfrac{1}{4}$, which is also a ratio of the stretch to the length of the spring.

If we assume the original length of the spring to be the whole quantity equal to one (1), then the stretch is equal to $\dfrac{1}{4}$ of the one (1) quantity.

The common fraction $\dfrac{1}{4}$ may be converted to a decimal fraction $.25 = \dfrac{25}{100}.$

ALGEBRA

A more convenient comparison of quantities is by per cent or by hundredths. If we assume the length of the spring as the whole quantity to be equal to 100 hundredths, then the stretch can be expressed in terms of 25 hundredths. In other words, we say, the stretch of the spring is 25 per cent of the original length of the spring, written 25%. The sign %, then, stands for hundredths or two decimal places. It should be noted that a per cent expressed $\frac{1}{4}\%$ means $\frac{1}{4}$ of 1%.

PROCEDURE FOR DETERMINING PERCENTAGES

The procedure for determining the required percentage is given in three steps.

Step 1. Express the relation between two quantities as a ratio in the form of a common fraction.

Step 2. Convert the common fraction to a decimal fraction.

Step 3. Multiply the decimal fraction by 100 and thus express the result in terms of per cent.

Examples:

1. $\frac{30}{60} = \frac{1}{2}, \frac{1}{2} = .50, .50 = 50\%$

2. $\frac{11}{33} = \frac{1}{3}, \frac{1}{3} = .33\frac{1}{3}, .33\frac{1}{3} = 33\frac{1}{3}\%$

Problems Involving the Finding of Percentages. A convenient method for finding percentages is suggested herewith. The method is adopted in engineering and shop computations because of its simplicity and practicability. It consists of expressing the known and the unknown data in the form of a direct or an inverse proportion, and, finally, the unknown value is found by methods previously explained.

Examples:

1. There are 36 employees normally engaged in a certain department. Assume 3 members are absent on a given date. What is the percentage of absences?
   The proportion may be shown in the following manner:

$$\frac{\text{Greater number of employees}}{\text{Smaller number of employees}} = \frac{\text{Greater percentage (100\%)}}{\text{Smaller percentage}},$$

$$\frac{36}{3} = \frac{100}{x}, \quad 36x = 3 \times 100, \quad x = \frac{\overset{1}{\cancel{3}} \times \overset{25}{\cancel{100}}}{\underset{\underset{3}{12}}{\cancel{36}}} = 8\frac{1}{3}\%.$$

The greater percentage is always a constant equal to 100 per cent; it represents the whole quantity, a percentage of which is the value required in the problem.

2. The inspection department rejected 17 bushings out of a lot of 340 bushings. What is the percentage of rejections?

$$\frac{\text{Greater number of bushings}}{\text{Smaller number of bushings}} = \frac{100 \text{ per cent}}{\text{Required percentage}},$$

$$\frac{340}{17} = \frac{100}{x}, \quad 340x = 100 \times 17, \quad x = \frac{100 \times \overset{1}{\cancel{17}}}{\underset{20}{\cancel{340}}} = 5\%.$$

3. A copper bar 14″ long is elongated .07″ due to a load to which the bar is subjected. What is the percentage of elongation?

$$\frac{\text{Greater length}}{\text{Smaller length}} = \frac{100 \text{ per cent}}{\text{Percentage of elongation}}$$

$$\frac{14}{.07} = \frac{100}{x}, \quad 14x = 100 \times .07, \quad x = \frac{100 \times \overset{.01}{\cancel{.07}}}{\underset{2}{\cancel{14}}} = \frac{1}{2}\%.$$

An interesting feature where percentage is involved is the method of calculating the percentage of profit or loss in purchase and sale of merchandise or equipment.

Examples:

1. A machine tool that was purchased by a dealer for $1,250 is resold for $1,425. What is the dealer's percentage of profit?

Usually in commercial transactions, the percentage of profit is calculated on the basis of the selling price. In this example, the selling

ALGEBRA

price is equal to $1,425, the profit is: $1,425 − $1,250 = $175, and the proportion may be shown as follows:

$$\frac{\text{Greater sum}}{\text{Smaller sum}} = \frac{100 \text{ per cent}}{\text{Percentage of profit}} \text{ or } \frac{\text{Selling price}}{\text{Actual profit}} = \frac{100 \text{ per cent}}{\text{Percentage of profit}}$$

$$\frac{1425}{175} = \frac{100}{x}, \; 1425x = 100 \times 175, \; x = \frac{100 \times 175}{1425} = \frac{17500}{1425} = 12.28\%.$$

2. A machine tool having a list price of $1,780 was sold at a discount of 15%. What price was paid for the machine tool?

The proportion may be shown in the following manner:

$$\frac{\text{Greater sum}}{\text{Smaller sum}} = \frac{100\%}{15\%} \text{ or } \frac{\text{List price}}{\text{Discount}} = \frac{100\%}{15\%}$$

The discount is equal to x.

$$\frac{1780}{x} = \frac{100}{15}, \; 100x = 1780 \times 15, \; x = \frac{1780 \times 15}{100} = \$267 = \text{discount}.$$

The final step is the calculation of the cost of the machine tool, 1780 − 267 = $1,513.

SOLUTION OF SIMPLE SIMULTANEOUS EQUATIONS

The algebraic equations of the preceding chapter were of the type involving only one unknown quantity. In this chapter we shall deal with equations in which more than one unknown quantity is involved.

Assume an equation in which two unknown quantities are involved, say $x + y = 14$. The solution of this equation as it stands is impossible, because we can substitute for x and y a number of values that will satisfy the equation. For example $x = 7$ and $y = 7$ will satisfy the equation; as will $x = 9$ and $y = 5$; $x = 10$ and $y = 4$; $x = 2$ and $y = 12$; and so on. An equation of this type is called an *indeterminate equation* and is of no use for practical purposes.

Now if we take in addition to the equation $x + y = 14$, another indeterminate equation, $x - y = 4$, in which the unknown quantities x and y are involved in such a way that a single pair of values of the quantities unknown satisfies both equations, we find that solution is possible. When a set of equations contain the same pair of values for the unknown, the equations are called *simultaneous equations*. Simultaneous equations are extensively employed in the solution of problems involving more than one unknown quantity. In the example given $x + y = 14$ and $x - y = 4$, we find that the values $x = 9$ and $y = 5$ will satisfy both equations. If the second equation is of such form that no pair of values that will satisfy both equations can be chosen, the equations are called *inconsistent* and cannot be solved.

In order to solve simultaneous equations, several methods may be employed to find the unknowns. All of these methods have one common feature, and that is that each makes possible the elimination of one unknown quantity by producing a new equation which contains only one unknown.

The methods used in the solution of simultaneous equations are: (1) The elimination of one unknown by addition or subtraction. (2) The elimination of one unknown by substitution. (3) The elimination of one unknown by comparison.

Elimination by Addition or Subtraction. Let us solve the foregoing example by the first method, that is, elimination by addition or subtraction.

Example 1: Solve for x and y in $x + y = 14$ and $x - y = 4$.

By addition:

$$x + y = 14$$
$$x - y = 4$$
$$2x = 18$$
$$x = 9$$
$$y = 14 - 9 = 5$$

By subtraction:

$$x + y = 14$$
$$x - y = 4$$
$$2y = 10$$
$$y = 5$$
$$x = 14 - 5 = 9$$

In some cases it is more convenient to eliminate the unknown only by addition or only by subtraction, whichever method is the more conducive to solution. It is well to note that when the unknowns we are eliminating are of opposite signs, we add, and we subtract when they are of like signs.

Example 2: Solve for x and y in $7x + 2y = 29$ and $28x - 9y = 48$.

In this example it is convenient to eliminate x by subtraction, but a preliminary operation is necessary, namely, the balancing of the coefficients of x in both equations. We may multiply each member of the first equation by 4 without disturbing the equality, thus making the coefficients of x alike in both equations.

$$\begin{array}{ll} 7x + 2y = 29 & 28x + 8y = 116 \\ \times 4 & - \\ \hline 28x + 8y = 116 & 28x - 9y = 48 \\ & \hline 17y = 68 \\ & y = \dfrac{68}{17} = 4 \end{array}$$

x may be solved now: $7x + (2 \times 4) = 29$, $7x + 8 = 29$, $x = \dfrac{29 - 8}{7} = 3$.

In some cases each equation must be multiplied or divided by different values in order to equalize the coefficients of the unknown which is chosen to be eliminated.

Elimination by Substitution. An illustration of the second method used in elimination of one unknown—namely, by substitution—is here given:

Example: Solve for x and y in $6x - 2y = 16$ and $35x + 5y = 210$.

In this case it is more convenient to solve for y in terms of x in any one equation and substitute the value of y in the second equation, thus eliminating one unknown.

Let us solve for y in the first equation.

$$6x - 16 = 2y, \quad y = \dfrac{6x - 16}{2} = 3x - 8$$

Substituting the value of y in the second equation gives:

$$35x + 5(3x - 8) = 210$$
$$35x + 15x - 40 = 210$$
$$50x = 250$$
$$x = 5$$

y can be solved in terms of x. We found y to equal $3x - 8$ and $x = 5$, therefore $y = (3 \times 5) - 8 = 7$.

Elimination by Comparison. The third method—elimination by comparison—consists of solving for one unknown in terms of the other in both equations. Then by comparing or equating both expressions, the other unknown may be solved.

Example: Solve for x and y in $3x - 7y = 3$ and $5x + 3y = 49$.

In the first equation, $x = \dfrac{7y + 3}{3}$.

In the second equation, $x = \dfrac{49 - 3y}{5}$.

$$\frac{7y + 3}{3} = \frac{49 - 3y}{5}$$

For clearance of fractions the new equation is multiplied by 15.

$$\left(\frac{7y + 3}{\cancel{3}}\right) \times \overset{5}{\cancel{15}} = \left(\frac{49 - 3y}{\cancel{5}}\right) \times \overset{3}{\cancel{15}}, \; 35y + 15 = 147 - 9y,$$

$$35y + 9y = 147 - 15, \; 44y = 132, \; y = \frac{132}{44} = 3.$$

x may be solved by considering any convenient expression involving x.

In the first given equation, $3x - 7y = 3$, $x = \dfrac{(7 \times 3) + 3}{3} = \dfrac{24}{3} = 8$.

EVALUATION OF FORMULAS AND METHODS OF SUBSTITUTION

Methods used to solve simple formulas and equations were shown in the preceding chapter, and various steps for simplification of algebraic expressions were explained in detail. The necessity for simplification was pointed out as the step preliminary to substitution of numerical values for algebraic symbols representing known or assumed quantities.

This chapter deals with formulas of all kinds, and a general procedure is indicated for solution, thus saving time and effort that would be wasted if an attempt were made to obtain the answer by unsure means.

The steps to be followed in solving an equation are enumerated below. The solution of the equation $\dfrac{x}{4} - \dfrac{x}{18} = \dfrac{x}{12} + 3$ will be illustrated with the steps.

Step 1. Simplification of the equation, which includes the adjustment of plus and minus signs encountered in grouping algebraic expressions, possible multiplication and division, the clearing of fractions, and any other preliminary operations that may seem appropriate upon inspection of the given expression. Various methods of simplification have been explained in detail previously with the exception of clearing of fractions. If both sides of an equation containing fractions are multiplied by the least common denominator of all given fractions, the equation can be cleared. In the equation given to be solved the least common denominator of all given fractions is 36. Multiplying both sides of the equation by 36 gives:

$$\frac{36x}{4} - \frac{36x}{18} = \frac{36x}{12} + (36 \times 3), 9x - 2x = 3x + 108.$$

Step 2. All terms containing the unknown must be transposed to the left member and all terms that do not contain the unknown must be transposed to the right member.

$$9x - 2x - 3x = 108$$

Step 3. Collect like terms.

$$4x = 108$$

Step 4. Divide both members of the equation by the coefficient of the unknown.

$$\frac{4x}{4} = \frac{108}{4}, x = 27.$$

Step 5. Test the results by substituting the value obtained for the unknown in the original equation.

$$\frac{27}{4} - \frac{27}{18} = \frac{27}{12} + 3, 6\frac{3}{4} - 1\frac{1}{2} = 2\frac{1}{4} + 3.$$

$$5\frac{1}{4} = 5\frac{1}{4}, \text{ proving the operation is correct.}$$

Example of Solving an Algebraic Equation: Solve the equation:
$$\frac{3(x-2)}{4} - \frac{2(x-3)}{6} = \frac{4x(14-5)}{15} - \left(3\frac{1}{2} + \frac{29}{30}\right)$$

Step 1. Clearing of fractions: The least common denominator of all given fractions is 60. Multiplying both sides of the given equation by 60, gives:

$$\frac{60 \times 3(x-2)}{4} - \frac{60 \times 2(x-3)}{6} = \frac{60 \times 4x(14-5)}{15} - 60 \times \left(3\frac{1}{2} + \frac{29}{30}\right), \ 45(x-2) - 20(x-3) = 16x(14-5) - 268.$$

Removing the parentheses and adjusting the signs:

$$45x - 90 - 20x + 60 = 224x - 80x - 268.$$

Step. 2. Transposing:

$$45x - 20x - 224x + 80x = 90 - 60 - 268.$$

Step 3. Collecting like terms:

$$-119x = -238, \ 119x = 238.$$

Step 4. Dividing both sides of the equation by the coefficient of x:

$$\frac{119x}{119} = \frac{238}{119}, \ x = 2.$$

Step 5. Proving the result by substituting the value of x in the given equation:

$$\frac{3(2-2)}{4} - \frac{2(2-3)}{6} = \frac{4 \times 2 \times 9}{15} - 4\frac{7}{15}, \ 0 - \left(-\frac{1}{3}\right) = 4\frac{12}{15} - 4\frac{7}{15},$$

$$\frac{1}{3} = 4\frac{12}{15} - 4\frac{7}{15}, \ \frac{1}{3} = \frac{1}{3},$$

proving the operation is correct.

The foregoing examples of methods used in solution of algebraic equations demonstrate the fact that various letters and symbols employed in formulas take only temporarily the place of the given figures which are substituted in the corresponding places. This operation is commonly referred to as the evaluation of the formula. The primary aim in evaluating a formula is to determine its value by substituting the known numbers for the letters and symbols, and, finally, performing the necessary operations.

ALGEBRA

Examples of Evaluating Formulas:

1. If we have to calculate the weight of a hollow cylinder made from a known metal, we can obtain the general formula for determining the weight of a hollow cylinder from a handbook or, by simple analysis, we may develop the required formula independently.

    The formula is $W = w\frac{\pi}{4}(D_0^2 - D_i^2)h$.

    W = weight of metal cylinder in pounds.
    w = unit weight = weight per cubic inch of given metal in pounds.
    π = a constant = 3.1416.
    D_0 = outside diameter of cylinder in inches.
    D_i = inside diameter of cylinder in inches.
    h = altitude of cylinder in inches.

    To evaluate this formula, we have to substitute given values for the letters to satisfy a given condition.

2. The formula used in problems dealing with strength of materials to evaluate the effect of a sudden application of load is:

    $$W(h + \Delta) = \frac{sA\Delta}{2}$$

    To evaluate the formula for the purpose of solving the unknown quantity (s), all the known values are substituted for the corresponding letters after the given formula is changed in such a way that s appears isolated on one side of the equation.

    Given formula: $W(h + \Delta) = \frac{sA\Delta}{2}$

    Clearing of fractions: $2W(h + \Delta) = sA\Delta$

    Dividing both sides by $A\Delta$: $\frac{2W}{A\Delta}(h + \Delta) = s$

    More convenient: $s = \frac{2W(h + \Delta)}{A\Delta}$

    Substituting values for all the letters on the right side solves the unknown s.

3. The formula used in design of hollow shafts is:

$$S_s = \frac{16Td_0}{\pi(d_0^4 - d_i^4)}$$

Assuming the unknown quantity is d_i, the evaluation of the formula must be preceded by isolating d_i to one side of the equation.

Given formula: $S_s = \dfrac{16Td_0}{\pi(d_0^4 - d_i^4)}$

Multiplying both sides by $\pi(d_0^4 - d_i^4)$:

$$S_s\pi(d_0^4 - d_i^4) = 16Td_0$$

Dividing both sides by $S_s\pi$: $d_0^4 - d_i^4 = \dfrac{16Td_0}{\pi S_s}$

Transposing d_0^4 to the right side: $-d_i^4 = -d_0^4 + \dfrac{16Td_0}{\pi S_s}$

Changing signs: $d_i^4 = d_0^4 - \dfrac{16Td_0}{\pi S_s}$

Value d_0 offers an opportunity to factor the right side of the equation, but it is best to leave it alone, as its form is fairly simple.

Finally: $d_i = \sqrt[4]{d_0^4 - \dfrac{16Td_0}{\pi S_s}}$

All known values are substituted for the letters below the radical sign, and d_i can be determined by performing the necessary operations with regard to the substituted values.

QUADRATIC EQUATIONS

A quadratic equation contains the square (but no higher power) of the unknown quantity. An equation containing only the square of the unknown is an *incomplete* or *pure quadratic*. For example, the expression $17x^2 - 12 = 5x^2 + 36$ is a pure quadratic.

An expression containing the square and the first power of the unknown is a *complete* or *affected quadratic*. For example, the expression $43x^2 + 10x - 33 = 3x + 153$ is an affected quadratic.

Solving a Pure Quadratic. When solving a pure quadratic equation, all the unknown quantities are collected on one side of the equation and the known quantities collected on the opposite side. The known quantities combined form a quantity that is divided by the coefficient of the combined unknown quantities, the square root of each side of the final equation being extracted. The previously given example is solved in the following manner.

Example: $17x^2 - 12 = 5x^2 + 36$

Collecting terms: $17x^2 - 5x^2 = 12 + 36$
Collecting terms: $12x^2 = 48$
Dividing by the coefficient 12: $x^2 = 4$
Extracting the square root: $x = \pm 2$

When the square of the unknown quantity is equal to a negative value, the square root cannot be found, therefore it is called an *imaginary* quantity and can only be indicated, for example, $x^2 = -12$, $x = \sqrt{-12}$.

Solving an Affected Quadratic. Affected or complete quadratic equations may be solved by factoring, by completing the square, or by the use of a formula.

1. *Solution by factoring.* Solving an equation in x by the use of factoring is based on the principle that a product of two or more numbers equals zero when at least one of the factors is zero. If the equation to be solved is not already given as equaling zero, we must first transpose all terms to one member and obtain zero as the other member.

Example: Solve for x in the expression $x^2 + 2x - 35 = 0$.

The term $2x$ may be separated into the sum of two terms: $x^2 + 7x - 5x - 35 = 0$

Grouping: $(x^2 + 7x) - (5x + 35) = 0$
Factoring: $x(x+ 7) - 5(x + 7) = 0$
$(x - 5)(x + 7) = 0$
If $x - 5 = 0$, then $x = 5$.
If $x + 7 = 0$, then $x = -7$.

The given equation has two solutions, namely, $x = 5$ and $x = -7$.
Substituting $x = 5$ in the equation:

$$5 \times 5 + 2 \times 5 - 35 = 0$$
$$25 + 10 - 35 = 0$$

Substituting $x = -7$ in the equation.

$$(-7) \times (-7) + 2 \times (-7) - 35 = 0$$
$$49 - 14 - 35 = 0$$

2. *Solution by completing the square.* When the left member of the equation cannot be factored by inspection, the method of completing the square is used.

Example: Solve for x in the expression $3x^2 - 9x - 54 = 0$.

All the unknowns are transposed to one side of the equation and the knowns to the other side:

$$3x^2 - 9x = 54$$

Both sides of the equation are divided by the coefficient of x^2, which is 3. (This is always done before completing the square unless the coefficient of x^2 is 1.)

$$x^2 - 3x = 18$$

The left-hand side of the equation is changed to a *perfect square* by adding the square of half the coefficient of the first power of the unknown to both sides of the equation:

$$x^2 - 3x + \left(\frac{3}{2}\right)^2 = 18 + \left(\frac{3}{2}\right)^2$$

$$\left(x - \frac{3}{2}\right)^2 = 20\frac{1}{4}$$

$$x - \frac{3}{2} = \pm \sqrt{20\frac{1}{4}} = \pm 4\frac{1}{2}$$

$$x = 1\frac{1}{2} \pm 4\frac{1}{2} = +6 \text{ or } -3$$

The given equation has two solutions, namely, $x = 6$ and $x = -3$. Substituting $x = 6$ in the equation:

$$3 \times 6 \times 6 - 9 \times 6 = 54$$

Substituting $x = -3$ in the equation:

$$3 \times (-3) \times (-3) - [9 \times (-3)] = 54$$
$$3 \times 9 - (-27) = 54$$
$$27 + 27 = 54$$

3. *Solution by the quadratic formula.* If all the terms of a complete quadratic equation are collected in the proper order on one side of the equal-

ALGEBRA

ity sign, the resulting expression is in the form of a trinomial. The first term includes the unknown to the second power with or without a numerical or literal coefficient. The second term includes the unknown to the first power with or without a coefficient. The third term is the known quantity.

The general form of a quadratic equation is:

$$ax^2 + bx + c = 0$$

The coefficients a and b, and the known value c, may be positive or negative.

If a given expression can be written in the general form of a quadratic equation, the square root which is the unknown x may be obtained from the following formula:

$$x = \frac{-b \pm \sqrt{b^2 - 4ac}}{2a}$$

Example 1: Solve for x in the expression $5x^2 + 3x = 26$.

Collecting all terms to one side: $5x^2 + 3x - 26 = 0$. Comparing with the general form of the quadratic equation, coefficient $5 = a$, coefficient $3 = b$, and $-26 = c$.

$$x = \frac{-b \pm \sqrt{b^2 - 4ac}}{2a} = \frac{-3 \pm \sqrt{3^2 - 4 \times 5 \times (-26)}}{2 \times 5} =$$

$$\frac{-3 \pm \sqrt{9 - (-520)}}{10} = \frac{-3 \pm \sqrt{529}}{10} = \frac{-3 \pm 23}{10} = +2 \text{ or } -2.6$$

Substituting $x = +2$ in the given equation:

$$(5 \times 4) + (3 \times 2) = 26$$
$$20 + 6 = 26$$

Substituting $x = -2.6$ in the given equation:

$$[5 \times (-2.6)^2] + [3 \times (-2.6)] = 26$$
$$(5 \times 6.76) - 7.8 = 26$$
$$33.8 - 7.8 = 26$$
$$26 = 26$$

Example 2: Solve for x in the expression $3x^2 - 12x = -9$.
Collecting all terms to one side: $3x^2 - 12x + 9 = 0$.

Comparing with the general form of the quadratic equation:

$$x = \frac{-b \pm \sqrt{b^2 - 4ac}}{2a} = \frac{-(-12) \pm \sqrt{(-12)^2 - (4 \times 3 \times 9)}}{2 \times 3} =$$

$$12 \pm \frac{\sqrt{144 - 108}}{6} = \frac{12 \pm \sqrt{36}}{6} = \frac{12 \pm 6}{6} = +3 \text{ or } +1$$

Substituting $x = +3$ in the given equation:

$$3 \times 3^2 - 12 \times 3 = -9$$
$$27 - 36 = -9$$
$$-9 = -9$$

Substituting $x = +1$ in the given equation:

$$(3 \times 1) - 12 = -9$$
$$3 - 12 = -9$$
$$-9 = -9$$

ALGEBRAIC PRINCIPLES APPLIED TO SHOP AND TOOLROOM PROBLEMS

Algebra is of great value in the solving of shop and toolroom problems. The problems given in this section are actual problems taken from the shop and they warrant careful study.

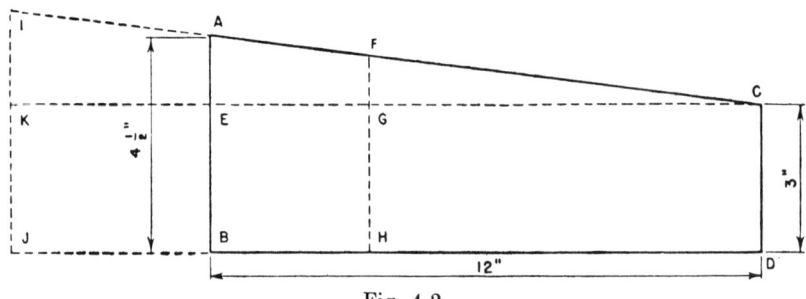

Fig. 4-2.

Simplified Rules for Solving Taper Problems. Shop problems involving tapers are numerous, and a simplified procedure for the purpose of solving such problems is explained in the following:

A part is said to be tapered, if it varies uniformly from end to end or for some definite distance. For instance, a rectangular steel plate may be machined to take the shape shown in Fig. 4-2.

ALGEBRA

The plate $ABCD$ is said to have $1\frac{1}{2}''$ taper per foot, which is obtained by subtracting $3''$ from $4\frac{1}{2}''$ and from the fact that the plate is $12''$ long.

If other plates are considered, such as $CDFH$ or $CDIJ$, they also have $1\frac{1}{2}''$ taper per foot, but the total amount of slope differs in each case, namely, AE, FG, and IK, though the rate of slope for all remains constant—$1\frac{1}{2}''$ per foot. Such tapers are employed in the shop when producing machine parts such as rams of shaper heads, gibs of machine tools, taper keys, cotters, etc.

Another type of taper is widely used in fitting hubs on axles, and for spindles, shafts, mandrels, arbors, cutter shanks, plugs, gages, etc. The accuracy of such machine members is of great importance, and the correct analysis and manipulation of taper formulas is essential. A part shaped in accordance with this type of taper decreases gradually in diameter from end to end or for some definite distance, and its shape is conical, as shown in Fig. 4-3.

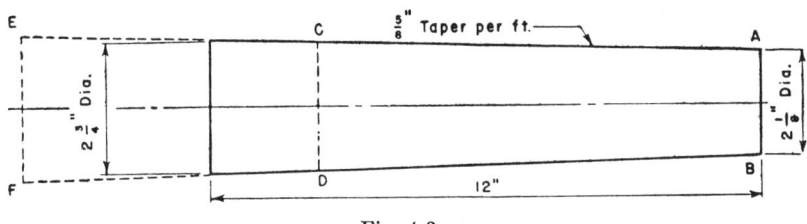

Fig. 4-3.

The part shown in Fig. 4-3 is said to be $\frac{5''}{8}$ taper per foot, i.e., its diameter varies $\frac{5''}{8}$ in $12''$ length. It should be noted that tapers of conical parts refer to diameters and not to one side, a condition that must be considered in calculating such tapers to obtain accurate results.

If the part shown in Fig. 4-3 is less than $12''$ long, such as $ABCD$, or over $12''$ long, such as $ABEF$, it is still $\frac{5''}{8}$ taper per foot, but the difference between the end diameters is either under or over $\frac{5''}{8}$. This difference is called the *actual amount of taper*. Examples that are to follow, show the relationship between the length and the actual amount of taper existing in taper parts.

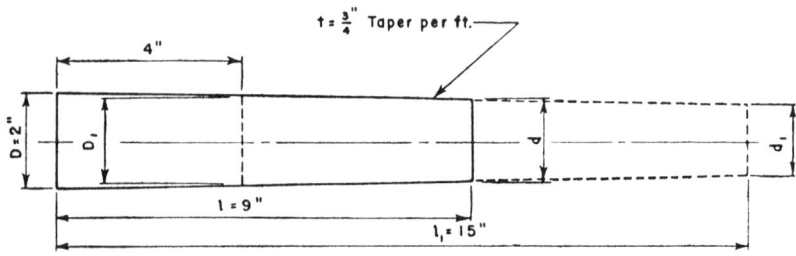

Fig. 4-4.

Let us consider a taper part shown in Fig. 4-4.
The values are:

D = large diameter in inches
d = small diameter in inches
l = length in inches
$D-d$ = actual amount of taper

All of the elements except the small diameter (d) are given. To calculate d, we reason as follows: If the actual amount of taper $D-d$ is determined, the rest of the problem is simple. $D-d$ must be smaller than $\frac{3''}{4}$, because it is the difference between the end diameters of a tapered part 9″ long, whereas $\frac{3''}{4}$ is the difference between the end diameters of an imaginary tapered part of the same slope measuring 12″ in length. Therefore, $D-d$ is to $\frac{3''}{4}$ as 9″ is to 12″.

$$\frac{D-d}{\frac{3}{4}} = \frac{9}{12}, \quad 12(D-d) = 9 \times \frac{3}{4}, \quad D-d = \frac{9 \times \frac{3}{4}}{12} = \frac{9 \times \overset{1}{\cancel{3}}}{\underset{4}{\cancel{4}} \times \cancel{12}} = \frac{9''}{16}.$$

$$D-d = \frac{9}{16}, \quad 2-d = \frac{9}{16}, \quad 2 - \frac{9}{16} = d, \quad d = 1\frac{7''}{16}. \text{ (Answer)}$$

The problem may be solved by the suggested method of direct proportion previously explained.

$$\frac{\text{Large length}}{\text{Small length}} = \frac{\text{Large amount of taper}}{\text{Small amount of taper}}$$

ALGEBRA

$$\frac{12''}{9''} = \frac{\frac{3}{4}''}{x}, \quad 12x = 9 \times \frac{3}{4}, \quad x = \frac{9 \times \frac{3}{4}}{12} = \frac{9}{16}'',$$

$$d = D - x = 2'' - \frac{9''}{16} = 1\frac{7''}{16}.$$

If the length of part in Fig. 4-4 is $l_1 = 15''$, what is the small diameter d_1? Using the suggested method of proportion, we have:

$$\frac{\text{Large length}}{\text{Small length}} = \frac{\text{Large amount of taper}}{\text{Small amount of taper}}$$

$$\frac{15''}{12''} = \frac{x}{.75''}, \quad 12x = 15 \times .75, \quad x = \frac{15 \times .75}{12} = .9375'' = \frac{15''}{16},$$

$$d_1 = D - x = 2'' - \frac{15''}{16} = 1\frac{1}{16}''. \text{ (Answer)}$$

If it is desired to calculate the diameter D_1 in Fig. 4-4 located $4''$ from the left end, the proportion may be set up as follows:

$$\frac{\text{Large length}}{\text{Small length}} = \frac{\text{Large amount of taper}}{\text{Small amount of taper}}$$

$$\frac{12''}{4''} = \frac{.75''}{x}, \quad 12x = 4 \times .75, \quad 12x = 3, \quad x = \frac{1''}{4},$$

$$D_1 = D - x = 2'' - \frac{1''}{4} = 1\frac{3''}{4}. \text{ (Answer)}$$

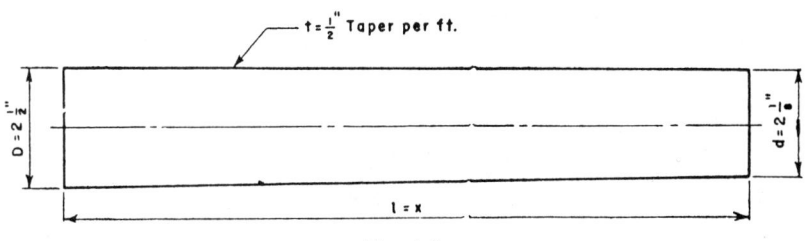

Fig. 4-5.

Another example involving tapers is illustrated in Fig. 4-5.

It is required to solve for the length l in Fig. 4–5. The proportion is:

$$\frac{\text{Large length}}{\text{Small length}} = \frac{\text{Large amount of taper}}{\text{Small amount of taper}}$$

From inspection of Fig. 4-5, it may be concluded that the actual amount of taper is equal to $2\frac{1}{2}'' - 2\frac{1}{8}'' = \frac{3}{8}''$, which is less than the $\frac{1}{2}''$ taper per foot, therefore, the actual length of the part must be proportionately smaller.

$$\frac{12''}{x} = \frac{.5''}{.375}, \quad .5x = 12 \times .375,$$

$$x = \frac{12 \times .375}{.5} = 9'', \quad l = 9''. \text{ (Answer)}$$

If l, d, and t are given in Fig. 4-5, the unknown D may be calculated by proportion as previously explained. In a similar manner, if D, d, and l are given, the unknown t may be calculated.

The constant length of $12''$ is always brought into the proportion. By inspection of given conditions it can be determined whether this constant is to be considered as the large or the small length.

A general formula may be developed for the solution of problems involving tapers. See Fig. 4-6.

The values are:

D = large diameter in inches
d = small diameter in inches
l = length in inches
t = taper per foot in inches
$12''$ = a constant
$D - d$ = actual amount of taper

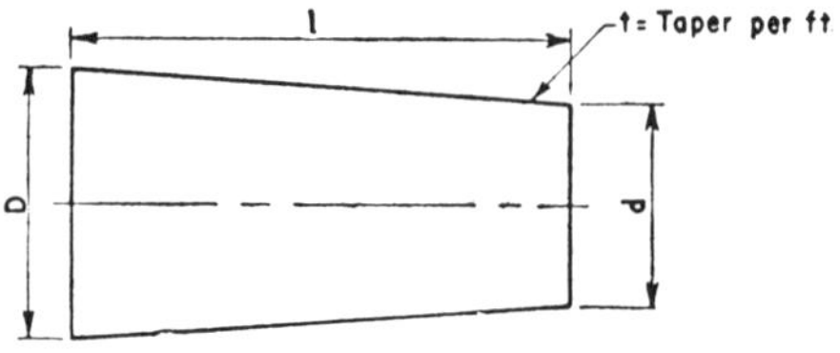

Fig. 4-6.

Then $\dfrac{D - d}{l}$ = amount of taper per inch length,

and $\dfrac{t}{12}$ = amount of taper per inch length.

Therefore, $\dfrac{D - d}{l} = \dfrac{t}{12}$, which is the general formula used for solution

ALGEBRA

of any one element in a problem dealing with tapers, provided the remaining elements are known.

The formula just given may be written $12(D-d) = tl$, which is a more convenient form to use when solving for an unknown. The four types of taper problems that can be solved by application of the general formula are:

1. $D = \dfrac{tl}{12} + d$

2. $d = D - \dfrac{tl}{12}$

3. $l = \dfrac{12(D - d)}{t}$

4. $t = \dfrac{12(D - d)}{l}$

The most commonly used tapers are the following:

1. Morse, about $\dfrac{5''}{8}$ taper per foot.

2. Brown & Sharpe, $\dfrac{1''}{2}$ taper per foot.

3. $\dfrac{3''}{4}$ taper per foot.

4. Taper pin standard, $\dfrac{1''}{4}$ taper per foot.

5. Jarno, $.6''$ taper per foot.

In the Jarno type, the number of the taper designates the principal dimensions. For example, a No. 8 Jarno taper is $.8''$ in diameter at the small end, $1''$ in diameter at the large end, and $8 \times \dfrac{1}{2} = 4''$ in length.

A simple formula may be developed for a Jarno taper.

Assume l = length of taper part in inches
D = diameter at large end in inches
d = diameter at small end in inches
N = number of Jarno taper

Then $D = N \times \frac{1''}{8}$

$d = N \times \frac{1''}{10}$

$l = N \times \frac{1''}{2}$

Locating Holes in a Drill Jig Plate. Numerous shop problems dealing with precision work can be solved with comparative ease by the application of algebraic principles explained and demonstrated in the foregoing examples.

A problem frequently encountered in toolroom work is accurate layout of holes in a drill jig plate by use of precision disks.

Let us assume a steel plate used on a jig in which three closely spaced holes are to be drilled and reamed. The project is illustrated in Fig. 4-7, and the problem is to calculate the diameters of three disks that can be placed on the plate in contact with each other in such a way that the disk centers will coincide with the axes of the holes to be drilled. Then the plate can be clamped to the lathe face plate, where each disk is indicated by means of a sensitive indicator supported on the tool slide and resting in the sharp V-center of the disk. The indicated disk is removed, and the hole in line with it is drilled and reamed. The same procedure is continued until all the disks are indicated, and each hole is drilled and

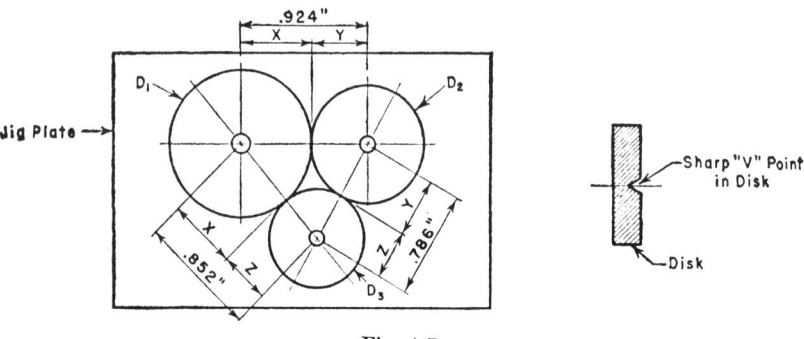

Fig. 4-7.

reamed in the proper order. The disks are usually held on the jig plate temporarily by means of wax or soft solder.

The center distances, as shown in Fig. 4-7, are given. The disk diam-

ALGEBRA

eters are calculated by developing suitable simultaneous equations involving two unknowns. The procedure is as follows:

D_1, D_2, and D_3 are the unknown diameters of the three disks.

$$x = \frac{D_1}{2}, \quad y = \frac{D_2}{2}, \quad \text{and} \quad z = \frac{D_3}{2}$$

Subtract entire lower expression from upper expression:

$$\begin{array}{r} x + y = .924 \\ - \\ \underline{x + z = .852} \\ y - z = .072 \end{array}$$

Add both expressions:

$$\begin{array}{r} y + z = .786 \\ + \\ \underline{y - z = .072} \\ 2y = .858 \\ y = \frac{.858}{2} = .429 \end{array}$$

But, $2y = D_2$, therefore, $D_2 = .858''$.

The remaining disks can be calculated in several different ways. Let us take the expressions involving y and z and subtract one from the other.

$$\begin{array}{r} y + z = .786 \\ - \\ \underline{y - z = .072} \\ 2z = .714 \\ z = \frac{.714}{2} = .357 \end{array}$$

But $2z = D_3$, therefore, $D_3 = .714''$.
$D_1 = 2x$, but $x = .924 - y = .924 - .429 = .495$.
Therefore, $D_1 = 2x = 2 \times .495 = .990''$.

In order to place the disks in proper relation to the jig plate, any given disk, such as D_1, must be properly located with respect to any two edges of the plate, say the upper and left-side edges. Accurate placement can be accomplished with the aid of precision gage blocks.

Locating Holes with Toolmaker's Buttons, Precision Disks, and Rollers. Toolmaker's buttons can be used for locating holes where the space between the holes is too great, and where the precision disks recommended in the preceding example would be impractical because of their large diameters. The method is illustrated in Fig. 4-8.

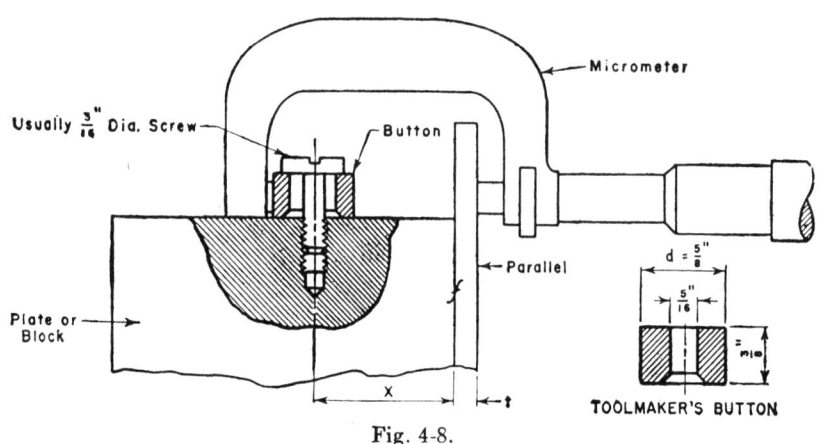

Fig. 4-8.

It is required to calculate the distance x which extends from the center of a particular hole to be drilled to a finished surface marked f (symbol for the word *finish*). A hole is drilled and tapped in the plate for a $\frac{3}{16}''$ diameter screw at an approximate distance, x, from the finished surface, f. A toolmaker's button is screwed onto the plate lightly, so it can be moved a short distance either away or toward the edge marked f, if required. The correct position of the button can be calculated from the sum of the required distance (x), half of the diameter (d) of the button, and the thickness (t) of the auxiliary parallel clamped to the plate. The micrometer spanning this over-all distance must give the required reading, and, by a slight displacement of the button in either direction, a correct micrometer reading can be obtained. This is possible inasmuch as the hole in the button is larger than the screw diameter, and the screw head is of sufficient size to overlap the large hole in the button. The preliminary fastening of the button onto the plate is usually accomplished by a very light pressure between the screw head and the plate surface. The button is now in a position to be marked by an indi-

ALGEBRA

cator that is fastened to the tool slide and is so designed that when the plate is clamped to the face plate of the lathe in the correct position the indicator will show a constant reading for all positions of the face plate throughout a complete revolution of the lathe spindle. The button is removed and a hole is drilled, reamed, or bored in the correct position.

The illustration shows only one distance x to be located. In practice, similar steps are taken to maintain other distances as, for example, the position of a hole in relation to another finished surface of the plate. This procedure can be repeated wherever holes are to be located.

The toolmaker's button shown in Fig. 4-8 is one of the common types used in the toolroom and inspection department. However, there are other sizes that can be used for this class of work, but the outside diameter of the button in each case must be precise.

This method is widely applied. Dies, jigs, fixtures, gages, and many tool set-ups can be checked effectively by methods similar to those used in the preceding illustrations.

Again, in Fig. 4-9, it is seen that holes can be located accurately by means of precision disks or rollers.

Given a circular plate, it is required to drill and ream in it three holes spaced at distances indicated in the illustration. The center hole may be indicated with the center of the plate. The hole to the right can be located by boring the plate to a suitable diameter and inserting proper disk in contact with this diameter, as shown. Assume the disk diameter is

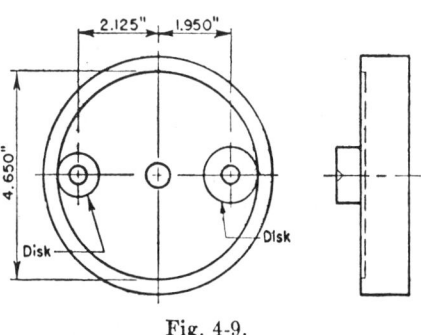

Fig. 4-9.

equal to .750″, then the bore diameter is equal to $2 \times \left(1.950 + \dfrac{.750}{2}\right) = 4.650''$. The bored hole is very shallow—just deep enough, in fact, to hold the disk in place. The disk is indicated, and the hole is drilled and reamed. The hole to the left is located by calculating a disk center to contact the bored hole as shown. The center of the disk must coincide with the center of the hole, and the diameter of the left disk is equal then to $2 \times \left(\dfrac{4.650}{2} - 2.125\right) = .400''$. The disk is indicated as shown in the foregoing examples and removed. The hole whose axis is a continuation of the disk axis is drilled and reamed. This method is practical when the difference in the dimensions locating both holes from the center of the plate is small, as it makes it possible to use disks of small diameters in both cases.

Machine Shop and Toolroom Problems. A problem frequently encountered in the shop is the right triangle and the inscribed circle, as shown in Fig. 4-10.

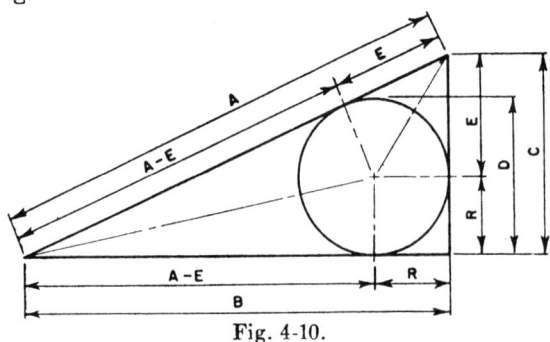

Fig. 4-10.

In practice, it may be necessary to find the diameter of a disk which is tangent to all sides of a given right triangle. The algebraic operations are as follows: A consists of $A-E$ and E; B consists of $A-E$ and R. By inspection of the figure it is evident that the length designated by $A-E$ is the same in both places, a principle that will be explained in Chapter V, "Geometry."

$$\begin{aligned} A - B &= E - R \\ + \\ \underline{A - C} &= \underline{A - E - R} \\ 2A - B - C &= A - 2R \end{aligned}$$

($A - E$ is common to both A and B.)

(E is common to both A and C.)

(Obtained by the addition of above expressions.)

$$2R = B + C - A,$$
$$\text{or } D = B + C - A$$

Another problem which involves ratios and proportions is illustrated in Fig. 4-11.

The given location of two holes is indicated in Fig. 4-11. It is required to find the horizontal distance "y" and the vertical distance "x" from the center of the disk.

From the proportion existing in similar

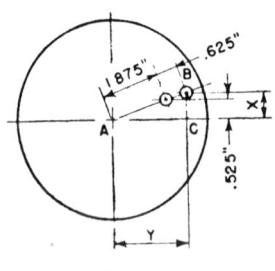

Fig. 4-11.

ALGEBRA

triangles (a principle that is treated in Chapter V, "Geometry"), the following expression can be written:

$$\frac{x}{1.875 + .625} = \frac{.525}{1.875}$$

$$\frac{x}{2.500} = \frac{.525}{1.875}$$

In any proportion, the product of the end terms is equal to the product of the mean terms:

$$1.875x = 2.500 \times .525$$

$$x = \frac{2.500 \times .525}{1.875} = \frac{1.3125}{1.875} = .700$$

Dimension "y" can be calculated from the right triangle ABC by the Pythagorean theorem that will be explained in Chapter V, "Geometry."

$$y = \sqrt{(1.875 + .625)^2 - (.700)^2} = \sqrt{6.25 - .49} = \sqrt{5.76} = 2.4$$

An interesting problem in the shop is presented by a problem illustrated in Fig. 4-12. In the problem, a right triangle ABC is given, and it is required to find another right triangle ADE whose area is twice that of the given triangle.

The area of the given triangle $ABC = \dfrac{ac}{2}$, as is explained in Chapter V, "Geometry."

The area of the required triangle $ADE = \dfrac{ed}{2}$.

The area of trapezoid $BCDE = \left(\dfrac{c+d}{2}\right)b$.

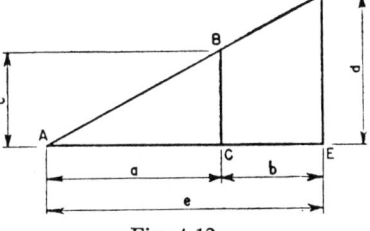

Fig. 4-12.

$$\frac{ac}{2} - \left(\frac{c+d}{2}\right)b = 0$$

$$\frac{ac}{2} + \left(\frac{c+d}{2}\right)b = \frac{ed}{2}$$

The following proportion may be set up from the similar triangles ABC and ADE, as explained in Chapter V, "Geometry":

$$\frac{a}{e} = \frac{c}{d}, \; ad = ce, \; c = \frac{ad}{e}$$

In $\dfrac{ac}{2} - \left(\dfrac{c+d}{2}\right)b = 0$, c is replaced by $\dfrac{ad}{e}$

$$\frac{aad}{2e} - \left(\frac{ad + ed}{2e}\right)b = 0$$

$$\frac{a^2d}{2e} - \frac{d(a+e)b}{2e} = 0$$

$a^2d - d(a+e)b = 0$, but $b = e - a$.

$a^2 - (a+e)b = 0$, and $a^2 - (a+e)(e-a) = 0$, therefore $a^2 - e^2 + a^2 = 0$.

Finally $2a^2 = e^2$ and $a = \sqrt{\dfrac{e^2}{2}} = \sqrt{\dfrac{1}{2}}\, e$, or $e = \sqrt{2a^2} = \sqrt{2}\,a$.

The results show the relationship that must exist between a and e in order to construct a right triangle whose area is twice the area of the given triangle.

A practical application of the quadratic equation is illustrated in Fig. 4-13, which shows a problem frequently met with in the toolroom or the inspection department. This problem consists of finding the correct diameter D of a cylindrical roller or disk that is used for checking the accuracy of a gage. If the roller or disk contacts the gage at positions A, B, and C, indicated in Fig. 4-13, it is concluded that the gage is correct.

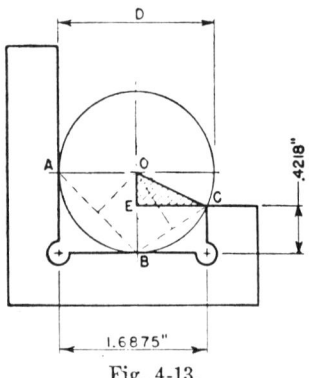

Fig. 4-13.

Procedure: In the right triangle OCE, the hypotenuse OC is the radius of the required roller, and is equal to $\dfrac{D}{2}$. In accordance with the principle common to all right triangles, the square of the hypotenuse is equal to

ALGEBRA

the sum of the squares of the other two sides of the right triangle. Therefore, $(OC)^2 = (OE)^2 + (CE)^2$.

But $OC = \dfrac{D}{2}$, $OE = \dfrac{D}{2} - .4218$, and $CE = 1.6875 - \dfrac{D}{2}$.

Hence $\left(\dfrac{D}{2}\right)^2 = \left(\dfrac{D}{2} - .4218\right)^2 + \left(1.6875 - \dfrac{D}{2}\right)^2$

Removing parentheses:

$$\dfrac{D^2}{4} = \dfrac{D^2}{4} - .4218D + .1779 + 2.8476 - 1.6875D + \dfrac{D^2}{4}$$

Collecting all terms to one side of the equation:

$$\dfrac{D^2}{4} - \dfrac{D^2}{4} - \dfrac{D^2}{4} + .4218D + 1.6875D - 3.0255 = 0$$

Simplifying:

$$-\dfrac{D^2}{4} + 2.1093D - 3.0255 = 0$$

Changing signs and clearing the fraction:

$$\left(4 \times \dfrac{D^2}{4}\right) - 4 \times (2.1093D) + (4 \times 3.0255) = 0$$

Putting the expression in the form of the quadratic equation: $ax^2 + bx + c = 0$, where $x = D$, $a = 1$, $b = -8.4372$ and $c = 12.1020$:

$$D^2 - 8.4372D + 12.1020 = 0$$

$$x = \dfrac{-b \pm \sqrt{b^2 - 4ac}}{2a}$$

$$D = \dfrac{-(-8.4372) \pm \sqrt{(-8.4372)^2 - (4 \times 1 \times 12.1020)}}{2 \times 1} = \dfrac{8.4372 \pm \sqrt{71.1863 - 48.408}}{2}$$

$$D = \dfrac{8.4372 \pm \sqrt{22.7783}}{2} = \dfrac{8.4372 \pm 4.7727}{2} = +6.6049 \text{ or } +1.8323$$

Of the two values obtained for D, the latter value is the logical one to use to suit the limitations of the problem. Therefore, the diameter of the roller or disk must be equal to $1.8323''$ in order to be in line contact with the gage at the points A, B, and C.

PRACTICE PROBLEMS APPLICABLE IN THE SHOP AND TOOLROOM

1. Find the ratio of 27 inches and 3 inches.
2. Find the ratio of 3 inches and 27 inches.
3. If the price of 7 drill bushings is $2.83, what is the cost of 13 drill bushings?
4. If 3 girls assemble 723 radio tubes per day, how many girls will be required to assemble 1,205 tubes per day? Assume the same kind of tubes and a constant rate of speed in performance of the task.
5. If gear A in Fig. 4-14 revolves 340 r.p.m., how many r.p.m. will gear B revolve?

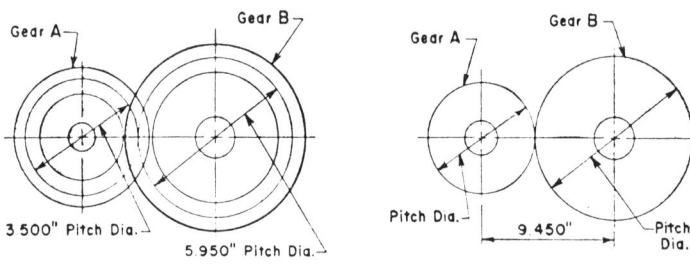

Fig. 4-14. Fig. 4-15.

6. If gear A rotates $3\frac{1}{2}$ times as fast as gear B, and the center distance between them is equal to $9.450''$, as shown in Fig. 4-15, what are the pitch diameters of each gear?

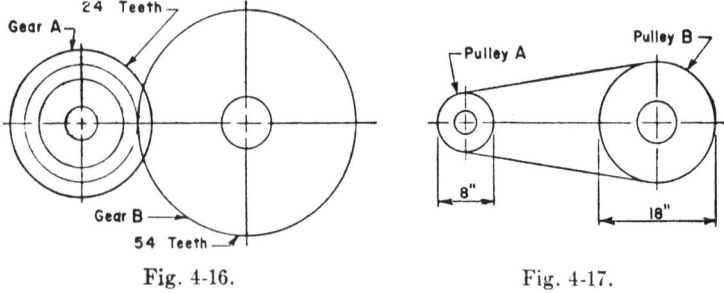

Fig. 4-16. Fig. 4-17.

7. If the speed of gear A in Fig. 4-16 is equal to 450 r.p.m. what is the speed of gear B?

ALGEBRA

8. What is the speed of pulley A in Fig. 4-17, if pulley B makes 128 r.p.m. ?
9. What is the magnitude of the force F necessary to lift the weight W in Fig. 4-18?

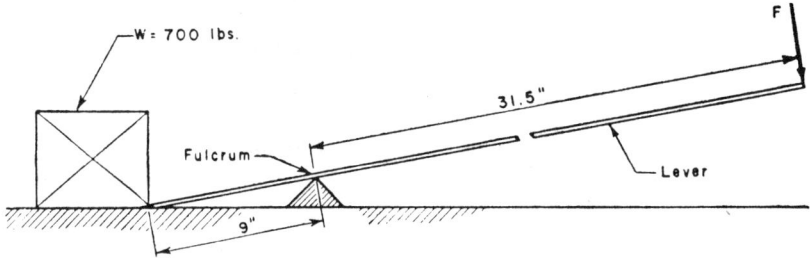

Fig. 4-18.

10. What is the speed of the driven shaft in Fig. 4-19 if the speed of the motor is 1,150 r.p.m.?
11. If a certain project is completed in the shop by 7 men in 18 days, how many men will be employed to complete a similar project in 14 days?
12. If a toolroom project is completed in scheduled time by 12 men working $7\frac{1}{2}$ hours per day, how many men must be employed to complete the same project in scheduled time working 10 hours per day?
13. 17 is what per cent of 85?
14. 13 is what per cent of 39?
15. 2.465 is what per cent of 14.5?
16. $\frac{27}{32}$ is what per cent of $2\frac{1}{4}$?
17. 22 is 20% of what number?
18. 30 is $3\frac{1}{3}$% of what number?
19. A firm purchased a desk which was listed at $110, subject to a discount of 15 per cent. What was paid for the desk?

Note: Cost = list price − discount.

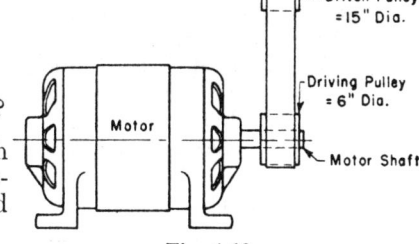

Fig. 4-19.

20. In a drafting room composed of 84 draftsmen, 7 draftsmen are absent. What per cent of the men are absent?
21. Change the following decimals to per cent: .13, .240, .374, .201, .0047.
22. Change the following per cent to decimals: 17, 1.4, 3.02, 51.004, .3072.
23. In the inspection of 743 roller-bearing races, 16 were rejected. What per cent of races was rejected?

24. In a shipment of 382 gears, 13 were rejected. What per cent of gears was accepted?
25. In casting cast iron, $\frac{1''}{8}$ per foot is allowed for shrinkage. What per cent of shrinkage occurs?
26. In casting aluminum, $\frac{5''}{32}$ per foot is allowed for shrinkage. What per cent of shrinkage occurs?
27. A machine-tool firm manufactures 170 drill presses per year, of which 137 are single-spindle machines and the remainder multiple-spindle machines. What per cent of multiple-spindle drill presses is built per year?
28. A dealer pays $6.35 for a workbench and sells it for $9.25. What per cent of profit is realized?
29. A machinist earns $2,340 per year and spends $2,125. What per cent of earnings does he save?
30. A company purchased a machine tool for $2,785. After being in operation for a certain length of time, the machine was sold to a dealer for $1,315. What per cent of depreciation resulted?
31. A tin-base alloy containing 91 per cent tin, 4.50 per cent copper, and 4.50 per cent antimony is used in aircraft bearings. How many pounds of each constituent are contained in a bearing weighing 23.42 pounds?
32. A magnesium-base alloy containing 10 per cent aluminum, .30 per cent zinc, .10 per cent manganese, .70 per cent silicon and the remainder magnesium is suitable for die-casting. How many pounds of each constituent are contained in a casting weighing 47.58 pounds?
33. What is the net price of a shipment of bolts, the list price of which is $72.86, subject to a discount of 35 per cent plus a further discount of 6 per cent for cash?
34. What per cent of the indicated horsepower is the actual effective horsepower of an engine, if the indicated horsepower is 24.8 and the actual horsepower 17.6?
35. A lot of screws is listed in the catalogue as follows: $37.32 list, less 40 per cent − 20 per cent − 10 per cent. What is the net price of the lot of screws?
36. If 4 men working $8\frac{1}{2}$ hours can fill a space in a freight car with iron castings weighing 7,300 pounds, how many working $6\frac{3}{8}$ hours will fill another space in the freight car with 10,950 pounds of iron castings, assuming that the efficiency of the men is constant and the castings are identical in kind?
37. If 7 men in a railway repair shop working 9 hours per day make 14 cars ready for use in 4 days, how many days will it take 5 men

working at 6 hours per day to complete 25 cars, assuming the efficiency of the men and the type of work are unchanged?

38. If 6 men on a farm working 10 hours per day 6 days per week can harvest 12,798 bushels of fruit in 3 weeks, how many weeks will be required to harvest 18,770.4 bushels of fruit by 8 men working 12 hours per day $5\frac{1}{2}$ days per week, assuming the efficiency of the men remains constant throughout the task?

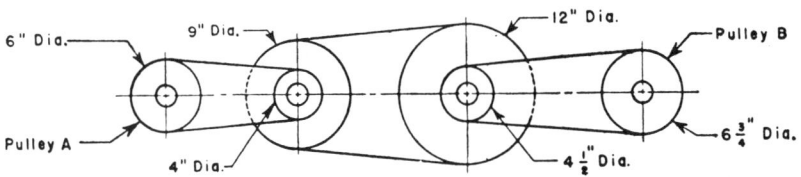

Fig. 4-20.

39. Find the speed of pulley B in Fig. 4-20, if the speed of pulley A is 750 r.p.m. and the diameters of the intermediate pulleys are as shown in the illustration.
40. Find the speed of pulley B in Fig. 4-20, if the speed of pulley A is 860 r.p.m. and the diameters of the intermediate pulleys are as shown.
41. Find the speed of pulley A in Fig. 4-20, if the speed of pulley B was known to be 940 r.p.m. and the diameters of the intermediate pulleys were unchanged from those shown in Fig. 4-20.
42. Assume the diameters of the pulleys in Fig. 4-20 are unchanged from those shown except for pulley A, whose diameter is not known and is to be determined. The speed of pulley A is given as 720 r.p.m. and that of pulley B as 360 r.p.m.

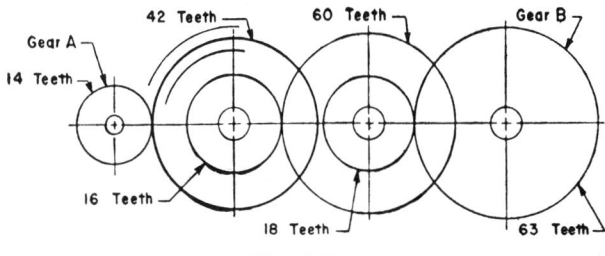

Fig. 4-21.

43. Find the speed of gear B in Fig. 4-21, if the speed of gear A is 820 r.p.m. and the number of teeth of the intermediate gears are as shown.

132   MATHEMATICS FOR INDUSTRY

44. Find the speed of gear B in Fig. 4-22, if the speed of gear A is 640 r.p.m. and the number of teeth of the intermediate gears are as shown.

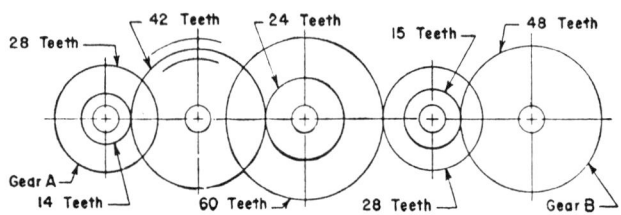

Fig. 4-22.

45. Solve $x - y = 6$ and $x + y = 16$ for x and y.
46. Solve $8x - 3y = 23$ and $40x + y = 163$ for x and y.
47. Solve $3x + 2y = 34$ and $4x - y = 27$ for x and y.
48. Solve $7x - 5y = 28$ and $5x + 2y = 59$ for x and y.
49. Solve $9x + 4y = 92$ and $5x - 2y = 30$ for x and y.
50. Solve $2.6x - y = 7$ and $4x + 3y = 38$ for x and y.
51. Solve $17x - 11y = 80$ and $9x + 6y = 66$ for x and y.
52. Solve $13x - 4y = 148$ and $5x + 3y = 66$ for x and y.
53. Solve $26x + 7y = 59$ and $25x - 3y = 47$ for x and y.

54. Solve $\dfrac{x}{3} + \dfrac{x}{15} = \dfrac{x}{21} + 2\dfrac{7}{15}$ for x.

55. Solve $\dfrac{y}{5} - \dfrac{y}{25} + \dfrac{y}{15} = 1\dfrac{61}{75}$ for y.

56. Solve $\dfrac{7(x+3)}{3} - \dfrac{4(x-2)}{18} = \dfrac{.5x(8-3)}{24} + \left(17 - 5\dfrac{5}{18}\right)$ for x.

57. Solve $\dfrac{4(x-2)}{7} + \dfrac{2(x+2)}{42} = \dfrac{3x(4+2)}{35} - \dfrac{11}{15}$ for x.

58. Solve $\dfrac{8(y+.75)}{6} - \dfrac{5(6y-7.5)}{2} + \dfrac{12y+13}{5} = \left(4.5 - \dfrac{15}{2}\right)$ for y.

59. What horsepower is necessary to raise a machine weighing $4\dfrac{1}{2}$ tons to a floor 32 feet high in $3\dfrac{1}{4}$ minutes?

Note: Horsepower $= \dfrac{W \times h}{t \times 33{,}000}$, where W = weight in pounds, h = height in feet, t = time in minutes, and 33,000 is a constant denoting the number of pounds lifted 1 foot in 1 minute which is the equivalent of 1 horsepower.

60. What horsepower is required to pump 200,000 barrels of water per

ALGEBRA

8-hour working day to a height of 35 feet? Assume 1 barrel of water is equal to 4.2 cubic feet, and 1 cubic foot of water weighs 62.5 pounds.

61. In an electric power machine, the horsepower formula is as follows: Horsepower equals $\dfrac{\text{Volts} \times \text{Amperes}}{746}$, where volts times amperes equals watts and 1 horsepower = 746 watts. What is the capacity of the machine in horsepower if the number of volts equals 220 and the number of amperes equals 300?

62. What horsepower is developed by a steam engine, having a cylinder bore 6″ in diameter and a stroke of $8\frac{1}{2}''$ at 350 r.p.m. of the crank, if the mean effective pressure is 110 pounds per square inch? The formula $H = \dfrac{PLAN}{33{,}000}$ is used for calculating the horsepower of a steam engine, where H = indicated horsepower, P = mean effective pressure of steam in pounds per square inch, L = length of stroke in feet, A = area of piston in square inches, and N = number of strokes of piston per minute.

Note: Care should be taken to express the length of the stroke in feet. Also, the area of the piston = $A = \dfrac{\pi}{4} D^2$, where D = bore diameter in inches.

63. What is the diameter of the cylinder bore of a steam engine that develops 60 horsepower, has a stroke of $9\frac{1}{4}''$ and crank rotation of 85 r.p.m.? The mean effective pressure is equal to 70 per cent of the boiler pressure, which is indicated as 75 pounds per square inch.

Note: Use the formula $H = \dfrac{PLAN}{33{,}000}$, as explained in Problem 62. The area A of the cylinder bore is calculated from the formula, then the diameter D of the cylinder bore is calculated from the area formula, which is $A = \dfrac{\pi}{4} D^2$.

64. The total torque in pound-inches transmitted by a plate friction clutch is expressed by the formula: $T = 2\pi fp \dfrac{r_o^3 - r_i^3}{3}$, where T = torque in pound-inches, f = coefficient of friction of the clutch facing = .35, p = unit pressure on the clutch facings = 30 pounds per square inch, r_o = outside radius of the clutch facing = 5″, r_i = inside radius of the clutch facing = 3″. Calculate the total torque in pound-inches.

65. In calculating the strength of a 3″ diameter transmission shaft, a value designated by the symbol α is used, which expresses the ratio of the maximum intensity of stress resulting from the axial

load to the average axial stress. The equation is expressed as follows:
$$\alpha = \frac{1}{1 - .0044\left(\frac{L}{K}\right)},$$
where L = length of shaft between supporting bearings = 72″, and k = radius of gyration of the shaft = $\frac{d}{4}$ = $\frac{\text{shaft diameter in inches}}{4}$. Calculate the value of α.

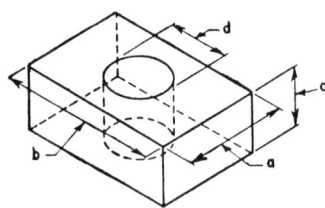
Fig. 4-23.

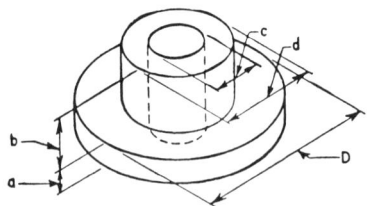
Fig. 4-24.

66. Calculate the weight of the steel block shown in Fig. 4-23, using the following formula: $W = wc\left(ab - \frac{\pi}{4}d^2\right)$, where W = the weight of the block in pounds, w = the weight of 1 cubic inch of steel = .28 pounds, a = 4.75″, b = 11.25″, c = 8.5″ and d = diameter of hole through the block = 2.125″.

67. Calculate the weight of the aluminum part shown in Fig. 4-24, using the following formula: $W = w\frac{\pi}{4}[D^2a + d^2b - c^2(a+b)]$,

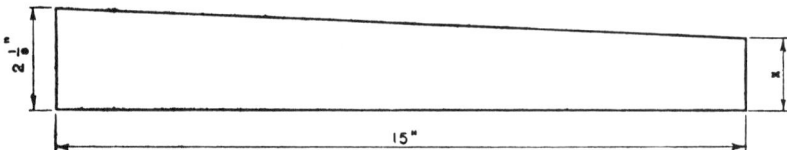
Fig. 4-25.

where W = the weight of the aluminum part in pounds, w = the weight of 1 cubic inch of aluminum = .098 pound, D = diameter of base = 7.5″, d = diameter of base = 4.25″, c = diameter of hole through the part = 1.87″, a = 4.75″, and b = 5.50″.

68. Calculate the dimension x in Fig. 4-25, if the taper per foot is equal to $\frac{1''}{2}$

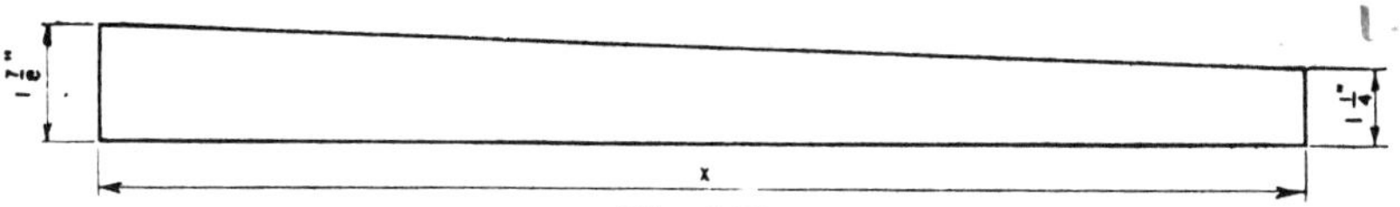

ALGEBRA

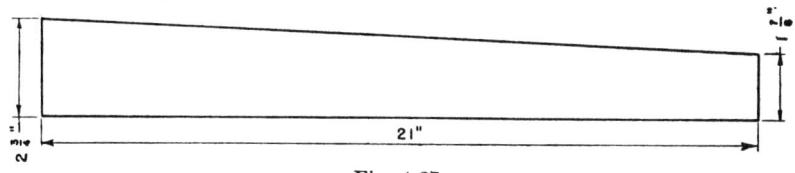

Fig. 4-26.

69. Determine the dimension x in Fig. 4-26, if the taper per foot is equal to $\dfrac{3''}{8}$.

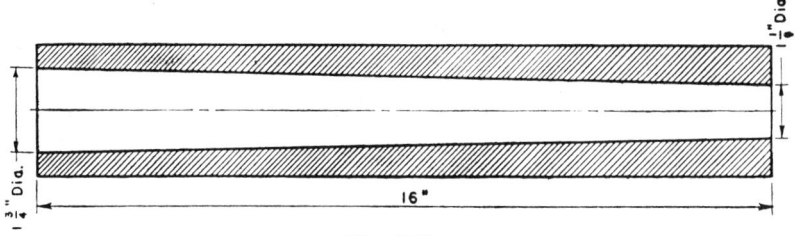

Fig. 4-27.

70. Determine the taper per foot as shown in Fig. 4-27.

Fig. 4-28.

71. Determine the taper per foot as shown in Fig. 4-28.

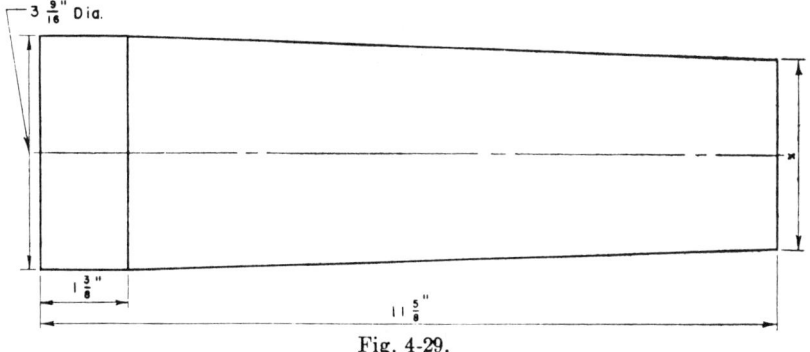

Fig. 4-29.

72. Calculate the diameter x in Fig. 4-29 if the taper per foot is equal to $\dfrac{3''}{4}$.

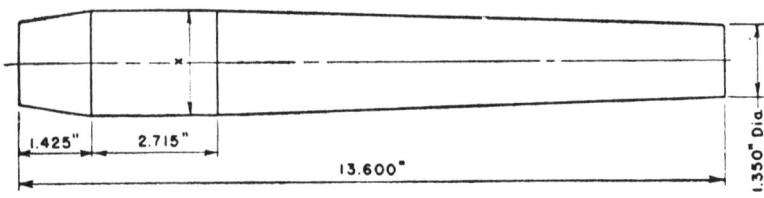

Fig. 4-30.

73. Calculate the diameter x in Fig. 4-30, if the taper per foot is equal to $\dfrac{11''}{16}$.

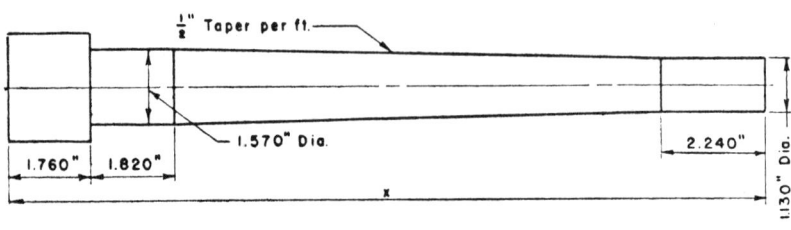

Fig. 4-31.

74. Calculate the length x of part shown in Fig. 4-31.

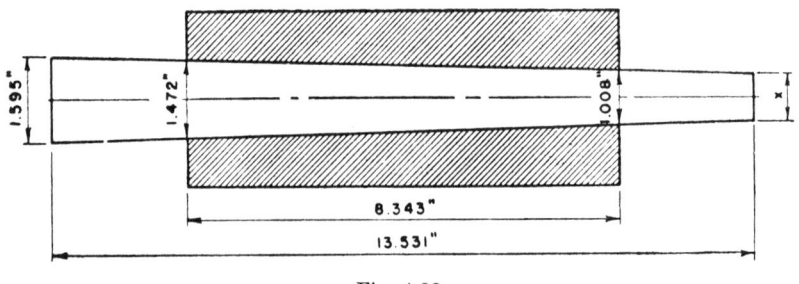

Fig. 4-32.

75. Calculate the dimension x in Fig. 4-32.

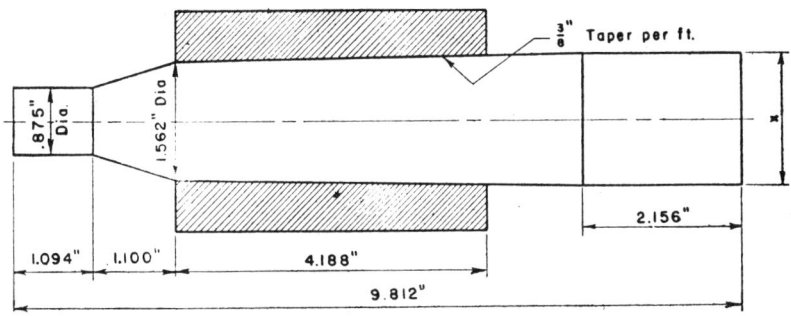

Fig. 4-33.

76. Calculate the diameter x in Fig. 4-33.

77. Calculate the distance x, if the small end of the taper plug gage in Fig. 4-34 is 1.360″ diameter instead of 1.375″ diameter, as shown.

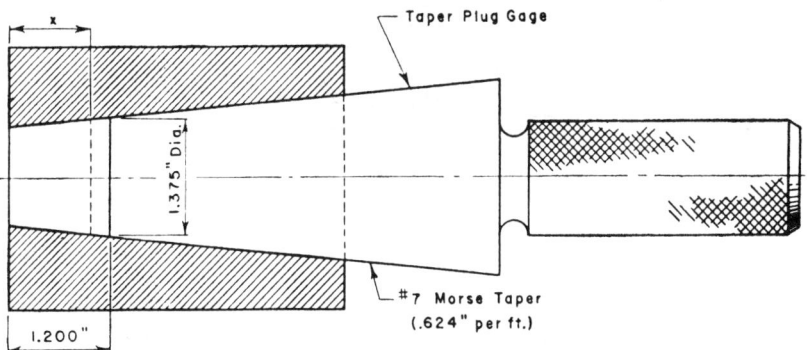

Fig. 4-34.

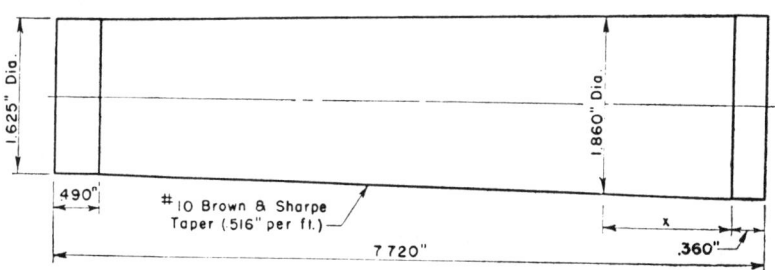

Fig. 4-35.

78. Determine the distance x in Fig. 4-35.

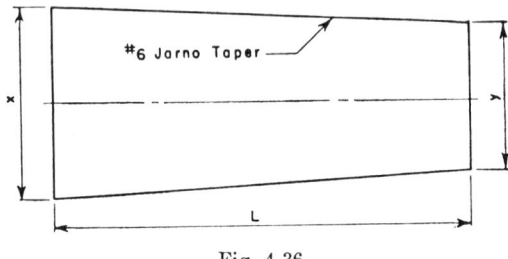

Fig. 4-36.

79. What are the diameters x and y and length L of the part shown in Fig. 4-36?

80. Find the center distance x between two disks, $\frac{7''}{8}$ diameter and $\frac{9''}{16}$ diameter, which are used for the purpose of constructing a gage that will check a taper $\frac{5''}{8}$ per foot. If the gage is correct, its jaws

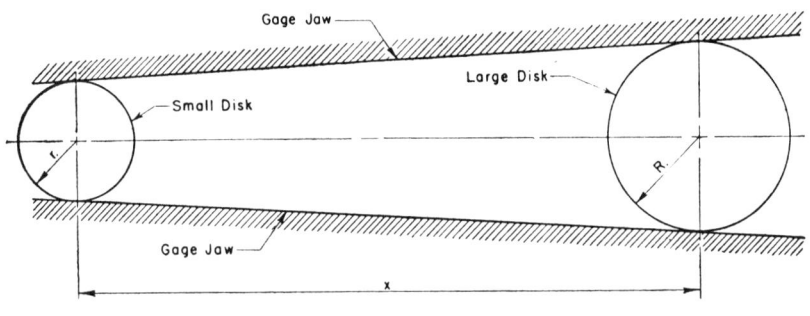

Fig. 4-37.

will touch both disks, as shown in Fig. 4-37. Use formula: $x = \frac{R-r}{t}\sqrt{1+t^2}$. R = radius of large disk, r = radius of small disk, t = taper per inch on one side.

81. What diameters must the disks be made of in Fig. 4-38, so that when the jaws are in contact with them and the distance L over

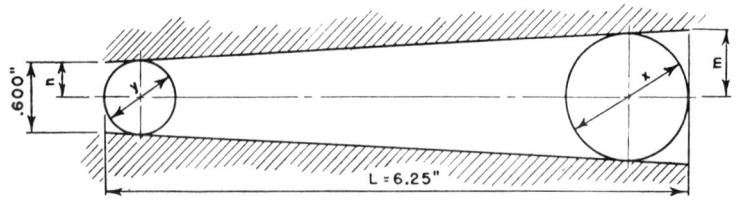

Fig. 4-38.

the disks measures 6.25″, the taper per foot will be $\frac{7''}{8}$? Use formulas:

$$x = \frac{2m}{L} [\sqrt{L^2 + (m - n)^2} - (m - n)]$$

$$y = \frac{2n}{L} [\sqrt{L^2 + (m - n)^2} + (m - n)]$$

82. A parking lot is in the form of a rectangle, one side of which is 39 feet longer than the other. Find the length of the sides, if the area is equal to 2,700 square feet.

Hint: Set up a quadratic equation and solve it by any convenient method.

Determine the value of x in the equations in Problems 83 to 88:

83. $3x^2 - 14x = -8$.
84. $7x^2 + 6x = 81$.
85. $8.4x^2 - 7.6x = 172$.
86. $12\frac{1}{2}x^2 + 8\frac{1}{3}x = 500$.
87. $4x^2 + 9x + 45 = 100$.
88. $20x^2 - 18x - 24 = 92$.
89. Find the radius of a circle inscribed in the right triangle as shown in Fig. 4-39.

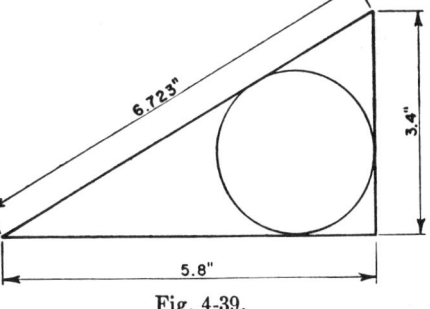

Fig. 4-39.

90. Find the diameters of 3 gage disks required in locating the centers of 3 holes before reaming, Fig. 4-40.

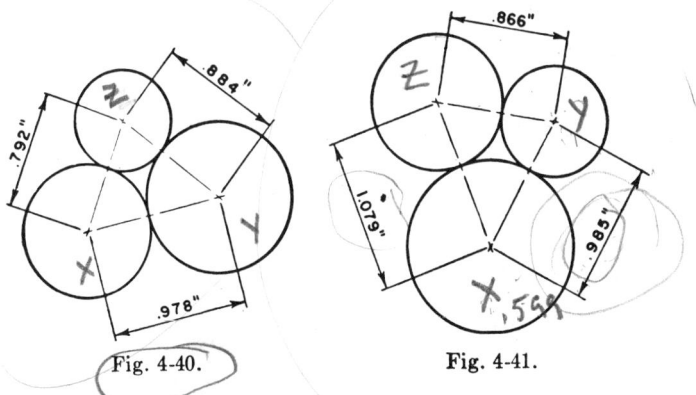

Fig. 4-40.   Fig. 4-41.

91. Find the diameters of 3 gage disks required in locating the centers of 3 holes before reaming, Fig. 4-41.

92. The total capacity of two unequal tanks is 368 gallons. When $\frac{1}{3}$ of the contents of the larger tank and $\frac{1}{5}$ of the contents of the smaller tank are taken out, the remaining contents in both tanks are the same. Determine the capacity of each tank.

93. The perimeter of a steel rectangular plate is 38 inches. If the length of the plate is reduced by $3\frac{1}{2}''$ and its width is increased by $1\frac{3}{4}''$, the area does not change. Calculate the dimensions of the plate, and determine its area.

94. A factory plot is of a rectangular shape. If its length is increased by 25 feet and the width is reduced by 25 feet, the area of the plot is changed to 8,400 square feet. If the length is reduced by 25 feet and the width is increased by 25 feet, the area of the plot is changed to 9,700 square feet. Calculate the length and width of the plot.

95. If a number is squared, the result is greater than the sum of the squares of its two halves by 72. Find the number.

96. If 3 is added to a number, its reciprocal is reduced by $\frac{1}{16}$. What is the number?

97. A train covered the distance of 840 miles at a certain speed. If the speed of the train had been slowed down by 15 miles per hour, the trip would have taken $2\frac{1}{2}$ hours longer. What is the speed of the train?

98. $S = c + \frac{cp}{100}$, find p.

99. $x = y\left(\frac{2t}{p+q} - 1\right)$, find t.

100. $t = 2\pi \sqrt{\frac{2Wa^2 + Mb^2}{Mgb}}$, find M.

101. $t = 2\pi \sqrt{\frac{l^2 + 3h^2}{3gh}}$, find l.

102. $S = \frac{6Fh}{ft^2}$, find t.

103. $F = Sf\frac{\pi y}{P}$, find y.

104. $F = \frac{SfY}{P}\left(\frac{600}{600 + V}\right)$, find V.

105. $F = \frac{SfY}{P}\left(\frac{78}{78 + \sqrt{V}}\right)$, find V.

106. $F = \dfrac{SfY}{P}\left(\dfrac{150}{200 + V} + .25\right)$, find V.

107. $f = .54 - \dfrac{140}{500 + V}$, find V.

108. $L = 2C + 1.57\,(D + d)$, find d.

109. $\dfrac{\pi d^2}{4} S_s = dt S_c$, find d.

110. $1\tfrac{3}{4}d + .1(p - d) = c$, find d.

111. $K = \dfrac{c}{c + e}$, find c.

112. Prove that $\dfrac{h}{w} = \dfrac{2S_s}{S_c}$, if $\dfrac{DLhS_c}{4} = \dfrac{DLwS_s}{2}$.

113. $\dfrac{2T}{DLw} = \dfrac{16T}{\pi D^3} \times \dfrac{1}{.75}$, find D.

Chapter V
GEOMETRY

The mathematical science of geometry is a very old science dating back to the Egyptians and Babylonians. However the worker in the shop and toolroom, as well as many workers in various fields, will find that a knowledge of geometry is a necessary aid in solving many problems. This chapter brings out some of the most important facts of geometry and the application of these facts to shop and toolroom problems.

Every material body must occupy a definite portion of space. Geometry is a mathematical science that deals with the space occupied by a material body.

Plane geometry is that branch of the science of mathematics which makes a study of the figures made by points and lines that lie in the same plane. Solid geometry is a study of solids or figures with the dimension of thickness as well as length and width.

PRINCIPLES OF PLANE GEOMETRY

Many terms used in geometry are briefly defined and explained here to enable the student to apply the principles of geometry with ease and confidence.

The boundary which separates a material body from surrounding space is called the *surface* of the body.

A *point* is the intersection of one part of a line from another part of the same line or the intersection of two or more lines. It is definitely positioned, but it does not have length, or width, or thickness.

A *line* is the boundary of a surface or the intersection of two or more surfaces. It has definite position, and has only one dimension, namely, length. A line is straight when it takes the same direction throughout; when the direction changes, the line is said to be "curved." A line is "broken" when it is composed of several connected straight lines.

A line may be generated by a point, whereas a surface is the result of generating a line.

In a plane surface, any two points can be connected by a straight line lying entirely in the surface. This condition does not exist in a curved surface.

Angles are formed by two straight lines which meet at a given point. The point at which the lines meet is the *vertex* of the angle, and the lines are called the *sides* of the angle.

GEOMETRY

An angle may be read by the single letter at the vertex, by a small letter written between the sides of the angle, or by three letters. When using three letters, one is at the vertex and one on each side. The letter at the vertex is always the middle letter. An angle *ABC* is shown in Fig. 5-1, where *B* is the vertex and *AB* and *BC* are the sides.

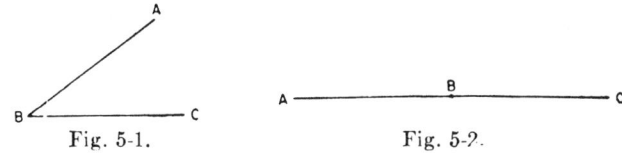

Fig. 5-1. Fig. 5-2.

A *straight angle* is one in which the sides extend in a straight line in opposite directions from the vertex, as shown in Fig. 5-2, where *AB* and *BC* are the sides of the angle and *B* the vertex.

Angles having a common side and a common vertex are called *adjacent angles*, as shown in Fig. 5-3, where *ABC* and *CBD* are adjacent angles whose common side is *BC* and common vertex *B*.

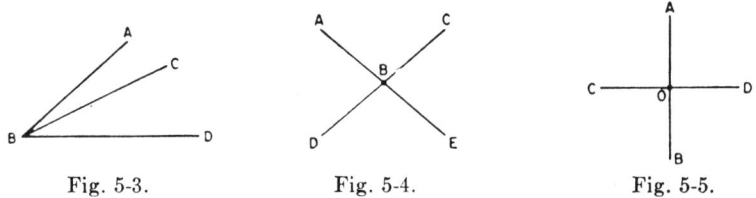

Fig. 5-3. Fig. 5-4. Fig. 5-5.

Vertical or *opposite angles* are formed by two lines extended in opposite directions and crossing each other at a certain point. These angles have a common vertex, as shown in Fig. 5-4, where angles *ABC* and *DBE*, or *ABD* and *CBE* are vertical or opposite angles having a common vertex *B*.

For brevity, angles are designated by the symbol ∠ or ⊿, as hereinafter shown.

When adjacent angles formed by intersected lines *AB* and *CD* are equal to each other, they are called *right angles*, as in Fig. 5-5, where angles *AOC*, *AOD*, *BOC*, and *BOD* are right angles.

Angular measurements are expressed in degrees, each degree is $\frac{1}{90}$ of a right angle. The right angle is taken as equal to 90°, and the straight angle as equal to 180°. Each degree is subdivided into smaller units called minutes ('), and is equal to 60 of these smaller units. Minutes are subdivided into still smaller units called seconds ("), and each minute is equal to 60 seconds.

An angle which is less than 90° is called an *acute* angle, and an angle greater than 90° is called an *obtuse* angle.

If the sum of two angles is equal to a right angle (90°), the two angles are *complementary* to each other.

If the sum of two angles is equal to a straight angle (180°), the two angles are *supplementary* to each other.

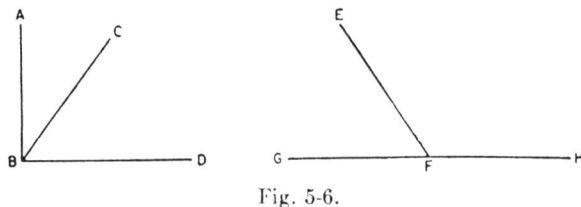

Fig. 5-6.

In Fig. 5-6, ∠ *ABC* and *CBD* are complementary, ∠*EFG* and *EFH* are supplementary.

Parallel lines are lines that lie in the same plane and are equally distant from each other at all points. In Fig. 5-7, *AB* and *CD* are parallel lines and are written *AB* ∥ *CD*.

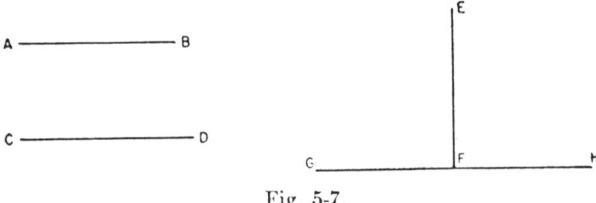

Fig. 5-7.

When two straight lines intersect and form right angles, they are called *perpendicular lines*. In Fig. 5-7, *EF* and *GH* are perpendicular lines and are written *EF* ⊥ *GH*.

PLANE GEOMETRIC FIGURES

Plane geometric figures are plane surfaces, i.e., surfaces which a straightedge may contact at all points. These surfaces are bounded by straight or curved lines. If the boundaries are straight lines, the figures are *polygons*, which are named in accordance with the number of sides or angles each presents.

A *triangle* is a plane figure bounded by three sides, such as ABC in Fig. 5-8. It is represented by the symbol △ or △.

The sum of all angles in any triangle is equal to 180°. The proof of this theorem will be presented later in this chapter.

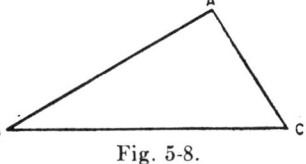
Fig. 5-8.

The sum of all angles in any polygon can be calculated with the aid of a simple formula: $S = 180° \times (N - 2)$, where S = sum of all angles in the polygon in degrees and N = number of sides of the polygon. For an example, the sum of all angles in an octagon is equal to $180° \times (8 - 2) = 180° \times 6 = 1080°$. The proof of this theorem will be presented later in this chapter.

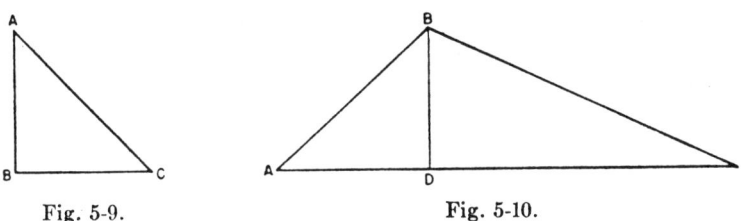

Fig. 5-9.   Fig. 5-10.

A *right triangle* has one right angle, as shown in Fig. 5-9, where $\angle ABC$ is a right angle. The *hypotenuse* is the side opposite the right angle. The other two sides are called the *base* and *altitude*. The base is the side upon which the triangle is supposed to stand.

In an *oblique triangle*, Fig. 5-10, none of the angles is a right angle.

In any triangle, say in Fig. 5-10, AC is called the base, and BD is the altitude or height.

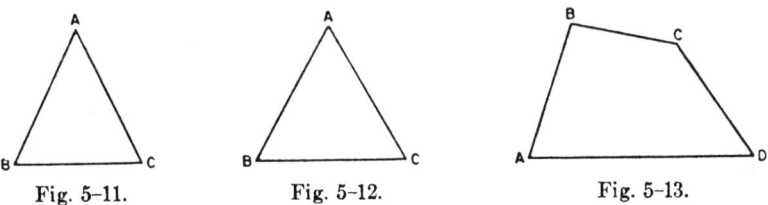

Fig. 5-11.   Fig. 5-12.   Fig. 5-13.

An *isoceles triangle* is a triangle having two equal sides, therefore the two angles opposite these sides are equal, as shown in Fig. 5-11, where sides AB and AC are equal, and $\angle ABC = \angle ACB$.

An *equilateral triangle*, Fig. 5-12, is a triangle that has all of its sides equal. Its angles are also equal. An *equiangular triangle*, Fig. 5-12, is a triangle having all of its angles equal. Its sides are also equal.

A *quadrilateral* is a four-sided polygon, Fig. 5-13. If the four sides and the four angles of a quadrilateral are equal, the figure is a *square*. If the

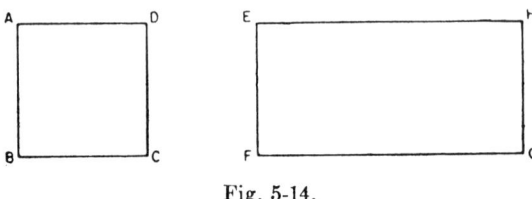

Fig. 5-14.

opposite sides are parallel and the four angles are right angles, the figure is a *rectangle*. In Fig. 5-14, *ABCD* is a square and *EFGH* a rectangle.

A quadrilateral which has both pairs of opposite sides parallel is a *parallogram*. The square and the rectangle shown in Fig. 5-14 are special

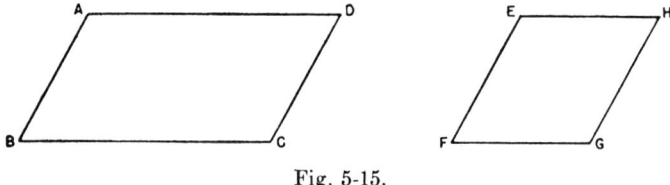

Fig. 5-15.

examples of parallelograms. Other parallelograms are shown in Fig. 5-15, in which *ABCD* is a *rhomboid* and *EFGH* a *rhombus*.

If only one pair of opposite sides are parallel, the quadrilateral is called a *trapezoid*, Fig. 5-16.

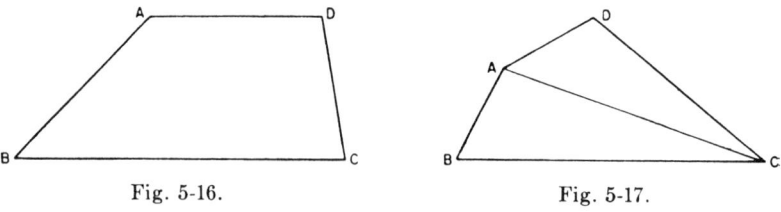

Fig. 5-16.   Fig. 5-17.

If none of the opposite sides are parallel, the quadrilateral is called a *trapezium*, Fig. 5-17.

In any polygon except a triangle, a line joining any two vertices that are not adjacent is a *diagonal*. In Fig. 5-17, line AC is a diagonal.

Other polygons, such as a *pentagon, hexagon, heptagon, octagon, nonagon*, and *decagon*, are shown in Fig. 5-18.

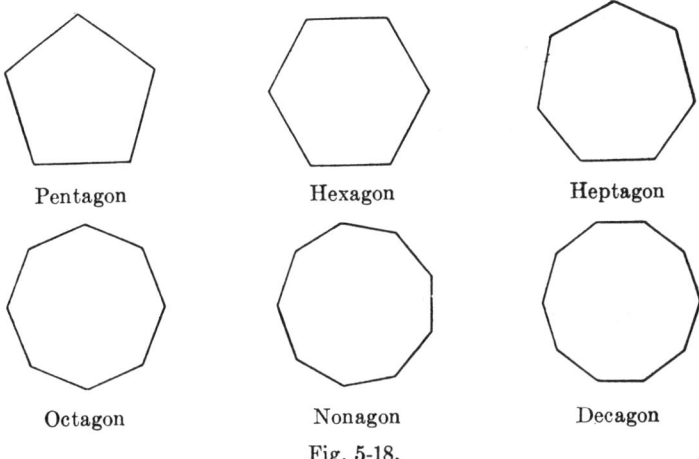

Fig. 5-18.

In a regular polygon, all sides are equal, and these geometric figures are also called equilateral polygons. The equilateral triangle and the square are also equiangular.

The sum of all sides of a polygon expressed as a linear dimension is called the *perimeter*.

Plane geometric figures bounded by curved lines will be treated later in this chapter.

AXIOMS AND THEOREMS

In the study of geometry certain statements are so obvious that we accept them without proof. Such fundamental truths are called *axioms*. Some authorities distinguish between axioms and *postulates*. Axioms refer to fundamental truths pertaining to quantities in general without special reference to geometry. Postulates are axioms that refer especially to geometry. In this text we shall not make this distinction. It is well to note, however, that Axiom 9 and 10 following are also called postulates.

Statements that need proving by the process of reasoning and by reference to suitable axioms and to previously proved statements are called *theorems* or *propositions*. For the sake of simplicity we have grouped the theorems into three groups: General Theorems, Triangle Theorems, and Circle Theorems. This enables the reader to refer easily to other theorems while proving the theorem at hand.

If a statement is a direct deduction or inference drawn from another statement, which may be an axiom or a demonstrated theorem, it is called a *corollary*.

Axioms. Some of the axioms commonly used in geometry have been mentioned in the study of algebra. These axioms, and others, are enumerated herewith:

1. When equal quantities are added to equal quantities, the sums are equal.

2. When equal quantities are subtracted from equal quantities, the remainders are equal.

3. When equal quantities are multiplied by equal quantities, the products are equal.

4. When equal quantities are divided by equal quantities, the quotients are equal.

5. Quantities or things that are equal to the same quantity or thing are equal to each other.

6. In a mathematical expression, any quantity may be replaced by another quantity, provided both quantities have the same value.

7. The whole of a quantity equals the sum of its parts.

8. The whole of a quantity is greater than any of its parts.

9. Only one straight line may be drawn between two given points.

10. Only one line can be drawn through a given point parallel to a given line.

A direct deduction or corollary may be drawn from Axiom 9, which can be stated as follows:

If two straight lines intersect, intersection can occur only at one point.

Another corollary may be drawn from Axiom 10, and its statement reads as follows:

Two lines parallel to a third line are parallel to each other.

GENERAL THEOREMS OR PROPOSITIONS

Some of the general theorems or propositions commonly used in geometry are enumerated in the paragraphs that follow.

Theorem 1. *A perpendicular drawn from a given external point to a*

GEOMETRY

given line is the only perpendicular line and the shortest line that can be drawn under the given conditions.

Given: Line AB, point C outside of it, and line $CD \perp$ to AB, as shown in Fig. 5-19.

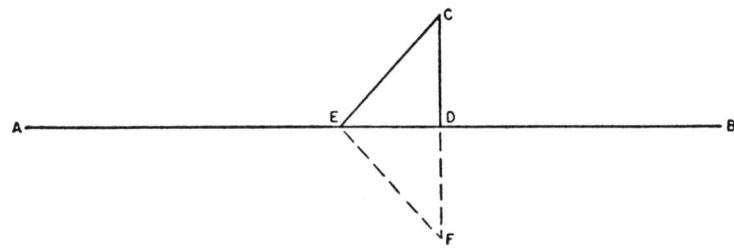

Fig. 5-19.

Required: To prove that line CD is the only perpendicular line and the shortest line that can be drawn from point C to line AB.

Proof: Assume any other line, such as CE, and prove that it is not $\perp$ to AB. Extend line CD downward and obtain line $DF =$ line CD. Connect points E and F by a straight line, EF. $\triangle CDE$ and EDF are produced in which ED is a common side, $CD = DF$, and $\angle CDE$ and EDF are both right angles and equal to each other. ∴ the two triangles are equal to each other (this will be proved later in this chapter), and $\angle CED$ and DEF are equal to each other. (The symbol ∴ represents the word "therefore" and will be used hereafter.)

Since $\angle CEF$ is not equal to 180°, $\angle CED$ is not equal to 90°, i.e., it is not a right angle, as $\angle CED =$ one-half $\angle CEF$. This development leads us to conclude that line CE is not $\perp$ to line AB. In like manner, any other line that does not produce a right angle with line AB is not perpendicular to it. ∴ line CD is the only perpendicular to line AB.

To prove that line CD is the shortest line that can be drawn under the given conditions, compare line CDF with line CEF. It is found that line CDF is shorter than line CEF (Axiom 9). Since $CD =$ half of CDF and $CE =$ half of CEF, line CD is shorter than line CE, written thus $CD < CE$ (Axiom 4). For the same reason, any other line drawn in the figure can be proved to be longer than the perpendicular line CD.

Note: The symbol $<$ represents the words "less than," and the symbol $>$ represents the words "greater than."

Theorem 2. *If two straight lines intersect, the opposite angles (also called vertical angles) are equal.*

Given: Two straight lines AB and CD intersecting each other at point E, as shown in Fig. 5-20.

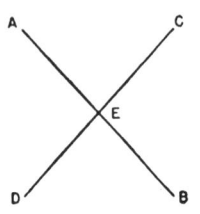

Fig. 5-20.

Required: To prove that the opposite $\angle AEC = BED$ and $AED = BEC$.

Proof: $\angle AEC$ and AED are supplementary to each other. $\angle BED$ and AED are supplementary to each other. $\therefore \angle AEC = \angle BED$.

It may be proved by the same reasoning that $\angle AED = \angle BEC$.

Theorem 3. *When two given lines lie in the same plane, and both are perpendicular to the same line, the given lines are parallel to each other.*

Given: Two lines AB and $CD \perp$ to line EF, as shown in Fig. 5-21.

Required: To prove that $AB \parallel CD$.

Proof: Assume that lines AB and CD are not parallel and meet at a point when extended far to the right of the figure. In that event there would be two perpendiculars from that point to the line EF, a circumstance which contradicts General Theorem 1, which states that only one perpendicular can be drawn from a given point to a given line. Therefore, we may conclude that lines AB and CD must be parallel, as they never can meet when they appear as they are shown in Fig. 5-21.

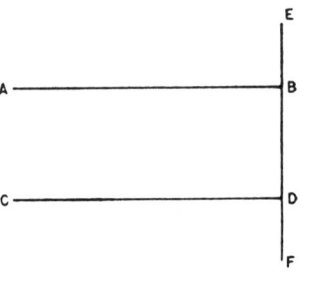

Fig. 5-21.

Theorem 4. *When two lines are parallel to each other, and one of these lines is perpendicular to a given line, the other line is also perpendicular to the given line.*

Given: Two lines AB and CD parallel to each other and line $AB \perp$ to a third line EF, as shown in Fig. 5-22.

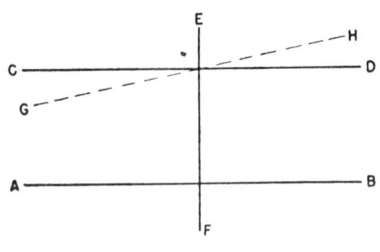

Fig. 5-22.

Required: To prove that CD is also $\perp$ to EF.

Proof: Assume that CD is not $\perp$ to EF, and that GH, as shown in Fig. 5-22, is $\perp$ to EF. In that event $GH \parallel AB$ in accordance with General Theorem 3, stating that two lines perpendicular to a third line are parallel to each other. But $AB \parallel CD$ in accord-

GEOMETRY

ance with the given conditions. ∴ GH and CD must occupy the same position; and since GH was assumed to be ⊥ to EF, CD is also ⊥ to EF (Axiom 10).

Theorem 5. *If two parallel lines are cut by a third line, the alternate-interior angles are equal.*

Given: Two parallel lines AB and CD cut by line EF, as shown in Fig. 5-23.

Required: To prove that $\angle c = \angle f$, and $\angle d = \angle e$.

Note: In Fig. 5-23, angles a, b, g, and h are called *exterior* angles, and angles c, d, e, and f *interior* angles. Angles of pairs a and h and b and g are called *alternate-exterior* angles. Angles of pairs c and f and d and e are called *alternate-interior* angles. Angles of pairs a and e, c and g, b and f, and d and h are called *corresponding* angles (often called pairs of *exterior-interior* angles).

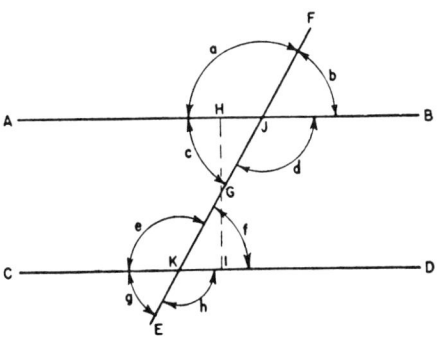

Fig. 5-23.

Proof: Locate the mid-point G in line JK and draw through it line $HI \perp AB$. This line is also ⊥ to CD, according to General Theorem 4. Consider right △GHJ and GIK; they are equal to each other because △GHJ and GIK are right angles and are equal to each other, △HGJ and IGK are opposite angles and are equal to each other (General Theorem 2), and lines GJ and GK are equal to each other by construction (G was taken as mid-point). ∴ these right triangles are equal to each other as proved later in this chapter, and all the corresponding elements in equal triangles are equal, hence $\angle c = \angle f$. Angles d and e are each equal to a straight angle (straight angle = 180°), less an angle c or f, i.e., angles d and e are supplementary to angles c and f in corresponding order, and therefore are equal to each other.

Corollary. If two parallel lines are cut by a third line, the alternate-exterior angles are equal.

The same reasoning used with alternate-interior angles applies to the alternate-exterior angles, a and h and b and g. Each of these angles is supplementary to a corresponding alternate-interior angle, and if these have been proved to be equal to each other, the alternate-exterior angles are equal in a corresponding order. Hence, $\angle a = \angle h$, being supplementary to equal △c and f, and $\angle b = \angle g$, being supplementary to equal △d and e.

Theorem 6. *If two parallel lines are cut by a third line, the corresponding (exterior-interior) angles are equal.*

Given: Two parallel lines, AB and CD, cut by a third line, EF, as shown in Fig. 5-24.

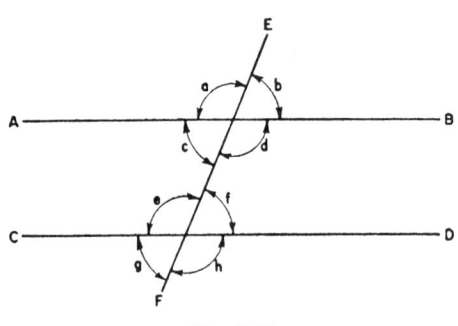

Fig. 5-24.

Required: To prove that the corresponding angles a and e, c and g, b and f, d and h, are equal to each other.

Proof: $\angle a = \angle d$, because opposite angles are equal to each other (General Theorem 2).

$\angle e = \angle d$, because alternate-interior angles are equal to each other (General Theorem 5).

∴ $\angle a = \angle e$ (Axiom 5). In a similar manner, the other angles can be proved equal to each other.

Theorem 7. *If a perpendicular line is drawn from the mid-point of a given line, all points on the perpendicular line are equidistant from the extremities of the given line.*

Given: Line AB and perpendicular CD drawn from the mid-point C, as shown in Fig. 5-25.

Required: To prove that any point, such as D, is equidistant from the extreme points A and B of the given line, i.e., $AD = BD$.

Proof: Draw lines AD and BD, forming two right triangles, ACD and BCD, in which line CD is common to both triangles, $AC = BC$ by construction, $\angle ACD = \angle BCD$ since they are right angles. ∴ $\triangle ACD = \triangle BCD$, and the corresponding elements in equal triangles are equal, hence $AD = BD$. In a similar manner, it can be proved that any other point, such as E, is equidistant from A and B, i.e., $AE = BE$, etc. Also $\angle ADC = \angle BDC$ and $\angle DAC = \angle DBC$.

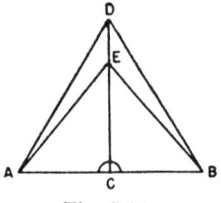

Fig. 5-25.

Theorem 8. *If the sides of two angles are parallel and point in the same direction, the angles are equal.*

GEOMETRY 153

Given: Two angles ABC and DEF with corresponding sides pointing in the same direction, as shown in Fig. 5-26. $AB \parallel DE$ and $BC \parallel EF$.

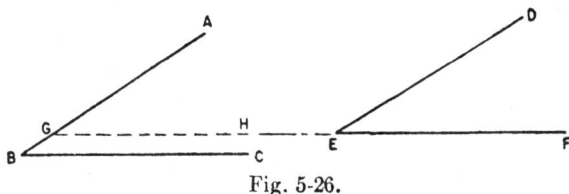

Fig. 5-26.

Required: To prove $\angle ABC = \angle DEF$.

Proof: Extend EF till it intersects AB at point G, forming $\angle AGH$.

$\angle ABC = \angle AGH$, because they are corresponding angles formed by parallel lines (General Theorem 6).

$\angle DEF = \angle AGH$, because they are corresponding angles formed by parallel lines (General Theorem 6).

$\therefore \angle ABC = \angle DEF$ (Axiom 5).

Corollary. If two of the parallel sides point in opposite directions, the angles are supplementary.

Theorem 9. *If the sides of one given angle are perendicular to the sides of the other given angle, the angles are equal, provided their sides all point in the same direction.*

Given: Two angles ABC and DEF with sides perpendicular to each other, as shown in Fig. 5-27. $AB \perp DE$ and $BC \perp EF$.

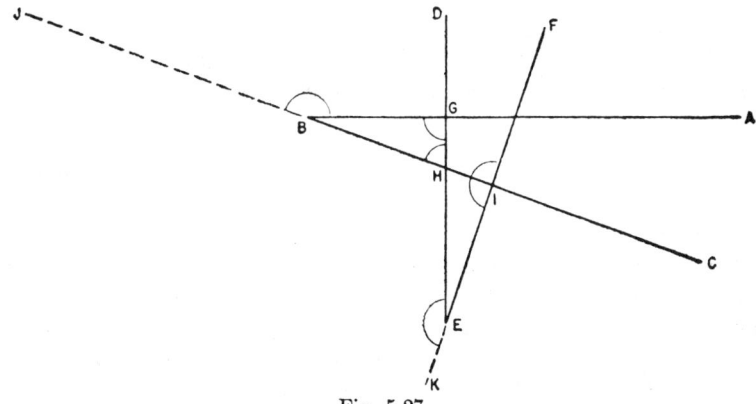

Fig. 5-27.

Required: To prove that $\angle ABC = \angle DEF$.

Proof: $\angle ABC$ is the complement of $\angle BHG$. $\angle DEF$ is the complement of $\angle EHI$. $\therefore \angle ABC = \angle DEF$, because $\angle BHG = \angle EHI$ as opposite angles (General Theorem 2).

Corollary. If the sides of one given angle are perpendicular to the sides of the other given angle, the angles are supplementary, provided their sides all point in opposite directions. For example, $\angle ABJ$ is supplementary to $\angle DEF$, and $\angle DEK$ is supplementary to $\angle ABC$.

Theorem 10. If several parallel lines are equidistant from each other, any line intersecting the parallels will form equal segments.

Given: Parallel lines AB, CD, EF, and GH intersected by line IJ and OP as shown in Fig. 5-28.

Required: To prove that segments $KL = LM = MN$ and segments $OR = RS = ST$.

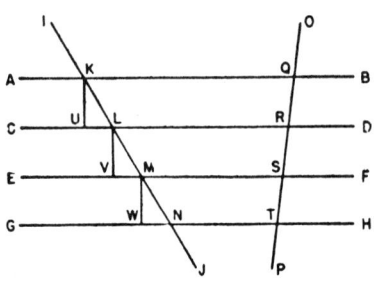

Fig. 5-28.

Proof: Drop perpendiculars KU, LV, and MW as shown in illustration, forming right $\triangle$ KUL, LVM, and MWN, which are equal to each other because sides KU, LV, and MW are equal by construction. $\angle KUL$, LVM, and MWN are right angles and $\angle LKU$, MLV, and NMW are equal as corresponding angles formed by parallel lines (General Theorem 6). In equal triangles the corresponding elements are equal, hence $KL = LM = MN$.

In like manner it can be proved that the segments QR, RS, and ST intercepted on line OP are all equal to each other.

THE TRIANGLE

The triangle constitutes one of the most important geometric figures. Upon it depend many principles of geometry and other branches of mathematics.

The various definitions pertaining to the triangle and its elements have been given, and some axioms applying to it have been presented from time to time. In this section, treatment of the triangle will include an enumeration of some of the most important theorems or propositions and, by application of their principles, we shall be able directly to solve problems frequently encountered in shop practice.

THEOREMS OR PROPOSITIONS RELATING TO THE TRIANGLE

Theorem 1. *The sum of the three angles in any triangle is equal to two right angles, or 180°.*

GEOMETRY 155

Given: Triangle ABC, shown in Fig. 5-29.
Required: To prove that the sum of $\angle a, b,$ and $c = 180°$.
Proof: Extend lines BC to F, AB to G, and draw $DE \parallel AC$.
$\angle c = \angle d$, because they are corresponding angles formed by parallel lines (General Theorem 6).

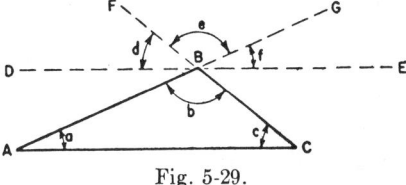

Fig. 5-29.

$\angle a = \angle f$, because they are corresponding angles formed by parallel lines (General Theorem 6).
$\angle b = \angle e$, since they are opposite angles (General Theorem 2).
The sum of angles d, e, and $f = 180°$ because the sum is a straight angle. ∴ the sum of equal $\angle a$, b, and $c = 180°$ (Axiom 6). Q.E.D. (This is an abbreviation of the Latin phrase "which was to be demonstrated.")

Corollary 1. If two angles of a triangle are equal respectively to two angles of another triangle, the third angles of both triangles are likewise equal, because the sum of all angles in each triangle equals 180°.

Corollary 2. In any right triangle, the acute angles are complementary to each other, because their sum equals 90°.

Note: The statement that the sum of all angles in a triangle equals 180° may be expressed by the formula $S = 180° \times (N - 2)$. In this formula, $S =$ sum of all angles in degrees and $N =$ number of sides in the figure. Applying the formula to the triangle we have $S = 180° \times (3 - 2) = 180°$.

The same formula applies to any polygon of N sides. This can be proved by dividing the polygon into a number of triangles. The number of triangles is always equal to the number of sides of the polygon, less 2. For example, in an octagon the number of triangles $= 8 - 2 = 6$, and the sum of all angles $= 180° \times 6 = 1080°$.

Theorem 2. *Any two triangles are equal, if any two sides and the included angle of one triangle are equal to the corresponding two sides and the included angle of the other triangle.*

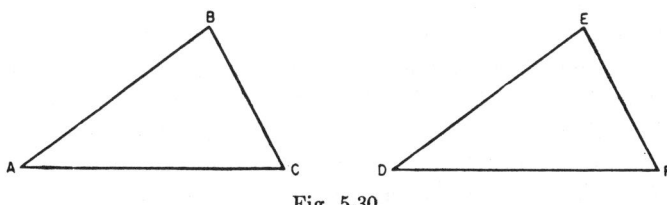

Fig. 5-30.

Given: Triangles ABC and DEF in Fig. 5-30. $AB = DE$, $AC = DF$, and $\angle BAC = \angle EDF$.
Required: To prove that the given triangles are equal.

156 MATHEMATICS FOR INDUSTRY

Proof: △*DEF* is superimposed upon △*ABC* in such a way that points *A* and *D*, lines *AB* and *DE*, and lines *AC* and *DF* coincide. This is possible because ∠*BAC* = ∠*EDF* and *AB* = *DE* and *AC* = *DF*. As points *B* and *E* must coincide, and points *C* and *F* must also coincide, line *BC* must coincide with line *EF* (Axiom 7). All the elements of one given triangle will coincide with all the corresponding elements of the other given triangle, ∴ the given triangles are equal to each other. The remaining elements of both given triangles are mutually equal to each other—side *BC* = side *EF*, ∠*ABC* = ∠*DEF*, and ∠*BCA* = ∠*EFD*.

Theorem 3. *Any two triangles are equal if any two angles and the included side in one triangle are equal to the corresponding two angles and the included side of the other triangle.*

Given: Triangles *ABC* and *DEF* in Fig. 5-31. ∠*BAC* = ∠*EDF*, ∠*BCA* = ∠*EFD* and side *AC* = side *DF*.

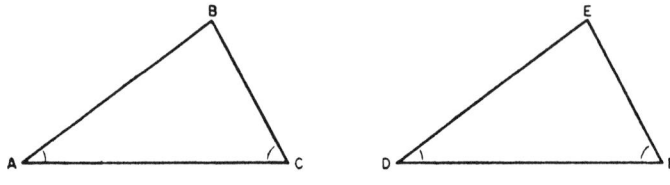

Fig. 5-31.

Required: To prove that the given triangles are equal.

Proof: △*DEF* is superimposed upon △*ABC* in such a way that points *A* and *D*, lines *AC* and *DF*, and points *C* and *F* coincide. Then line *AB* will coincide with line *DE* because ∠ *BAC* and *EDF* are equal to each other. For the same reason, line *CB* will coincide with line *EF*, ∠ *BCA* and *EFD* being equal to each other. Points *B* and *E* must coincide according to the corollary to Axiom 9, which states that two straight lines can intersect at only one point.

Corollary. Triangle Theorem 3 applies similarly where two angles and any side of one triangle are equal to the corresponding two angles and the side of the other triangle. The truth of the statement is obvious: since the sum of the three angles in any triangle is equal to 180°, then if two angles of one triangle are equal to two angles of the other triangle, the third angle of each of the triangles will be of the same magnitude.

Theorem 4. *When one side of a triangle is extended, an exterior angle is formed, which is equal to the sum of the two opposite interior angles inside the triangle.*

Given: Triangle ABC and side AC extended to D, thus forming exterior $\angle BCD$, as shown in Fig. 5-32.

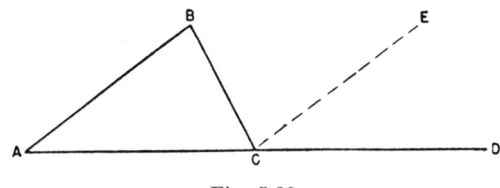

Fig. 5-32.

Required: To prove that $\angle BCD =$ the sum of opposite $\triangle ABC$ and BAC.

Proof: Draw line $CE \parallel AB$.
$\angle ECD = \angle BAC$ (General Theorem 6).
$\angle BCE = \angle ABC$ (General Theorem 5).
$\angle BCD = \angle BCE + \angle ECD = \angle ABC + \angle BAC$ (Axiom 1 and Axiom 7).
$\therefore \angle BCD = \angle ABC + \angle BAC$, Q.E.D (Axiom 5).

Theorem 5. *If two angles of a triangle are equal, the sides opposite them are equal and the triangle is isosceles.*

Given: Triangle ABC and $\triangle BAC$ and ACB are equal, as shown in Fig. 5-33.

Required: To prove that the sides AB and BC are equal.

Proof: Draw BD bisecting $\angle ABC$ and forming $\triangle ABD$ and CBD that are equal to each other because $\angle BAC = \angle ACB$ as given, $\angle ABD = \angle CBD$ by construction, and BD is a common side (Triangle Theorem 3). In equal triangles, the corresponding elements are equal. $\therefore AB = BC$, and furthermore, the given triangle ABC is an isosceles triangle.

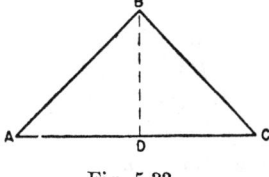

Fig. 5-33.

Corollary 1. The angles at the base of an isosceles triangle are equal.

Corollary 2. In an isosceles triangle, a perpendicular line dropped from the vertex upon the base, bisects the base and the angle at the vertex. AC in Fig. 5-33 is the base and $\angle ABC$ is the vertex.

Corollary 3. If instead of two equal sides, the triangle has three equal sides, it is an equilateral triangle, and it is also an equiangular triangle, i.e., it has three equal angles.

Theorem 6. *If the three sides of one triangle are equal respectively to the three sides of another triangle, the triangles are equal.*

Given: Triangles ABC and DEF with the corresponding sides equal to each other, as shown in Fig. 5-34.

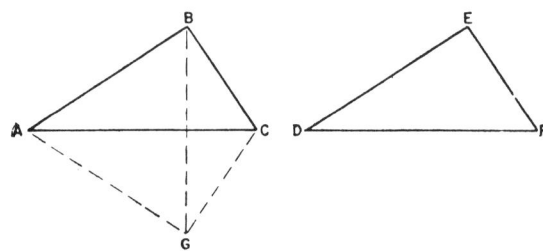

Fig. 5-34.

Required: To prove that the triangles are equal.

Proof: Attach $\triangle DEF$ to $\triangle ABC$ in such manner that their sides AC and DF coincide but point E falls opposite point B at G. Draw line BG, forming two isosceles triangles, BAG and BCG.

In $\triangle BAG$, $\angle ABG = \angle AGB$ and in $\triangle BCG$, $\angle CBG = \angle BGC$ (Corollary 1 to Triangle Theorem 5).

But, $\angle ABC = \angle ABG + \angle CBG$, and $\angle AGC = \angle DEF = \angle AGB + \angle BGC$.

∴ $\angle ABC = \angle DEF$, because if parts are equal, their sums are equal, and $\triangle ABC = \triangle DEF$ according to Triangle Theorem 2.

Theorem 7. *In any triangle, except equilateral or equiangular triangles, there are at least two unequal angles and two unequal sides, and the longer side lies opposite the greater angle.*

Given: Triangle ABC in which $\angle BCA > \angle BAC$, as shown in Fig. 5-35.

Required: To prove that $AB > BC$.

Proof: Draw line CD making $\angle ACD = \angle DAC$. The constructed $\triangle ACD$ is an isosceles triangle, and $AD = CD$.

$CD + DB > BC$, because the straight line BC is the shortest distance between points B and C.

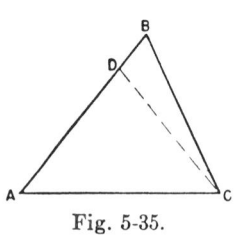

Fig. 5-35.

∴ $AD + DB > BC$, because $AD = CD$ (Axiom 6). Hence, $AB > BC$.

Theorem 8. *If a line cuts two sides of a triangle and is drawn parallel to the third side of the triangle, the sides cut by the line are divided proportionally.*

Given: Triangle ABC and line $DE \parallel AC$, as shown in Fig. 5-36.

Required: To prove that $\dfrac{AD}{DB} = \dfrac{CE}{EB}$.

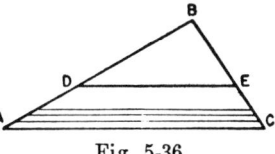

Fig. 5-36.

Proof: Assume that line AD consists of a certain number of very small units, say x units, then line DB will consist of y units, and a proportion between the lines will assume the form: $\dfrac{AD}{DB} = \dfrac{x}{y}$.

If lines are drawn through the small divisions, representing the selected units on AD and DB, parallel to AC, these lines will intersect CE and EB at the same number of points, namely, x and y.

Then $\dfrac{CE}{EB} = \dfrac{x}{y}$, according to General Theorem 10.

Hence, $\dfrac{AD}{DB} = \dfrac{CE}{EB}$ (Axiom 5).

Corollary. A corollary may be drawn to the above theorem, stating that the sides of the triangle are proportional to the portions cut by the line $DE \parallel AC$, as shown in Fig. 5-36.

$\dfrac{AB}{DB} = \dfrac{CB}{EB}$, because $\dfrac{AD + DB}{DB} = \dfrac{CE + EB}{EB}$. Also $\dfrac{AB}{AD} = \dfrac{CB}{CE}$.

Theorem 9. *Two triangles are similar if the three angles of one are equal to the three angles of the other. Furthermore, in similar triangles, the corresponding sides are proportional.*

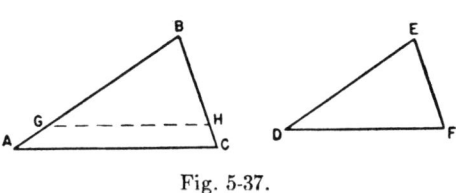

Fig. 5-37.

Given: Triangles ABC and DEF with angles of one equal to the angles of the other, as shown in Fig. 5-37.

Required: To prove that the triangles are similar and that $\dfrac{AB}{DE} = \dfrac{AC}{DF} = \dfrac{BC}{EF}$.

Proof: Place $\triangle DEF$ over $\triangle ABC$ in such a way that points E and B coincide. DE is placed over AB and EF over BC.

Then $\triangle BGH = \triangle DEF$ and $\angle BGH = \angle BAC$ by construction; likewise $\angle BHG = \angle BCA$, hence GH is parallel to AC and, according to Triangle Theorem 8, the following proportions may be written:
$$\frac{AB}{GB} = \frac{AC}{GH} = \frac{BC}{BH}.$$

But, $GB = DE$, $GH = DF$, and $BH = EF$, therefore $\frac{AB}{DE} = \frac{AC}{DF} = \frac{BC}{EF}$.

$\triangle ABC$ is similar to $\triangle DEF$, or, as symbolized, $\triangle ABC \sim DEF$, because their corresponding angles are equal to each other and their corresponding sides are proportional. (The symbol $\sim$, as indicated, represents the words "similar to.")

Note: Any two polygons are similar if their angles are equal to each other and their corresponding sides are proportional.

Corollary 1. For triangles to be similar, only two angles of one must be equal to two angles of the other, because their third angles will automatically be equal, as the sum of all angles in any triangle equals 180°.

Corollary 2. For right triangles to be similar, only one acute angle in one triangle must be equal to one acute angle in the other triangle, as all right triangles have one right angle, i.e., an angle which equals 90°.

Corollary 3. Two triangles are similar if their respective sides are parallel, because their corresponding angles are equal.

Corollary 4. Two triangles are similar if their sides are respectively perpendicular, because the angles formed in each triangle by such construction are equal to each other, as previously proved.

Corollary 5. Similar triangles are constructed by drawing perpendiculars from two points on one side of an angle to the other side of the angle, as shown in Fig. 5-38.

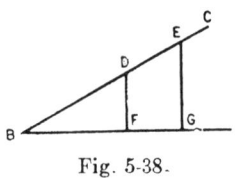
Fig. 5-38.

If $DF \perp AB$ and $EG \perp AB$, then $DF \parallel EG$. $\angle BFD = \angle BGE$, $\angle BDF = \angle BEG$ (General Theorem 6), and $\angle ABC$ is common to $\triangle DFB$ and EGB. $\therefore \triangle DFB \sim \triangle EGB$.

Theorem 10. *If, in a right triangle, a line is drawn from the vertex of the right angle perpendicular to the hypotenuse, two right triangles are formed which are similar to each other and to the given triangle.*

Given: **Right triangle** ABC and line BD drawn perpendicular to hypotenuse AC, as shown in Fig. 5-39.

Required: To prove that $\triangle ABD \sim \triangle BDC$, $\triangle ABD \sim \triangle ABC$, and $\triangle BDC \sim \triangle ABC$.

Proof: $\triangle ABD \sim \triangle ABC$, because $\angle BAC$ is common to both triangles, and right triangles are similar if both have an equal angle besides the right angle (Corollary 2 of Triangle Theorem 9).

For the same reason, $\triangle BDC \sim \triangle ABC$, $\therefore ABD \sim \triangle BDC$, because both are similar to $\triangle ABC$ (Axiom 5).

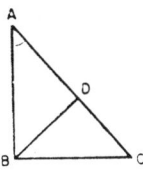

Fig. 5-39.

Corollary 1. In any right triangle the line drawn perpendicular to the hypotenuse is the mean proportional between the portions of the hypotenuse on either side of the perpendicular.

In Fig. 5-39, $\triangle ABD \sim \triangle BDC$, and their corresponding sides are proportional, hence, $\dfrac{AD}{BD} = \dfrac{BD}{DC}$.

Corollary 2. In a right triangle, if a perpendicular is drawn from the vertex of the right angle to the hypotenuse, each side adjacent to the right angle is the mean proportional between the hypotenuse and the portion of the hypotenuse adjacent to that side.

In Fig. 5-39, $\triangle BDC \sim \triangle ABC$, and their corresponding sides are proportional, hence, $\dfrac{AC}{BC} = \dfrac{BC}{DC}$.

$\triangle ABD \sim \triangle ABC$, and their corresponding sides are proportional, hence, $\dfrac{AC}{AB} = \dfrac{AB}{AD}$.

Corollary 3. In Fig. 5-39, it is evident from the similarity of the right triangles ABD and BDC that $\angle BAD = \angle DBC$ and $\angle ABD = \angle ACB$.

The Pythagorean Theorem. *In any right triangle the square of the hypotenuse is equal to the sum of the squares of the two other sides.*

This theorem or proposition has wide application in the solution of shop problems. It is one of the oldest mathematical propositions known and yet is one of the most widely used in all branches of engineering.

Given: Right triangle ABC, as shown in Fig. 5-40.

Required: To prove the square of the hypotenuse, AB, is equal to the sum of the squares of the two other sides, AC and BC.

Proof: Draw $CP \perp AB$ and prolong it to meet FH at G.
Draw CF and BE.

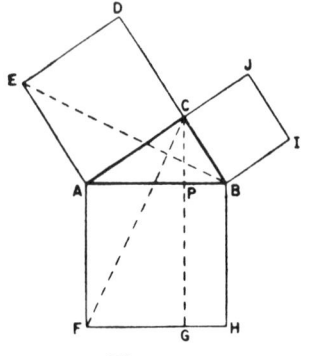

Fig. 5-40.

Since ∠ ACD and ACB are right angles, DCB is a straight line. Similarly, ACJ is a straight line.

In △ABE and AFC, $AF = AB$ and $AE = AC$.

Also ∠BAE = ∠CAF, since each consists of a right angle plus the angle BAC. Therefore △ABE = △AFC.

But the area of rectangle $AFGP$ = 2 times the area of △AFC, since both have the same base AF and the same altitude FG (or AP).

Likewise, the area of square $ACDE$ = twice the area of △ABE, since each has the same base AE and a common altitude AC.

Therefore area of rectangle $AFGP$ = area of square $ACDE$.

It may be shown similarly that area of rectangle $BPGH$ = area of square $CBIJ$.

Therefore, $AFGP + BPGH = ACDE + CBIJ$. That is, the square of the hypotenuse, AB, is equal to the sum of the squares of the two other sides, AC and BC.

Projection-Type Problems. Frequently, in shop problems, we speak of projections of lines upon other lines. In such cases the projection of one line upon another line is that portion of the second line contained between perpendiculars drawn to it from the ends of the first line. For example, in Fig. 5-41, the projection of AB (line 1) on CD (line 2) is EF, and the projection of AB (line 1) on GH (line 2) is AI.

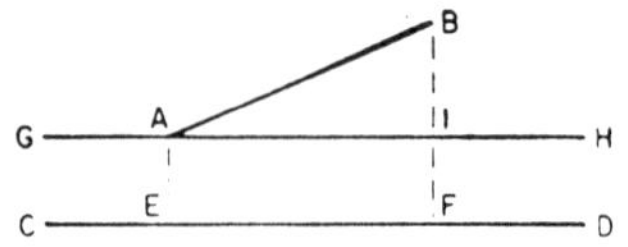

Fig. 5-41.

In some projection problems, the triangle involved is an oblique triangle instead of a right triangle. In oblique triangles both angles at the base may be acute, or one of the angles at the base may be obtuse.

Triangle ABC, shown in Fig. 5-42, is an oblique triangle with both angles at the base acute. For brevity, let us designate line AB as c, line BC as a, and line AC as b.

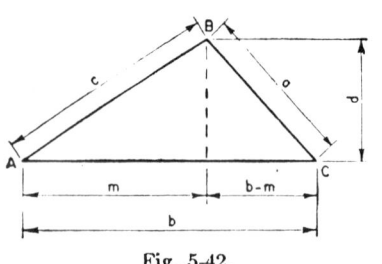

Fig. 5-42.

GEOMETRY

The projection of line c upon line b is line m, and the length of line m is expressed by the formula: $m = \dfrac{c^2 + b^2 - a^2}{2b}$.

The proof of this formula is as follows:

$c^2 + b^2 - a^2 = m^2 + d^2 + b^2 - [d^2 + (b - m)^2] = m^2 + d^2 + b^2 - d^2 - (b - m)^2 = m^2 + b^2 - b^2 + 2bm - m^2$.

$c^2 + b^2 - a^2 = 2bm$.

$\therefore m = \dfrac{c^2 + b^2 - a^2}{2b}$.

Triangle ABC, shown in Fig. 5-43, is an oblique triangle with one of the angles at the base obtuse. The proof of the formula $m = \dfrac{c^2 - b^2 - a^2}{2b}$ is as follows:

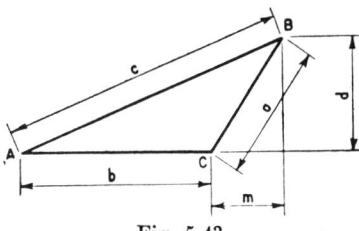

Fig. 5-43.

$c^2 - b^2 - a^2 = (b + m)^2 + d^2 - b^2 - m^2 - d^2 = b^2 + 2bm + m^2 - b^2 - m^2$

$c^2 - b^2 - a^2 = 2bm$.

$\therefore m = \dfrac{c^2 - b^2 - a^2}{2b}$.

THE CIRCLE

The circle occurs more frequently than any other geometric figure in practical problems of engineering. Cylinders, pistons, shafts, pipes, wires, pins, steam boilers, structural columns, etc., involve the circle.

The *circle* may be defined as a plane figure bounded by a curved line, the *circumference*. Every point on the circumference is equidistant from a point within called the *center*.

The *diameter* of a circle is a straight line drawn through the center connecting two points on the circumference. It is the widest distance across the circle.

The *radius* is a straight line drawn from the center of the circle to a point on the circumference. All radii of the same circle are equal and their length is always one half that of the diameter.

A part of the circumference is a circular *arc*, such as arc AB in Fig. 5-44.

If the circular arc $= \dfrac{1}{360}$ of the circumference, it is an arc of 1°.

The angle subtended by such an arc is an angle of 1°. The angle AOB is subtended by the arc AB, and if the arc $AB = N$ degrees, the angle AOB is N degrees.

The straight line AB joining the ends of the arc is called a *chord*, and it subtends its arc.

The area included between the arc and the chord is called a *segment*.

The area included between an arc and two radii drawn to the extremities of the arc is called a *sector*. In Fig. 5-44, the area between the radii AO, BO, and arc AB is the sector AOB.

When two or more circles have the same center, they are called *concentric* circles.

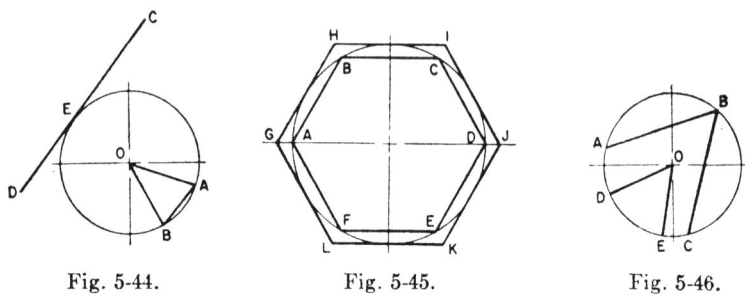

Fig. 5-44.   Fig. 5-45.   Fig. 5-46.

When a straight line of unlimited length touches the circumference at only one point but does not go through it, this line is *tangent* to the circumference, and the point is the *point of tangency*. CD is tangent to the circumference, and E is the point of tangency in Fig. 5-44.

When a polygon is inside the circle and has its vertices on the circumference, the polygon is *inscribed* in the circle. Fig. 5-45 shows a polygon, $ABCDEF$, inscribed in a circle. In the same figure, polygon $GHIJKL$ is *circumscribed* about the circle. All sides of the circumscribed polygon are tangent to the circle.

A central angle is an angle whose vertex is at the center of the circle and whose sides are radii of the circle.

If an angle is drawn inside the circle with its vertex on the circumference of the circle, the angle is called an *inscribed* angle, as shown in Fig. 5-46, where $\angle ABC$ is an inscribed angle, and arc AC is its intercepted arc. (Central and inscribed angles are said to intercept the arc between their sides.) Angle DOE, Fig. 5-46, is the central angle and arc DE is its intercepted arc.

GEOMETRY

THEOREMS OR PROPOSITIONS RELATING TO THE CIRCLE

Theorem 1. *If two central angles are equal in the same circle or in equal circles, their subtended arcs are also equal, and conversely equal arcs in the same circle or in equal circles have equal central angles.*

Given: Equal circles, as shown in Fig. 5-47, and two equal central angles, AO_1B and CO_2D.

Required: To prove arcs AB and CD are equal.

Proof: Superimpose one circle upon the other, so that their centers O_1 and O_2 coincide, and radius O_2D falls on radius O_1B. Since $\angle AO_1B = \angle CO_2D$, radius O_2C will coincide with radius O_1A, and point C will fall on point A. Arc CD will follow the outline of arc AB and will coincide with it because all points of each arc are the same distance from the center. If the conditions of the problem are reversed, the same reasoning for proving the proposition can be applied.

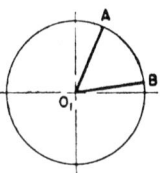
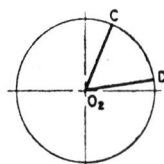

Fig. 5-47.

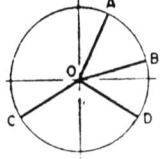

Fig. 5-48.

Corollary. If the two central angles are not equal, their subtended arcs are not equal, but the central angles are proportional to their subtended arcs, as shown in Fig. 5-48. The proportion is set up in this manner:

$$\frac{\angle AOB}{\angle COD} = \frac{\text{arc } AB}{\text{arc } CD} = \frac{AB}{CD} \quad (\frown \text{ is the symbol for an arc.})$$

Theorem 2. A line drawn tangent to a given circle is perpendicular to the radius drawn to the point of tangency; conversely a line perpendicular to a radius at its extremity is tangent to the circle.

Given: Circle and line AB tangent to the circle at point C.

Required: To prove that tangent line AB is perpendicular to radius OC, as shown in Fig. 5-49.

Proof: OC is the shortest line from the center of the circle to line AB, because any other line, such as OD, has its extremity, D, fall outside of the circle. $\therefore AB \perp OC$. If the conditions of the problem are reversed, the same reasoning will prove the proposition.

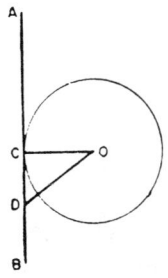

Fig. 5-49.

Corollary. Any line perpendicular to the tangent at the point of tangency must pass through the center of the circle.

Theorem 3. If two lines are drawn tangent to a circle from an outside point, these lines are equal to each other, and the angles formed by these lines and a line joining the outside point and the center of the circle are also equal.

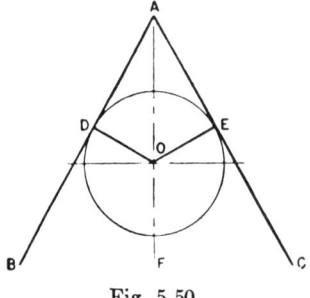

Fig. 5-50.

Given: Two lines AB and AC drawn tangent to a circle, as shown in Fig. 5-50. D and E are the points of tangency.

Required: To prove that lines AD and AE are equal, and that $\angle OAD = \angle OAE$.

Proof: Construct two right $\triangle OAD$ and OAE by drawing radii OD and OE.

These triangles are equal, because $\angle ODA = \angle OEA$ (both are right angles), OA is a common side, and $OD = OE$ as radii of the same circle. In equal triangles corresponding elements are equal. $\therefore AD = AE$ and $\angle OAD = \angle OAE$.

Corollary. A line joining the center of a circle and an outside point, bisects the angle between the tangents drawn from this point. For example, $\angle BAC$ is bisected by line AF going through the center of the circle.

Theorem 4. If an angle is inscribed in a circle, it is equal to one half the number of degrees of its intercepted arc.

Given: Angle ABC inscribed in a circle.

Required: To prove that $\angle ABC$ is measured by one half of its intercepted arc, AC, as shown in Fig. 5-51.

Proof: Draw lines BD, OA, and OC.

$\angle COD = \angle OCB + \angle OBC$ (Triangle Theorem 4).

$\angle AOD = \angle OAB + \angle OBA$ (Triangle Theorem 4).

$\angle AOC = \angle COD + \angle AOD = \angle OCB + \angle OBC + \angle OAB + \angle OBA$ (Axiom 5).

$\angle AOC = 2\,(\angle OBC + \angle OBA)$, because $\angle OCB = \angle OBC$ $\therefore$ $\triangle OBC$ is isosceles and because $\angle OAB = \angle OBA$, $\triangle OBA$ is isosceles.

$$\angle OBC + \angle OBA = \frac{\angle AOC}{2}.$$

But, $\angle ABC = \angle OBC + \angle OBA$, and $\angle AOC$ is measured by its intercepted arc AC, $\therefore$ $\angle ABC$ is measured by one half of its intercepted arc AC.

Corollary 1. $\angle ABC = \dfrac{\angle AOC}{2}$ in Fig. 5-51, thus an inscribed angle is always equal to one half of the central angle, both angles being subtended by the same arc.

Corollary 2. An inscribed angle based on the diameter of the given circle is a right angle since the center angle is a straight angle, i.e., it equals 180°. See Fig. 5-52. The inscribed angle $ABC = 90°$, because $\angle AOC = 180°$ and $\angle ABC$ is measured by half of the intercepted arc AC. Any other angle, such as $\angle ADC$, $\angle AEC$, or $\angle AFC$, equals 90°.

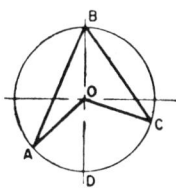

Fig. 5-51.

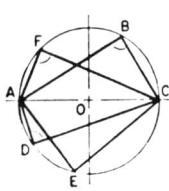

Fig. 5-52.

Corollary 3. In the same circle all inscribed angles subtending the same or equal arcs are equal to each other.

Theorem 5. If a circle is inscribed in a right triangle, its diameter is equal to the sum of the two sides of the triangle, less the hypotenuse.

Given: Right triangle ABC and a circle inscribed in it, as shown in Fig. 5-53.

Required: To prove that the diameter of the circle $= AB + BC - AC$. (This proposition was discussed in Chapter IV on the basis of an algebraic treatment. Here it is treated from the geometric point of view.)

Proof: Find the points of tangency D, E, and F by dropping perpendiculars from the center of the circle O upon all sides of the given triangle.

$AD = AF$ (Circle Theorem 3).
$EC = FC$ (Circle Theorem 3).
$OD = OE = OF = BD = BE$ by construction.
$AB - AD = BD$, $\therefore AB - AF = BD$ (Axiom 2).
$BC - EC = BE$, $\therefore BC - FC = BE$ (Axiom 2).

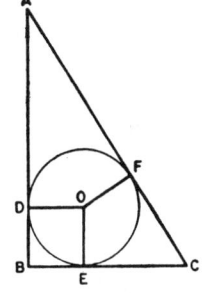

Fig. 5-53.

Adding the two expressions $AB - AF = BD$ and $BC - FC = BE$:

$AB + BC - (AF + FC) = BD + BE$ (Axiom 1).
$AB + BC - AC =$ diameter of inscribed circle (Axiom 6).

Theorem 6. Whenever two circles or circular arcs are tangent to each other externally, the point of tangency and both centers lie in a straight line.

Given: Two circular arcs tangent to each other at point C, as shown in Fig. 5-54.

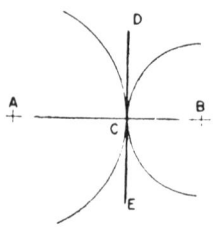

Fig. 5-54.

Required: To prove that centers A and B and point of tangency C lie on a straight line.

Proof: Draw line of tangency DE. $DE \perp AB$ according to previous Circle Theorem 2 which states that a tangent to a circle is perpendicular to the radius of the circle.

$\therefore$ $\angle ACD$ is a right angle, so is $\angle BCD$, hence $\angle ACD + \angle BCD = 180° =$ a straight angle, and line ACB is a straight line, that is, points A, B, and C are all in one straight line.

Corollary. Whenever two circles or circular arcs are tangent to each other on the inside, the point of tangency and both centers lie in a straight line. See Fig. 5-55. A and B are the centers of the given circular arcs and C is the point of tangency.

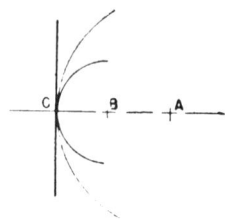

Fig. 5-55.

Theorem 7. If a perpendicular is dropped from the center of a circle upon any chord in the circle, the chord is bisected and so is the subtended arc.

Given: Chord AB and perpendicular OC going through the center, as shown in Fig. 5-56.

Required: To prove that line $AC =$ line BC, and $\overset{\frown}{AD} = \overset{\frown}{DB}$.

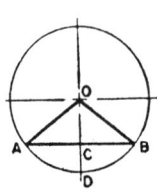

Fig. 5-56.

Proof: Draw radii OA and OB to complete two right triangles OCA and OCB, which are equal to each other, because OC is common to both, $\angle OCA = \angle OCB$ as right angles, and $OA = OB$ as radii of the same circle. In equal triangles corresponding elements are equal, $\therefore AC = BC$, $\angle AOC = \angle BOC$, $\overset{\frown}{AD} = \overset{\frown}{DB}$ (Circle Theorem 1).

Corollary. If line OC in Fig. 5-56 is extended through the circle, it will coincide with the diameter of the circle; thus we may state that when the diameter of a circle is perpendicular to a given chord, it will bisect the chord and the subtended arc.

Theorem 8. When two chords drawn through the circle intersect

GEOMETRY

within the circle, the point of intersection divides each chord into two parts or segments in such a way that the product of the two segments of one chord is equal to the product of the two segments of the other chord.

Given: Two chords AB and CD intersected at point E in the circle, as shown in Fig. 5-57.

Required: To prove that $AE \times EB = CE \times ED$.

Proof: Draw lines AC and BD to complete triangles AEC and BED. These triangles are similar, because $\angle AEC = \angle BED$ as opposite angles (Triangle Theorem 2), $\angle ACD = \angle ABD$ as angles subtended

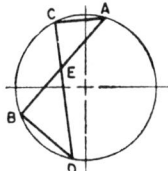

Fig. 5-57.

by the same arc (Corollary 3 of Circle Theorem 4), and the third pair of angles must be equal to each other as the sum of all angles in any triangle = 180°. In similar triangles, corresponding elements are proportional, $\therefore \dfrac{CE}{EB} = \dfrac{AE}{ED}$, and $AE \times EB = CE \times ED$.

Theorem 9. If a triangle is inscribed in a circle, the product of any two sides of the triangle is equal to the product of the diameter of the circle and the altitude upon the third side of the inscribed triangle.

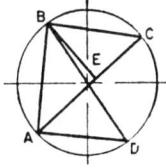

Fig. 5-58.

Given: Triangle ABC inscribed in circle with $BE \perp AC$ and diameter BD, as shown in Fig. 5-58.

Required: To prove that $AB \times BC = BD \times BE$.

Proof: Draw line AD to construct $\triangle BAD$ which is similar to right triangle BEC, because $\angle BAD$ is a right angle (Corollary 2 of Circle Theorem 4), $\therefore$ $\angle BAD = \angle BEC$, $\angle ACB = \angle ADB$ as angles subtended by the same arc AB (Corollary 3 of Circle Theorem 4), and the third pair of angles are equal to each other.

In similar triangles, corresponding elements are proportional $\therefore$ $\dfrac{AB}{BE} = \dfrac{BD}{BC}$, and $AB \times BC = BD \times BE$.

Theorem 10. When two lines cut through a circle and intersect outside of the circle, the included angle between these two lines is measured by half of the difference of the intercepted arcs.

Given: Two intersected lines AB and AC, usually called secants, which cut through the circle, as shown in Fig. 5-59, and intersect at A outside of the circle.

Required: To prove that $\angle BAC$ is measured by $\dfrac{\overset{\frown}{FG} - \overset{\frown}{DE}}{2}$.

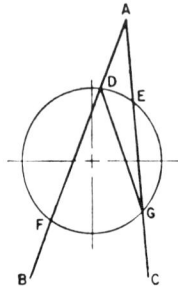

Fig. 5-59.

Proof: Draw line DG.

$\angle FDG$ is measured by half of arc FG (Circle Theorem 4).

$\angle DGE$ is measured by half of arc DE (Circle Theorem 4).

$\angle FDG = \angle BAC + \angle DGE$ (Triangle Theorem 4) and then $\angle BAC = \angle FDG - \angle DGE$.

$\therefore$ $\angle BAC$ is measured by one-half arc FG − one-half arc DE, or $\dfrac{\overset{\frown}{FG} - \overset{\frown}{DE}}{2}$ (Axiom 6).

Corollary. If one or both of the intersecting lines is tangent to the given circle, the included angle between these two lines is also measured by half of the difference of the intercepted arcs. This is shown respectively in Figs. 5-60 and 5-61.

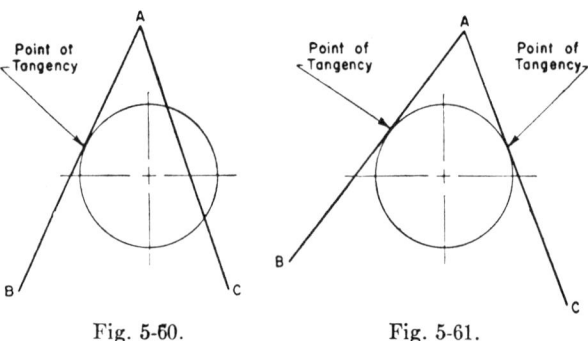

Fig. 5-60.   Fig. 5-61.

CONSTRUCTION PROBLEMS RELATING TO THE CIRCLE

Problem 1. Construct a circle circumscribed about a triangle.
Given: Triangle ABC, as shown in Fig. 5-62.
Required: To circumscribe a circle about $\triangle ABC$.
Procedure: Bisect line AB at point D, and erect $DE \perp AB$.
Bisect line AC at point F, and erect $FG \perp$ to AC.

GEOMETRY

These two perpendiculars intersect at point O, which is the center of the circumscribed circle. The radius of the circle is any line from the center O to any of the three vertices of the given triangle, such as OA, OB, and OC.

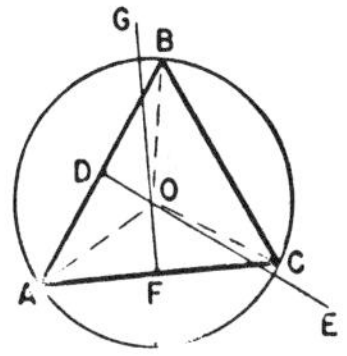

Proof: $\triangle AOD = \triangle BOD$, because $\angle BDO = \angle ADO$ as right angles, DO is a common side, $AD = DB$, $\therefore AO = OB$.

In the same manner, it can be proved that $AO = OC$, $\therefore$ point O is equidistant from points A, B, and C, and a circle drawn from point O as a center with a radius equal to any of the lines OA, OB, and OC will pass through the vertices of the triangle.

Fig. 5-62.

It should be noted in conclusion that three points in any case determine a circle.

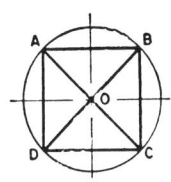

Fig. 5-63.

Problem 2. Construct a square inscribed in a circle.
Given: A circle, as shown in Fig. 5-63.
Required: To inscribe a square $ABCD$ in the given circle.
Procedure: Draw two diameters AC and BD perpendicular to each other.
Connect points A, B, C, and D by lines AB, BC, CD, and DA, and figure $ABCD$ becomes the required square.

Proof: $\angle BAD = \angle ADC = \angle DCB = \angle CBA = 90°$, because each is an inscribed angle and is measured by half of its subtended arc, which is a semicircle (Corollary 2 of Circle Theorem 4).

Problem 3. Construct a regular hexagon inscribed in a given circle.
Given: A circle, as shown in Fig. 5-64.
Required: To inscribe a regular hexagon $ABCDEF$ in the given circle.
Procedure: From any point on the circumference, such as A, draw an arc with a radius equal to that of the given circle, intersecting the circumference at point B. Construct $\triangle AOB$, which is an equilateral triangle since all the sides are equal. It is also an equiangular triangle.

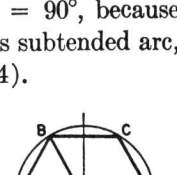

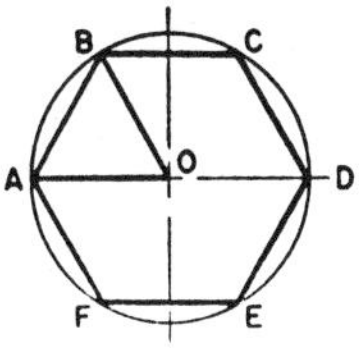

Fig. 5-64.

$\angle AOB = 60°$ (Triangle Theorem 1), or $\angle AOB = \dfrac{360°}{6}$, hence six equilateral triangles equal to $\triangle AOB$ can be constructed as shown in Fig. 5-64.

Lines AB, BC, CD, DE, EF, and FA are all equal to each other, and the constructed figure is a regular hexagon.

PRACTICE PROBLEMS APPLICABLE IN THE SHOP AND TOOLROOM

1. Determine $\angle ACB$ in right $\triangle ABC$, if $\angle CAB = 37°$, see Fig. 5-65.
2. Determine $\angle ACD$ in Fig. 5-65.
3. Determine $\angle DCE$ in Fig. 5-65.
4. Determine $\angle CAF$ in Fig. 5-65.

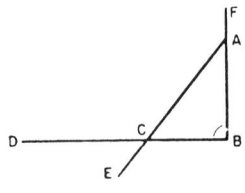

Fig. 5-65.

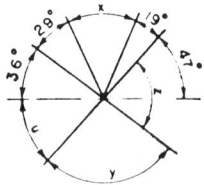

Fig. 5-66.

5. Find the altitude of $\triangle ABC$ in Fig. 5-65, if its base is AC.
6. Determine x in Fig. 5-66.
7. Determine y in Fig. 5-66.
8. Determine z in Fig. 5-66.
9. Determine u in Fig. 5-66.
10. What is the complement of x in Fig. 5-66?
11. What is the supplement of y in Fig. 5-66?
12. What is the complement of z in Fig. 5-66?
13. What is the complement of u in Fig. 5-66?

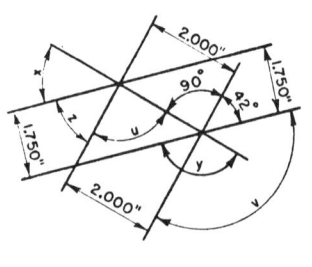

Fig. 5-67.

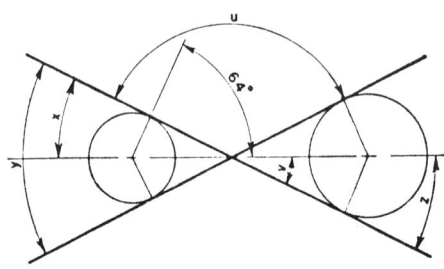

Fig. 5-68.

14. Determine x in Fig. 5-67.
15. Determine y in Fig. 5-67.
16. Determine z in Fig. 5-67.
17. Determine u in Fig. 5-67.
18. Determine v in Fig. 5-67.

GEOMETRY

19. Determine x in Fig. 5-68.
20. Determine y in Fig. 5-68.
21. Determine z in Fig. 5-68.
22. Determine u in Fig. 5-68.
23. Determine v in Fig. 5-68.

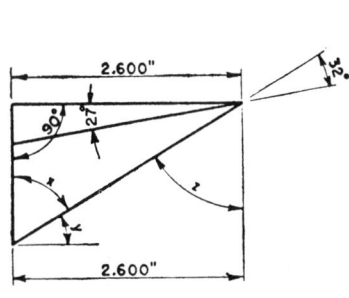

Fig. 5-69.   Fig. 5-70.

24. Determine x in Fig. 5-69.
25. Determine y in Fig. 5-69.
26. Determine z in Fig. 5-69.
27. Determine x in Fig. 5-70.
28. Determine y in Fig. 5-70.
29. Determine z in Fig. 5-70.
30. Determine u in Fig. 5-70.
31. Determine v in Fig. 5-70.
32. Find the points of tangency of circular arcs A, B, C, and D, as shown in Fig. 5-71.

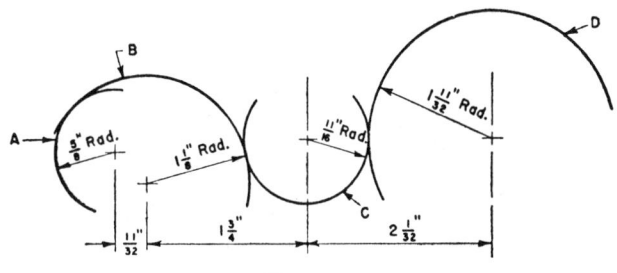

Fig. 5-71.

33. Determine x in Fig. 5-72.
34. Determine y in Fig. 5-72.

35. Determine z in Fig. 5-72.
36. Determine u in Fig. 5-72.
37. Determine v in Fig. 5-72.
38. Determine x in Fig. 5-73.
39. Determine y in Fig. 5-73.
40. Determine z in Fig. 5-73.
41. Determine u in Fig. 5-73.
42. Determine v in Fig. 5-73.
43. Determine x in Fig. 5-74.
44. Determine y in Fig. 5-74.
45. Determine z in Fig. 5-74.
46. Determine u in Fig. 5-74.
47. Determine v in Fig. 5-74.
48. Determine m in Fig. 5-74.
49. Determine n in Fig. 5-74.

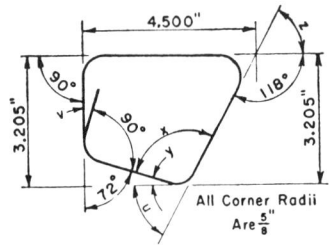

Fig. 5-72.

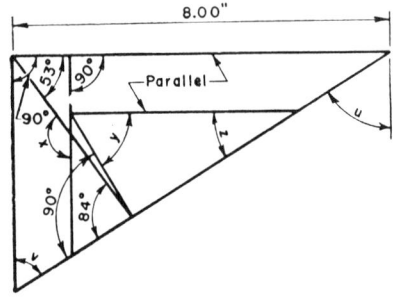

Fig. 5-73.

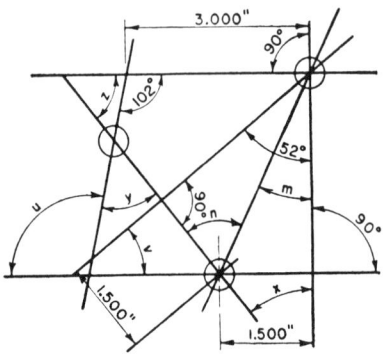

Fig. 5-74.

50. Determine x in Fig. 5-75.
51. Determine y in Fig. 5-75.

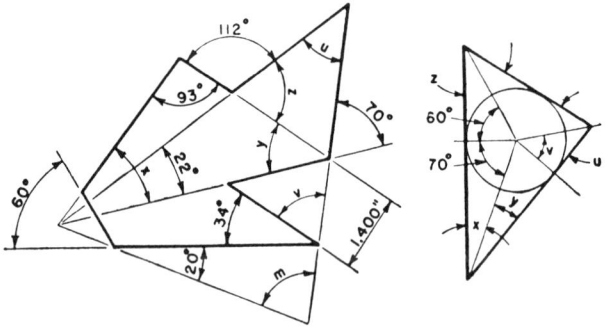

Fig. 5-75.

Fig. 5-76.

GEOMETRY

52. Determine z in Fig. 5-75.
53. Determine u in Fig. 5-75.
54. Determine v in Fig. 5-75.
55. Determine m in Fig. 5-75.
56. Determine x in Fig. 5-76.
57. Determine y in Fig. 5-76.
58. Determine z in Fig. 5-76.
59. Determine u in Fig. 5-76.
60. Determine v in Fig. 5-76.
61. Determine x in Fig. 5-77.
62. Determine y in Fig. 5-77.
63. Determine z in Fig. 5-77.
64. Determine x in Fig. 5-78.
65. Determine y in Fig. 5-78.
66. Determine z in Fig. 5-78.
67. Determine u in Fig. 5-78.
68. Determine v in Fig. 5-78.
69. Locate the point of tangency between line AB and arc C in Fig. 5-78.

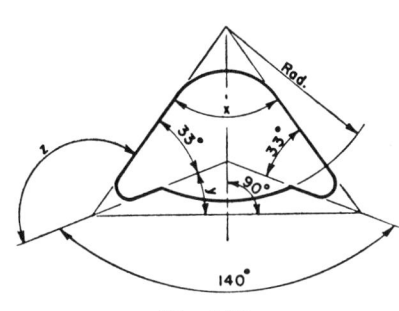

Fig. 5-77.

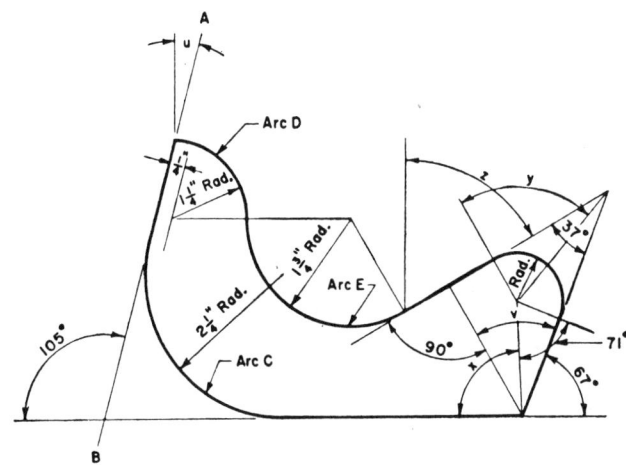

Fig. 5-78.

70. Locate the point of tangency between arcs D and E in Fig. 5-78.
71. Determine x in Fig. 5-79.
72. Determine y in Fig. 5-79.
73. Determine z in Fig. 5-79.
74. Determine u in Fig. 5-79.
75. Determine x in Fig. 5-80.
76. Determine y in Fig. 5-80.
77. Determine z in Fig. 5-80.

78. Determine u in Fig. 5-80.
79. Determine v in Fig. 5-80.

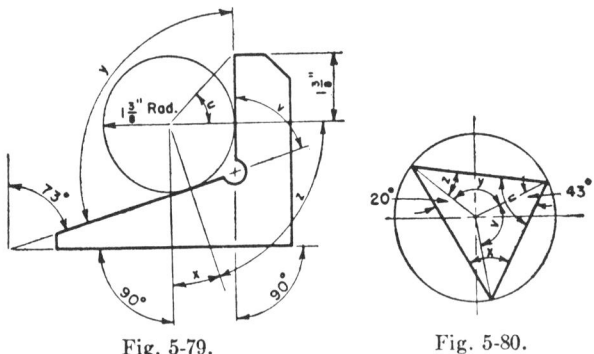

Fig. 5-79. Fig. 5-80.

80. Determine x in Fig. 5-81.
81. Determine y in Fig. 5-81.
82. Determine z in Fig. 5-81.

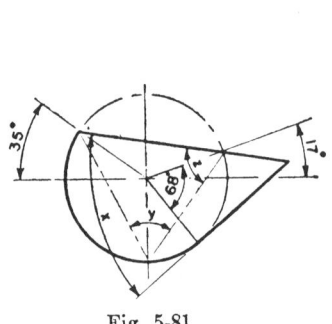

Fig. 5-81.

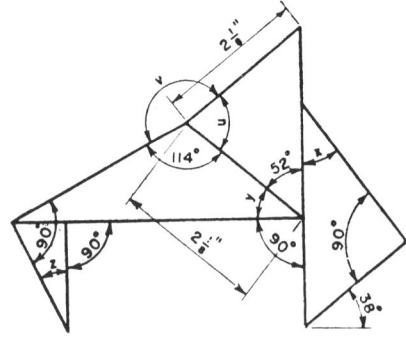

Fig. 5-82.

83. Determine x in Fig. 5-82.
84. Determine y in Fig. 5-82.
85. Determine z in Fig. 5-82.
86. Determine u in Fig. 5-82.
87. Determine v in Fig. 5-82.
88. Find the length of AB if the length $AC = 3\frac{3}{4}''$ and the length of $BC = 2\frac{1}{2}''$, as shown in Fig. 5-83.

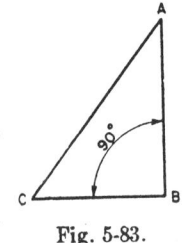

Fig. 5-83.

GEOMETRY

89. Assume the length of each leg of the right triangle in Fig. 5-83 is equal to $2\frac{7}{8}''$. Calculate the length of the hypotenuse.

90. Assume the length of the hypotenuse in Fig. 5-83 is equal to $5\frac{3}{4}''$. What would be the length of each leg of the right triangle if both legs were equal to each other?

91. Assume the length of leg BC in Fig. 5-83 is $1\frac{1}{2}$ times that of leg AB, and the hypotenuse is equal to $6\frac{1}{2}''$. Calculate the length of each leg.

92. Calculate the distance x in Fig. 5-84.
93. Calculate the radius in Fig. 5-85.

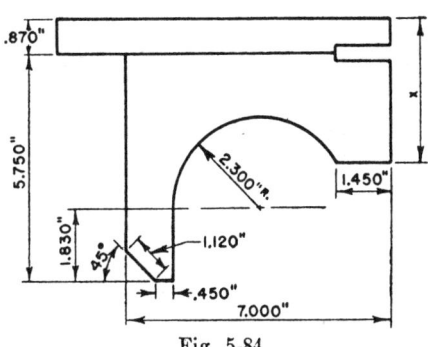

Fig. 5-84.

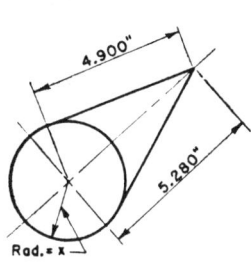

Fig. 5-85.

94. Calculate the distance x in Fig. 5-86, which is drawn symmetrical about its vertical center line.
95. Calculate the distance x in Fig. 5-87.

Hint: Use the projection formula.

96. Calculate the distance y in Fig. 5-87.
97. Calculate the altitude h in Fig. 5-87.

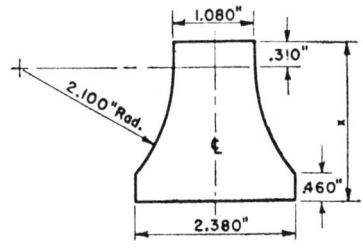

Fig. 5-86.

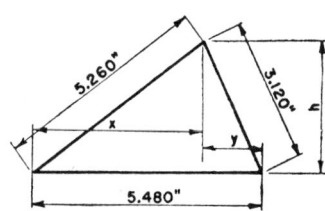

Fig. 5-87.

178   MATHEMATICS FOR INDUSTRY

98. Calculate the distance x in Fig. 5-88.
99. Calculate the altitude h in Fig. 5-88.
100. Calculate x in Fig. 5-89.
101. Calculate the dimension y in Fig. 5-89.

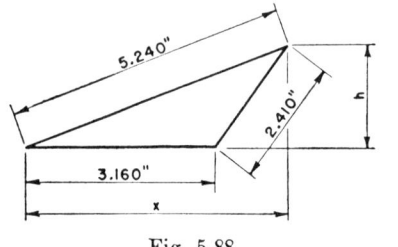

Fig. 5-88.   Fig. 5-89.

102. Determine x in Fig. 5-90.
103. Calculate $\angle x$ in Fig. 5-91, if the length l of circular arc $DE = 4.200''$ and $AB \perp BC$.

Hint: Calculate $\angle y$ by proportion. $\dfrac{\text{Arc } l}{\text{circumference}} = \dfrac{\angle y}{360°}$.

Substituting numerical values: $\dfrac{4.200}{2\pi R} = \dfrac{\angle y}{360°}$; $\dfrac{4.200}{2 \times 3.1416 \times 2.25} = \dfrac{\angle y}{360°}$.

Taking cross products: $2 \times 3.1416 \times 2.25 \times y = 360° \times 4.200$.

$$\angle y = \frac{360° \times 4.200}{2 \times 3.1416 \times 2.25} = 106° 57'.$$

$\angle BCD = \angle CBD$, because $CD = BD = 2\frac{1}{4}''$ and $\triangle BCD$ is an isosceles.
But $\angle CBD = 90° - 72° = 18°$. $\therefore \angle BCD = 18°$ and $\angle x = \angle y + 18° = 124° 57'$.

104. Find the length of the circular arc AC, if its radius $= 1.8''$ and the angle subtended by the arc $= 66°$, Fig. 5-92.

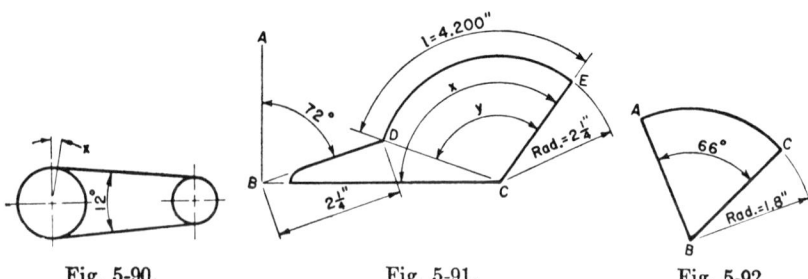

Fig. 5-90.   Fig. 5-91.   Fig. 5-92.

GEOMETRY

105. Determine x in Fig. 5-93.
106. Calculate the length l of the circular arc shown in Fig. 5-94.

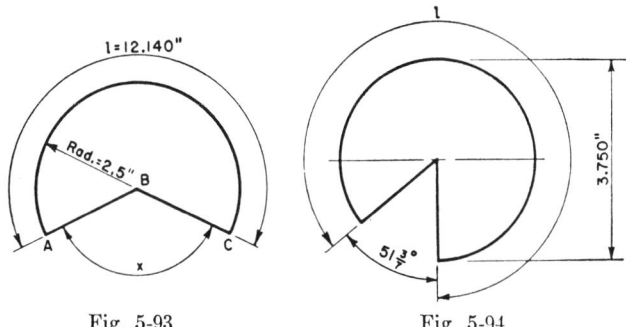

Fig. 5-93.   Fig. 5-94.

107. Determine x in Fig. 5-95.
108. Determine y in Fig. 5-95.
109. Determine z in Fig. 5-95.
110. Calculate the length of line OA in Fig. 5-96.
111. Calculate the length of line OB in Fig. 5-96.
112. Calculate the length of line OC in Fig. 5-96.

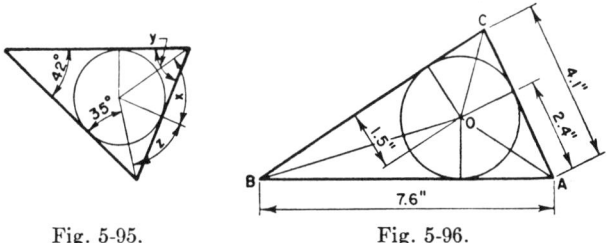

Fig. 5-95.   Fig. 5-96.

113. Calculate the radii R and r in Fig. 5-97.

Hint: In any right triangle, where one of the angles $= 30°$, the side opposite the 30° angle is equal to one half of the hypotenuse.

114. Calculate the radii R and r in Fig. 5-98.

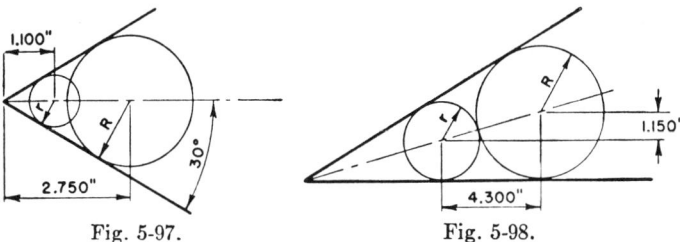

Fig. 5-97.   Fig. 5-98.

115. Determine x in Fig. 5-99.
116. Determine y in Fig. 5-99.
117. Determine z in Fig. 5-99.
118. Determine u in Fig. 5-99.
119. Calculate the diameter of a circle that can be inscribed in the right triangle in Fig. 5-100.

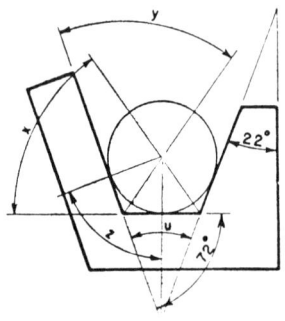

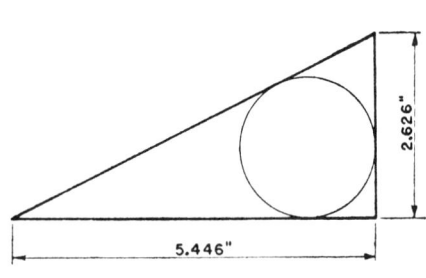

Fig. 5-99. Fig. 5-100.

120. Calculate the dimension h in Fig. 5-101.
121. Calculate the dimension c in Fig. 5-101.

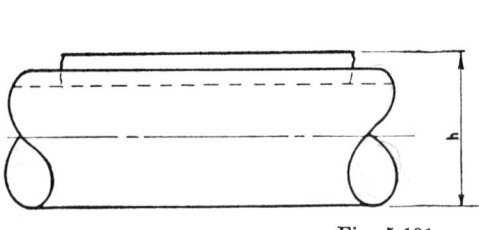
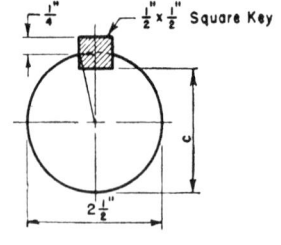

Fig. 5-101.

122. Find the diameter of the circle passing through the holes shown in Fig. 5-102.
123. Calculate the radius R in Fig. 5-103.
124. Determine x in Fig. 5-103.
125. Determine y in Fig. 5-103.
126. Determine z in Fig. 5-103.

GEOMETRY

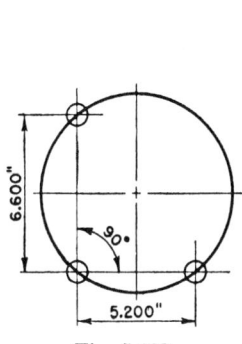

Fig. 5-102.

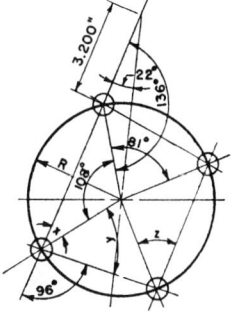

Fig. 5-103.

127. Determine x in Fig. 5-104, if the length of arc $l = 3.442''$.
128. Determine y in Fig. 5-104.
129. Determine x in Fig. 5-105.
130. Determine y in Fig. 5-105, if the length of arc $l = 5.100''$.

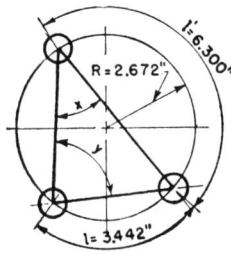

Fig. 5-104.

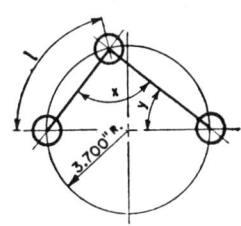

Fig. 5-105.

131. Calculate the radius R, as shown in Fig. 5-106.
132. Calculate the minor axis x of the ellipse shown in Fig. 5-107.

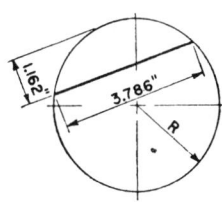

Fig. 5-106.

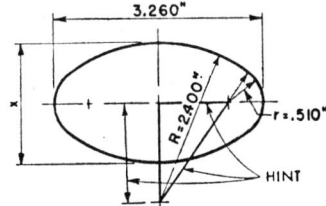

Fig. 5-107.

133. Find the radius R of a circle that passes through the three holes shown in Fig. 5-108.

134. Find the distance x across the flats of the hexagon shown in Fig. 5-109.

135. Find the side x of the square shown in Fig. 5-110.

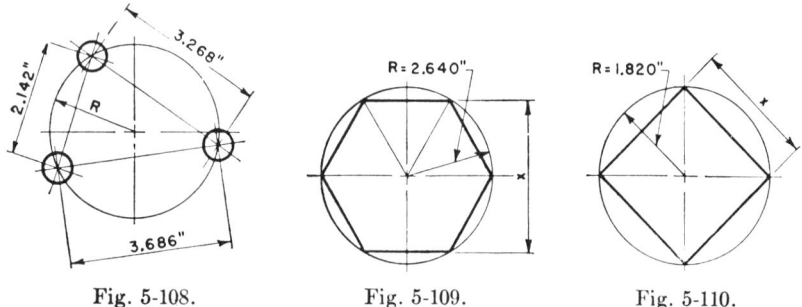

Fig. 5-108.   Fig. 5-109.   Fig. 5-110.

136. Calculate the distance x shown in Fig. 5-111.

137. Find the distance M, which is the measurement over the precision pins inserted in the respective holes shown in Fig. 5-112.

138. Find the diameter of a circle inscribed in the equilateral triangle shown in Fig. 5-113.

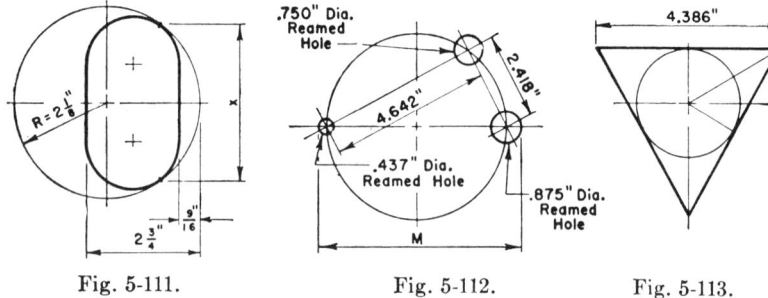

Fig. 5-111.   Fig. 5-112.   Fig. 5-113.

Hint: Each angle in an equilateral triangle = 60°, and in a right triangle having one acute angle = 30°. The side opposite this angle is equal to one half of the hypotenuse.

CHAPTER VI

GEOMETRY—CONTINUED

This chapter and the preceding chapter should be regarded as one unit dealing with the study of geometry. The preceding chapter presented the important theorems of plane geometry. This chapter deals with the measurement of plane figures and problems in construction. Then follows the fundamentals of solid geometry.

THE PROPERTIES OF THE CIRCLE

Some of the important properties of the circle have been discussed in the preceding chapter in connection with the properties of other geometric figures. This has been done because the circle is an essential part of many geometric construction problems and for this reason it is almost impossible to isolate the study of the circle.

The most important properties of the circle are illustrated in Fig. 6-1. They are enumerated as follows:

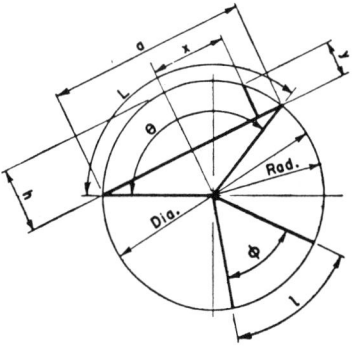

Fig. 6-1.

1. $\dfrac{\text{Circumference of a circle}}{\text{Diameter of the circle}} = \pi = 3.1416$. If the length of the circumference and the diameter of a circle are carefully measured, the ratio $\dfrac{\text{circumference}}{\text{diameter}}$ will be found to have a fixed value, whatever size of circle may be selected. This constant ratio has been given the symbol π (which is the Greek letter pi) and has a value of 3.1416. The solution of many geometric problems is based upon this ratio. Its wide use becomes evident in this and subsequent chapters of this text.

2. Circumference $= \pi D$. The circumference of a circle is equal to its diameter multiplied by π.
3. Circumference $= 2\pi R$. The circumference of a circle is equal to its radius multiplied by 2π.

4. Diameter = $\dfrac{\text{circumference}}{\pi}$ or diameter = circumference × .3183 (a constant derived from $\dfrac{1}{\pi}$.)

5. Diameter of a circle, whose circumference is equal to the perimeter of a square, is equal to a side of the square × 1.27324 (a constant derived from $\dfrac{4}{\pi}$).

6. Side of a square, whose perimeter is equal to the circumference of a circle, is equal to the diameter of the circle × $\dfrac{\pi}{4}$ or the diameter of a circle × .7854 (a constant derived from $\dfrac{\pi}{4}$).

7. Side of a square inscribed in a circle = diameter of circle × .7071.
8. Diameter of a circle circumscribed about a square = side of square × 1.4142.
9. Area of a circle = πR^2 or $\dfrac{\pi}{4}D^2$ or = $.7854 D^2$ (.7854 is a constant derived from $\dfrac{\pi}{4}$).

10. Side of a square having the same area as a given circle = diameter of given circle × .886.
11. The following general formulas were derived from Fig. 6-1:

$$D = \dfrac{a^2 + 4h^2}{4h},$$ where a = length of chord, and h = height of segment, and D = diameter.

$$R = \dfrac{a^2 + 4h^2}{8h},$$ where R = radius.

$$a = 2\sqrt{2hR - h^2}$$

$$h = R - \dfrac{1}{2}\sqrt{4R^2 - a^2}$$

$$h = R + y - \sqrt{R^2 - x^2},$$ where x and y are variables depending upon the location.

$$x = \sqrt{R^2 - (R + y - h)^2}$$

$$y = h - R + \sqrt{R^2 - x^2}$$

$$L = \frac{\pi R \theta^\circ}{180} = .01745 R \theta^\circ, \text{ where } L = \text{length of arc and } \theta = \text{angle in degrees.}$$

$$\theta = \frac{180^\circ L}{\pi R} = 57.2958 \frac{L}{R}$$

The lengths of circular arcs in a circle are proportional to the central angles which they intercept. For example, in Fig. 6-1, the length of the circular arc L is proportional to the central angle θ, and, as previously stated, the circular arc is measured by the central angle. Any other circular arc in the same circle, such as l, is in the same proportion to arc L as their corresponding central angles ϕ and θ. The relationship may be expressed thus, $\frac{l}{L} = \frac{\phi}{\theta}$.

In a given circle, the length of any circular arc may be calculated if its central angle is known. For example, if in Fig. 6-1 angle $\phi = 60^\circ$ and the diameter of the circle = $2''$, the length of arc

$$l = \frac{\pi D \times 60}{360} = \frac{\pi \times \cancel{2}^1 \times \cancel{60}^1}{\cancel{360}_3} = \frac{3.1416}{3} = 1.0472''.$$

In the same circle, if angle $\theta = 120^\circ$, the length of arc $L = 2l = 1.0472 \times 2 = 2.0944''$.

The principle just illustrated is applied in solution of many shop problems. For example, the length of the circular pitch of a gear is equal to the pitch diameter of the gear $\times \pi$, divided by the number of teeth in the gear. The gear teeth are equally spaced, therefore the central angles spacing the teeth and their corresponding circular arcs or circular pitches are equal, and each circular pitch, as explained before, is obtained by dividing the circumference of the pitch circle by the number of teeth in the gear.

A few practice problems concluding Chapter V are solved by application of the principle just explained, and other problems found throughout the text likewise depend for their correct solution upon application of this principle.

GEOMETRIC PROBLEMS OF CONSTRUCTION

Some of the basic fundamentals of geometric construction are illustrated by the methods employed in the solution of the following problems. The tools necessary for the solution of these problems are a few simple drawing instruments, such as a rule, compass, and dividers.

Problem 1. *To bisect a given straight line.* See Fig. 6-2, where AB is the given line.

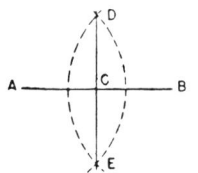

Fig. 6-2.

With ends A and B of the given line as centers and radii equal to any distance that is greater than half the length of line AB, draw arcs which intersect at points D and E. A straight line, DE, drawn through these points bisects the given line AB at C and is perpendicular to AB.

Problem 2. *To bisect a given circular arc.* See Fig. 6-3, where AB is the given circular arc.

Join points A and B. With points A and B of the given arc as centers and radii equal to any distance that is greater than half the length of line AB, draw arcs which intersect at points C and D. A straight line, CD, drawn through these points bisects the given arc AB at point E.

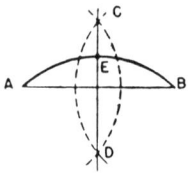

Fig. 6-3.

Problem 3. *To bisect a given angle.* See Fig. 6-4, where ABC is the given angle.

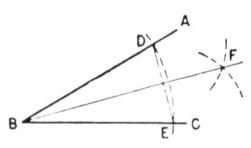

Fig. 6-4.

With B as a center and any radius, draw an arc which intersects lines AB and BC at D and E, respectively. With D and E as centers and radii equal to any distance greater than half of the length of line DE, draw arcs which intersect at point F. A straight line drawn through points B and F bisects the given angle.

Problem 4. *To draw a line parallel to a given line at a given distance from the line.* See Fig. 6-5, where AB is the given line and AC is the given distance from it.

With A and B as centers and radii equal to AC, draw circular arcs DE and FG. Draw line CH tangent to arcs DE and FG. Line CH is the required line parallel to given line AB. Lines AC and BH are perpendicular to given line AB and to the newly drawn line CH.

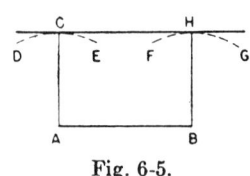

Fig. 6-5.

Problem 5. *To draw, through a given point, a line parallel to a given line.* See Fig. 6-6, where AB is the given line and C the given point.

With C as a center and any convenient radius such as CA, draw arc AE. With A as a center and the same radius, draw arc CD. With A as a center and a radius equal to CD, draw arc FG, intersecting arc AE at point E. Draw a line through points E and C, which is parallel to given line AB.

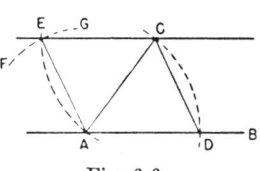

Fig. 6-6.

Problem 6. *To erect a perpendicular to a given line at a given point.* See Fig. 6-7, where AB is the given line and C the given point.

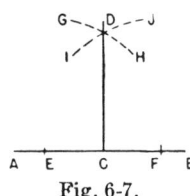

Fig. 6-7.

On line AB lay off from point C distance to right and left such as CE and CF, which lines are equal to each other. With E and F as centers and with radii greater than EC, draw arcs GH and IJ, which intersect at point D. A line drawn through points C and D is perpendicular to given line AB at given point C.

Problem 7. *To erect a perpendicular to a given line from a given outside point.* See Fig. 6-8, where AB is the given line and C the given outside point.

With C as a center and any radius greater than the distance between point C and line AB, draw arc DE, intersecting AB at points D and E. With D and E as centers and radii greater than half the length of line DE, draw arcs FG and HI, intersecting at point J. A line drawn through points C and J is perpendicular to given line AB.

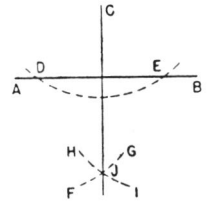

Fig. 6-8.

Problem 8. *To erect a perpendicular to a given line at one of its extreme points.* See Fig. 6-9, where AB is the given line and A one of its extremities.

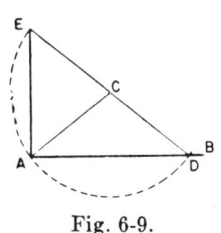

Fig. 6-9.

With any point C as a center outside of line AB and with the distance CA as a radius, draw arc DAE. Join DC and extend line to intersect arc at point E, thus forming diameter DCE. A line drawn through points A and E is perpendicular to given line AB.

Problem 9. *To locate the center of a given circular arc.* See Fig. 6-10, where ABC is the given arc.

Choose any three points on the given arc, such as D, B and E. Bisect arcs BD and BE, using method described in Problem 2. Lines FG and HI bisect arcs BD and BE. These lines extended intersect at point O, which is the required center of the given circular arc.

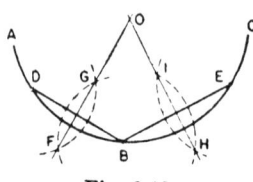

Fig. 6-10.

Problem 10. *To draw a square on a given line.* See Fig. 6-11, where AB is the given line.

Erect a perpendicular AC to given line AB at point A. Lay off on AC a length AD equal to AB. With centers B and D and radii equal to AB, draw arcs EF and GH, intersecting at point I. Draw lines DI and BI to complete the required square $ABID$.

A rectangle may be constructed in a similar manner by making $AD >$ or $< AB$.

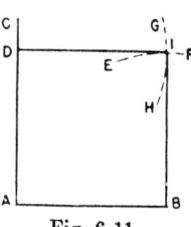

Fig. 6-11.

Problem 11. *To construct a 45° angle at a point on a given line.* See Fig. 6-12, where AB is the given line and A is the point on it.

Erect a perpendicular CE to line AB at any convenient point, say C, using method described in Problem 6. Lay off on the perpendicular a length CF equal to length AC. Join points A and F by a straight line AF, which forms a 45° angle with given line AB at point A.

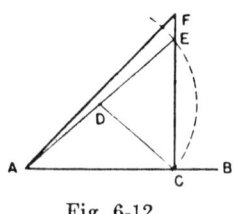

Fig. 6-12.

Problem 12. *To construct a 60° angle at a point on a given line.* See Fig. 6-13, where AB is the given line and A is the point on it.

With A as a center and any radius draw arc DC. With C as a center and with the same radius, draw arc EF, intersecting arc CD at point G. Draw line AG through points A and G, forming a 60° angle with the given line AB at point A.

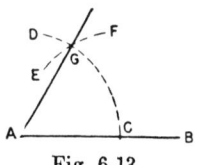

Fig. 6-13.

Problem 13. *To construct a 30° angle at a point on a given line.* See Fig. 6-14, where AB is the given line and A is the point on it.

Construct a 60° angle, CAB, at point A on line AB by the method described in Problem 12. Bisect $\angle CAB$, using the method described in Problem 3. $\angle CAE = \angle EAB = 30°$.

Note that if a perpendicular is dropped from point C upon given line AB, angle ACF is also equal to 30°, since right triangles ACF and

GEOMETRY

DCF are equal to each other, and angle *ACD* of equilateral triangle *ACD* is bisected by line *CF*. Therefore $\angle ACF = \angle DCF = 30°$.

Problem 14. *To construct an angle at a point on a given line equal to a given angle.* See Fig. 6-15, where *AB* is the given line, *A* is the point, and $\angle CDE$ is the given angle.

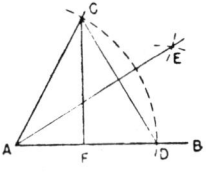

Fig. 6-14.

With *D* as a center and any radius, draw arc *FG*, intersecting the sides of the given angle at points *H* and *I*. With *A* as a center and with the same radius *DI*, draw arc *JK*. With *K* as a center and radius *HI*, draw arc *LM*, intersecting arc *JK* at point *N*. Draw line *AN* through points *A* and *N*, thus forming angle *NAB*, which is equal to given angle *CDE*.

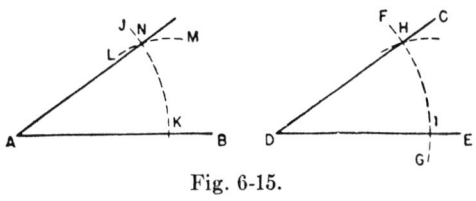

Fig. 6-15.

Problem 15. *To inscribe a hexagon in a given circle.* See Fig. 6-16.

Draw the diameter *AB* through the center *O* of the given circle. With *A* and *B* as centers and radii equal to the radius of the given circle, strike arcs above and below points *A* and *B*, intersecting the circle at points *C*, *D*, *E*, and *F*. Join all these points with straight lines, thus forming the required hexagon *ACEBFD* inscribed in the given circle.

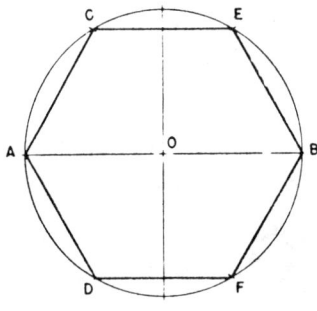

Fig. 6-16.

Problem 16. *To circumscribe a hexagon about a given circle.* See Fig. 6-17.

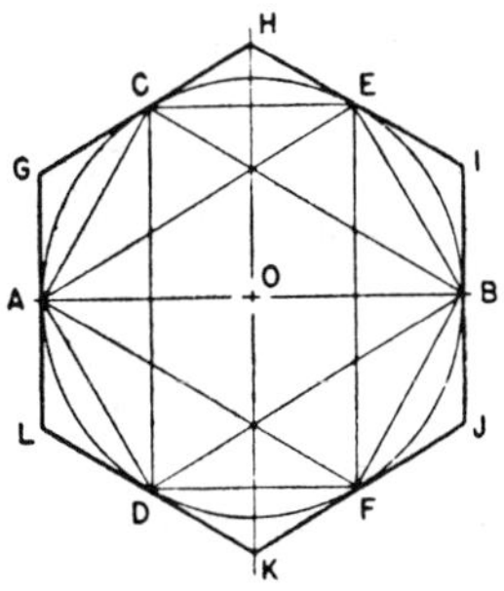

Method 1. If lines are drawn tangent to the given circle at points *A*, *C*, *E*, *B*, *F*, and *D* in Fig. 6-16, the required hexagon *GHIJKL* is constructed as shown in Fig. 6-17.

Method 2. Draw lines *AE*, *EF*, *AF*, *CD*, *CB*, *BD*. Draw lines *GH*, *IJ*, *KL*, *GL*, *HI*, *KJ* parallel to the above lines in the respective order, and obtain the required hexagon *GHIJKL*, as shown in Fig. 6-17.

Fig. 6-17.

Problem 17. *To construct a cycloid.* See Fig. 6-18. (A cycloid is a curve that is developed as follows: A circle of given diameter rolls on a straight line, and a given point on the circumference of this circle will form the required curve.)

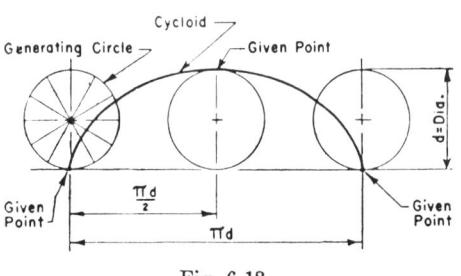

Fig. 6-18.

The length of the cycloid curve = $4 \times d$, where d = diameter of forming circle. The area bounded by the curve and the base line = 3 times the area of the forming circle.

Problem 18. *To construct an ellipse.* See Fig. 6-19, where AB and CD are the given axes.

Method 1. With O as a center and OA as a radius, draw arc AE. Divide OA into a convenient number of parts: OH, HG, GF, and FA, and draw parallel lines through points H, G, and F, intersecting arc AE at K, J, and I.

Draw line BP parallel and equal to OD, and join points O and P. Lay out

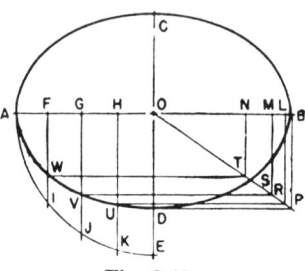

Fig. 6-19.

on OB the following lengths: $OL = HK$, $OM = GJ$, and $ON = FI$. Draw lines LR, MS, and NT parallel to BP, and lay out their equivalents HU, GV, and FW in the respective order. Join points D, U, V, W, and A by a smooth curve which represents one quarter of the ellipse. The remaining portion of the required ellipse may now be constructed from the curve just obtained.

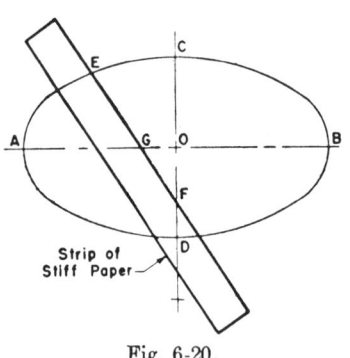

Fig. 6-20.

Method 2. A simple method often used for rapid construction of an ellipse is illustrated in Fig. 6-20 and described as follows:

AB is the major or transverse axis and CD the minor or conjugate axis. Lay off on a strip of stiff paper the distance $EF = AO$ and

$EG = CO$. Move the strip of paper to the right and to the left in such a manner that point G always travels on the major axis AB, and point F travels on the minor axis CD. One of the points is obtained as shown in the figure. In a similar manner a number of points may be determined through which the required ellipse is traced.

Problem 19. *To construct a helix on the surface of a cylinder.* See Fig. 6-21. (A helix is the curve formed on any cylinder by a point moving around and along the surface of the cylinder as an ordinary screw thread. The distance measured parallel to the axis, traversed by the point in one revolution, is the *lead*.)

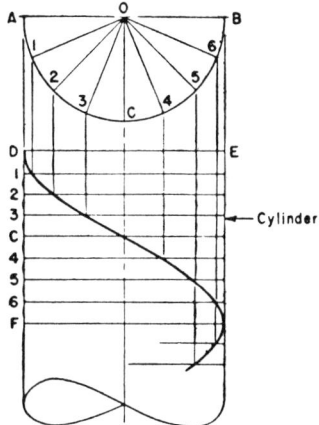

Fig. 6-21.

Divide the semicircumference into any number of equal parts, say eight. A–1 = 1–2 = 2–3, etc. Lay off DF = half the lead and divide DF into the same number of equal parts, namely, eight. D–1 = 1–2 = 2–3, etc.

Project points 1, 2, 3, etc., from the semicircumference downward and draw horizontal lines through points 1, 2, 3, etc., on the cylinder. Where the horizontal lines cross the correspondingly labeled vertical lines, points of intersection are marked, through which a smooth curve is traced. This curve is half of the required helix. The other half may be obtained in a similar manner, except that it will appear hidden back of the cylinder.

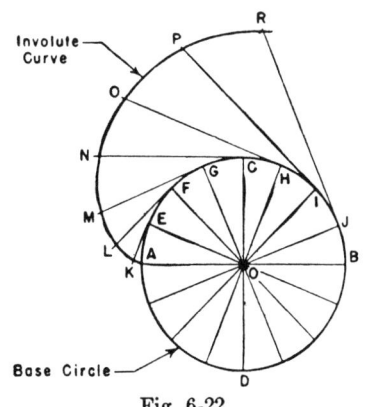

Fig. 6-22.

Problem 20. *To construct an involute curve.* See Fig. 6-22, where $ABCD$ is the base circle. (An involute curve is the path of a point on a cord as it unwinds from a line, polygon, or circle.)

Divide the circumference of the base circle into any number of equal parts, say sixteen. $AE = EF = FG$, etc.

Draw tangent lines at points E, F, G, etc. Lay off EK = length of arc AE, FL = length of arc AF, GM = length of arc AG, etc. Trace curve through points A, K, L, M, etc., which points describe the required involute curve.

Problem 21. *To divide a given line into any number of equal parts.* See Fig. 6-23, where AB is the given line to be divided into seven equal parts.

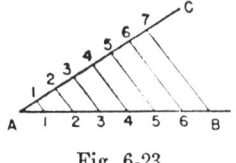

Fig. 6-23.

Draw line AC inclined to AB, and set off on line AC seven equal spaces, such as A-1 = 1-2 = 2-3, etc.

Connect point 7 with point B, and draw lines through points 6, 5, 4, 3, 2, and 1 parallel to 7-B. These lines divide given line AB into seven equal parts, such as A-1 = 1-2 = 2-3, etc.

Problem 22. *To draw a circular arc through two given points with a given radius.* See Fig. 6-24, where A and B are the given points and CD the given radius.

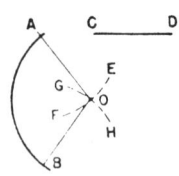

Fig. 6-24.

With A and B as centers and CD as the radius, draw arcs EF and GH, respectively. These arcs intersect at point O, which is the center of the required circular arc. With O as a center and CD as a radius, draw the required arc AB.

Problem 23. *To draw a circle passing through three given points.* See Fig. 6-25, where A, B, and C are the given points.

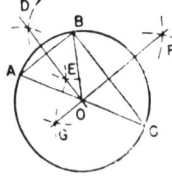

Fig. 6-25.

Bisect lines AB and BC by the method described in Problem 1 and extend bisectors DE and FG until they intersect at point O, which is the center of the required circle. With O as a center and any radius, such as OA, draw the required circle.

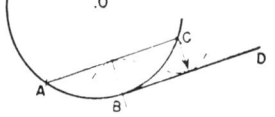

Fig. 6-26.

Problem 24. *To draw a line tangent to a given circular arc through a given point on the arc.* See Fig. 6-26, where ABC is the given arc and B the given point on the arc.

With B as a center, set off equal segments BA and BC on the given circular arc. Draw line AC. Draw line BD parallel to AC, using method described in Problem 4. Line BD is tangent to given circular arc.

GEOMETRY

Problem 25. *To draw tangents to a given circle from a given point outside the circle.* See Fig. 6-27, where point A is given outside the circle.

With A as a center and AO as a radius, draw arc FG. With O as a center and the diameter of the given circle as a radius, draw arcs HI and JK, intersecting arc FG at points B and C. Draw lines OB and OC, intersecting the given circle at points D and E. Draw lines from A through D and E, thus forming the required tangent lines.

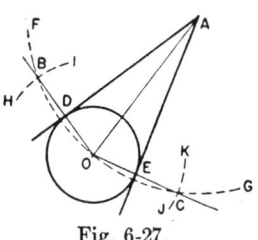

Fig. 6-27.

Problem 26. *To draw between two inclined lines a series of circles touching the inclined lines and touching each other.* See Fig. 6-28, where AB and CD are the given inclined lines.

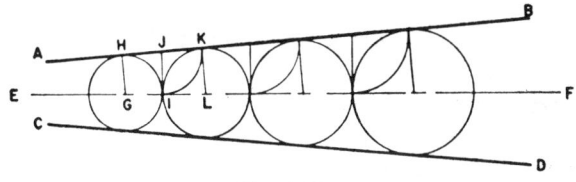

Fig. 6-28.

Bisect the angle formed by the two inclined lines. EF is the bisector. Choose any point G on line EF and draw GH perpendicular to AB. Draw a circle with G as a center and GH as a radius touching both given lines and intersecting line EF at I. Draw IJ perpendicular to EF, and with J as a center and JI as a radius, draw circular arc IK, intersecting AB at point K. Draw KL perpendicular to AB, and with L as a center and LK as a radius, draw a circle that will touch line AB at K and the first circle at I. As many circles may be drawn between the given inclined lines as may be required to satisfy the conditions, following the procedure given.

Problem 27. *To construct a triangle on a given base, the length of the sides being given.* See Fig. 6-29, where AB is the given base and AC and BC are the lengths of the sides of the required triangle.

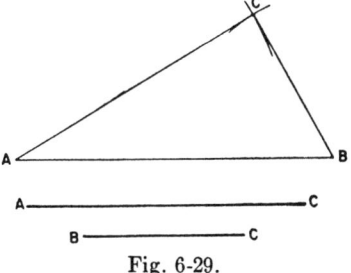

Fig. 6-29.

With A as a center and AC as a radius, strike an arc. With B as a center and BC as a radius, strike another arc, intersecting the first arc at point C. Draw lines AC and BC to complete the required triangle.

Problem 28. *To construct a parallelogram when two adjacent sides and the included angle are given.* See Fig. 6-30, where the sides AB and AD and $\angle DAB$ are given.

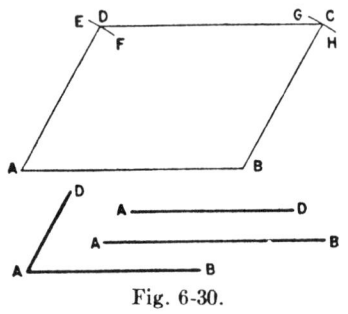

Fig. 6-30.

Draw the base of the parallelogram equal to AB. With A and B as centers and a radius equal to AD, draw arcs EF and GH. Construct given angle DAB, intersecting arc EF at point D. Draw $BC \parallel AD$, intersecting arc GH at point C. Draw line CD, thus completing the required parallelogram.

Problem 29. *To circumscribe a circle about a given triangle.* See Fig. 6-31, where triangle ABC is given.

Bisect line AC at point D and line BC at point E. Erect perpendiculars at points D and E, intersecting at point O, which is the center of the required circle. OA, or OB, or OC is the radius.

It should be noted that Problem 29 is similar to Problem 23, Fig. 6-25, where the three given points are the vertices of the given triangle in Fig. 6-31.

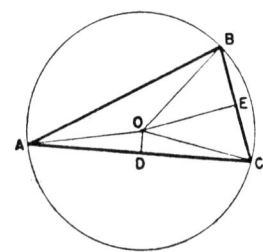

Fig. 6-31.

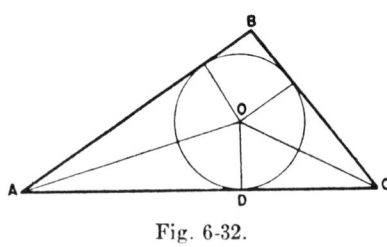

Fig. 6-32.

Problem 30. *To inscribe a circle in a given triangle.* See Fig. 6-32, where ABC is the given triangle.

Bisect $\angle BAC$ and draw the bisector AO.

Bisect $\angle BCA$ and draw the bisector CO.

The point of intersection O is the center of the required circle. Drop a perpendicular OD, which is the radius of the required circle.

GEOMETRY

Problem 31. *To circumscribe a circle about a given square.* See Fig. 6-33, where $ABCD$ is the given square.

Draw diagonals AC and BD, intersecting at point O. With O as a center and OA as a radius, draw the required circle.

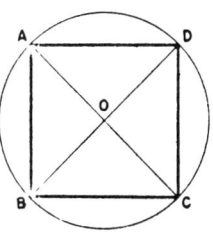

Fig. 6-33.

Note that if it is required to inscribe a square in a given circle, draw any two diameters in the given circle perpendicular to each other. The points at which the diameters cut the circle are the corners of the inscribed square.

Problem 32. *To inscribe a circle in a given square.* See Fig. 6-34, where $ABCD$ is the given square.

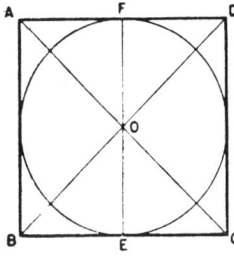

Fig. 6-34.

Draw diagonals AC and BD, intersecting at point O. Drop the perpendicular OE upon BC, and with O as a center and OE as a radius, draw the required circle.

Note that if it is required to circumscribe a square about a given circle, draw any two diameters perpendicular to each other and extend them beyond the circle. Bisect $\angle BOC$ and draw the bisector OE. Bisect $\angle AOD$ and draw the bisector OF. Draw $BC \perp OE$ and $AD \perp OF$. Join the intersection points A, B, C, and D and complete the required square.

Problem 33. *To inscribe a pentagon in a given circle.* See Fig. 6-35, where $ACBD$ is the given circle.

Draw any two diameters perpendicular to each other, such as AB and CD, intersecting at the center of the circle O. Bisect AO at E. With E as a center and EC as a radius, draw arc CF, intersecting AB at F. With C as a center and CF as a radius, draw arc FG, intersecting the given circle at G. Draw CG, which is one side of the pentagon. With C as a center and the same radius CF, draw an arc that intersects the given circle at H. Draw CH, which is another side of the pentagon. Points I and J are obtained in a similar manner and the required pentagon is drawn in.

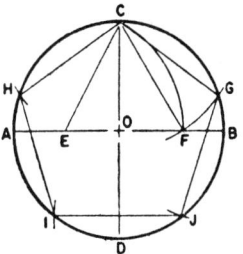

Fig. 6-35.

Problem 34. *To construct an octagon on a given line.* See Fig. 6-36, where AB is the given line.

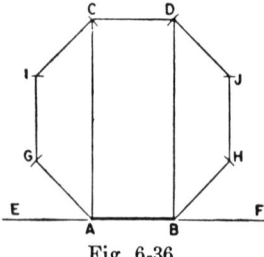
Fig. 6-36.

Draw lines AC and BD perpendicular to AB. Extend line AB to E and F respectively.

Bisect right angles CAE and DBF and draw the bisectors AG and BH. Make lines AG and BH equal to AB.

Draw lines GI and HJ parallel to AC and make them equal to AB. With I and J as centers and AB as radii, draw arcs that intersect AC and BD at points C and D. Draw lines IC, JD, and CD to complete the required octagon.

Problem 35. *To inscribe an octagon in a given circle.* See Fig. 6-37, where $ACBD$ is the given circle.

Draw two diameters AB and DC at right angles to each other, passing through the center of the given circle O.

Bisect arcs AC, CB, BD, and DA at E, F, H, and G respectively.

Join all the points on the circumference to obtain the required octagon.

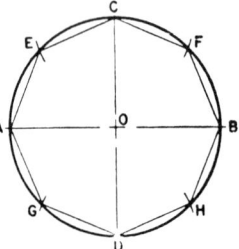
Fig. 6-37.

Problem 36. *To circumscribe an octagon about a given circle.* See Fig. 6-38.

Draw two diameters AD and BC at right angles to each other and extend them beyond the given circle.

Draw the square $ABCD$ about the given circle.

Draw lines GH and KL perpendicular to AD. Draw lines EF and IJ perpendicular to BC. These lines are to be tangent to the given circle.

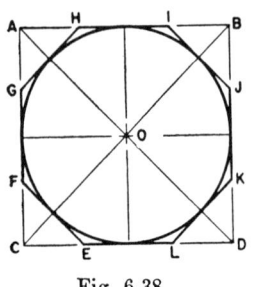
Fig. 6-38.

Draw in the required octagon $EFGHIJKL$.

Problem 37. *To construct an octagon in a given square.* See Fig. 6-39, where $ABCD$ is the given square.

Draw the diagonals AD and BC, intersecting at O. With B as a center and BO as a radius, draw arc EOF, intersecting the given square at points E and F. With A, C, and D as centers and the same radius BO, draw arcs intersecting the given square at points G, H, I, J, K, and L. Connect all the points of intersection to obtain the required octagon $EHLFJKGI$.

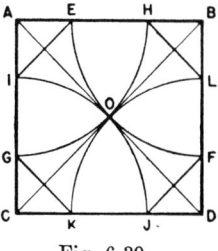

Fig. 6-39.

AREAS OF PLANE GEOMETRIC FIGURES

The ability to compute areas of surfaces is very important in shop problems when the distribution of acting loads over definite areas must be determined, or when the bearing areas of machine elements are to be calculated or the weights of certain structural details are to be estimated.

Since some surfaces are complex by nature they are divided into a number of elementary geometric figures, the areas of which are calculated in accordance with the following description, and the sum of the areas is then computed, giving the area of the surface in question.

At various intervals general rules are stated which may be of value to the reader.

1. *The area of a triangle is equal to one-half the product of the base and altitude.* The area of a right triangle $= \dfrac{ah}{2}$. See Fig. 6-40.

2. The area of an equilateral triangle $= \dfrac{ah}{2} = a^2 \dfrac{\sqrt{3}}{4} = .433a^2$. The altitude $h = \dfrac{a}{2}\sqrt{3} = .866a$. See Fig. 6-41.

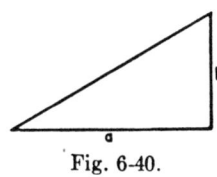
Fig. 6-40.

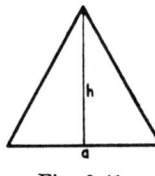
Fig. 6-41.

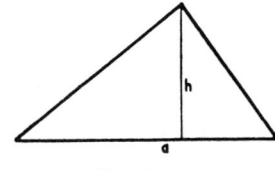
Fig. 6-42.

3. The area A of any triangle $= \dfrac{ah}{2}$. See Fig. 6-42. $h = \dfrac{2A}{a}$.

4. *The area of a square is equal to the square of its side.* The area A of a square $= a^2$. See Fig. 6-43. Also $A = .5b^2$; $b = \sqrt{2}\,a = 1.414a$.
5. *The area of a rectangle is equal to the product of its base and its altitude.* The area A of a rectangle $= ah$. See Fig. 6-44. Also $b = \sqrt{a^2 + h^2}$; $h = \sqrt{b^2 - a^2}$.
6. *The area of a parallelogram is equal to the product of its base and its altitude.* The area A of a parallelogram $= ah$. See Fig. 6-45. (A general parallelogram shown in the figure is also a rhomboid. If the

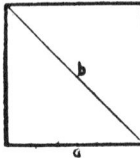
Fig. 6-43.

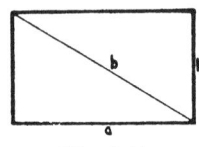
Fig. 6-44.

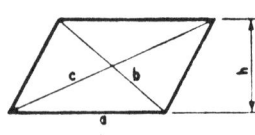
Fig. 6-45.

opposite sides are parallel and all sides are equal, the figure is a rhombus.) In a rhombus $b^2 + c^2 = 4a^2$, and $A = \dfrac{bc}{2}$.

7. *The area of a trapezoid is equal to the product of the altitude and one-half the sum of the bases.* The area A of a trapezoid $= \dfrac{a+b}{2}\,h$. See Fig. 6-46.

8. *The area of a trapezium is obtained by dividing the quadrilateral into parts and then finding the sum of the areas of these parts.* The area A of a trapezium $= \dfrac{(c+d)b + ec + fd}{2}$. See Fig. 6-47.

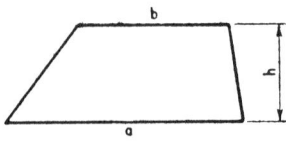
Fig. 6-46.

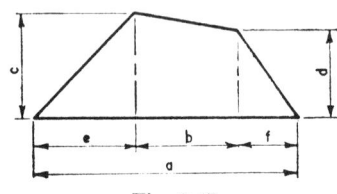
Fig. 6-47.

9. *The area of a regular polygon is equal to half the product of its apothem (distance from its center to any side) and perimeter.* The area A of a regular polygon is also $N \times$ area of each triangle, where $N =$ number of triangles. See Fig. 6-48. $A = \dfrac{Nar}{2} = \dfrac{Na}{2}\sqrt{R^2 - \dfrac{a^2}{4}}$.

GEOMETRY

10. **The area of a circle equals π times the square of the radius or one-fourth of π times the square of the diameter.** The area A of a circle $= \pi R^2 = \dfrac{\pi D^2}{4} = .7854 D^2$. See Fig. 6-49.

11. The area A of a hollow circle $= \dfrac{\pi}{4}(D^2 - d^2) = .7854(D^2 - d^2)$. Also $A = \pi(R^2 - r^2) = \pi \dfrac{D + d}{2}(R - r) = \pi(R + r)(R - r)$. See Fig. 6-50. A hollow circle is also called an *annulus*.

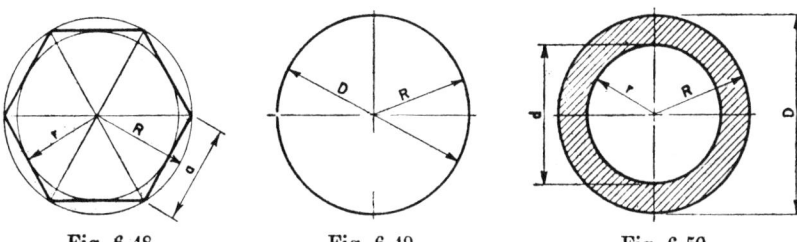

Fig. 6-48.   Fig. 6-49.   Fig. 6-50.

12. The area A of a sector of a circle $= \dfrac{\pi \theta R^2}{360} = .008727 \theta R^2 = \dfrac{LR}{2}$. See Fig. 6-51. Also $L = \dfrac{\pi R \theta}{180} = .01745 R\theta$, where $\theta =$ number of degrees in all these figures.

13. The area A of a segment of a circle $= \dfrac{R^2}{2}\left(\dfrac{\pi \theta}{180} - \sin \theta\right)$. See Fig. 6-52. If θ is greater than 90°, $A = \dfrac{R^2}{2}\left[\dfrac{\pi \theta}{180} - \sin(180 - \theta)\right]$. (The mean-

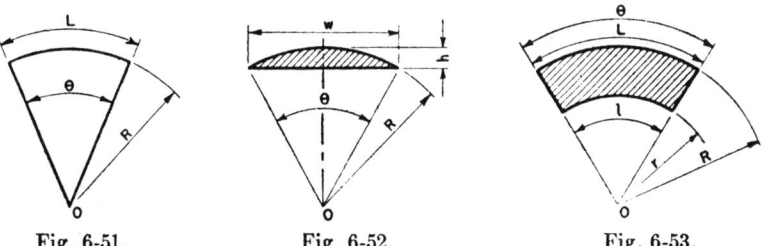

Fig. 6-51.   Fig. 6-52.   Fig. 6-53.

ing of sin is explained in Chapter VII.) An approximate method: $A = \dfrac{h^3}{2w} + \dfrac{2}{3}wh$, where $w =$ length of chord.

14. The area A of a sector shown shaded in Fig. 6-53 $= \dfrac{\pi\theta(R^2 - r^2)}{360}$.

   Also $A = \dfrac{R - r}{2}(L + l)$.

15. A *fillet* is used on a pattern or casting. The area A of a fillet $= .215R^2$. See Fig. 6-54.

16. A *parabola* is the curve of intersection of a cone of revolution with a plane parallel to one of the elements of the cone. The area A of a parabola $= \dfrac{2}{3}ah$. See Fig. 6-55.

17. The area A of an ellipse $= \pi ab$. See Fig. 6-56. The perimeter of an ellipse $= \pi(a + b)$ approximately.

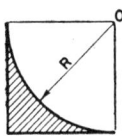 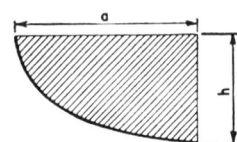 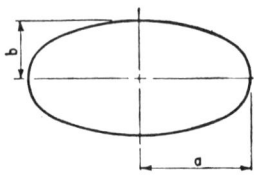

Fig. 6-54.   Fig. 6-55.   Fig. 6-56.

18. The area A of an irregular plane surface is estimated in the following manner. See Fig. 6-57. Draw line O—O and divide the given surface along this line into an even number (N) of parallel and small strips of width indicated by W, the heights and ordinates of which are

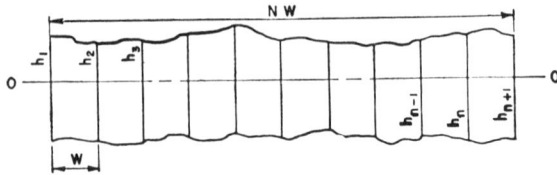

Fig. 6-57.

h_1, h_2, $h_3 \cdots h_{n-1}$, h_n, and h_{n+1}. Then the approximate area of the given surface A = sum of ordinates $\times$ width W. For a more accurate result, consider the contours between *three* ordinates as parabolas, then the area

$$A = \dfrac{W}{3}[h_1 + h_n + 1 + 4(h_2 + h_4 + h_6 \cdots + h_n) + 2(h_3 + h_5 + h_7 \cdots + h_{n-1})].$$

GEOMETRY

SOLID GEOMETRY

Whereas plane surfaces are surfaces with only two dimensions, solids are objects with three dimensions—length, breadth, and thickness. In this text we shall deal mainly with the areas and volumes of these solids. As an aid to the reader the various solids shall be defined when they are first mentioned.

AREAS AND VOLUMES OF SOLIDS

Area is the surface content of any figure and is expressed in square units, such as square inches, square feet, etc. The capacity of a solid figure measured by cubic units, that is, cubic inches, cubic feet, etc., is called the volume.

For the sake of simplicity the following equivalents will be used in all the formulas:

V = volume
S = total surface area
S_l = lateral surface area
A_b = area of the base
A_r = area of the right section
A_t = area of the top
L = slant height
L_g = slant height between centers of gravity of areas
P_b = perimeter of the base
P_r = perimeter of the right section
P_t = perimeter of the top
h = height or altitude
h_g = vertical distance between centers of gravity of areas
r = radius
d = diameter
w = length of chord

1. A *cube* is a solid whose six faces are equal squares. The volume of a cube = $V = a^3$. The total area of the surface of a cube = $S = 6a^2$. See Fig. 6-58.
2. A *prism* is a solid whose ends (called bases) are equal, similar, and parallel polygons and whose sides (called lateral faces) are parallelograms. A *rectangular prism* is a prism whose bases are rectangles. A *parallelepiped* is a prism whose bases are parallelograms. The volume of a rectangular prism or parallelepiped = $V = abc$. The total area of the surface = $S = 2(ab + bc + ac)$. See Fig. 6-59.

3. A *right prism* is one whose lateral faces are perpendicular to the bases. A *regular prism* is a right prism whose base is a regular polygon. The volume of a right regular prism $= V = A_b h$, where A_b equals area of the base. The area of the lateral surface $S_l =$ perimeter of base $\times$ height $= P_b h$. See Fig. 6-60. The total area of the surface $= S = S_l + 2A_b$, where $S_l =$ lateral surface area and $A_b =$ area of the base.

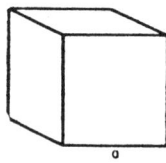
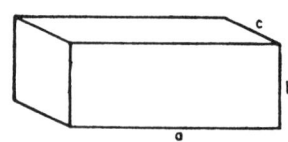
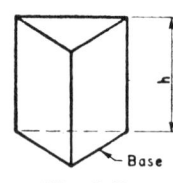

Fig. 6-58.   Fig. 6-59.   Fig. 6-60.

4. The volume of any prism $= V = A_r L$, where $A_r =$ area of the right section and $L =$ slant height. See Fig. 6-61. (It may also be said that the volume of any prism is equal to the product of its base and altitude.) $S_l = P_r L$, where $P_r =$ perimeter of right section.

$$S = S_l + 2A_b.$$

5. The *frustum* of a solid is the part formed by cutting off the top of the solid by a plane parallel to the base. The volume of a frustum of a prism $= V = A_b h_g = A_r L_g$, where $h_g =$ vertical distance between centers of gravity of areas and $L_g =$ slant height between centers of gravity of areas. See Fig. 6-62.

6. A *pyramid* is a solid whose base is a polygon and whose sides are triangles which meet at a common point to form the vertex of the

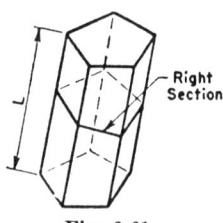
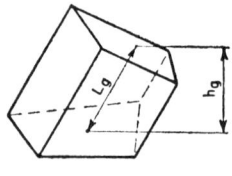
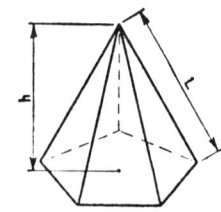

Fig. 6-61.   Fig. 6-62.   Fig. 6-63.

pyramid. A *right pyramid*, or a *regular pyramid* is one whose base is a regular polygon and the sides equal isoceles triangles. The volume of a right regular pyramid $= V = \frac{1}{3}A_b h$, where h = vertical height or altitude. $S_l = \frac{1}{2}P_b L$. See Fig. 6-63.

7. A *cone* is a solid whose base is a circle and whose surface tapers from the base to a point called the vertex. The same formulas used for the pyramid apply to a regular cone, shown in Fig. 6-64, where $A_b = \frac{\pi}{4}d^2$ and $P_b = \pi d$.

8. The volume of any pyramid or cone $= V = \frac{1}{3} A_b h$. See Fig. 6-65.

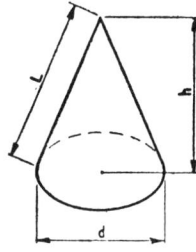

Fig. 6-64.

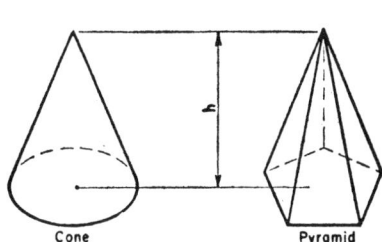

Fig. 6-65.

9. The volume of a frustum of a right regular pyramid or cone $= V = \frac{1}{3}h(A_b + A_t + \sqrt{A_b A_t})$. See Fig. 6-66. $S_l = \frac{1}{2}L(P_b + P_t)$, where P_t = perimeter of the top. $S = S_l + A_b + A_t$.

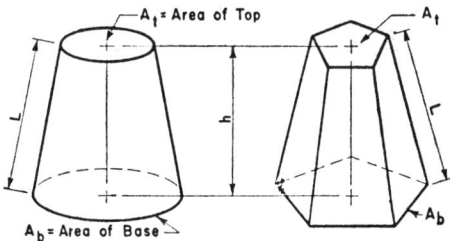

Fig. 6-66.

10. The volume of a frustum of a general pyramid or cone $= V = \frac{1}{3}h(A_b + A_t + \sqrt{A_b A_t})$. See Fig. 6-67. Note that the top and bottom surfaces are parallel.

11. A *right circular cylinder* is a right cylinder (one whose side is perpendicular to its base) whose bases are circles. The volume of a right circular cylinder $= V = \dfrac{\pi}{4}d^2h$. See Fig. 6-68. $S_l = \pi dh$ and $S = \pi d(r + h)$.

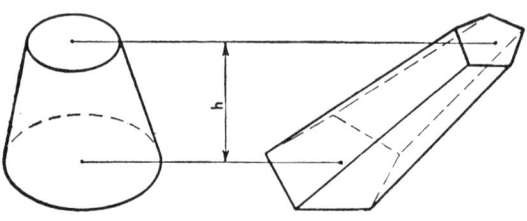
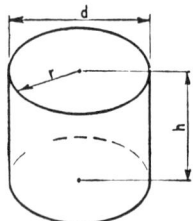

Fig. 6-67.   Fig. 6-68.

12. The volume of any cylinder $= V = A_b h = A_r L$, where A_r area of right section. (A right section of a cylinder is a section made by a plane perpendicular to all its elements.) $S_l = P_b h = P_r L$ and $S = S_l + 2A_b$. See Fig. 6-69.

13. The volume of a frustum of any cylinder $= V = \dfrac{1}{2} A_r (L_1 + L_2)$. See Fig. 6-70, where L_1 and $L_2 =$ respective slant heights. Also $V = A_b h_g$.

14. The volume of a frustum of a right circular cylinder $=$

$$V = \frac{\pi d^2}{8}(h_1 + h_2).$$

See Fig. 6-71, where h_1 and h_2 are the respective heights.

$$S_l = \frac{\pi d}{2}(h_1 + h_2), \quad \text{and} \quad S = S_l + A_b + A_t.$$

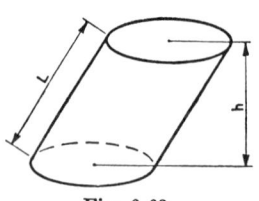
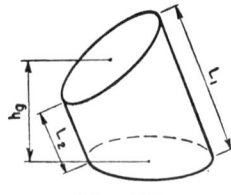
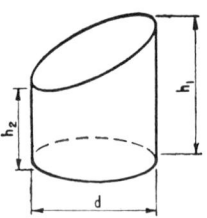

Fig. 6-69.   Fig. 6-70.   Fig. 6-71.

GEOMETRY

15. A *sphere* is a solid bounded by a curved surface every point of which is equally distant from a point within called the center. The volume of a sphere = $V = \frac{4}{3}\pi r^3$ and $S = 4\pi r^2$. See Fig. 6-72.

16. A *spherical sector* is a portion of the volume of a sphere generated by the revolution of a circular sector about a diameter of the circle. The volume of a spherical sector = $V = \frac{2}{3}\pi r^2 h$. $S = \frac{\pi r}{2}(4h + w)$, where w = the length of the chord. See Fig. 6-73.

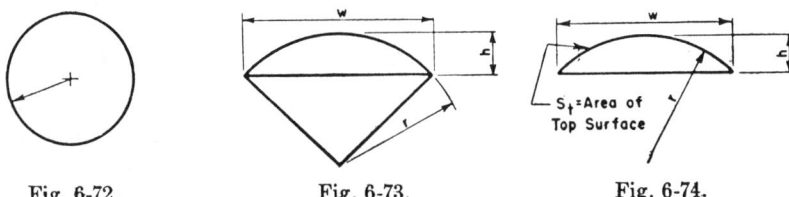

Fig. 6-72.   Fig. 6-73.   Fig. 6-74.

17. A *spherical segment* is a portion of the volume of a sphere included between two parallel planes. Fig. 6-74 shows the segment of one base where one of the bounding planes touches the sphere. The volume of this spherical segment = $V = \frac{\pi}{3}h^2(3r - h)$. Also $V = \frac{\pi}{24}h(3w^2 + 4h^2)$. See Fig. 6-74. $S_t = 2\pi rh = \frac{\pi}{4}(4h^2 + w^2)$.

18. A *spherical zone* is the portion of the surface of the sphere between two parallel planes. The volume of figure shown = $V = \frac{\pi h}{24}(3w^2 + 3a^2 + 4h^2)$. See Fig. 6-75. $S_l = 2\pi rh$, and
$$S = \frac{\pi}{4}(8rh + w^2 + a^2).$$

19. A *torus* (also called an *anchor ring*) is a ring formed from a cylinder bent into a circular form. The volume of a torus = $V = 2\pi^2 r^2 R$, and $S = 4\pi^2 rR$. See Fig. 6-76, where R = mean radius, r = radius of circular section and S = total area of surface of torus.

20. The volume of a hollow cylinder = $V = V_1 - V_2$. Also $V = \pi h(R^2 - r^2)$, or $V = \pi ht(D - t)$. See Fig. 6-77, where V_1 = volume of outside cylinder and V_2 = volume of cylindrical hole. R = large

radius, r = small radius, D = large diameter, t = thickness of wall. See Fig. 6-77.

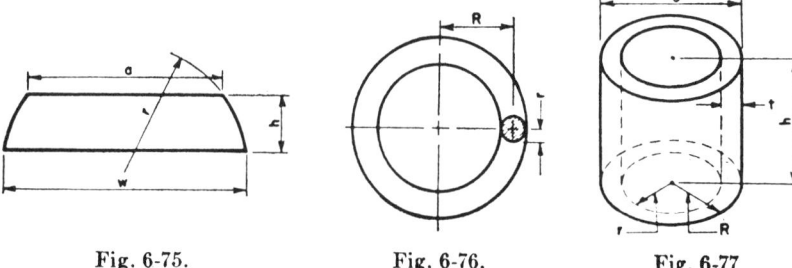

Fig. 6-75.   Fig. 6-76.   Fig. 6-77.

METHODS AND SHORT CUTS IN ESTIMATING WEIGHTS

In order to estimate weights of castings, forgings and other machine elements we must determine their volumes. Many machine elements are composed of parts, the volumes of which may be found by methods described in the foregoing examples. Some minute portions of the machine elements may be different from the so-called *standard* geometric solids. In that case, a close approximation is assumed to correspond to nearly similar geometric solids. The partial volumes are all added together and expressed in cubic inches. The weights of commonly used materials per cubic inch may be found in Table 8 of the Appendix and the total weight of the part in question computed therefrom.

Illustrative Examples. A few illustrative examples follow and the methods used are explained in detail. The volumes of geometric solids are obtained from the preceding section on volumes.

Example 1. It is required to compute the weight of a wrought-iron rivet, as shown in Fig. 6-78.

The body of the rivet consists of a right cylinder, whose diameter = $1\frac{1}{8}''$ and height = $3\frac{3}{4}''$. The head of the rivet is a frustum of a right cone.

The volume of the body = $V = \frac{\pi}{4}d^2h$ = .7854 × (1.125)² × 3.75 = 3.729 cu. in. The volume of the head =

$V = \frac{1}{3}h(A_b + A_t + \sqrt{A_b A_t})$ =

$\frac{.562 \times [.7854 \times (1.125)^2 + .7854 \times (2.031)^2 +}{3}$

$\sqrt{.7854 \times (1.125)^2 \times .7854(2.031)^2}$] = 1.395 cu. in.

GEOMETRY

The total volume of solid material = 3.729 + 1.395 = 5.124 cu. in.
The weight of 1 cu. in. of wrought iron = .2834 lb., in accordance with Table 8 of the Appendix.
The weight of the rivet shown in Fig. 6-78 = .2834 × 5.124 = 1.452 lb.

Note: The reader is advised to use the slide rule for computations similar to those in Example 1, if the degree of accuracy desired is satisfactory for the requirements of the job. For precision calculations, computation must be carried out in the conventional manner. The reader may use Table 2 of the Appendix, giving numbers, squares, square roots, etc., or more comprehensive tables found in standard reference books such as *Machinery's Handbook*, *American Machinist Handbook*, and *Marks' Handbook*, etc.

Example 2. Compute the weight of the steel rivet shown in Fig. 6-79.
The body of the rivet is a right cylinder, and its volume = $V = \frac{\pi}{4} d^2 h$ = .7854 × (1.25)² × 5.125 = 6.290 cu. in.

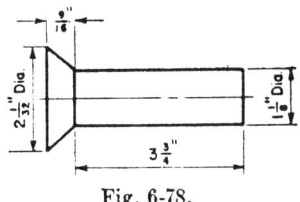

Fig. 6-78.

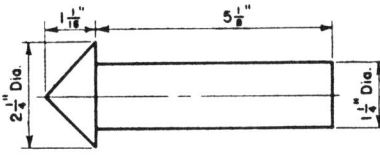

Fig. 6-79.

The head of the rivet is a regular cone, and its volume = $V = \frac{1}{3} A_b h$ =

$\frac{1}{3}$ × .7854 × (2.25)² × 1.062 = 1.408 cu. in.

The total volume of the solid material = 6.290 + 1.408 = 7.698 cu. in.

The weight 1 cu. in. of steel = .2816 lb. (See Table 8 of the Appendix.)
The weight of the rivet shown in Fig. 6-79 = .2816 × 7.698 = 2.168 lb.

Example 3. Compute the weight of the hollow copper rivet shown in Fig. 6-80.
The body of the rivet is a right cylinder, partly hollow.

The volume of the solid cylinder = $V = \frac{\pi}{4} d^2 h$ = .7854 × (.75)² × 2.625 = 1.160 cu. in.

The volume of the cylindrical hollow portion $= V = .7854 \times (.5625)^2 \times 2.25 = .559$ cu. in.

The volume of the solid material in the body $= 1.160 - .559 = .601$ cu. in.

The volume of the rivet head $=$ the volume of a spherical segment.

$V = \dfrac{\pi}{3}h^2(3r - h)$, where $r = \dfrac{3''}{4}$ and $h = \dfrac{5''}{8}$. Then $V = \dfrac{3.1416}{3} \times (.625)^2 \times [(3 \times .75) - .625] = 1.0472 \times .3906 \times (2.250 - .625) = .665$ cu. in.

The total volume of solid material in the rivet $= .601 + .665 = 1.266$ cu. in.

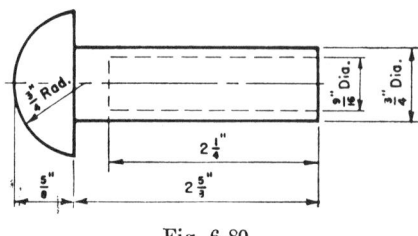

Fig. 6-80.

The weight of 1 cu. in. of copper $= .3184$ lb. (Table 8, Appendix).

The weight of the rivet shown in Fig. 6-80 $= .3184 \times 1.266 = .403$ lb.

Example 4. Find the weight of the aluminum ring shown in Fig. 6-81.

The volume of the solid material $= V = 2\pi^2 r^2 R$, a formula for the volume of a torus, where $R = 2\dfrac{13''}{32}$ and $r = \dfrac{9''}{32}$. Then

$$V = 2\pi^2 r^2 R = 2 \times (3.1416)^2 \times (.28125)^2 \times 2.40625 =$$

$$2 \times 9.8696 \times .0791 \times 2.40625 = 3.752 \text{ cu. in.}$$

The weight of 1 cu. in. of aluminum $= .0924$ lb.

The weight of the ring in Fig. 6-81 $= .0924 \times 3.752 = .347$ lb.

Example 5. Compute the weight of the cast-iron gear blank shown in Fig. 6-82.

When computing the weights of machine elements of complex shape, the object is divided into a number of elementary parts, and the volume of each is estimated. Some approximation is inevitable, but with experience one can compute the volume and then the weight of the object with fairly close accuracy.

Step 1. The volume of the rim is computed first; it is a hollow cylinder and $V = \pi ht(D - t) = 3.1416 \times 1.625 \times 1.250(16 - 1.250) = 94.126$ cu. in.

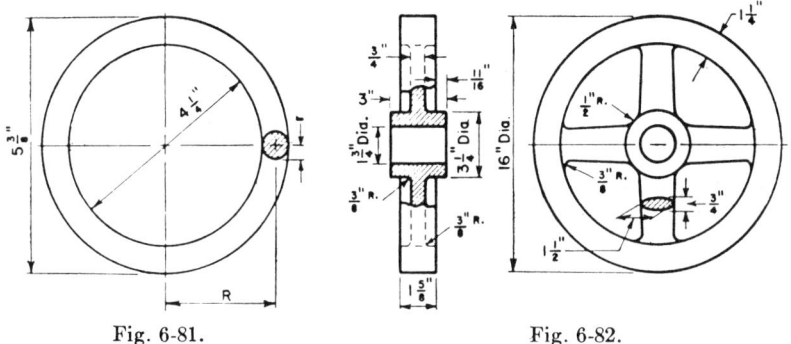

Fig. 6-81.   Fig. 6-82.

Step 2. The volume of the hub, which is a hollow cylinder, is equal to
$V = \pi ht(D - t) =$

$$3.1416 \times 3 \times \left(\frac{3.25 - 1.75}{2}\right) \times \left(3.25 - \frac{3.25 - 1.75}{2}\right) =$$

$$3.1416 \times 3 \times .75 \times 2.50 = 17.672 \text{ cu. in.}$$

Step 3. The arms are of elliptical section, and the average area of the cross section is approximately at the middle of the arm length, since the arms taper. The area of an ellipse, in accordance with Fig. 6-56, is
$A = \pi ab = 3.1416 \times \dfrac{1.5}{2} \times \dfrac{.75}{2} = 3.1416 \times .75 \times .375 = .883$ sq. in.

The volume of one arm without the fillets $= Al$, where $A = .883$ sq. in., and $l =$ length of arm.

$$l = \frac{16}{2} - 1.25 - \frac{3.25}{2} = 8 - 1.25 - 1.625 = 5.125''.$$

$V = Al = .883 \times 5.125 = 4.525$ cu. in. = volume of one arm. Volume of four arms $= 4.525 \times 4 = 18.100$ cu. in.

Step 4. The area of a fillet $= .215R^2$. Looking at the right-side view: There are eight fillets of $\dfrac{1''}{2}$ radius. Their total area $= 8 \times .215 \times .5^2 = .430$ sq. in. There are eight fillets of $\dfrac{3''}{8}$ radius. Their total area $=$

$8 \times .215 \times .375^2 = .240$ sq. in. The combined areas $= .430 + .240 = .670$ sq. in. The length of each fillet $= \frac{3''}{4}$. The combined volumes of the fillets $= .670 \times .75 = .503$ cu. in.

Looking at the left-side view: There are sixteen fillets of $\frac{3''}{8}$ radius and approximately $1\frac{1''}{2}$ long. The combined areas of the fillets $= 16 \times .215R^2 = 16 \times .215 \times .375^2 = .482$ sq. in. The combined volumes of the fillets $= .482 \times 1.5 = .723$ cu. in.

Step 5. The total volume of solid material in the entire object is equal to:

$$\begin{array}{r} 94.126 \\ 17.672 \\ 18.100 \\ .503 \\ .723 \\ \hline 131.124 \text{ cu. in.} \end{array}$$

Step 6. The weight of 1 cu. in. of cast iron, in accordance with Table 8 of the Appendix, is equal to .26 lb.

The weight of the part shown in Fig. 6-82 $= 131.124 \times .26 = 34.09$ lb., say 34 lb.

A slide rule is very convenient for these computations if a certain degree of accuracy can be sacrificed for the sake of speed. (A complete description of the slide rule and instruction in its use can be found in Chapter XI.)

A convenient way to estimate the weight of a complex casting is to weigh the wooden pattern used for the casting and then compute the weight of the casting itself by determining the specific gravities of the materials of both object and pattern. (See Table 8 of the Appendix.)

For example, if the pattern is of white pine and weighs 7 lb. 3 oz., and steel is to be used for the casting, the procedure would be:

The specific gravity of steel $= 7.80$. The specific gravity of white pine $= .50$. (In accordance with Table 8 of the Appendix.) The weight of the steel casting $=$ the weight of the pattern $\times$ the ratio of the specific gravities $= 7\frac{3}{16} \times \frac{7.80}{.50} = 112.12$ lb., say 112 lb.

Another method used in computing the weight of the casting is to estimate the volume of the pattern. In accordance with tables found in

GEOMETRY

standard handbooks, the weight of 1 cu. in. of white pine = .018 lb. The weight of the pattern = $7\frac{3}{16}$ lb. Therefore the volume of the pattern = $\frac{7.187}{.018}$ = 399 cu. in.

The volume of the casting, disregarding the shrinkage, is also 399 cu. in. approximately.
The weight of 1 cu. in. of steel = .2816 lb.
The total weight of the casting = .2816 × 399 = 112.358 lb. This checks favorably with the result obtained by means of the ratio of the specific gravities of the materials.

Note: These methods apply to solid castings only. When computing the volumes of cores, which require core boxes and core prints on the patterns, methods described in handbooks employing various factors may be used, although the results obtained are only approximate.

SPECIFIC GRAVITIES OF VARIOUS MATERIALS

We know that some materials weigh more than others, yet the volumes of the respective materials of a given group of materials may be identical. In one of the foregoing examples a comparison was made between a steel casting and its wooden pattern. The volumes of both objects were the same, yet the weight of the steel casting was 112.12 lb. and that of the wooden pattern was 7.187 lb. We speak of this variation as the ratio of the densities of the two materials, in this case, $\frac{112.12}{7.187}$ = 15.6.

We also noted that the weight or density of 1 cu. in. of steel is .2816 lb., and the weight of 1 cu. in. of white pine (the material of the pattern) is .018 lb.; therefore the steel is $\frac{.2816}{.018}$ = 15.6 times the weight of the white pine. This ratio—15.6—is the specific gravity of the steel if white pine is chosen as the standard to which another material is referred. For convenience, the standard usually used for solids and liquids is water. Water is the substance of most common reference chiefly because, in the metric system, the weight of 1 cubic centimeter of water under standard conditions is 1 gram, and because water is a substance familiar to all owing to its abundance. All substances are referred to water, but air or hydrogen gas is used as a standard of comparison when a ratio between volumes of gases is to be established.

Example 1. The specific gravity of white pine is .5. This is a ratio between the weight of the white pine and the weight of an equal volume of water. In other words, water is twice as heavy as white pine.

Example 2. The specific gravity of steel is 7.80. This shows that steel is 7.80 times as heavy as water.

By comparing all substances with water as a standard, we may be able to compute the volume of any material, if the volume of some other material is known.

Specific gravities and weights of commonly used materials are given in Table 8 of the Appendix.

GEOMETRIC PRINCIPLES APPLIED IN SOLUTION OF SHOP PROBLEMS

A number of carefully chosen problems is here presented to demonstrate the application of geometric principles.

Problem 1. Find the diameter of a circle, the center of which is undetermined.

The contour of the object may be a circular arc whose diameter or radius is to be determined. Perhaps it is desired to replace a broken disk or pulley and the center cannot be physically located. See Fig. 6-83.

Procedure: Scribe a convenient chord C and measure its length.

Bisect this chord and obtain $\frac{C}{2}$.

Measure the height (h) of this chord.

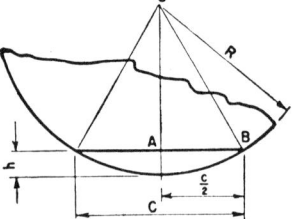

Fig. 6-83.

Solve for R, which is the radius of the right triangle OAB.

$OA = R - h$, and $(R - h)^2 + \left(\frac{C}{2}\right)^2 = R^2$.

$$R^2 - 2hR + h^2 + \frac{C^2}{4} =$$

$$R^2 - 2hR + h^2 + \frac{C^2}{4} = 0.$$

$$2hR = h^2 + \frac{C^2}{4}$$

$$R = \frac{h^2 + \frac{C^2}{4}}{2h},$$

and $D = \dfrac{h^2 + \frac{C^2}{4}}{h}$,

or $D = h + \dfrac{C^2}{4h}$, where D = diameter.

Problem 2. For a hollow portion of a detail, find the diameter of its circular arc profile. The method used is shown in Fig. 6-84.

Procedure: Place a convenient straightedge of known length C, as shown in Fig. 6-84.

Bisect C and obtain $\dfrac{C}{2}$.

Measure the height h of this chord.

Solve for R in the same manner as in Fig. 6-83, and finally $D = h + \dfrac{C^2}{4h}$, where D = diameter of circular arc in problem.

Problem 3. The methods indicated in Problems 1 and 2 for finding the required diameters are theoretically correct, but inaccuracies resulting from difficulties in scribing a correct chord, and unavoidable errors in accurately measuring the height of the chord, can be responsible for incorrect results. In such case a more precise method, using precision measuring tools, is applied in solution of the problem, as illustrated in Fig. 6.85.

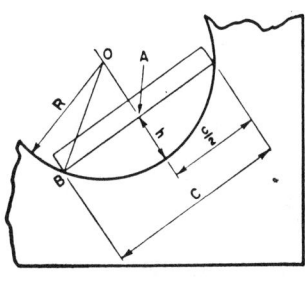
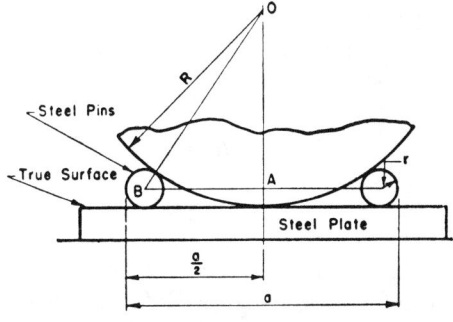

Fig. 6-84.   Fig. 6-85.

Procedure: Place two precision steel pins of a convenient diameter, as shown, in contact with the object and the steel flat.

Measure with micrometer or precision gage blocks the distance a.

Solve for R in the right triangle OAB, where $OB = R + r$, $AB = \dfrac{a}{2} - r$, and $OA = R - r$.

$$OB^2 = AB^2 + OA^2$$

$$(R + r)^2 = \left(\dfrac{a}{2} - r\right)^2 + (R - r)^2$$

$$R^2 + 2Rr + r^2 = \dfrac{a^2}{4} - ar + r^2 + R^2 - 2Rr + r^2$$

$$4Rr = \dfrac{a^2}{4} - ar + r^2$$

$$R = \dfrac{\dfrac{a^2}{4} - ar + r^2}{4r} = \dfrac{a^2 - 4ar + 4r^2}{16r} = \dfrac{a^2 - 4r(a - r)}{16r}$$

$$D = \dfrac{a^2 - 4r(a - r)}{8r}$$

Problem 4. A precision method used in finding the radius or diameter of an internal circular arc is illustrated in Fig. 6-8G.

Procedure: Three precision steel disks of convenient diameter are placed in contact with the internal contour of the part and in contact with each other. Disks of the same diameter are chosen simply as a matter of convenience.

A measurement M is obtained by means of precision tools.

Consider the right triangle ABC:

$$b = \dfrac{M}{2} - r$$

$$a = \sqrt{(2r)^2 - b^2} = \sqrt{4r^2 - b^2}$$

Dimension a can also be measured as shown in the figure and compared with the calculated dimension a.

GEOMETRY

Consider the right triangle OAC:

$$(R - r)^2 = (R - r - a)^2 + b^2$$

$$R^2 - 2Rr + r^2 = R^2 - 2Rr - 2aR + 2ar + a^2 + r^2 + b^2$$

$$2aR = 2ar + a^2 + b^2$$

$$\therefore R = \frac{2ar + a^2 + b^2}{2a}$$

$$D = \frac{2ar + a^2 + b^2}{a}, \text{ where}$$

$D =$ diameter of the circular arc.

If the internal portion of the object is spherical, precision steel balls must be employed instead of disks, since the balls make point contact instead of line contact with the spherical shape of the part, whose radius and diameter are to be determined.

Note: The setup shown in Fig. 6-86 is cumbersome, and it is difficult to maintain the contact at all points and to take an accurate measurement M. However, mathematical analysis of the construction is rational.

Problem 5. A problem that is common in machine shop practice is shown in Fig. 6-87. It is required to calculate the diameter of a circle in which four disks or balls of known diameter are to fit as shown.

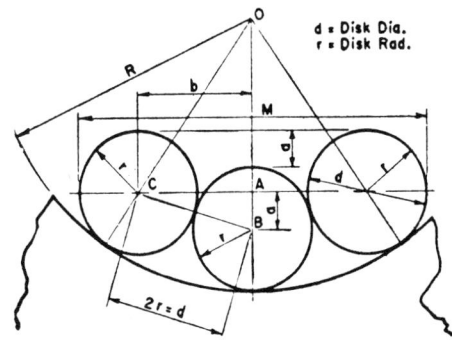

Fig. 6-86.

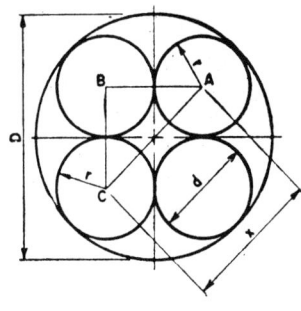

Fig. 6-87.

Procedure: Consider right triangle ABC:

$$AB = BC = 2r = d$$
$$AC = \sqrt{AB^2 + BC^2} = \sqrt{2d^2} = 1.4142d$$
$$D = AC + r + r = 1.4142d + d = 2.4142d$$

Problem 6. In the problem illustrated in Fig. 6-88, a circle is inscribed in a given right triangle, ABC. The lengths of the hypotenuse and of one of the legs of the triangle are known. It is required to find the diameter, d, of the inscribed circle.

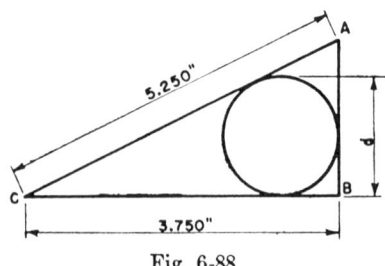

Fig. 6-88.

Procedure: The diameter of a circle inscribed in a right triangle is equal to the sum of the lengths of both legs of the triangle, less the length of the hypotenuse.

$$d = AB + BC - AC$$

But $AB = \sqrt{(AC)^2 - (BC)^2} = \sqrt{(5.250)^2 - (3.750)^2} = 3.674$

$$d = 3.674 + 3.750 - 5.250 = 2.174$$

Problem 7. The problem illustrated in Fig. 6-89 is common in automobile body and airplane sheet metal panels and other parts. Dies and forms for obtaining the proper shape of these parts must be constructed with a high degree of accuracy. These shapes are often similar to those of circular arcs, and if the radii of the arcs are unusually large, or scribing of the arcs is not convenient, the following method of obtaining the required arc may be applied.

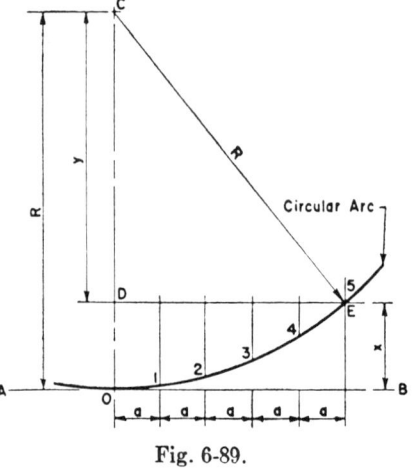

Fig. 6-89.

A series of points is obtained through which a smooth curve approaching the circular arc is traced. Of course, the more points obtained, the smoother the curve.

This method of obtaining a circular arc is used frequently in the toolroom and inspection department, where an enlarged template of a circular arc is drawn and the object itself is compared by means of projection comparators.

Procedure: Draw line AB and choose any point O at which the line will be tangent to the required circular arc of given radius R.

Step off from point O any convenient distances a on line AB, and erect perpendiculars at these points.

To locate points on the required arc, say point 5, consider the right triangle CDE, in which CE is the radius (R) and DE is equal to $5 \times a$. Hence $Y = \sqrt{R^2 - (5a)^2}$ and $X = R - Y$, thus locating point 5 a distance x above line AB.

In a similar manner, any point on the required curve may be found by constructing a suitable right triangle. The accuracy of the curve depends upon the number and size of distances a chosen. The closer are the lines of division, the more accurate will be the traced curve.

The spacing of the distances may be unequal. For convenience, they are spaced uniformly in the illustration, but if some additional points are desired, the procedure is similar to that described for the problem of Fig. 6-89. As a matter of fact, it is recommended that points farther away from the O point be spaced closer together.

Problem 8. The problem illustrated in Fig. 6-90 is often encountered in layout work, when it is required to construct an approximate length of a given circular arc, CD.

Procedure: Draw line AB tangent to arc at point C.

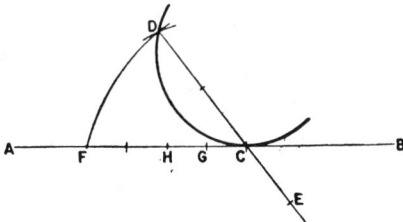

Fig. 6-90.

Extend chord DC downward and set distance $CE = \dfrac{DC}{2}$.

With E as a center and ED as a radius, strike arc DF, intersecting line AB at F. Line CF is equal to the length of arc CD—the percentage of error being negligible.

If it is required to find a circular arc on a given circle equal in length to a given line CF, divide given line into four equal parts CG, GH, etc. With G as a center and GF as a radius, draw an arc, intersecting the given circle at point D. Thus arc CD is found.

Problem 9. The problem illustrated in Fig. 6-91 shows how to draw a common tangent to two given circles A and B.

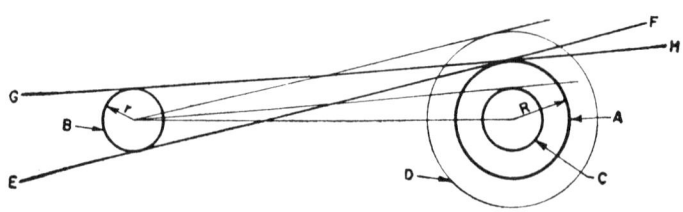

Fig. 6-91.

Procedure: Draw circle C with a radius equal to $R - r$ and circle D with a radius equal to $R + r$, concentric with the given circle A.

Draw tangent lines from the center of circle B to circles C and D.

Draw lines EF and GH parallel to these tangent lines, and thereby find required lines tangent to both given circles.

Problem 10. The problem illustrated in Fig. 6-92 shows a method by which a circle is drawn through one given point A, touching two given circles with centers K and L, respectively.

Procedure: Draw two lines, DG and DH, tangent to both given circles and intersecting at point D.

Draw any line DM from point D, intersecting the given circles at points B, C, E, and F.

Draw a circle through three points B, F, and A by a previously shown method.

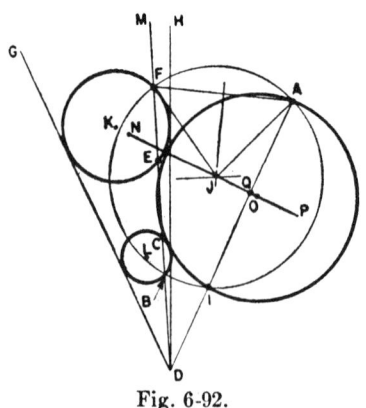

Fig. 6-92.

The center of this circle is at J, and the circle intersects line DA at point I.

Draw a circle through points A and I, touching both given circles. The center O of this circle must lie on line NP. Line NP also goes through the center J and is perpendicular to AI at its mid-point, Q.

GEOMETRY

AI is a chord in both circles: center J and center O. Therefore both of these centers must lie on the line NP perpendicular to chord AI at its mid-point.

PRACTICE PROBLEMS APPLICABLE IN THE SHOP AND TOOLROOM

1. Find the circumference of a circle whose diameter is equal to $2.12''$.
2. Find the circumference of a circle whose diameter is equal to $1\frac{17''}{64}$.
3. Find the circumference of a circle, in feet, whose diameter is equal to $2'\ 3\frac{1''}{4}$.

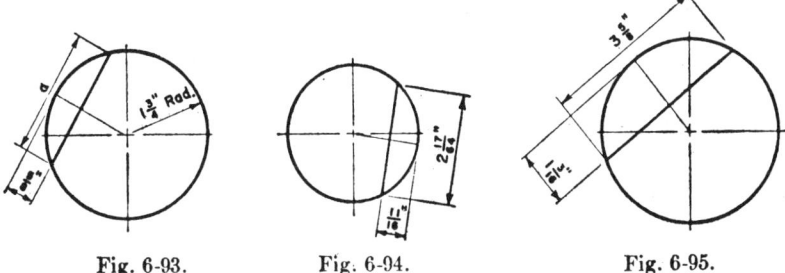

Fig. 6-93.　　　Fig. 6-94.　　　Fig. 6-95.

4. Find the diameter of a circle whose circumference is equal to $13\frac{3''}{8}$.
5. Find the diameter of a circle whose circumference is equal to $3'\ 4\frac{1''}{2}$.
6. Find the length of chord a in Fig. 6-93.
7. Find the radius of the circle in Fig. 6-94.
8. Find the diameter of the circle in Fig. 6-95.

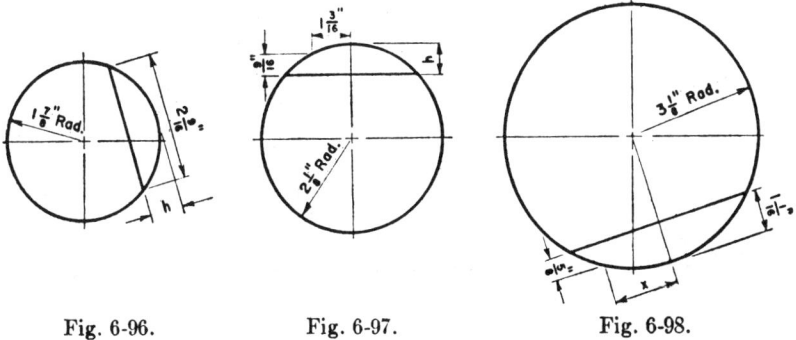

Fig. 6-96.　　　Fig. 6-97.　　　Fig. 6-98.

9. Find the height h of the chord in Fig. 6-96.
10. Find the height h of the chord in Fig. 6-97.
11. Find the distance x in Fig. 6-98.

12. Find the distance y in Fig. 6-99.
13. Find the length L of the arc in Fig. 6-100.

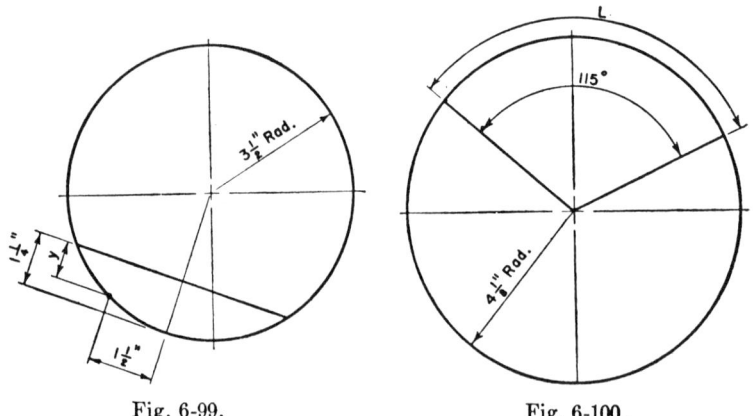

Fig. 6-99. Fig. 6-100.

14. Find the angle θ in Fig. 6-101.
15. Find the length l of the arc in Fig. 6-101, if $\angle \phi = 50°$.
16. Bisect line AB in Fig. 6-102 by geometric construction. Choose any length.

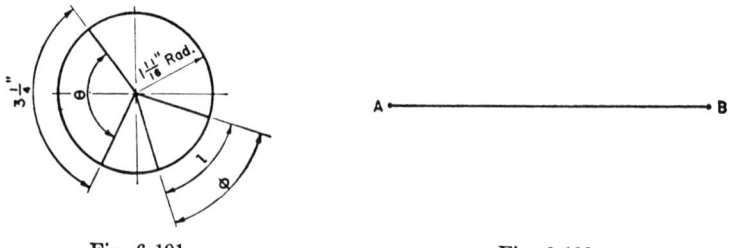

Fig. 6-101. Fig. 6-102.

17. Bisect the circular arc AB in Fig. 6-103. Scale: Half size.

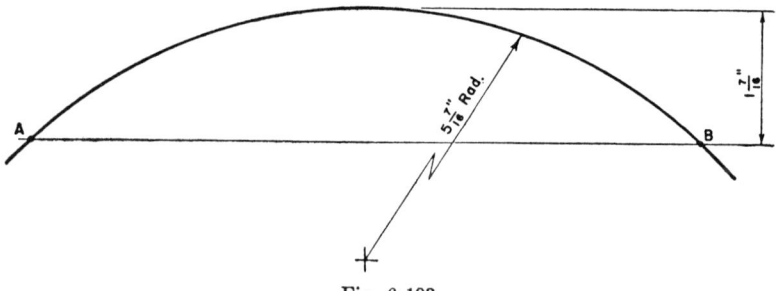

Fig. 6-103.

GEOMETRY

18. Bisect angle ABC in Fig. 6-104. Scale: Half size.

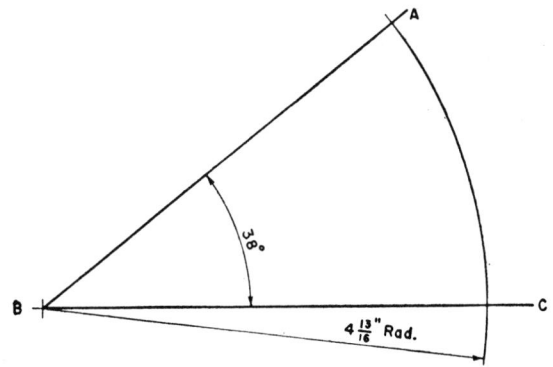

Fig. 6-104.

19. Draw a line parallel to line AB, $2\frac{1}{16}''$ from it, as shown in Fig. 6-105. Scale: Half size. Use two methods.

20. Erect a perpendicular to line AB at point C, as shown in Fig. 6-106.

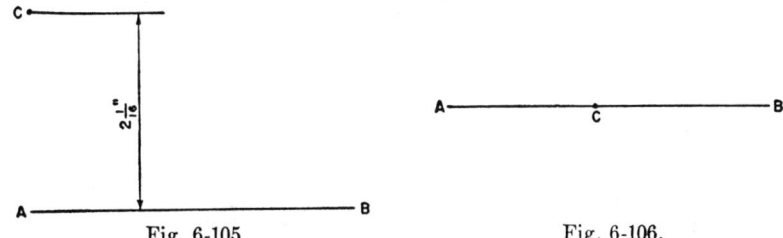

Fig. 6-105.   Fig. 6-106.

21. Erect a perpendicular to line AB from point C, as shown in Fig. 6-107. Scale: Half size.

22. Erect a perpendicular to line AB at its extremity A, as shown in Fig. 6-108.

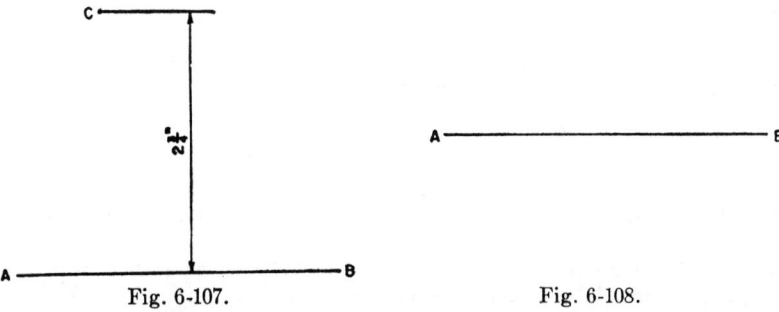

Fig. 6-107.   Fig. 6-108.

23. Locate the center of circular arc *ABC*, as shown in Fig. 6-109. Choose any circular arc.
24. Construct a square whose side is equal to $1\frac{1}{4}''$. Scale: Full size.
25. Construct a 45° angle.
26. Construct a 60° angle.
27. Construct a 30° angle.
28. Construct an angle geometrically on line *DE* at point *D* equal to ∠*ABC* in Fig. 6-110.

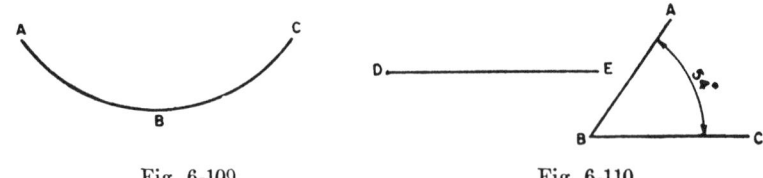

Fig. 6-109.   Fig. 6-110.

29. Inscribe a hexagon in a circle whose diameter is equal to $4\frac{7}{16}''$. Scale: Half size.
30. Circumscribe a hexagon about the circle in Problem 29.
31. Construct an ellipse whose major and minor axes are $3\frac{7}{16}''$ and $1\frac{13}{16}''$, respectively. Scale: Full size.
32. Construct the ellipse in Problem 31 by means of a strip of stiff paper. Scale: Full size.
33. Construct a helix on the surface of a cylinder whose diameter is equal to $1\frac{5}{8}''$ and the lead = $2\frac{1}{8}''$. Scale: Full size.
34. Construct an involute curve whose base circle = $1\frac{5}{8}''$ diameter. Use half of the base circle for the development of the involute. Scale: Full size.
35. Divide a line $4\frac{27}{64}''$ long into seven equal parts by geometric construction, and check the divisions with a rule. Scale: Full size.
36. Draw a circular arc through any two given points with a radius equal to $1\frac{9}{16}''$. Scale: Full size.
37. Draw a circle passing through any three given points.
38. Draw a tangent line to the circular arc through a given point *A* on this arc, as shown in Fig. 6-111. Scale: Full size.
39. Draw two tangent lines to the circle shown in Fig. 6-112 from the given point *A*. Scale: Full size.

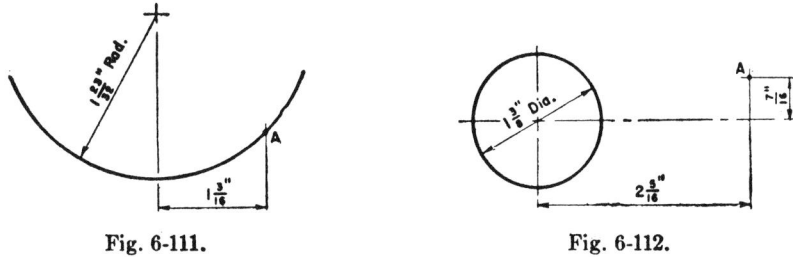

Fig. 6-111. Fig. 6-112.

40. Draw three circles between the two given inclined lines shown in Fig. 6-113. The circles are to be tangent to the given lines and to

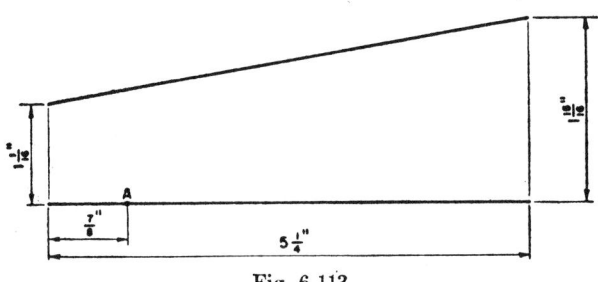

Fig. 6-113.

each other. The first circle is tangent to one of the given lines at point A. Scale: Full size.
41. Construct a triangle on base AB, as shown in Fig. 6-114. The sides of the triangle are to be AC and BC. Scale: Full size.
42. Construct a parallelogram in Fig. 6-115 if the sides and one of the angles are given. Scale: Full size.

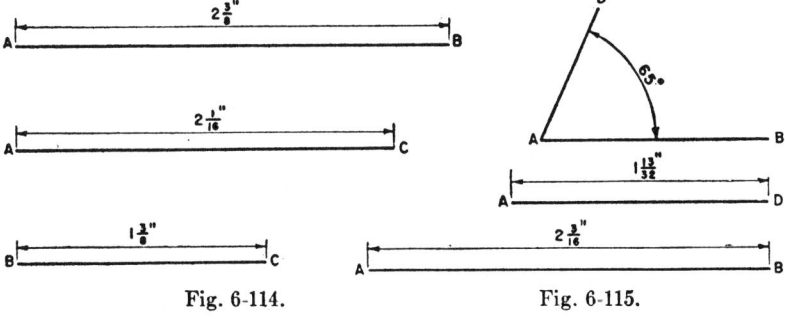

Fig. 6-114. Fig. 6-115.

43. Circumscribe a circle about the triangle ABC shown in Fig. 6-116. Scale: Full size.
44. Inscribe a circle in the triangle ABC shown in Fig. 6-117. Scale: Full size.

45. Circumscribe a circle about the square shown in Fig. 6-118. Scale: Full size.

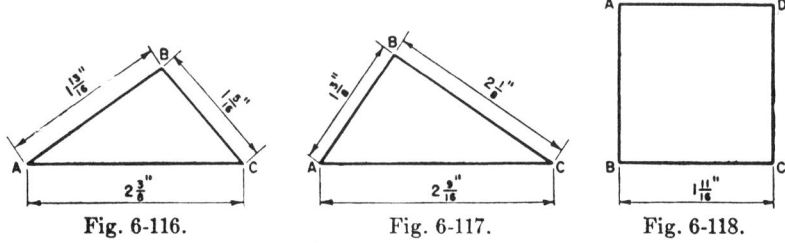

Fig. 6-116.   Fig. 6-117.   Fig. 6-118.

46. Inscribe a circle in the square shown in Fig. 6-119. Scale: Full size.
47. Circumscribe a square about the circle shown in Fig. 6-120. Scale: Full size.
48. Inscribe a pentagon in the circle shown in Fig. 6-121. Scale: Full size.

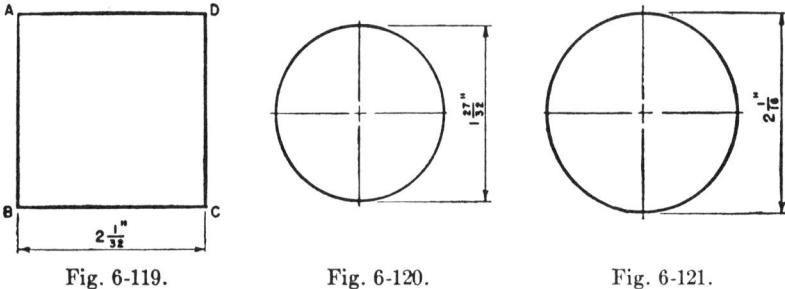

Fig. 6-119.   Fig. 6-120.   Fig. 6-121.

49. Construct an octagon on the base shown in Fig. 6-122. Scale: Full size.
50. Inscribe an octagon in the circle shown in Fig. 6-123. Scale: Full size.
51. Circumscribe an octagon about the circle shown in Fig. 6-124. Scale: Full size.

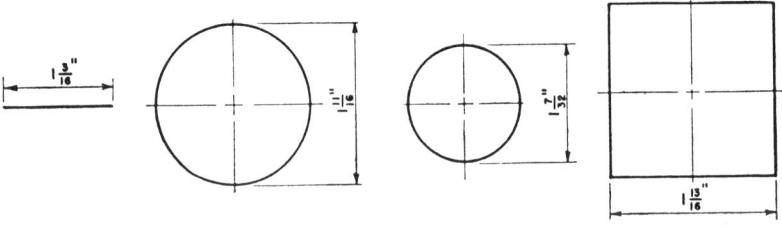

Fig. 6-122.   Fig. 6-123.   Fig. 6-124.   Fig. 6-125.

GEOMETRY

52. Construct an octagon in the square shown in Fig. 6-125. Scale: Full size.
53. Find the area of a triangle whose base = $3\frac{7}{32}''$ and altitude = $1\frac{13}{16}''$.
54. Find the altitude and area of an equilateral triangle whose base = $2\frac{5}{16}''$.
55. Find the area of a square whose side = $1\frac{17}{64}''$.
56. Find the area of a rectangle whose base = $4\frac{3}{16}''$ and altitude = $2\frac{7}{32}''$.
57. Find the area of a parallelogram whose base = $6\frac{1}{4}''$ and altitude = $3\frac{5}{8}''$.

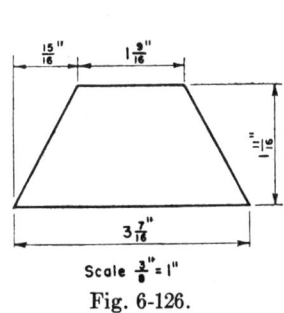

Fig. 6-126.

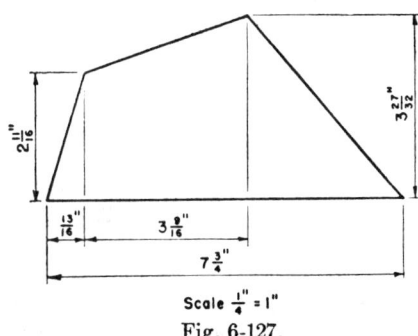

Fig. 6-127.

58. Find the area of the trapezoid shown in Fig. 6-126.
59. Find the area of the trapezium shown in Fig. 6-127.
60. Find the area of a regular hexagon whose side = $1\frac{5}{8}''$.
61. Find the area of a circle whose diameter = $3\frac{17}{32}''$.
62. Find the area of the hollow circle shown in Fig. 6-128.
63. Find the area of the sector shown in Fig. 6-129. Also find the length L of the circular arc.
64. Find the area of the segment shown in Fig. 6-130. Use the approximate method.

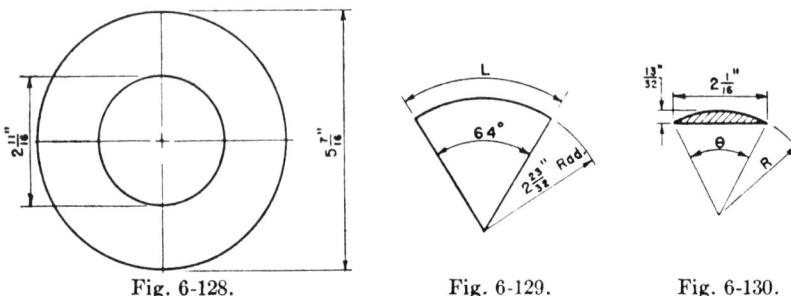

Fig. 6-128. Fig. 6-129. Fig. 6-130.

Note: Make a full-size layout and determine radius R by scaling and angle θ by use of the protractor.

65. Find the area of the sector shown shaded in Fig. 6-131.
66. Find the area of the fillet shown in Fig. 6-132.
67. Find the area of the parabola shown in Fig. 6-133

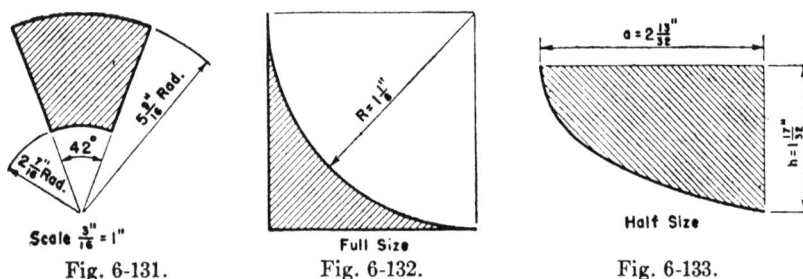

Fig. 6-131. Fig. 6-132. Fig. 6-133.

68. Find the perimeter and area of the ellipse shown in Fig. 6-134.
69. Find the volume and the total area of the surface of the cube shown in Fig. 6-135.
70. Find the volume and total area of the surface of the parallelepiped shown in Fig. 6-136.

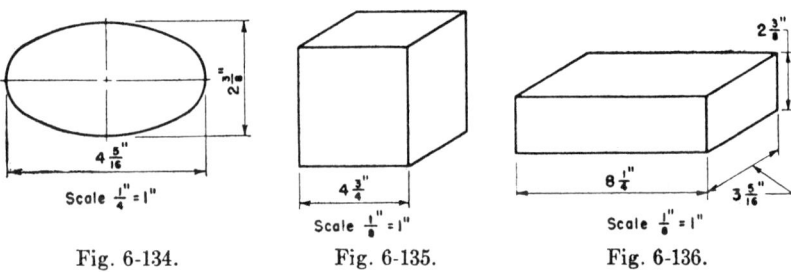

Fig. 6-134. Fig. 6-135. Fig. 6-136.

GEOMETRY

71. Find the volume and total area of the surface of the right regular prism shown in Fig. 6-137. The base of the prism is an equilateral triangle whose side $= 2\frac{7}{16}''$.
72. Find the area of the right section and the volume of the prism shown in Fig. 6-138.

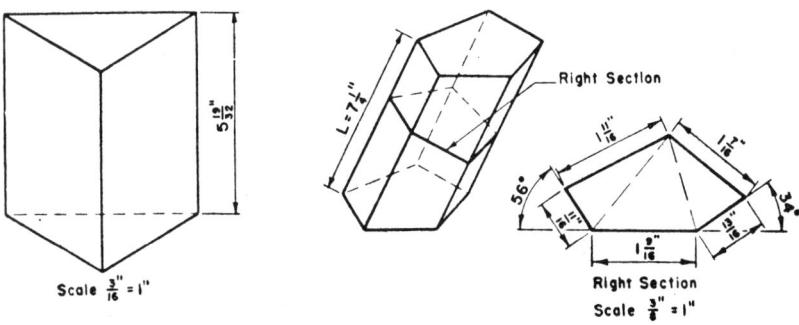

Fig. 6-137.   Fig. 6-138.

Note: Calculate the area of the right section by making a full-size layout, then estimate the area of each triangle by scaling its base and its altitude.

73. Find the volume of the cone shown in Fig. 6-139.
74. Find the volume of the frustum of the cone shown in Fig. 6-140.
75. Find the volume and total area of the surface of the cylinder shown in Fig. 6-141.

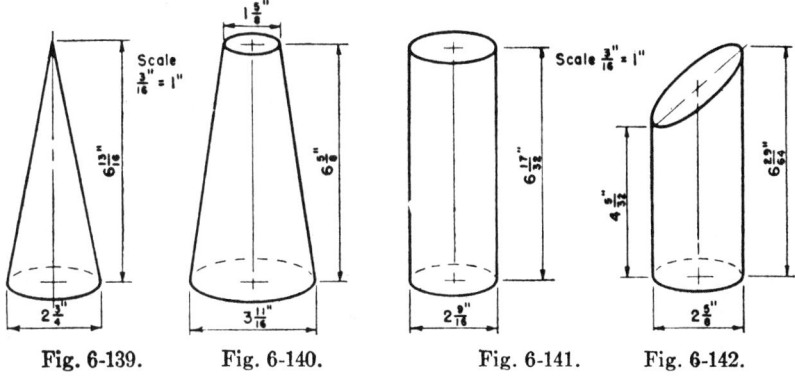

Fig. 6-139.   Fig. 6-140.   Fig. 6-141.   Fig. 6-142.

76. Find the volume of the frustum of the right circular cylinder shown in Fig. 6-142.
77. Find the volume of a sphere whose diameter $= 1\frac{7}{8}''$.

78. Find the volume of the spherical sector shown in Fig. 6-143.
79. Find the volume of the spherical segment shown in Fig. 6-144.
80. Find the volume of the spherical zone shown in Fig. 6-145.

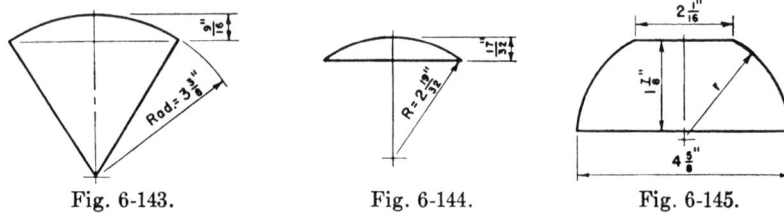

Fig. 6-143.   Fig. 6-144.   Fig. 6-145.

81. Find the volume of the torus shown in Fig. 6-146.
82. Find the volume of the solid material in the hollow cylinder shown in Fig. 6-147.

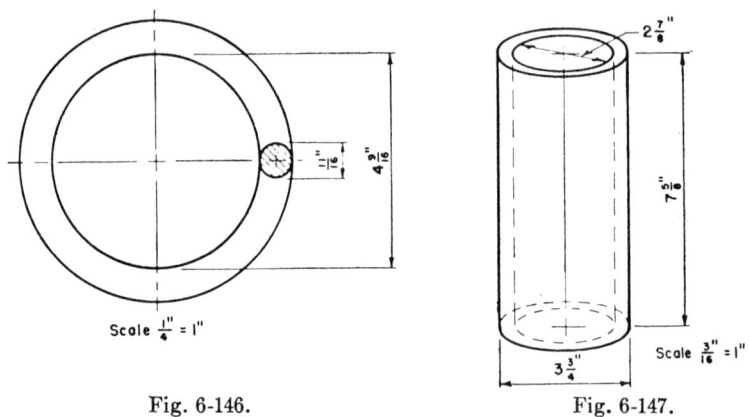

Fig. 6-146.   Fig. 6-147.

83. Estimate the weight of the copper rivet shown in Fig. 6-148.
84. Estimate the weight of the aluminum detail shown in Fig. 6-149.

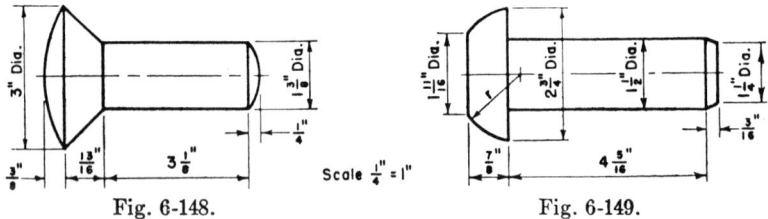

Fig. 6-148.   Fig. 6-149.

85. Estimate the weight of the hollow rivet made from wrought iron and shown in Fig. 6-150.

GEOMETRY

86. Find the weight of the steel part shown in Fig. 6-151.

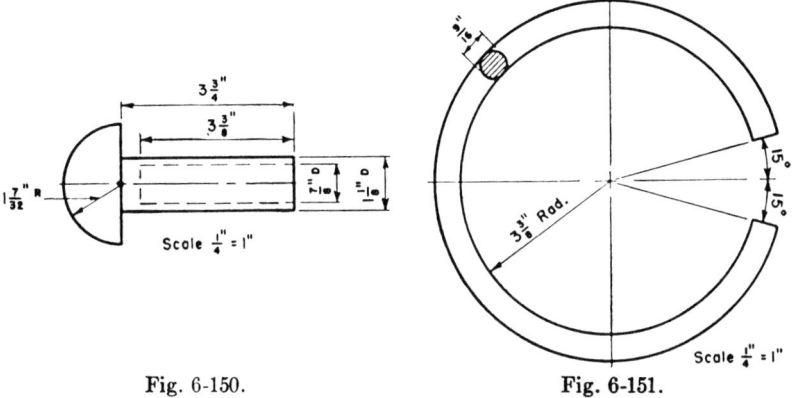

Fig. 6-150.   Fig. 6-151.

87. Find the weight of the cast-iron wheel shown in Fig. 6-152.

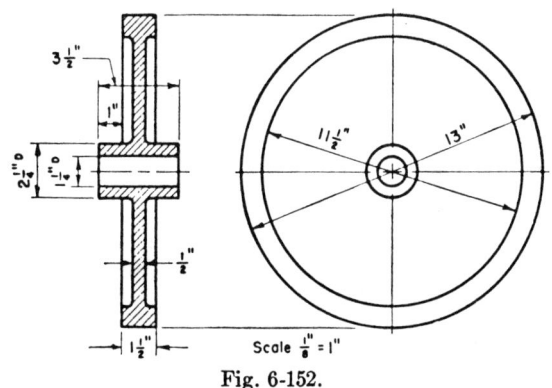

Fig. 6-152.

88. Find the minor axis of the ellipse in Fig. 6-153.

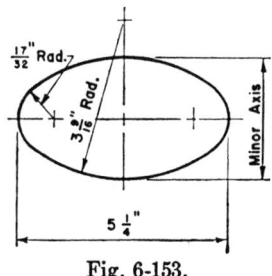

Fig. 6-153.

89. Determine graphically the diameter of the bolt circle in Fig. 6-154.

230 MATHEMATICS FOR INDUSTRY

90. Calculate the diameter of the disk whose portion is shown in Fig. 6-155.

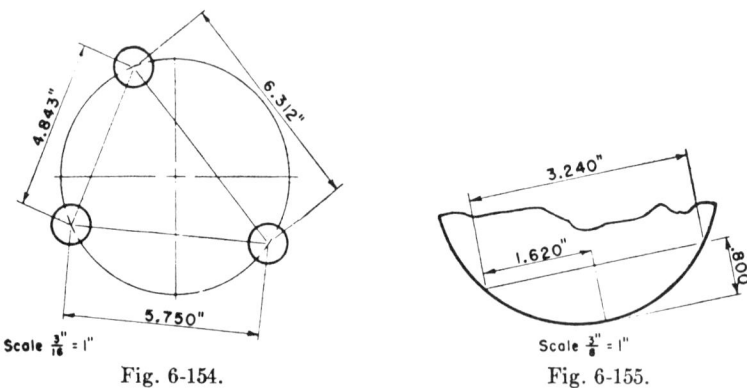

Fig. 6-154. Fig. 6-155.

91. Calculate the diameter of the bored hole in casting shown partially in Fig. 6-156.

92. Calculate the weight of the steel plate shown in Fig. 6-157.

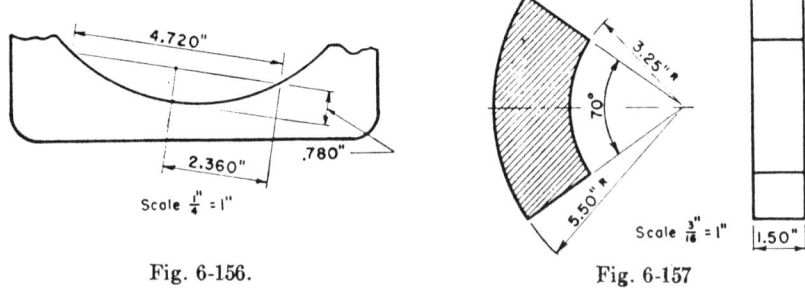

Fig. 6-156. Fig. 6-157.

93. Calculate the diameter of the disk, the set-up of which is shown in Fig. 6-158.

94. Calculate the diameter of the part set up as shown in Fig. 6-159.

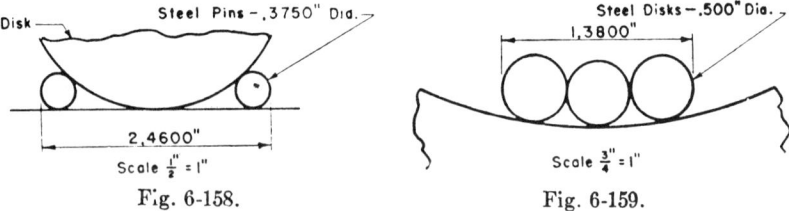

Fig. 6-158. Fig. 6-159.

95. Construct a triangle in which the bisector of the 52° angle opposite the base = $3\frac{11''}{16}$, and the altitude = 3″.

GEOMETRY

96. Locate the heights of the following points above line AB, as shown in Fig. 6-160: Points 1, 2, 3, 4, 5, and 6.

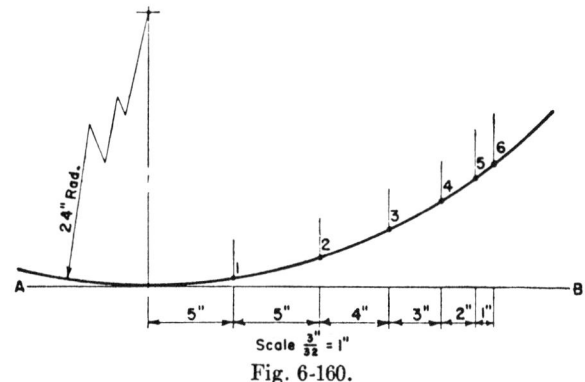

Fig. 6-160.

97. Construct the approximate length of arc CD on line AB, as shown in Fig. 6-161. Scale: Full size.

98. Draw a common tangent to the circles shown in Fig. 6-162. Scale: Full size.

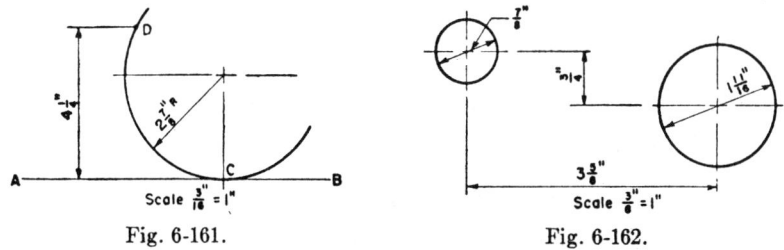

Fig. 6-161.   Fig. 6-162.

99. Draw a circle through the given point A, touching the two given circles, as shown in Fig. 6-163. Scale: Full size.

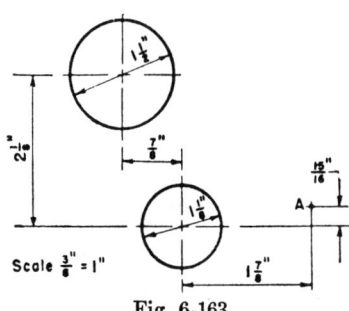

Fig. 6-163.

CHAPTER VII
TRIGONOMETRY

Trigonometry is a study of a branch of mathematics based on geometry but using the tools of algebra. Numbers are used, as in arithmetic, and algebraic equations are constructed. Many geometric problems can be solved by means of trigonometry with the smallest expenditure of time and effort.

The word *trigonometry* is derived from the Greek and means "triangle" and "measurement." From this it would seem that trigonometry is only the study of triangles. However, the subject also includes many other investigations involving angles.

The practical application of trigonometry is simple, and large use is made of tables of functions giving various quantities. These quantities are chosen to meet definite conditions and are multiplied by dimensions specified on the working drawings, thus enabling the mechanic or draftsman to use the quantities in a convenient manner.

Much of the information conveyed by the draftsman is difficult to apply as processing gets under way in the shop, therefore trigonometry is found a useful tool in that it offers the simplest solution of many troublesome problems. It is not, however, to be looked upon as a miracle worker, as definite information is necessary before attempting solution of the problem by any means whatsoever.

TRIGONOMETRIC FUNCTIONS AND TABLES

It has been shown in preceding chapters dealing with geometry that certain relations exist between the sides of a triangle. In this chapter it will be shown that angles are also related to the sides of a triangle. When constructing various angles or solving triangles or other geometric figures, the relations between the angles and sides of a triangle enter the problem as factors to be considered.

Trigonometric functions are the functions of an angle. We shall explain the trigonometric functions and show how they can be employed to solve the right triangle.

Let us draw an angle of less than 90°, such as ABC in Fig. 7-1. Erect a perpendicular from any point D on AB to BC, where it intersects at point E. A right triangle (DBE) is formed in which a certain ratio exists between DE and BE, written $\dfrac{DE}{BE}$.

TRIGONOMETRY

If another point, F, is taken on line AB and a perpendicular, FG, is dropped on BC, another right triangle, FBG, is formed, similar to triangle DBE. From the similarity of the two right triangles, the following can be stated: $\dfrac{FG}{BG} = \dfrac{DE}{BE}$.

A number of points may be taken on AB and corresponding perpendiculars dropped on BC, creating several ratios equal to $\dfrac{DE}{BE}$, as long as angle ABC is not changed from the value shown in Fig. 7-1.

Therefore, the ratio $\dfrac{DE}{BE}$ is a function of angle ABC and is called the *tangent* of angle ABC—designated tan $\angle ABC$.

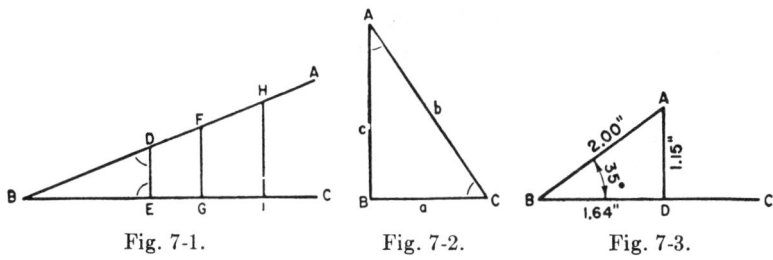

Fig. 7-1.   Fig. 7-2.   Fig. 7-3.

In a similar way, ratios between other sides of the right triangle DBE can be formed that will be equal to corresponding ratios of similar right triangles, such as FBG.

For example, the ratio $\dfrac{DE}{BD} = \dfrac{FG}{BF} = \dfrac{HI}{BH}$, since right triangles DBE, FBG, and HBI are similar.

The ratio $\dfrac{DE}{BD}$ is another function of angle ABC and is called the *sine* of angle ABC—designated, sin $\angle ABC$.

Since a triangle has three sides, six ratios can be taken and designated as functions of a particular angle, such as angle ABC. Another six functions of angle BDE may be taken in triangle DBE. Note $\angle BDE$ = complement $\angle ABC$. Of course, the functions of right angle DEB are not practical as it appears in Fig. 7-1.

The trigonometric relations or functions of an angle in a right triangle are enumerated as follows: sine, cosine, tangent, cotangent, secant, and cosecant, and abbreviated, sin, cos, tan, cot, sec, and csc.

For convenience, the angles are designated by capital letters, and the sides of the triangle opposite the angles are designated by corresponding small letters. It is also helpful to designate the sides by special names. For example, in the triangle ABC of Fig. 7-2, side a is the opposite side with respect to angle A, side c is the adjacent side with respect to angle A, and side b is the hypotenuse as usual.

The six trigonometric functions of angle A and angle C of Fig. 7-2 are as follows:

$$\sin A = \frac{a}{b} = \frac{\text{opposite side}}{\text{hypotenuse}} \qquad \sin C = \frac{c}{b}$$

$$\cos A = \frac{c}{b} = \frac{\text{adjacent side}}{\text{hypotenuse}} \qquad \cos C = \frac{a}{b}$$

$$\tan A = \frac{a}{c} = \frac{\text{opposite side}}{\text{adjacent side}} \qquad \tan C = \frac{c}{a}$$

$$\cot A = \frac{c}{a} = \frac{\text{adjacent side}}{\text{opposite side}} \qquad \cot C = \frac{a}{c}$$

$$\sec A = \frac{b}{c} = \frac{\text{hypotenuse}}{\text{adjacent side}} \qquad \sec C = \frac{b}{a}$$

$$\csc A = \frac{b}{a} = \frac{\text{hypotenuse}}{\text{opposite side}} \qquad \csc C = \frac{b}{c}$$

It can be noted from the foregoing tabulation that the cosecant of an angle is the reciprocal of its sine, the secant of an angle is the reciprocal of its cosine, and the cotangent of an angle is the reciprocal of its tangent.

Note: In a right triangle, both acute angles are complementary to each other. Since the sum of all angles in any triangle is equal to 180°, the sum of both acute angles is equal to 90°.

Functions and Cofunctions. In the right triangle of Fig. 7-2, it can be seen that the sine of the acute angle A and the cosine of acute angle C are the same. We also noted that angle A and angle C are complementary to each other. The term *cofunction* refers to the function of the complementary angle and the word *cosine* is a contraction from "sine of the complement."

The six trigonometric functions may be grouped in pairs: sine and cosine; tangent and cotangent; secant and cosecant. Not only is the sine of one acute angle equal to the cosine of its complement, but the tangent of one acute angle is equal to the cotangent of its complement and the secant of one angle is equal to the cosecant of its complement. In other words, in each pair either function is a cofunction of the other and the function of one acute angle is equal to the cofunction of its complement. For example, the sin of 40° = cos of 50°; the tan of 35° = cot of 55°; the sec of 52° = csc of 38°.

Trigonometric Functions Found by Construction and Measurement. Trigonometric functions can be found by construction and measurement. For example, let us find the sine, cosine, and tangent of 35°.

Construct an angle $ABC = 35°$, as shown in Fig. 7-3. Measure a convenient distance $AB = 2.0''$, and drop a perpendicular AD on BC. Measure lines AD and BD, and find them to be $1.15''$ and $1.64''$, respectively.

Then the sin of 35° = $\dfrac{AD}{AB} = \dfrac{1.15}{2.00} = .575$

the tan of 35° = $\dfrac{AD}{BD} = \dfrac{1.15}{1.64} = .70$

the cos of 35° = $\dfrac{BD}{AB} = \dfrac{1.64}{2.00} = .82$

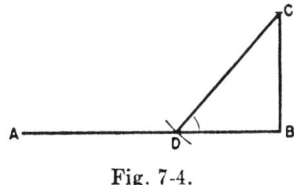

Fig. 7-4.

In a similar manner, the required angle may be constructed if one of its functions is known. For example, if the sine of an angle is $.575''$, a line AB is drawn, as shown in Fig. 7-4.

Erect a perpendicular, BC, at one of the extremities of line AB. Lay off $BC = .575''$, and with C as a center and a radius equaling $1''$, draw an arc intersecting AB at point D. Join C and D and obtain the required angle $CDB = 35°$.

Trigonometric Functions Found by Tables. The foregoing methods for finding trigonometric functions and constructing angles are not fully reliable with respect to accuracy, and, in laying out the desired distances, precision is essential. Even with great care, the accuracy of the results obtained may, at best, be within two decimal places. This being the case, the prepared tables containing trigonometric functions of all angles within five or more decimal places constitute a distinct aid. Application of these functions in solving triangles is here presented.

Note: Tables containing Natural Trigonometric Functions can be found in the Appendix of this text.

In the tables of natural trigonometric functions the angles in **degrees** are indicated at the top of each page, beginning with 0° and 179° on the first page and concluding with 44° and 135° on the last page. The angles in degrees are also indicated at the bottom of each page, beginning with 90° and 89° on the first page and concluding with 134° and 45° on the last page. The extreme left column of each page shows the minutes from 0 to 60 to be used with angles less than 45°, and from 90° up to 135°; the extreme right column shows the minutes from 0 to 60 to be used with angles greater than 45° and up to 90°, and from 135° up to 180°. Note that the minute column at the left starts at the top and the one at the right starts at the bottom.

The reason for arranging the tables as explained above is that the function of an acute angle is equal to the cofunction of its complement, therefore, for every function there is an equal cofunction, and it is only necessary to prepare tables for angles up to 45°.

The function of any angle greater than 45° is equal to the cofunction of its complement, which is less than 45°, and if the sine column reads from top to bottom, the same column will read cosine from bottom to top. Identical reasoning applies to the other functions, such as the cosine, tangent, etc.

For example, if the sine of 32° 16′ = .53386, the cosine of 57° 44′ = .53386. Thirty-two (32) degrees is found at the top of the page, and 16′ in the extreme left column reading downward. Fifty-seven (57) degrees is found at the bottom of the same page, and 44′ in the extreme right column reading upward.

Summing up, then, the trigonometric functions of all angles up to 45° are found in the columns reading downward, and those of all angles greater than 45° are contained in the columns reading upward, except for angles between 90° and 180°, which are placed as just described.

Examples (from the tables of Natural Trigonometric Functions in the Appendix):
1. Find the tangent of 18° 46′. This angle is less than 45°, and its tangent is found in the column of tangents reading downward in line with 46′ in the first column to the left, above which the figure 18° appears. The tangent of 18° 46′ = .33978.
2. The sine of 41° 17′ = .65978.
3. The sine of 67° 27′ is read from the bottom of the *Sine* column in line with 27′ in the extreme right column of the page, at the bottom of which is written 67°. The sine of 67° 27′ = .92354.
4. The tan of 56° 48′ = 1.5282.
5. The cos of 29° 38′ = .86921.
6. The cot of 61° 54′ = .53395.

TRIGONOMETRY 237

7. The sec of 28° 11′ = 1.1345.
8. The csc of 74° 24′ = 1.0382.
9. The sec of 51° 7′ = 1.5930.

If the trigonometric function of the angle is known, the angle itself may be determined from the tables, the process being merely reversed.

Examples:
1. If the tangent of an angle is .37455, the angle proper may be found by locating the function in the column headed *Tan* on the page listing functions of 20°. If in doubt as to whether **the required** angle is smaller or greater than 45°, the given function **may be located** in either the *Tan* or the *Cotan* column. If it is found in **the *Cotan*** column, reading downward, it is evident that the required **angle is** greater than 45°, and the column reading *Cotan* downward is read upward as *Tan*.
2. If the sine of an angle is .88444, this function may be located in the column of cosines, reading downward, which is an indication that the required angle is greater than 45°. This column reading upward is the *Sine* column, and the angle corresponding to this function is 62° 11′.
3. If the secant of an angle is 1.1282, the angle proper is 27° 35′.
4. If the cotangent of an angle is .56347, the angle proper is 60° 36′.

Interpolation. The tables of trigonometric functions in the Appendix, like the tables in many handbooks and reference books, are compiled for angles expressed in degrees and minutes. By a process called *interpolation* the same set of tables may be used to find functions of angles measured to seconds. Conversely, a fairly accurate angle may be computed with reference to the given function. By using the process of interpolation we assume that the change in the function is proportional to the change in the angle. Although this is not true if a large change is made—if an angle is doubled, its sine does not double—it is very nearly true for small changes in the angle.

Example 1: If it is required to compute the sine of 24° 37′ 24″, we find in the table the sine of 24° 38′, which is equal to .41681. The next step consists of finding the sine of 24° 37′, which is equal to .41654. The difference between the functions of the two angles is .41681 − .41654 = .00027.

The given angle is 24° 37′ 24″, and the sine of this angle can be computed or interpolated by adding to .41654 the amount equal to

$$.000\cancel{27}^{9} \times \frac{\cancel{24}^{6}}{\cancel{60}_{\cancel{20}_{5}}} = \frac{.00054}{5} = .00011,$$

238   MATHEMATICS FOR INDUSTRY

since $\frac{24}{60}$ is the ratio between 24 seconds and 60 seconds and .00027 is the increase in the function of the two angles which differ by 1 minute. Hence the sine of 24° 37′ 24″ = .41654 + .00011 = .41665.

Example 2: The tan of 36° 42′ 51″ is equal to the tan of 36° 42′ + the increase in the function corresponding to 51″. This can be shown as follows:

$$\tan 36° 43' = .74583$$
$$\tan 36° 42' = .74538$$
$$\overline{.00045}$$

$$.00045 \times \frac{51}{60} = \frac{.00153}{4} = .00038$$

$$\tan 36° 42' 51'' = .74538 + .00038 = .74576$$

The tangent of this angle may be computed by subtracting from the tangent of 36° 43′ an amount equal to

$$.00045 \times \frac{60-51}{60} = .00045 \times \frac{9}{60} = .00007; .74583 - .00007 = .74576.$$

This value is the same as that calculated by the previous method.

Example 3: The cot of 24° 17′ 23″ is found as follows:

$$\cot 24° 17' = 2.2165$$
$$\cot 24° 18' = 2.2147$$
$$\overline{.0018}$$

In this example the cotangent decreases with the increase in the angle, hence the cotangent of 24° 17′ 23″ is smaller than the cotangent of 24° 17′ by the amount of $.0018 \times \frac{23}{60} = .0007$.

$$\cot 24° 17' 23'' = 2.2165 - .0007 = 2.2158$$

Note: The cofunctions, such as cos, cot, and csc decrease with the increase in the angle, and conversely, increase with the decrease in the angle.

Example 4: Find the sec of 78° 28′ 41″.

$$\sec 78° 29' = 5.0087$$
$$\sec 78° 28' = 5.0015$$
$$\overline{.0072}$$

$$.0072 \times \frac{41}{60} = .0049$$

$$\sec 78° 28' 41'' = 5.0015 + .0049 = 5.0064$$

Note: The tan and cot of 45° = 1.0000.

TRIGONOMETRY

Trigonometric Functions Found by Graphical Method. A graphical method illustrating each trigonometric function is shown in Fig. 7-5.

A circle is drawn with a radius equal to 1 (1 in., 1 ft., 1 cm., or any other unit of measurement). The six functions of angle α can be found directly by measuring the lengths of the following lines: $AB, OB, DC, EF, OC,$ and OE.

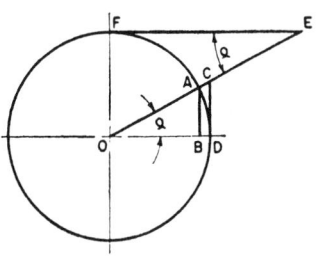

Fig. 7-5.

Since trigonometric functions are ratios of the sides of a right triangle containing the required angle, and ratios may be expressed in the form of indicated divisions or common fractions, the above lines AB, OB, etc., can be written in the form of numerators, and the denominators in each case are the lengths of the radius of the circle, which is always equal to 1.

Tabulation of all functions is as follows:

$$\sin \alpha = \frac{AB}{OA} = \frac{AB}{1} = AB$$

$$\cos \alpha = \frac{OB}{OA} = \frac{OB}{1} = OB$$

$$\tan \alpha = \frac{CD}{OD} = \frac{CD}{1} = CD$$

$$\cot \alpha = \frac{FE}{OF} = \frac{FE}{1} = FE$$

$$\sec \alpha = \frac{OC}{OD} = \frac{OC}{1} = OC$$

$$\csc \alpha = \frac{OE}{OF} = \frac{OE}{1} = OE$$

$$\angle OEF = \angle EOD = \angle \alpha$$

Of course, the graphic method for finding the functions shown in Fig. 7-5 is rather cumbersome, and accuracy of the results depends upon precision of measurement of the respective lines representing corresponding functions. But the necessity for getting a clear picture of the principle and magnitude of each function is realized quickly by analysis of Fig. 7-5.

SOLUTION OF PROBLEMS INVOLVING RIGHT TRIANGLES

Every triangle has six parts or elements—three sides and three angles. If three of the elements are given, one of which is a side, the other three elements can be found. The solution of the triangle is the finding of the unknown elements.

Any triangle can be solved by the use of trigonometric functions if enough information is available, but in this section we shall take up first the solution of problems directly involving right triangles and then problems which do not always appear in the form of right triangles.

Solution of Right Triangles. The information needed for solution of a right triangle will consist, at the least, of two known elements, since there is always one known angle equal to 90°. These two known, or given, elements must include at least one side, because if no side of the triangle is known, definite solution of the problem is not possible.

The methods of solution of right triangles may be divided into two classes: (1) triangles in which two sides are known; (2) triangles in which any one side and any one angle are known.

Method used when two sides are known. The following example illustrates the method used when two sides of the triangle are known.

Given: A right triangle, ABC, as shown in Fig. 7-6, in which the length of $AC = 6.324''$ and the length of $BC = 2.817''$.

Required: To find the length of AB and the angles of A and C.

Procedure: The length of AB may be found by a geometric method shown before, namely, $AB = \sqrt{(AC)^2 - (BC)^2}$, but the use of trigonometric functions makes solution of the problem less cumbersome. Before solving for the length of AB, the angle A or C must be found.

In order to find $\angle A$, we must know any one of its several functions. In this case, BC is known, and it is the side opposite $\angle A$. AC is the hypotenuse, and it is also known. Thus, the ratio of the side opposite and the hypotenuse may be found, and this ratio is the sine of $\angle A$. The expression is written: $\dfrac{BC}{AC} = \dfrac{2.817}{6.324} = .44545 = \sin \angle A$.

The nearest angle corresponding to sin .44545 is 26° 27'. The sin 26° 27' = .44542 (from the tables of functions), and it is close to the sine computed. However, if the result must be precise, the angle may be interpolated by the method shown previously.

TRIGONOMETRY

sin 26° 28' = .44568   computed sin = .44545
sin 26° 27' = .44542   sin 26° 27' = .44542
difference = .00026   difference   = .00003

The angle increment to be added to 26° 27' = $\frac{3}{26} \times 60 = \frac{90}{13} = 7''$, or very near. The required angle $A = 26° 27' 7''$.

$\angle C$ = complement of $\angle A = 90° - 26° 27' 7'' = 63° 32' 53''$.

The length of AB may be computed as follows:

AB is the side adjacent to $\angle A$. Therefore, $\frac{AB}{AC} = \frac{AB}{6.324} = \cos \angle A = \cos 26° 27' 7'' = \cos 26° 27' -$ (the value due to the increment 7''). Cos 26° 27' 7'' = .89532 − (.89532 − .89519) × $\frac{7}{60}$ = .89532 − $\left(.0001 \times \frac{7}{60}\right)$ = .89532 − .00001 = .89531. $\frac{AB}{6.324}$ = .89531. $AB = 6.324 \times .89531 = 5.66194$.

Because of the fact that AC and BC are expressed within three decimal places, AB may also be expressed within three decimal places, namely, 5.662''.

Method used when any one side or any one angle are known. The example that follows demonstrates the method of solving a right triangle in which any one side and any one angle of the triangle are given. See Fig. 7-7.

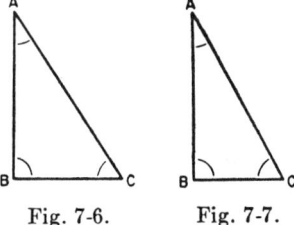

Fig. 7-6.   Fig. 7-7.

Given: Assume $AB = 7.832''$ and $\angle A = 34° 47'$.

Required: Find $\angle C$ and the lengths of BC and AC.

Procedure: $\angle C$ = complement of $\angle A = 90° - 34° 47' = 55° 13'$.

$\frac{AC}{AB} = \sec A, \frac{AC}{7.832} = \sec 34° 47' = 1.2175$. $AC = 7.832 \times 1.2175 = 9.53546$, say 9.535'', since AB is also within three decimal places.

$$\frac{BC}{AB} = \tan A$$

$$\frac{BC}{7.832} = \tan 34° 47' = .69459$$

$$BC = 7.832 \times .69459 = 5.440''$$

The example illustrated in Fig. 7-7 may be solved by taking functions different from those indicated. For instance, AC can be found by taking the ratio of $\dfrac{AB}{AC}$, which is the cos of angle A, and $AC = \dfrac{AB}{\cos A} \cdot \dfrac{AB}{AC}$ may be taken as the sin of angle C, and $AC = \dfrac{AB}{\sin C}$.

There is in most cases more than one combination that can be used for setting up the ratios necessary for solving the unknown sides and angles of the given right triangle. The example illustrated in Fig. 7-6 may be solved by selecting ratios other than those demonstrated.

A reliable check on the result is obtained by employing different combinations in the manner suggested, and if the value secured for each is about the same, we conclude that the computation is correct. If for any reason a calculated result is in doubt, the triangle can be laid out to scale and the figure obtained can be compared with the data available. Anything wrong will show up immediately in the drawing. One must bear in mind that the hypotenuse is the longest dimension in a right triangle, that the smallest side is opposite the smallest angle, and that an approximate comparison of the angles in the triangle can be made with the familiar angles of 30°, 45°, 60°, and 90°.

Problems Not Appearing in Form of Right Triangles. Problems in actual practice do not always appear in the form of right triangles. The geometric figures and construction may be quite complex, and the ability to solve problems by the use of trigonometric functions depends upon the degree of skill with which one is able to resolve a given problem into right triangles.

The following problems are but a few of the vast number of problems that occur in practice. They have been carefully selected to include the more important and significant features.

Problem 1. See Fig. 7-8.

Given: The bolt circle diameter of the eight (.3125") equally spaced holes is 7.2500".

Required: Accuracy of the spacing and location of the reamed holes is checked by means of steel precision pins inserted in two adjacent holes. A measurement (M) is taken across the two pins and compared with the corresponding dimension, which is computed by applying the principles of trigonometry.

Procedure: $\angle \alpha$ is a central angle subtended by an arc equal to $\dfrac{1}{8}$ of the circumference, therefore it is equal to 45°.

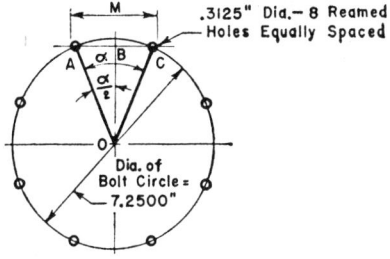

Fig. 7-8.

AC is the corresponding chord and AB is half of the chord.

$\triangle OBA$ is a right triangle in which OA, the hypotenuse, is equal to the radius of the bolt circle $= \dfrac{7.250}{2} = 3.625''$.

$$\frac{\alpha}{2} = \frac{45}{2} = 22.5°$$

$$\frac{AB}{OA} = \frac{AB}{3.625} = \sin\frac{\alpha}{2} = \sin 22.5° = .38268$$

$$AB = 3.625 \times .38268 = 1.3872$$

$$M = 2 \times \left(AB + \frac{.3125}{2}\right) = (2 \times AB) + .3125 =$$

$$(2 \times 1.3872) + .3125 = 3.0869''$$

Problem 2. See Fig. 7-9.

Given: Assume a cylindrical bar of $3\dfrac{1}{4}''$ diameter cut into a hexagon.

Required: To obtain the diameter (D) across the sides.

Procedure: In right triangle OAB, $OB = $ radius $= \dfrac{3.250}{2} = 1.625$.

$$\angle AOB = 30°$$

$$\frac{OA}{OB} = \frac{OA}{1.625} = \cos 30° = .86603$$

$$OA = 1.625 \times .86603 = 1.407$$

$$D = 2 \times OA = 2 \times 1.407 = 2.814''$$

Problem 3. See Fig. 7-10. This problem appears difficult at first but by application of trigonometric principles it can readily be solved.

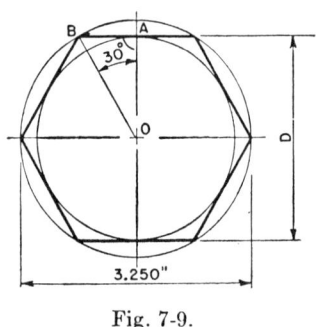

Fig. 7-9.

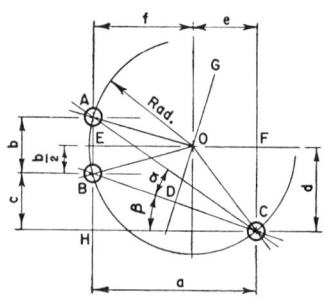

Fig. 7-10.

Given: Three holes spaced as shown: $a = 7.200''$, $b = 2.350''$, and $\angle \alpha = 15° 34'$.

Required: Find dimension c and $\angle \beta$.

Procedure: Bisect BC and draw a perpendicular DG from its mid-point, D. Bisect AB and draw a perpendicular, EF, from its mid-point, E. These two perpendiculars intersect at point O, which is used as a center for the circular arc passing through the centers of the given holes.

From the geometric principle previously explained, $\angle AOB = 2 \angle ACB$, since $\angle AOB$ is a central angle and $\angle ACB$ has its vertex, C, on the circumference of the circular arc, and both angles are subtended by the same arc, AB. $\angle AOE = \angle \alpha = 15° 34'$.

In right triangle AOE, $AE = \dfrac{b}{2} = \dfrac{2.350}{2} = 1.175''$ and $\angle AOE = 15° 34'$, $\therefore \dfrac{AE}{AO} = \sin \angle AOE$ and $AO = \dfrac{AE}{\sin 15° 34'} = \dfrac{1.175}{.26836} = 4.378'' = $ radius.

In the same right triangle, $\dfrac{OE}{AO} = \cos \angle AOE = \cos 15° 34'$, hence $OE = f = AO \times \cos 15° 34' = 4.378 \times .96332 = 4.217''$.

a is given and is equal to $7.200''$, and $e = a - f = 7.200 - 4.217 = 2.983''$.

In right triangle OFC, $OF = e = 2.983$ and $OC = $ radius $= 4.378$, hence

$$d = \sqrt{(OC)^2 - e^2} = \sqrt{(4.378)^2 - (2.983)^2} =$$

$$\sqrt{19.166884 - 8.898289} = \sqrt{10.268595} = 3.204''.$$

$$c = d - \frac{b}{2} = 3.204 - \frac{2.350}{2} = 3.204 - 1.175 = 2.029'', \text{ which is the}$$
required dimension.

In right triangle BCH, dimension $c = 2.029''$ and dimension $a = 7.200''$, hence $\tan \beta = \dfrac{c}{a} = \dfrac{2.029}{7.200} = .28181''$, and $\angle \beta = 15° 44'$, which is the required angle.

Problem 4. See Fig. 7-11.

Given: Fig. 7-11, which is a portion of a shop drawing showing a machine detail.

Required: The detail can be made in accordance with the information on the drawing, but in order to make the machine fixture for the given

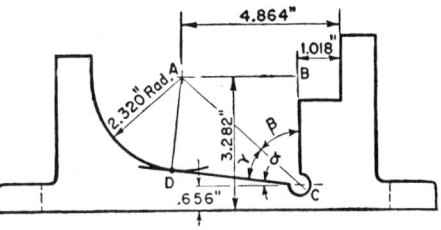

Fig. 7-11.

detail, the toolmaker needs to know the value of $\angle \alpha$ precisely. The tool inspector also needs the angle for precision checking of the completed fixture before it goes to the manufacturing division of the plant.

Procedure: In right triangle ABC, $AB = 4.864 - 1.018 = 3.846''$, $BC = 3.282 - .656 = 2.626''$. Hence $\dfrac{AB}{BC} = \tan \beta = \dfrac{3.846}{2.626} = 1.4646$, and $\angle \beta = 55° 40' 33''$ by interpolation. In the same right triangle, $\dfrac{AC}{AB} = \csc \beta = \dfrac{AC}{3.846} = 1.2108$, and $AC = 3.846 \times 1.2108 = 4.6567$.

In right triangle ACD, $AD = $ Rad. $= 2.320''$ and $AC = 4.6567''$, hence $\dfrac{AD}{AC} = \sin \angle \gamma = \dfrac{2.320}{4.6567} = .49827$, and $\angle \gamma = 29° 52' 55''$.

Finally, $\angle \alpha = 90° - (\beta + \gamma) = 90° - (55° 40' 33'' + 29° 52' 55'') = 4° 26' 32''$.

Problem 5. There are shop problems that can be solved rapidly by application of a simple geometric principle, as illustrated in Fig. 7-12, which shows a common linkage used in automobile and airplane mechanical controls such as clutches and brakes.

Given: F is the frame and bracket G is fastened to it. H is a lever, which pivots about point A in bracket G and is pinned to link I. Link I is pinned to slide J, which moves up and down along the straight line DE situated at a fixed distance (6.250″) from point A. The total movement or stroke of slide J is equal to 2.250″. For good design, the length of lever H is so computed that link I swings a distance, a, equally to the right and left of line DE.

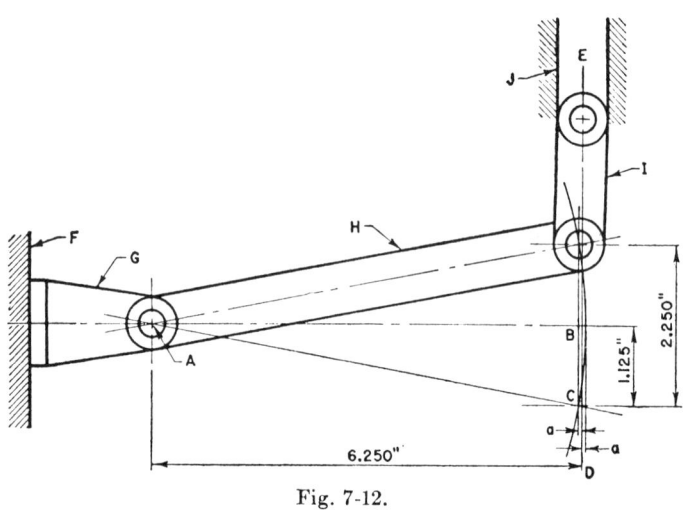

Fig. 7-12.

Required: Find the length of the lever H.

Procedure: Bisect the stroke and obtain the right triangle ABC, in which AC is the length of lever H and $= 6.250 + a$, $AB = 6.250 - a$, and $BC = \dfrac{2.250}{2} = 1.125''$. In this right triangle we have:

$$(AC)^2 = (AB)^2 + (BC)^2$$
$$(6.250 + a)^2 = (6.250 - a)^2 + (1.125)^2$$
$$(6.250)^2 + 12.500a + a^2 = (6.250)^2 - 12.500a + a^2 + (1.125)^2$$
$$12.500a + 12.500a = (1.125)^2$$
$$25a = 1.265625$$
$$a = \frac{1.265625}{25} = .0506''$$

The length of lever $H = AC = 6.250 + .0506 = 6.3006''$.

TRIGONOMETRY

RELATIONS BETWEEN TRIGONOMETRIC FUNCTIONS

The relations existing among the trigonometric functions of angles are very useful in solving many problems. This is true because the relations found to exist among the functions of angles can be used conveniently to find quantities that cannot be determined otherwise.

Note: Alpha (α) denotes any angle.

Relations between Functions of One Angle. The following formulas show the fundamental relations between functions of one angle. The formulas are often called the *fundamental identities*.

1. $\sin \alpha = \dfrac{1}{\csc \alpha}$.

2. $\csc \alpha = \dfrac{1}{\sin \alpha}$.

3. $\cos \alpha = \dfrac{1}{\sec \alpha}$.

4. $\sec \alpha = \dfrac{1}{\cos \alpha}$.

5. $\tan \alpha = \dfrac{1}{\cot \alpha}$.

6. $\cot \alpha = \dfrac{1}{\tan \alpha}$.

7. $\sin^2 \alpha + \cos^2 \alpha = 1$.
8. $\sec^2 \alpha = 1 + \tan^2 \alpha$.
9. $\csc^2 \alpha = 1 + \cot^2 \alpha$.

10. $\tan \alpha = \dfrac{\sin \alpha}{\cos \alpha}$.

11. $\cot \alpha = \dfrac{\cos \alpha}{\sin \alpha}$.

Functions of Complementary and Supplementary Angles. The formulas which follow show the relations existing between the functions of an angle and the functions of its complement and supplement.

1. $\sin(90° - \alpha) = \cos \alpha$.
2. $\cos(90° - \alpha) = \sin \alpha$.
3. $\tan(90° - \alpha) = \cot \alpha$.
4. $\cot(90° - \alpha) = \tan \alpha$.
5. $\sec(90° - \alpha) = \csc \alpha$.
6. $\csc(90° - \alpha) = \sec \alpha$.
7. $\sin(180° - \alpha) = \sin \alpha$.
8. $\cos(180° - \alpha) = -\cos \alpha$.
9. $\tan(180° - \alpha) = -\tan \alpha$.
10. $\cot(180° - \alpha) = -\cot \alpha$.
11. $\sec(180° - \alpha) = -\sec \alpha$.
12. $\csc(180° - \alpha) = \csc \alpha$.

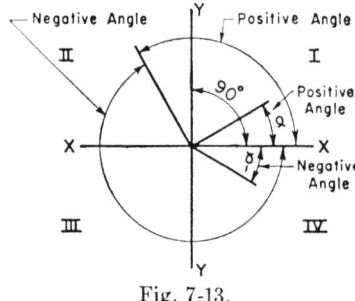
Fig. 7-13.

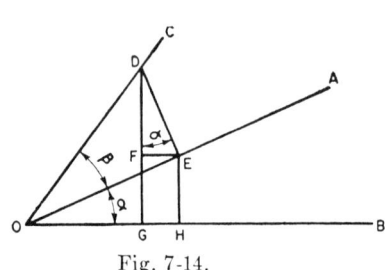
Fig. 7-14.

Functions of Negative Angles. Fig. 7-13 illustrates the meaning of negative angles. The four quadrants are obtained by drawing the X and Y axes and are designated I, II, III, and IV.

The formulas for the functions of a negative angle are:
1. $\sin(-\alpha) = -\sin \alpha$.
2. $\cos(-\alpha) = \cos \alpha$.
3. $\tan(-\alpha) = -\tan \alpha$.
4. $\cot(-\alpha) = -\cot \alpha$.
5. $\sec(-\alpha) = \sec \alpha$.
6. $\csc(-\alpha) = -\csc \alpha$.

TRIGONOMETRIC FUNCTIONS OF MORE THAN ONE ANGLE

There are times when the solution of a problem depends upon the ratio or function of more than one angle, especially in higher mathematics, and, for this purpose, some formulas expressing these ratios are enumerated in this section.

The validity of each formula may be tested by substituting real values instead of α and β.

TRIGONOMETRY

Functions of the Sum and Difference of Two Angles. The following formulas express the functions of the sum or difference of two angles in terms of functions of individual angles. They are called the *addition and subtraction formulas*.

Proof will be given for the first formula and formulas 2, 3, 4, 5, and 6 can be proved by referring to Fig. 7-14 and using the same procedure as that used for proving formula 1.

1. Sin $(\alpha + \beta) = \sin \alpha \cos \beta + \cos \alpha \sin \beta$.

   See Fig. 7-14. Note that $\angle \alpha$ is less than 90°, $\angle \beta$ is less than 90°, $\angle \alpha + \angle \beta$ is less than 90°.

   Proof: Draw line DE perpendicular to AO, and DG perpendicular to OB. $\angle EDF = \angle \alpha$ by construction.

   $\text{Sin } (\alpha + \beta) = \dfrac{DG}{OD} = \dfrac{EH + DF}{OD}$, since $EH = FG$.

   $\text{Sin } (\alpha + \beta) = \dfrac{EH}{OD} + \dfrac{DF}{OD}$. But $EH = OE \times \sin \alpha$, and $DF = DE \times \cos \alpha$, since $\triangle EOH$ and $\triangle DEF$ are right triangles. Substituting the values above, $\sin (\alpha + \beta) = \dfrac{OE \times \sin \alpha}{OD} + \dfrac{DE \times \cos \alpha}{OD}$. But $\dfrac{OE}{OD} = \cos \beta$, and $\dfrac{DE}{OD} = \sin \beta$.

   $\therefore \sin (\alpha + \beta) = \sin \alpha \cos \beta + \cos \alpha \sin \beta$.

   Test: The validity of this formula may be tested by substituting real values. An example is: assume $\angle \alpha = 18°$ and $\angle \beta = 24°$. Then $\sin (\alpha + \beta) = \sin \alpha \cos \beta + \cos \alpha \sin \beta$.
   Sin 42° = (sin 18° × cos 24°) + (cos 18° × sin 24°).
   .66913 = (.30902 × .91354) + (.95106 × .40674) = .28230 + .38683.
   The remaining formulas may be tested in the same manner.

2. Sin $(\alpha - \beta) = \sin \alpha \cos \beta - \cos \alpha \sin \beta$.
3. Cos $(\alpha + \beta) = \cos \alpha \cos \beta - \sin \alpha \sin \beta$.
4. Cos $(\alpha - \beta) = \cos \alpha \cos \beta + \sin \alpha \sin \beta$.

5. Tan $(\alpha + \beta) = \dfrac{\tan \alpha + \tan \beta}{1 - \tan \alpha \tan \beta}$.

6. Tan $(\alpha - \beta) = \dfrac{\tan \alpha - \tan \beta}{1 + \tan \alpha \tan \beta}$.

Functions of Double Angles. The double-angle formulas express the functions of twice an angle in terms of functions of the angle. (Refer to Fig. 7-14.) They are as follows:

Note: $\sin^2 \alpha = (\sin \alpha)^2$.

1. $\sin 2\alpha = 2 \sin \alpha \cos \alpha$.
2. $\cos 2\alpha = \cos^2 \alpha - \sin^2 \alpha = 1 - 2\sin^2 \alpha = 2\cos^2 \alpha - 1$.
3. $\tan 2\alpha = \dfrac{2 \tan \alpha}{1 - \tan^2 \alpha}$.

Functions of Half Angles. The following formulas can be used to find the functions of half of an angle.

1. $\sin \alpha = \pm \sqrt{\dfrac{1 - \cos 2\alpha}{2}}$.

2. $\cos \alpha = \pm \sqrt{\dfrac{1 + \cos 2\alpha}{2}}$.

3. $\tan \alpha = \pm \sqrt{\dfrac{1 - \cos 2\alpha}{1 + \cos 2\alpha}} = \dfrac{\sin 2\alpha}{1 + \cos 2\alpha} = \dfrac{1 - \cos 2\alpha}{\sin 2\alpha}$.

Product Formulas. It is sometimes desirable to transform the sum or difference of sines or cosines to a product. The formulas below are product formulas. The first four formulas are useful for expressing a product of two functions as the sum or difference of two functions. The last four formulas express the sum or difference of two functions as a product.

1. $\sin \alpha \cos \beta = \dfrac{1}{2} \sin(\alpha + \beta) + \dfrac{1}{2} \sin(\alpha - \beta)$.

2. $\cos \alpha \sin \beta = \dfrac{1}{2} \sin(\alpha + \beta) - \dfrac{1}{2} \sin(\alpha - \beta)$.

3. $\cos \alpha \cos \beta = \dfrac{1}{2} \cos(\alpha + \beta) + \dfrac{1}{2} \cos(\alpha - \beta)$.

4. $\sin \alpha \sin \beta = -\dfrac{1}{2} \cos(\alpha + \beta) + \dfrac{1}{2} \cos(\alpha - \beta)$.

5. $\sin \alpha + \sin \beta = 2 \sin \dfrac{1}{2}(\alpha + \beta) \cos \dfrac{1}{2}(\alpha - \beta)$.

6. $\sin \alpha - \sin \beta = 2 \cos \frac{1}{2}(\alpha + \beta) \sin \frac{1}{2}(\alpha - \beta)$.

7. $\cos \alpha + \cos \beta = 2 \cos \frac{1}{2}(\alpha + \beta) \cos \frac{1}{2}(\alpha - \beta)$.

8. $\cos \alpha - \cos \beta = -2 \sin \frac{1}{2}(\alpha + \beta) \sin \frac{1}{2}(\alpha - \beta)$.

SOLUTION OF PROBLEMS INVOLVING OBLIQUE TRIANGLES

The principles and methods for solving right triangles previously discussed are applied when the given problems can be reduced to right triangles directly; but, in many cases, the problems can be reduced first to oblique triangles, then to right triangles by dropping perpendiculars from the vertexes upon the opposite sides. The operations performed in solving right triangles have been explained and it is presumed that the reader has gained familiarity with the method. Reduction of oblique triangles to right triangles is often too cumbersome, however, and shorter processes for solving the obliques directly are here offered.

Any oblique triangle can be solved with the aid of trigonometric functions if three elements are given and at least one of the elements is a side.

Problems that can be reduced to oblique triangles are, for convenience, usually divided into four groups, choice depending upon the types of the given elements.

1. When any two angles and any one side are given.
2. When any two sides and the angle opposite one of the sides are given.
3. When any two sides and the angle included are given.
4. When all of the sides are given.

A few simple formulas (law of cosines, law of sines, projection formulas, and cotangent formulas) will be derived and, with their aid, solution of the oblique triangles will be faster and less cumbersome than reducing these triangles to right triangles and resolving them individually.

Law of Sines. *In any triangle, right or oblique, any side is proportional to the sine of its opposite angle.*

Proof: See Fig. 7-15.

Designate the sides of the given triangle, ABC, by a, b, and c, and the angles opposite these sides by A, B, and C, respectively. Draw the perpendicular h and obtain two right triangles, ABD and BCD.

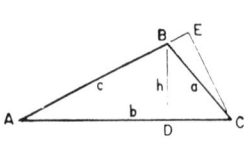

Fig. 7-15.

In right $\triangle ABD, \dfrac{h}{c} = \sin A$, hence $h = c \sin A$.

In right $\triangle BCD, \dfrac{h}{a} = \sin C$, hence $h = a \sin C$.

$\therefore c \sin A = a \sin C$, and $\dfrac{c}{\sin C} = \dfrac{a}{\sin A}$.

If a perpendicular, CE, is drawn from the vertex, C, the following can be proved: $\dfrac{a}{\sin A} = \dfrac{b}{\sin B}$.

Summing up: $\dfrac{a}{\sin A} = \dfrac{b}{\sin B} = \dfrac{c}{\sin C}$. This relation is known as the Law of Sines, and from the law may be written the three equations:

$$\dfrac{a}{\sin A} = \dfrac{b}{\sin B}$$

$$\dfrac{a}{\sin A} = \dfrac{c}{\sin C}$$

$$\dfrac{b}{\sin B} = \dfrac{c}{\sin C}$$

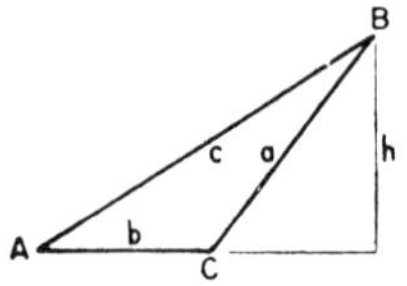

Fig. 7-16. 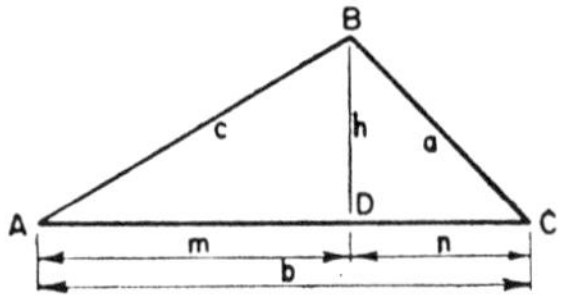   Fig. 7-17.

Each one of these equations contains four elements of the given triangle, and if any three elements are known, the fourth element may be found readily.

Example: In any equation, such as $\dfrac{a}{\sin A} = \dfrac{b}{\sin B}$, $a = \sin A \times \dfrac{b}{\sin B}$, $b = \sin B \times \dfrac{a}{\sin A}$, $\sin A = \dfrac{a \sin B}{b}$, and $\sin B = \dfrac{b \sin A}{a}$.

It can be proved that the same relation exists in the triangle ABC shown in Fig. 7-16, where $h =$ altitude drawn from vertex B to base AC.

$$\frac{a}{\sin A} = \frac{b}{\sin B} = \frac{c}{\sin C}$$

Law of Cosines. *In any triangle, the square of any side is equal to the sum of the squares of the other two sides, less double the product of these two sides multiplied by the cosine of the angle included between them.*

Proof: See Fig. 7-17.

Draw the perpendicular, h, and obtain two right triangles, ABD and BCD.

In right $\triangle BCD$, $a^2 = h^2 + n^2$.

But $n = b - m$, and since $m = c \cos A$, $n = b - c \cos A$, and $n^2 = (b - c \cos A)^2 = b^2 - 2bc \cos A + c^2 \cos^2 A$.

Also, in $\triangle BCD$, $h = c \sin A$, and $h^2 = c^2 \sin^2 A$.

Equation $a^2 = h^2 + n^2$ can be changed by substituting for h^2 and n^2 the equivalent values just obtained.

$\therefore a^2 = h^2 + n^2 = c^2 \sin^2 A + b^2 - 2bc \cos A + c^2 \cos^2 A$.

Factoring: $a^2 = b^2 + c^2 (\sin^2 A + \cos^2 A) - 2bc \cos A$.

Since $\sin^2 A + \cos^2 A = 1$, $a^2 = b^2 + c^2 - 2bc \cos A$.

In a similar manner, $b^2 = a^2 + c^2 - 2ac \cos B$ and $c^2 = a^2 + b^2 - 2ab \cos C$. This relation is known as the Law of Cosines.

Each one of these equations contains four elements of the given triangle, and if any three elements are known, the fourth element may be found readily.

Example: In any equation, such as $a^2 = b^2 + c^2 - 2bc \cos A$, $\cos A = \dfrac{b^2 + c^2 - a^2}{2bc}$.

If the equation involves angle B, $\cos B = \dfrac{a^2 + c^2 - b^2}{2ac}$.

If the equation involves angle C, $\cos C = \dfrac{a^2 + b^2 - c^2}{2ab}$.

Application of Law of Sines and Law of Cosines. Now we are prepared to treat each of the four groups of problems of the oblique triangles, previously mentioned, with the aid of the formulas derived from the laws of sines and cosines.

In solving the required triangle, a suitable formula involving three known elements of the triangle is selected, leaving one unknown element to be determined. Numerical values are substituted wherever possible, and the value of the unknown element is calculated.

A layout by which results may be checked is made, or another suitable formula is selected and the unknown element is solved by the new formula. If the results in both cases are identical, it is proved that they are correct.

Group 1. *Any two angles and any one side are given.*

Given: See Fig. 7-18, showing $\triangle ABC$, in which the length of $c = 3.480''$, $\angle A = 23° 30'$, and $\angle C = 64° 36'$.

Required: $\angle B$, length of a, and length of b.

Procedure: $\angle B = 180° - (\angle A + \angle C) = 180° - (23° 30' + 64° 36') = 180° - 88° 06' = 91° 54'$.

A problem of this type can be solved by using the Law of Sines, since all angles and one side are known. Therefore, any of the remaining sides may be determined.

$$\frac{a}{\sin A} = \frac{c}{\sin C}, \text{ and } a = \frac{\sin A \, c}{\sin C}.$$

Substituting the three known values in the preceding formula:

$$a = \frac{\sin 23° 30' \times 3.480}{\sin 64° 36'} = \frac{.39875 \times 3.480}{.90333} = 1.536''$$

$$\frac{b}{\sin B} = \frac{c}{\sin C}, \text{ and } b = \frac{\sin B \, c}{\sin C}$$

$$b = \frac{\sin 91° 54' \times 3.480}{\sin 64° 36'} = \frac{.99981 \times 3.480}{.90333} = 3.852''$$

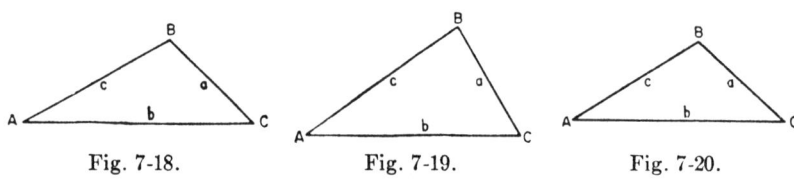

Fig. 7-18.   Fig. 7-19.   Fig. 7-20.

The Law of Cosines is not suitable for the foregoing problem, since only one side of the given triangle is known, and upon examination of the formulas, at least two sides of the triangle must be given in order to be able to solve its remaining unknown side. However, the Law of Cosines may be used to check the results obtained by means of the equations of the Law of Sines.

Group 2. *Any two sides and the angle opposite one of the sides are given.*

Given: See Fig. 7-19 showing $\triangle ABC$, in which the length of $a = 2.864''$, the length of $b = 4.228''$, and $\angle B = 82° 24'$.

Required: $\angle A$, $\angle C$, and the length of c.

Procedure: A problem of this type can be solved by using first the Law of Sines, since two sides of the given triangle and one angle are known. Therefore, the sine of $\angle A$ may be determined, and the third angle, C, may be calculated. Finally, by the Law of Sines, the remaining side c is calculated.

$$\frac{a}{\sin A} = \frac{b}{\sin B}$$

Then $\sin A = \dfrac{a \sin B}{b} = \dfrac{2.864 \times \sin 82° 24'}{4.228} = \dfrac{2.864 \times .99121}{4.228} =$ .67143 and $A = 42° 10' 40''$.

$C = 180° - (A + B) = 180° - (42° 10' 40'' + 82° 24') = 55° 25' 20''$.

By the Law of Sines we have $\dfrac{b}{\sin B} = \dfrac{c}{\sin C}$.

$c = \dfrac{\sin C\, b}{\sin B} = \dfrac{\sin 55° 25' 20'' \times 4.228}{\sin 82° 24'} = \dfrac{.82336 \times 4.228}{.99121} = 3.512''$.

With the elements given in the problem of Group 2, it is possible to find one triangle to meet the given conditions, two triangles, or no triangle at all. Any of these conditions can be tested by a layout of the triangle in accordance with the given data.

Group 3. *Any two sides and the angle included between them are given.*

Given: See Fig. 7-20 showing $\triangle ABC$, in which the length of $a = 5.274''$, the length of $b = 9.836''$, and $\angle C = 23° 47'$.

Required: The length of C and the values of angles A and B.

Procedure: A problem of this type cannot be solved immediately by the Law of Sines, because the given data does not include a single ratio between any of the sides of the given triangle & the sine of its opposite angle.

Therefore the Law of Cosines must be tried first, since the given data consists of the elements required for application of the Law of Cosines

$$c^2 = a^2 + b^2 - 2ab \cos C$$

$$c = \sqrt{a^2 + b^2 - 2ab \cos C}$$

$$= \sqrt{(5.274)^2 + (9.836)^2 - (2 \times 5.274 \times 9.836 \times .91508)}.$$

$$c = \sqrt{94.93955} = 9.744''$$

Now, by application of the Law of Sines:

$$\frac{c}{\sin C} = \frac{a}{\sin A}$$

$$\frac{9.744}{\sin 23° 47'} = \frac{5.274}{\sin A}$$

$$\sin A = \frac{5.274 \times \sin 23° 47'}{9.744} = \frac{5.274 \times .40328}{9.744} = .21827$$

$A = 12° 36'$ (to the nearest minute)

$B = 180° - (A + C) = 180° - (12° 36' + 23° 47') = 143° 37'$

Group 4. *When all the sides are given.*

Given: Fig. 7-21, showing $\triangle ABC$, in which the length of $a = 4.146''$, that of $b = 6.264''$, and that of $c = 8.786''$.

Required: The values of angles A, B, and C.

Procedure: The Law of Sines cannot be applied immediately, since no single ratio between any side of the triangle and the sine of its opposite angle is known. Therefore, applying the Law of Cosines for the first step would enable us to find one of the angles of the given triangle.

$a^2 = b^2 + c^2 - 2bc \cos A$, $(4.146)^2 = (6.264)^2 + (8.786)^2 - (2 \times 6.264 \times 8.786 \times \cos A)$.

$2 \times 6.264 \times 8.786 \cos A = (6.264)^2 + (8.786)^2 - (4.146)^2$.

$$\cos A = \frac{(6.264)^2 + (8.786)^2 - (4.146)^2}{2 \times 6.264 \times 8.786} =$$

$$\frac{39.237696 + 77.193796 - 17.189316}{110.071008}.$$

Then, $\cos A = \dfrac{99.242176}{110.071008} = .90162$ and $A = 25° 38'$.

TRIGONOMETRY

Now, by applying the Law of Sines, we have:

$$\frac{a}{\sin A} = \frac{b}{\sin B}$$

$$\frac{4.146}{\sin 25° 38'} = \frac{6.264}{\sin B}$$

$$\frac{4.146}{.43261} = \frac{6.264}{\sin B}$$

$$4.146 \sin B = .43261 \times 6.264$$

$$\sin B = \frac{.43261 \times 6.264}{4.146} = .65361$$

$$\angle B = 40° 49' \text{ (to nearest minute)}$$

$$\angle C = 180° - (A + B) = 180° - (25° 38' + 40° 49')$$

$$= 180° - 66° 27' = 113° 33'$$

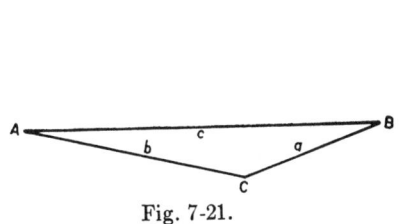

Fig. 7-21.

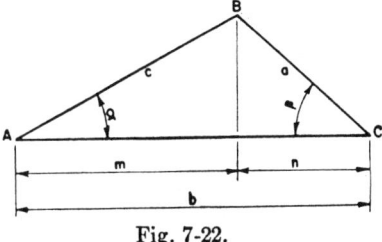

Fig. 7-22.

Projection Formulas. Oblique triangle problems of Group 4, in which three sides of the given triangle are given and it is required to find the values of the angles, may be solved by means of so-called projection formulas. The following problems illustrate the use of these formulas.

Problem 1. See Fig. 7-22.

Given: $\triangle ABC$, whose sides $a = 4.582''$, $b = 6.874''$, and $c = 6.246''$.
Required: The values of angles A, B, and C.
Procedure: In accordance with a previously given theorem of geometry, the projection of side c upon base b is m, and $m = \dfrac{c^2 + b^2 - a^2}{2b}$.

$$\text{Cos } \alpha = \frac{m}{c} = \frac{c^2 + b^2 - a^2}{2bc} = \frac{(6.246)^2 + (6.874)^2 - (4.582)^2}{2 \times 6.874 \times 6.246}, \cos$$

$$\alpha = \frac{65.269668}{85.870008} = .76009 \text{ and } \alpha = 40° 32' \text{ (to nearest minute)}.$$

Beta (β) can be determined in the same manner, that is, by considering n as the projection of side a upon the base b. Hence, $n = \dfrac{a^2 + b^2 - c^2}{2b}$.

$$\text{Cos } \beta = \frac{n}{a} = \frac{a^2 + b^2 - c^2}{2ab} = \frac{(4.582)^2 + (6.874)^2 - (6.246)^2}{2 \times 4.582 \times 6.874}$$

$$\cos \beta = \frac{29.234084}{62.993336} = .46408 \text{ and } \beta = 62° 21'$$

$\angle ABC = 180° - (\alpha + \beta) = 180° - (40° 32' + 62° 21') = 77° 7'$

Problem 2. A type of oblique triangle (ABC) whose sides are known is shown in Fig. 7-23. This triangle may be solved by means of projection formulas.

Given: $a = 4.827''$, $b = 5.212''$, and $c = 8.142''$.
Required: The values of angles A, B, and C.
Procedure: The value of p in accordance with the projection theorem of geometry is: $p = \dfrac{c^2 - b^2 - a^2}{2b}$.

$$\text{Cos } \gamma = \frac{p}{a} = \frac{c^2 - b^2 - a^2}{2ab} = \frac{(8.142)^2 - (5.212)^2 - (4.827)^2}{2 \times 4.827 \times 5.212}$$

$$\text{Cos } \gamma = \frac{66.292164 - 27.164944 - 23.299929}{50.316648}$$

$$= \frac{15.827291}{50.316648} = .31455$$

$\gamma = 71° 40'$

$\beta = 180° - 71° 40' = 108° 20'$

$$\text{Cos } \alpha = \frac{b + p}{c} = \frac{5.212 + \left(\dfrac{15.827291}{10.414}\right)}{8.142} = \frac{6.73035}{8.142} = .82662$$

$\alpha = 34° 15'$

$\angle ABC = 180° - (\alpha + \beta) = 180° - (34° 15' + 108° 20')$
$\qquad\quad = 180° - 142° 35' = 37° 25'$

TRIGONOMETRY

Problem 3. A useful geometric relation existing in any triangle is shown in Fig. 7-24, and its employment in solution of oblique triangles

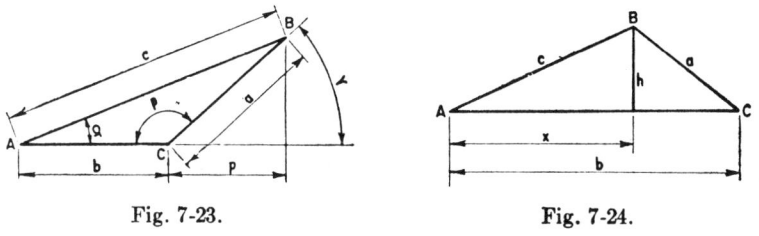

Fig. 7-23.   Fig. 7-24.

whose three sides are known is convenient, enabling one rapidly to determine the unknown angles.

Procedure:

$$h^2 = c^2 - x^2$$

$$h^2 = a^2 - (b - x)^2 = a^2 - b^2 + 2bx - x^2$$

$$c^2 - x^2 = a^2 - b^2 + 2bx - x^2$$

$$c^2 - a^2 + b^2 = 2bx$$

$$x = \frac{c^2 - a^2 + b^2}{2b}$$

$$\frac{x}{c} = \cos A$$

$$\frac{b - x}{a} = \cos C$$

$$B = 180° - (A + C)$$

Cotangent Formulas. Oblique triangles in which any side and its two adjacent angles are known may be solved by means of cotangent formulas.

Problem 1. Oblique triangle ABC as shown in Fig. 7-25.
Given: $b = 6.826''$, $A = 39° 26'$, and $C = 56° 47'$.
Required: a, c, and B.
Procedure:

$$B = 180° - (A + C) = 180° - (39° 26' + 56° 47')$$

$$= 180° - 96° 13' = 83° 47'$$

$$\frac{m}{h} = \cot A$$

$$\frac{n}{h} = \cot C$$

$$\frac{m + n}{h} = \cot A + \cot C. \text{ But } m + n = b$$

$$\therefore h = \frac{b}{\cot A + \cot C} = \frac{6.826}{\cot 39° 26' + \cot 56° 47'}$$

$$= \frac{6.826}{1.2160 + .65480} = 3.6487''$$

c may be calculated from the following relationship:

$$\frac{c}{h} = \csc A$$

$$\frac{c}{3.6487} = \csc 39° 26' = 1.5743$$

$$c = 3.6487 \times 1.5743 = 5.744''$$

$$\frac{a}{h} = \csc C = \csc 56° 47' = 1.1953$$

$$a = 3.6487 \times 1.1953 = 4.361''$$

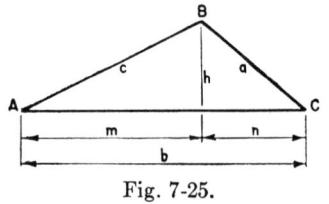

Fig. 7-25.

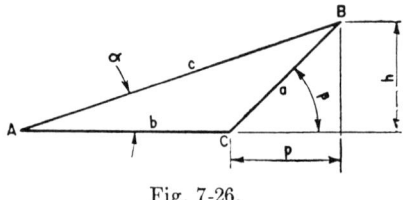

Fig. 7-26.

Problem 2. Another type of oblique triangle, Fig. 7-26, a side and two adjacent angles of which are known, may be solved by applying the cotangent formulas.

Given: $b = 5.735''$, $A = \alpha = 24° 51'$, and $C = 103° 23'$.

TRIGONOMETRY

Required: a, c, and B.
Procedure:

$$B = 180° - (A + C) = 180° - (24° \, 51' + 103° \, 23')$$
$$= 180° - 128° \, 14' = 51° \, 46'$$
$$\beta = 180° - C = 180° - 103° \, 23' = 76° \, 37'$$

$$\frac{b + p}{h} = \cot \alpha$$

$$\frac{p}{h} = \cot \beta$$

$$\frac{(b + p) - p}{h} = \cot \alpha - \cot \beta$$

$$\frac{b}{h} = \cot 24° \, 51' - \cot 76° \, 37'$$

$$h = \frac{b}{\cot \alpha - \cot \beta} = \frac{5.735}{\cot 24° \, 51' - \cot 76° \, 37'}$$

$$= \frac{5.735}{2.1592 - .23793}$$

$$h = \frac{5.735}{1.92127} = 2.985''$$

c may be calculated from the following relationship:

$$\frac{c}{h} = \csc \alpha = \csc 24° \, 51'$$

$$\frac{c}{2.985} = 2.3796$$

$$c = 2.985 \times 2.3796 = 7.103''$$

$$\frac{a}{h} = \csc \beta = \csc 76° \, 37' = 1.0279$$

$$a = h \times 1.0279 = 2.985 \times 1.0279 = \mathbf{3.068''}$$

ANGLES FREQUENTLY USED IN ENGINEERING PRACTICE

The following angles occur so frequently in engineering practice and machine shop work that it is appropriate to list their common functions:

$$\tan 45° = 1$$

$$\cot 45° = 1$$

$$\sin 45° = \frac{\sqrt{2}}{2} = .70711$$

$$\cos 45° = \frac{\sqrt{2}}{2} = .70711$$

$$\tan 30° = \frac{\sqrt{3}}{3} = .57735$$

$$\cot 30° = \sqrt{3} = 1.7320$$

$$\sin 30° = .5000$$

$$\cos 30° = .86603$$

$$\tan 60° = 1.7320$$

$$\cot 60° = .57735$$

$$\sin 60° = .86603$$

$$\cos 60° = .5000$$

$$\tan 90° = \infty \quad (\infty \text{ signifies } \textit{infinity})$$

$$\cot 90° = 0$$

$$\sin 90° = 1$$

$$\cos 90° = 0$$

SHOP TRIGONOMETRY

One can see from the foregoing discussion that applications of trigonometry in shop work are extremely useful and, at the same time, are very simple, consisting for the most part of taking proper values from standard tables of trigonometric functions and multiplying them by dimensions that appear on working drawings and specifications, thus obtaining values and dimensions that afford greater convenience to the shop man.

Angles designated on drawings are often converted to linear dimensions, as these are more conveniently handled by shop men. The draftsman, however, is not always able to put simplified dimensions on the drawings, as other users of the shop drawings prefer that designations be made in conformance with conventional practice.

In large manufacturing plants it is the practice of the engineering department not to duplicate the information that appears on the working drawings for the reason that changes in design of the product and introduction of new methods of manufacture frequently necessitate changes on the drawings and, consequently, there is the danger that dimensions, if repeated, will be overlooked as changes are made, thus creating confusion and possibly causing errors in the manufacturing division that can add up to serious loss.

In order to solve practical problems by means of trigonometry, there is a certain sequence of operations one must follow. A simple layout or sketch should be made of the given conditions, such as distances and angles, and also of the required distance and angles. To solve an unknown dimension or angle, a right or oblique triangle containing ample information and the required dimension or angle should be looked for. Frequently such triangles are not available, and auxiliary lines are drawn to form triangles that contain the unknown element.

The auxiliary lines are often chosen from lines already given on the drawing or are supplied by lines drawn parallel or perpendicular to the given lines. Lines drawn through the centers of circles on the drawing or through the centers of new circles tangent to the existing circles may constitute the auxiliary lines that form the triangles containing the unknown elements.

There is a possibility of drawing auxiliary lines through the vertices in a given circle or a circle tangent to the given circle, thus forming the triangle with sufficient data, and enabling one to solve for the unknown element.

In the event of not being able to construct a triangle which offers a sufficient number of known elements, another triangle is formed which contains, among other elements, an element (or its equivalent) included in the first triangle. Perhaps, more than one auxiliary triangle is required before the unknown element can be determined.

A good demonstration of a problem that would satisfy the conditions just described is illustrated in Fig. 7-27.[1]

---

[1] Keal and Leonard, *Technical Mathematics*, Vol. III (New York: John Wiley & Sons, 1923).

264   MATHEMATICS FOR INDUSTRY

Given: A 3″ diameter circle with 2″ and 1″ diameter circles inscribed on the 3″ diameter.

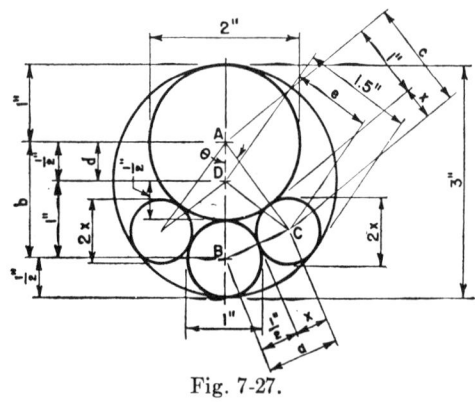

Fig. 7-27.

Required: What is the diameter of a circle tangent to all three given circles?

Note: Two such circles may be determined.

Procedure: By drawing auxiliary lines through the centers of the given and required circles, triangles ABC and ACD are formed.

Applying the Law of Cosines to triangle ABC:

$$\cos\theta = \frac{b^2 + c^2 - a^2}{2bc}$$

$$\cos\theta = \frac{(1.5)^2 + (1+x)^2 - (.5+x)^2}{2 \times 1.5 \times (1+x)}$$

$$= \frac{2.25 + 1 + 2x + x^2 - .25 - x - x^2}{2(1.5 + 1.5x)}$$

$$\cos\theta = \frac{3+x}{3+3x}$$

Applying the Law of Cosines to triangle ACD:

$$\cos\theta = \frac{c^2 + d^2 - e^2}{2cd}$$

$$= \frac{(1+x)^2 + .5^2 - (1.5-x)^2}{2 \times (1+x) \times .5}$$

$$\cos\theta = \frac{1 + 2x + x^2 + .25 - 2.25 + 3x - x^2}{1+x}$$

$$= \frac{-1 + 5x}{1+x}$$

from $\triangle ABC$, $\cos\theta = \dfrac{3+x}{3+3x}$

from $\triangle ACD$, $\cos \theta = \dfrac{-1 + 5x}{1 + x}$

$$\dfrac{3 + x}{3 + 3x} = \dfrac{-1 + 5x}{1 + x}$$

$$(3 + x)(1 + x) = (3 + 3x)(-1 + 5x)$$

$$3 + x + 3x + x^2 = -3 - 3x + 15x + 15x^2$$

$$3 + 3 = 8x + 14x^2$$

$$14x^2 + 8x = 6$$

$$7x^2 + 4x = 3$$

$$7x^2 + 4x - 3 = 0$$

This is a quadratic equation of the form:

$$ax^2 + bx + c = 0$$

where $x = \dfrac{-b \pm \sqrt{b^2 - 4ac}}{2a}$

$$x = \dfrac{-4 - \sqrt{4^2 - [4 \times 7 \times (-3)]}}{2 \times 7}$$

$$= \dfrac{-4 \pm \sqrt{16 + 84}}{14} = \dfrac{-4 \pm \sqrt{100}}{14} = \dfrac{-4 \pm 10}{14}$$

$$x = -1 \text{ and } x = \dfrac{6}{14}$$

Of the two possible values given by the solution just presented, only the value of $x = \dfrac{6}{14} = \dfrac{3}{7}$ is seen to fit the problem.

The diameters of the two circles tangent to the three given circles are $\dfrac{3}{7} \times 2 = \dfrac{6}{7} = .8671''$ each.

The remainder of this chapter will be devoted to solution of problems dealing with tapers, angles, sine bars, dovetails, and numerous machine and tool elements.

Problems Involving Tapers. Numerous cylindrical machine and tool details are made conical or, as explained in Chapter IV, are tapered.

By the use of trigonometry, tapers can often be calculated conveniently, especially when the angles included between sides of the details are specified, or when the included angles are required to correspond to other given information.

Problem 1. See Fig. 7-28.

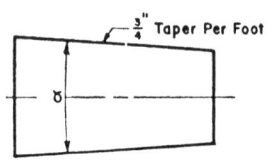

Fig. 7-28.

Given: Assume a detail, as shown in Fig. 7-28, having a $\frac{3''}{4}$ taper per foot.

Required: To calculate the included angle α.

Procedure: The information given in the figure is sufficient for calculating angle α, since this angle is constant for any length of the detail, and the diameters at the ends may be of any value so long as the difference between them is in the same proportion to $\frac{3''}{4}$ as the length of the part is to 12″.

Problem 2. See Fig. 7-29.

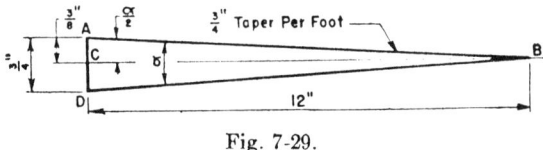

Fig. 7-29.

Given: Imagine the part illustrated by Fig. 7-29 having the taper and the included angle similar to Fig. 7-28.

Required: To calculate the included angle α.

Procedure: In right triangle ABC, $\tan \frac{\alpha}{2} = \frac{3/8}{12} = \frac{.375}{12} = .03125$, $\frac{\alpha}{2} = 1° 47' 24''$. $\alpha = 3° 34' 48''$.

A simple formula can be derived for solving the included angle, if the taper per foot is known.

$$\tan \frac{1}{2}\alpha = \frac{\text{taper per foot}}{24}$$

Note: Do not use the formula: $\tan \alpha = \frac{\text{taper per foot}}{12}$, because the tangent of an angle is not proportional to the angle itself, except in the case of very small angles, when the angles and their functions may be assumed to be proportional.

TRIGONOMETRY

If the included angle is given, the taper may be calculated: Taper per foot $= 24 \times \tan \frac{1}{2} \alpha$.

Problem 3. See Fig. 7-30.
Given: Assume a taper illustrated in Fig. 7-30.
Required: The included angle α.
Procedure: Draw BC parallel to the axis of the part. In the right triangle ABC, $\angle ABC$ is equal to $\frac{\alpha}{2}$ by construction,

$$AC = \frac{2.084 - 1.246}{2} = .419$$

$$\tan \frac{\alpha}{2} = \frac{AC}{7.370} = \frac{.419}{7.370} = .05685$$

$$\frac{\alpha}{2} = 3° \; 15' \; 15''; \; \alpha = 6° \; 30' \; 30''.$$

Problems Involving the Sine Bar. One of the most useful yet simple devices in the machine shop for measuring angles is the sine bar. There are many designs of sine bars, but the simplest one of all is illustrated

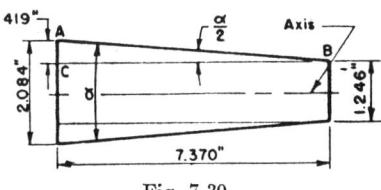
Fig. 7-30.

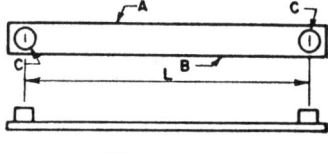
Fig. 7-31.

in Fig. 7-31. The bar consists of a steel straightedge, whose sides A and B are smooth, flat, and square. Two rollers, C, of the same diameter, are attached to the bar. The rollers are perfectly round and smooth and are spaced $5''$ or $10''$ apart, hence the designation of the bar as a $5''$ or $10''$ sine bar, or whatever the dimension indicated.

The device, consisting of the bar and attached rollers, is machined and ground to a tolerance as close as $.0001''$, or even closer. The distance L, in particular, must be accurate. The dimension L in the $5''$ sine bar is $5.0000''$, with a total tolerance not exceeding $.0001''$, or better.

When measuring or checking an angle or part, the sine bar is clamped to an angle plate and set up in such a way that one roller is situated above the other at a distance calculated to permit checking of the angle. See Fig. 7-32.

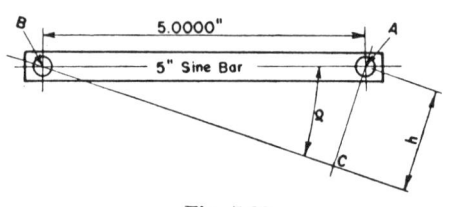

Fig. 7-32.

Problem 1. See Fig. 7-32.

Given: Distance between the rollers is 5.0000″.

Required: To set up the 5″ sine bar to an angle $\alpha = 30°$.

Procedure: The distance between the rollers (5.0000″) is used as a hypotenuse, AB, of the right triangle, ABC. The height, h, is calculated from the following relationship:

$$\frac{h}{AB} = \sin \alpha$$

$$\frac{h}{5.0000} = \sin 30° = .5000$$

$$h = 5.0000 \times .5000 = 2.5000''$$

Specially prepared tables found in handbooks give direct values of h for 5″ and 10″ sine bars for any desired angle.

The setup in Fig. 7-32 can be used as a gage to measure or check a 30° angle.

Problem 2. A practical example, showing the method employed in checking a taper machine detail or a taper plug gage by means of a sine

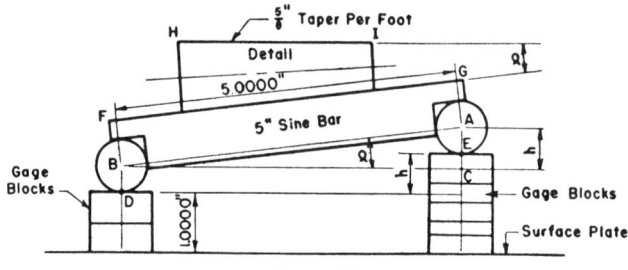

Fig. 7-33.

bar, is illustrated in Fig. 7-33. A 5″ sine bar of a commonly used type is employed in this setup.

Given: $\dfrac{5''}{8}$ taper per foot detail.

TRIGONOMETRY

Required: To check the accuracy of the taper, which is the same as checking the correctness of the included angle.

Procedure: It is good practice to place one end of the roller of the sine bar over a gage block instead of on the surface plate. The gage block may be of any thickness, a thickness of 1″ being considered convenient.

The rollers are of the same diameter, therefore the same right triangle is formed whether the hypotenuse is taken through the centers of the rollers A and B or through the contact points D and E.

To obtain proper height of the right roller above the surface plate, the necessary total thickness of gage blocks must be equal to $h + 1.0000″$.

To calculate h, the included angle α is determined from the previously derived formula for finding an included angle when the taper is known.

$$\tan \frac{1}{2}\alpha = \frac{\text{taper per foot}}{24} = \frac{5/8}{24} = \frac{.625}{24} = .02604$$

$$\frac{1}{2}\alpha = 1° 29' 30''$$

$$\alpha = 2° 59'$$

Now $\dfrac{h}{AB} = \sin \alpha$, since AB is parallel to DE and also to FG, and BC is parallel to the top of the surface plate and also to HI.

$$\frac{h}{AB} = \sin \alpha$$

$$\frac{h}{5.0000} = \sin \alpha = \sin 2° 59' = .05204$$

$$h = 5.0000 \times .05204 = .2602$$

The thickness of gage blocks at the right end of the sine bar is equal to $1.0000 + .2602 = 1.2602″$.

The sine bar now is set to the angle that corresponds to the correct taper of the detail and is firmly clamped to an angle plate that is rigidly supported on the surface plate. The gage blocks may be removed.

The checking of the taper detail consists of testing its upper edge, HI, by means of a height gage or a dial indicator. If HI is found to be of equal height above the surface plate throughout—that is, if it is parallel to the surface plate—angle α of the detail will be the same as the angle formed by the sine bar.

Problem 3. See Fig. 7-34.
Given: 5" sine bar.
Required: Determine an angle by means of a sine bar.
Procedure: Support the part to be checked on the surface plate, as shown in Fig. 7-34, and place the bar over the part with the rollers up-

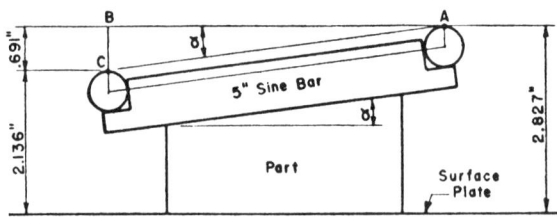

Fig. 7-34.

ward. Measure the distances from the surface plate to the top of the roller at each end of the bar. Assume these distances to be 2.827" and 2.136".

Construct the right triangle ABC, whose hypotenuse, AC, is equal to 5.0000" and side, BC, is equal to $2.827 - 2.136 = .691"$. $\therefore \sin \alpha = \frac{BC}{AC} = \frac{.691}{5.0000} = .13820$ and $\alpha = 7° 56' 37"$.

Problems Involving Precision Disks. A practical problem involving trigonometry is illustrated in Fig. 7-35. The shape of a V-block is checked

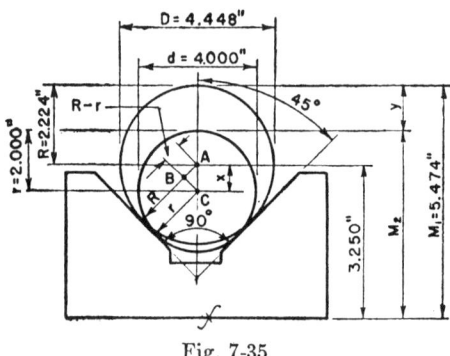

Fig. 7-35.

with the aid of a steel precision ball or disk of known diameter. The setup consists of inserting the ball or disk in the V-groove and taking a precise measurement from the bottom of the finished block to the top of the ball or disk. This measurement must correspond to that specified on the working drawing. Any deviation from the specification will indicate the degree of error.

Frequently a ball or disk of the diameter specified is not available at the time of checking and it becomes necessary to substitute one of a different diameter. The problem, then, consists of calculating the distance from the bottom of the block to the top of the ball or disk substituted.

Assume D is the specified diameter of ball or disk and is equal to 4.448″.

d is the diameter of the available ball or disk and is equal to 4.000″.

R is the radius of the specified ball or disk.

r is the radius of the available ball or disk.

M_1 is the measurement from the bottom of the block to the top of the specified ball or disk and is equal to 5.474″.

M_2 is the measurement from the bottom of the block to the top of the available ball or disk and is to be calculated.

In the right triangle ABC, $\dfrac{x}{R-r} = \sec 45° = 1.4142$, $x = 1.4142(R-r) = 1.4142(2.224 - 2.000) = 1.4142 \times .224 = .31678$, say .317.

By inspection of Fig. 7-35, $y = R - r + x = .224 + .317 = .541$, but $M_2 = M_1 - y = 5.474 - .541 = 4.933″$.

The included angle of the V-groove in Fig. 7-35 is shown as 90°. But any included angle of the V-groove may be checked by the foregoing method and the secant of half of the included angle is taken instead of 45°.

Problems Involving Angles and Inscribed Circles. The trigonometric relationships between commonly used angles and inscribed circles are shown in the next three figures.

In Fig. 7-36, $\angle GCH$ is equal to 90° and $\triangle CEF$ is a right triangle whose sides CE and EF are equal to the radius of the inscribed circle.

$CF = B$ and is the hypotenuse of the right $\triangle CEF$

$B^2 = R^2 + R^2 = 2R^2$

$B = \sqrt{2R^2} = 1.4142R$

$A = B - R = 1.4142R - R = .4142R$

272 MATHEMATICS FOR INDUSTRY

In Fig. 7-37, $\angle GCH$ is equal to 60° and $\triangle CEF$ is a right triangle whose side EF is equal to the radius R of the inscribed circle and the hypotenuse $CF = B$.

$$\frac{R}{B} = \sin 30° = .5000$$

$$R = .5000 B$$

$$2R = B$$

$$\therefore A = B - R = R$$

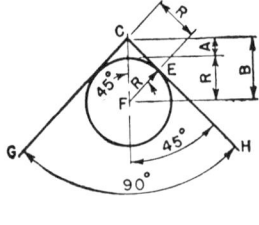
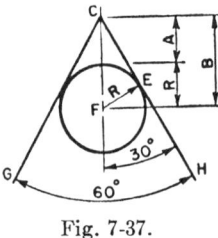
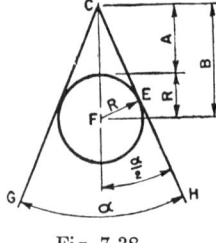

Fig. 7-36. Fig. 7-37. Fig. 7-38.

In Fig. 7-38, $\angle GCH$ is any angle α and $\triangle CEF$ is a right triangle whose side EF is equal to the radius R of the inscribed circle and the hypotenuse $CF = B$.

$$\frac{R}{B} = \sin \frac{\alpha}{2}$$

$$B = \frac{R}{\sin \frac{\alpha}{2}}$$

$$A = B - R = \frac{R}{\sin \frac{\alpha}{2}} - R = R\left(\frac{1}{\sin \frac{\alpha}{2}} - 1\right)$$

Problems Involving Angles and Circular Arcs. The next two figures illustrate two problems involving angles and circular arcs.

Problem 1. A typical toolroom problem is shown in Fig. 7-39, where an angle α formed by a line tangent to two similar circular arcs and a horizontal line is to be determined.

Given: Line EF tangent to

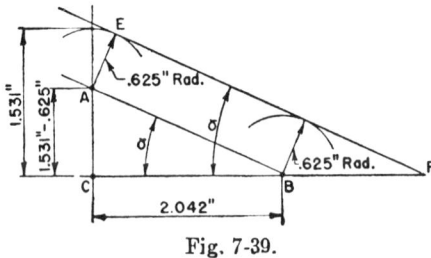

Fig. 7-39.

two circular arcs drawn with a .625″ radius, and the location of the centers of both arcs.

Required: To find $\angle \alpha$ formed by the tangent line EF and the horizontal line CF.

Procedure: Draw line AB through the centers of the given circular arcs, forming $\angle ABC = \alpha$, since $AB \parallel EF$.

In right $\triangle ABC$, $AC =$ given height $1.531″ -$ the length of radius R, which is $.625″$.

$$AC = 1.531 - .625 = .906$$

$$\frac{AC}{CB} = \tan \alpha$$

$$\frac{.906}{2.042} = \tan \alpha = .44368$$

$$\alpha = 23° \ 55' \ 34''$$

Problem 2. A more complex problem, dealing with an unknown angle formed by a line tangent to two given circular arcs and a horizontal base line is shown in Fig. 7-40.

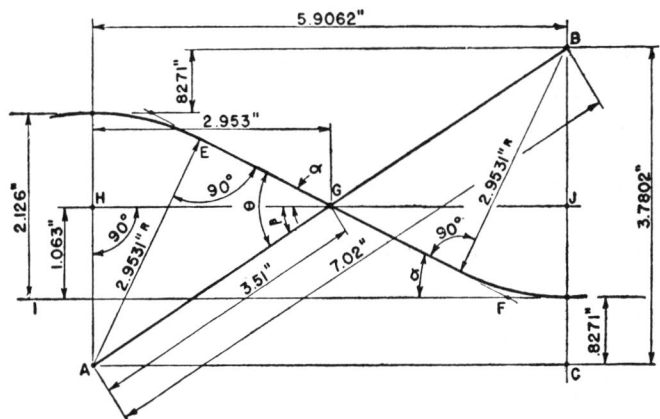

Fig. 7-40.

Given: The location of the centers A and B of both circular arcs drawn with a given radius $2.9531″$.

Required: To find $\angle \alpha$ formed by the tangent line EF and the horizontal base line, IF.

Procedure: As previously explained, the triangle containing the unknown element (in this case $\angle \alpha$) is not always presented in the given layout. Therefore, auxiliary lines are drawn either through the known centers or circular arcs or at any convenient location so long as a triangle is obtained. By means of the triangle it is possible to determine the unknown angle—not immediately, but in subsequent steps.

In Fig. 7-40 such a triangle, ABC, is constructed, and it is a right triangle in which AB is the hypotenuse, and its length $= \sqrt{(AC)^2 + (BC)^2} = \sqrt{(5.9062)^2 + (3.7802)^2} = \sqrt{49.216} = 7.02''$.

G is the mid-point of AB, and $AG = \dfrac{AB}{2} = \dfrac{7.02}{2} = 3.51''$; G is also the mid-point of HJ, and $HG = \dfrac{HJ}{2} = \dfrac{5.9062}{2} = 2.953''$.

In right $\triangle AEG$, $\dfrac{AE}{AG} = \sin\theta = \dfrac{2.9531}{3.51} = .8413$ and $\theta = 57°\ 17'$.

In right $\triangle AGH$, $\dfrac{AH}{HG} = \tan\beta = \dfrac{1.063 + .8271}{2.953} = \dfrac{1.8901}{2.953} = .6400$.

$\beta = 32°\ 37'$.
$\alpha = \theta - \beta = 57°\ 17' - 32°\ 37' = 24°\ 40'$.

Problems Involving Taper Plug Gages. Taper plug gages may be checked for accuracy of the angle of the taper and other dimensions with the aid of precision rollers and application of trigonometric principles. Two methods for checking taper plug gages will be given.

Method 1. A method commonly used in gage laboratories is shown in Fig. 7-41.

Given: Taper plug gage and equipment consisting of precision rollers and parallels or precision blocks.

Required: To check the accuracy of the angle of taper and the diameters at both ends of the gage.

Procedure: The procedure will be given in three steps.

Step 1. Check the angle of taper on one side, which is designated by α.

Place two rollers of the same diameter in contact with sides of gage and the smooth surface, as shown in Fig. 7-41. Obtain measurement M_1.

Remove rollers and place them in contact with block of known height and with sides of gage, as shown in the illustration, Fig. 7-41. Obtain another measurement M_2.

Construct right $\triangle JKL$, as shown, and calculate $\angle \alpha$.

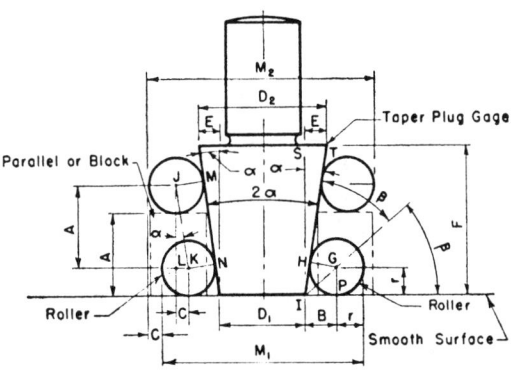

Fig. 7-41.

In this triangle, $\angle KJL = \alpha$, since $JM = KN =$ radius r of rollers, JK is parallel to MN and JL is parallel to the axis of the gage.

By inspection of the figure, it is evident that $LK = C = \dfrac{M_2 - M_1}{2}$, and $A =$ height of block. Therefore, $\tan \alpha = \dfrac{C}{A} = \dfrac{M_2 - M_1}{2A}$.

Angle α may be obtained from the tables of trigonometric functions. The included angle of the taper is 2α.

Step 2. Construct right $\triangle GIP = \triangle GIH$, in which $\angle GIP = \angle GIH = \beta$.

In right $\triangle GIP$, $\beta = \dfrac{90° - \alpha}{2}$, and $\dfrac{B}{r} = \cot \beta$. Therefore $B = r \cot \beta$.

Now the diameter of the gage at the small end $= D_1 = M_1 - 2B - 2r$.

Step 3. Construct right $\triangle IST$, in which $\dfrac{E}{F} = \tan \alpha$.

Angle α is known, F is the length of the gage and may be measured, $\therefore E = F \times \tan \alpha$.

Finally, $D_2 = D_1 + 2E$.

The values obtained in the manner related are now compared with those in the specification, and thus it is determined whether the gage is accurate within the allowable tolerances. If no specifications are available, the actual measurements of the gage are obtained and recorded.

Method 2. Taper ring gages may be checked in the manner just described with few exceptions. The method of checking these gages is

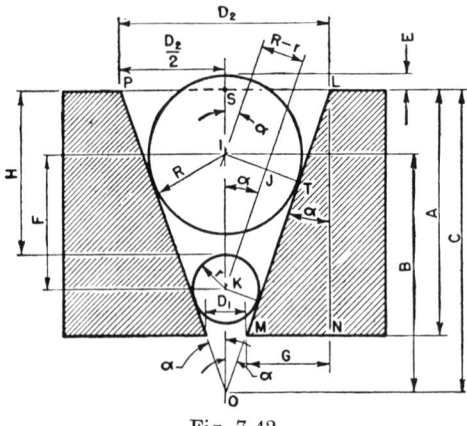

Fig. 7-42.

shown in Fig. 7-42. The equipment consists of two steel balls of different diameters conveniently selected to fit the taper gage, as shown in the illustration.

The method used for checking a taper ring gage is like that applied in checking taper members, which method consists of constructing a triangle containing sufficient data to permit calculation of the angle of the required taper.

Procedure: The procedure is in three steps.

Step 1. Insert a steel ball of convenient diameter into the ring gage. The size of the ball is determined by the diameter of the gage at the small end.

A measurement (H) is taken from the top of the smaller ball to the top surface of the gage.

The smaller ball is removed and a larger ball of convenient diameter is inserted into the gage, and a measurement E is taken from the top of the larger ball to the top surface of the gage.

A right triangle, IJK, is constructed between the centers of the two balls. In this triangle, $\angle IJK$ is equal to $90°$, $IJ = R - r$, and the hypotenuse, IK, is designated by F.

Note that R is the radius of the large ball and r the radius of the small one. F may be calculated as follows:

$$F = H + r + E - R$$

$$\sin \alpha = \frac{R - r}{F}$$

α may be found from the tables of trigonometric functions.

Step 2. Construct right $\triangle OTI$, in which $\angle IOT = \alpha$, since $JK \parallel OL$. $IT = R$ = radius of larger ball. The hypotenuse, IO, is designated B.

$$\therefore \frac{R}{B} = \sin \alpha \text{ and } B = \frac{R}{\sin \alpha}.$$

Now, in right $\triangle OSL$, $OS = C = B + R - E$.

$$SL = \frac{D_2}{2}, \text{ and } \frac{\frac{D_2}{2}}{C} = \tan \alpha.$$

$$\therefore \frac{D_2}{2} = C \times \tan \alpha, \text{ and } D_2 = 2C \times \tan \alpha.$$

Step 3. Construct right $\triangle LMN$, where $\angle MLN = \alpha$, since $LN \parallel SO$. In this triangle, LN is the thickness of the gage and is designated A, hence $\frac{MN}{LN} = \tan \alpha$, or $\frac{G}{A} = \tan \alpha$, and $G = A \times \tan \alpha$.

Finally, $D_1 = D_2 - 2G$.

Note that the larger ball may be of such diameter that its top is below the upper surface of the gage. It is good practice to select balls of such diameter that the distance, F, between centers is great enough to form a right triangle of reasonable proportions containing the angle α. The included angle of the taper is 2α.

In all taper gages the principal elements to check are the angle and the diameters at both ends.

It is also good practice in checking taper gages and other machine and tool parts to select several sets of precision rollers, pins, or balls, each of a different diameter. The results obtained from the use of these varied sets are averaged down to a usable value, the accuracy of which depends upon the number of tests performed.

Problems Involving Dovetails. A practical application of trigonometry in the shop is well illustrated in the setup used in the toolroom and inspection department for the purpose of checking dovetails. Two methods will be given, one for checking external dovetails and one for checking internal dovetails.

Method 1. Fig. 7-43 shows a method used in checking an external dovetail. The equipment consists of a pair of rollers or disks of a convenient diameter and a suitable measuring instrument or tool. The

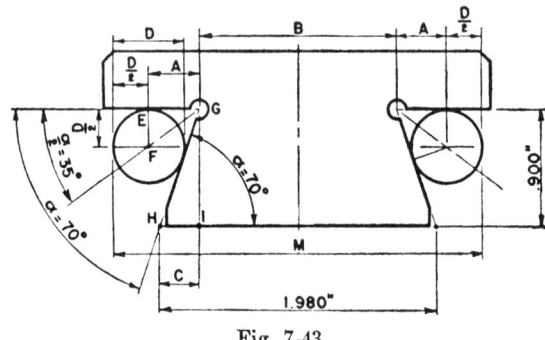

Fig. 7-43.

rollers are placed in contact with the dovetail surface and a measurement is taken across them, which procedure constitutes a basis for calculations that will determine the accuracy of the tested part.

Given: Assume that the principal dimensions of the dovetail shown in Fig. 7-43 are to be checked. These dimensions are the included angle α and the distance 1.980″, from which the location B of the vertices of $\angle \alpha$ on each side of the center line may be determined.

Required: To check principal dimensions of dovetail shown in Fig. 7-43.

Procedure: The rollers for this setup are .500″ in diameter, and the measurement (M) taken across them will indicate how accurate the dovetail is by comparing this measurement with the value required. The calculation is performed in the following manner:

In right $\triangle GEF$, $\angle EGF = \dfrac{\alpha}{2} = 35°$, $EF = \dfrac{D}{2} = \dfrac{.500}{2} = .250″$.

$\therefore EG = A = \dfrac{D}{2} \times \cot \dfrac{\alpha}{2} = .250 \times \cot 35° = .250 \times 1.4281 = .3570$.

In right $\triangle GIH$, $GI = 900″$, $\angle GHI = \alpha = 70°$. $\therefore HI = C = GI \times \cot \alpha = .900 \times \cot 70° = .900 \times .36397 = .3276$.

$B = 1.980 - 2C = 1.980 - (2 \times .3276) = 1.980 - .655 = 1.325$.

$M = 1.325 + 2A + D = 1.325 + (2 \times .357) + .500 = 1.325 + .714 + .500 = 2.539″$.

Method 2. The method of checking an internal dovetail is shown in Fig. 7-44.

Given: The basic dimensions specified on the working drawing are 4.500″, and the included angle $\alpha = 70°$.

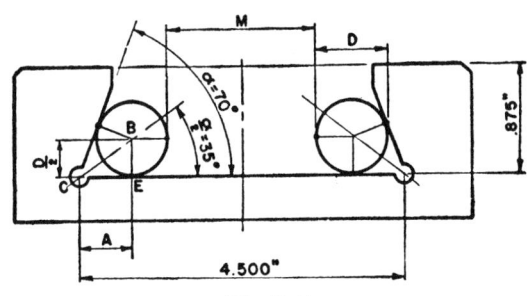

Fig. 7-44.

Required: To check the internal dovetail shown in Fig. 7-44.

Procedure: The rollers are .500″ in diameter, and a measurement (M) taken across them is compared with a corresponding dimension which is calculated as follows:

In right $\triangle BEC$, $\angle BCE = \dfrac{\alpha}{2} = 35°$ and $BE = \dfrac{D}{2} = .250″$. $\therefore CE = A = \dfrac{D}{2} \times \cot \dfrac{\alpha}{2} = .250 \times 1.4281 = .3570$.

$M = 4.500 - (2A + D) = 4.500 - (2 \times .357) - .500 = 4.500 - .714 - .500$.

$M = 3.286″$.

Additional Tool Design Problems. An interesting problem in tool design is illustrated in Fig. 7-45, wherein it is required to determine the length of a line that is tangent to two circles of different diameters.

Also, in tool and gage making and in precision inspection, it is frequently necessary to construct a temporary taper gage. If two disks of given diameter are chosen

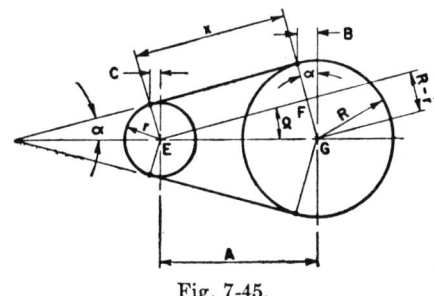

Fig. 7-45.

and set the necessary distance apart, a straightedge placed in contact with the disks will be seen to form half of the included angle of the required taper. Thus a taper gage is constructed with which given parts may be checked. The setup is shown in Fig. 7-46.

Problem 1. See Fig. 7-45.

Given: Two disks, of $\frac{3''}{4}$ and $1\frac{5''}{8}$ diameter, respectively, the center distance between disks equaling $A = 1\frac{11''}{16}$, as shown in Fig. 7-45.

Required: To find the dimension x, which is the length of the line tangent to the given disks.

Procedure: Draw line EF parallel to x and construct right $\triangle EFG$, in which $\angle \alpha$ can be calculated, since $FG = R - r$ and $EG = A = 1\frac{11''}{16} = 1.6875''$. $R = 1\frac{5}{8} \div 2 = \frac{13}{16} = .8125$. $r = \frac{3}{4} \div 2 = \frac{3}{8} = .375$.

$$\text{Sin } \alpha = \frac{R-r}{A} = \frac{.8125 - .375}{1.6875} = \frac{.4375}{1.6875} = .25925.$$

Angle $\alpha = 15° 1' 32''$.

In the same triangle, $EF = x = A \cos \alpha = 1.6875 \times \cos 15° 1' 32'' = 1.6875 \times .96581 = 1.630''$.

If desired, the dimension $B = R \times \sin \alpha = .8125 \times .25925$.
$$B = .2106''.$$
$C = r \times \sin \alpha = .375 \times .25925 = .0972''.$

Problem 2. The method of constructing a taper gage is shown in Fig. 7-46.

Given: Two disks of $\frac{1''}{2}$ and $\frac{3''}{4}$ diameter, respectively.

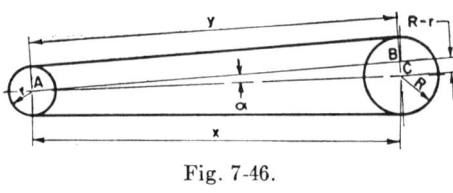

Fig. 7-46.

Required: To construct a gage $\frac{3''}{4}$ taper per foot by spacing the two given disks the necessary distance apart.

Procedure: In right $\triangle ABC$ of Fig. 7-46, $BC = R - r = \frac{3/4}{2} - \frac{1/2}{2} = .375 - .250 = .125.$

Tan $\alpha = \dfrac{.75}{12 \times 2} = .03125$, according to the foregoing rule for finding the tangent of half of the included angle if the taper per foot is known. Angle $\alpha = 1° 47' 24''$.

In the same triangle, the dimension x is the hypotenuse, $\therefore \sin \alpha = \dfrac{BC}{x} = \dfrac{R - r}{x}$.

But $\sin \alpha = \sin 1° 47' 24'' = .03123$, and $R - r = .125$, hence $.03123 = \dfrac{.125}{x}$, and $x = \dfrac{.125}{.03123} = 4.003''$.

If the dimension y is desired, $y = x \times \cos 1° 47' 24''$, $y = 4.003 \times .99951 = 4.000''$.

Problem 3. See Fig. 7-47.

Given: A toolroom problem dealing with a forming tool is illustrated in Fig. 7-47, which partially depicts a stellite tool that must be ground to the desired contour by use of a grinding wheel.

Required: The starting point, A, on the uncut material must be calculated in order to bring the grinding wheel up to that point. The location of point A from a definite point C is designated by the distance x.

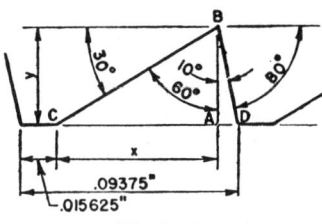

Fig. 7-47.

Procedure: In right $\triangle BAC$, $\dfrac{AB}{AC} = \dfrac{y}{x} = \cot 60° = .57735$. $y = .57735x$.

In right $\triangle BAD$, $\dfrac{y}{AD} = \dfrac{y}{.09375 - .015625 - x} = \dfrac{y}{.078125 - x} = \cot 10° = 5.6713$.

$y = (5.6713 \times .078125) - 5.6713x = .4430 - 5.6713x$.

Equating both values of y:

$$.57735x = .4430 - 5.6713x$$

$$.57735x + 5.6713x = .4430$$

$$6.2486x = .4430$$

$$x = \dfrac{.4430}{6.2486} = .071$$

PRACTICE PROBLEMS APPLICABLE IN THE SHOP AND TOOLROOM

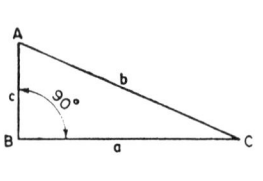
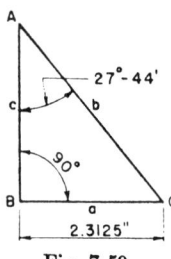

Fig. 7-48.   Fig. 7-49.   Fig. 7-50.

1. What is the tan of ∠ C in Fig. 7-48, if $a = 1$?
2. What is the cot of ∠ C in Fig. 7-48, if $c = 1$?
3. What is the sin of ∠ C in Fig. 7-48, if $b = 1$?
4. What is the cos of ∠ C in Fig. 7-48, if $b = 1$?
5. What is the sec of ∠ C in Fig. 7-48, if $a = 1$?
6. What is the csc of ∠ C in Fig. 7-48, if $c = 1$?
7. Find the tan of 21° 39'.
8. Find the cot of 68° 21'.
9. Find the sin of 57° 18'.
10. Find the cos of 32° 42'.
11. Find the sec of 38° 51'.
12. Find the csc of 51° 9'.
13. Find the tan of 51° 18'.
14. Find the cot of 42° 38'.
15. Find the sin of 59° 13'.
16. Find the cos of 18° 46'.
17. Find the sec of 63° 37'.
18. Find the csc of 7° 58'.
19. Find the angle whose tan = .33557.

Note: The statement in Problem 19 may be expressed as follows: The required angle = $\tan^{-1}$ .33557. $\tan^{-1}$ is read arc tangent. $\tan^{-1}$ .33557 represents the angle whose tan = .33557. In a similar manner, a problem may be stated in form of $\sin^{-1}$, $\cot^{-1}$, etc.

20. Find the angle whose cot = .48055.
21. Find the angle whose sin = .47869.
22. Find the angle whose cos = .80108.
23. Find the angle whose sec = 1.5666.
24. Find the angle whose csc = 2.6260.
25. Find the angle whose sin = .29973 to the nearest second.
26. Find the angle whose cos = .58484 to the nearest second.
27. Find the angle whose sec = 1.4481 to the nearest second.
28. Find the angle whose tan = .57013 to the nearest second.
29. Find the angle whose cot = .37936 to the nearest second.

TRIGONOMETRY

30. Find the angle whose csc = 2.8651 to the nearest second.

31. In Fig. 7-49, $A = \tan^{-1} \dfrac{a}{c}$. Find A, if $a = 2\dfrac{1}{4}''$ and $c = 3\dfrac{5}{16}''$.

32. In Fig. 7-49, $C = \tan^{-1} \dfrac{c}{a}$. Calculate C if a and c have the same values as those in Problem 31.

Note: C is a supplement of A, which can be used as a test of the accuracy of the answer to this problem.

33. In Fig. 7-49, $C = \sin^{-1} \dfrac{c}{b}$. Find C, if $b = 1''$ and $c = .7681''$.

34. Find the length of b in Fig. 7-50.
35. Find the angle C in Fig. 7-50.
36. Find the length of c in Fig. 7-50.

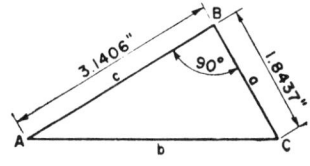
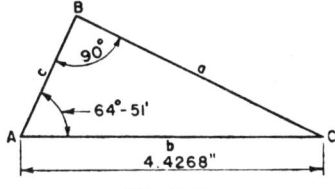

Fig. 7-51.   Fig. 7-52.

37. Find the angles A and C in Fig. 7-51.
38. Find the length of b in Fig. 7-51.
39. Find the length of a in Fig. 7-52.
40. Find the length of c in Fig. 7-52.
41. Find the angle C in Fig. 7-52.

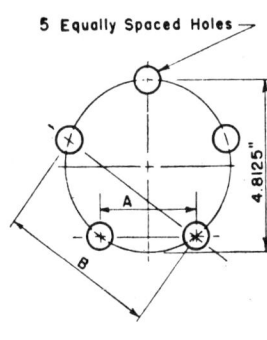
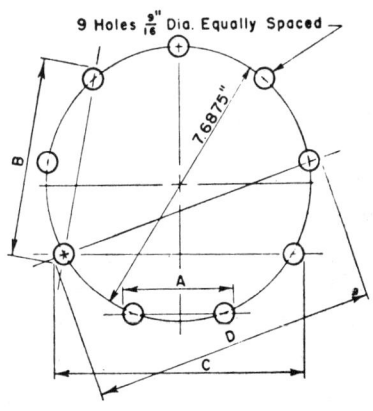

Fig. 7-53.   Fig. 7-54.

42. Determine the distance A in Fig. 7-53.
43. Determine the distance B in Fig. 7-53.
44. Determine the distance A in Fig. 7-54.
45. Determine the distance B in Fig. 7-54.
46. Determine the distance C in Fig. 7-54.
47. Determine the distance D in Fig. 7-54.

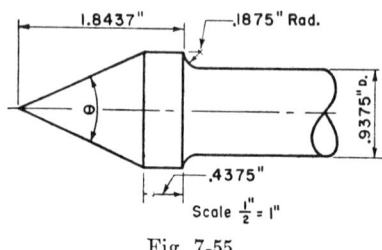

Fig. 7-55.

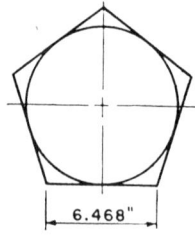

Fig. 7-56.

48. Find the included angle θ in Fig. 7-55 to the nearest minute.
49. A regular pentagon is shown in Fig. 7-56. Find the radius of the inscribed circle.

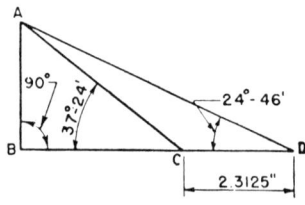

Fig. 7-57.

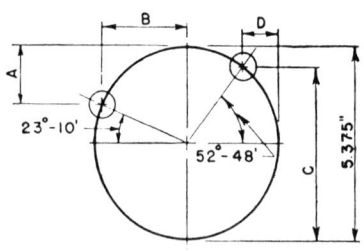

Fig. 7-58.

50. Find the lengths of AB and BC in Fig. 7-57.

Hint: In $\triangle ABC$, $\cot 37° 24' = \dfrac{BC}{AB}$, and $BC = AB \times \cot 37° 24'$. In $\triangle ABD$, $\cot 24° 46' = \dfrac{BC + 2.3125}{AB}$, and $BC = (AB \times \cot 24° 46') - 2.3125$. Equate both values of BC and solve for AB.

51. Find the dimensions A, B, C, and D in Fig. 7-58.
52. Find the length of the open belt in Fig. 7-59.

Hint: E and F are points of tangency. HGI is a right triangle in which $\sin \alpha = \dfrac{GI}{HI} = \dfrac{R - r}{C}$. In the same triangle HG may be calculated. Length of arc $EJ = \dfrac{2\pi r \times \alpha}{360}$, and that of arc $FK = \dfrac{2\pi R \times \alpha}{360}$.

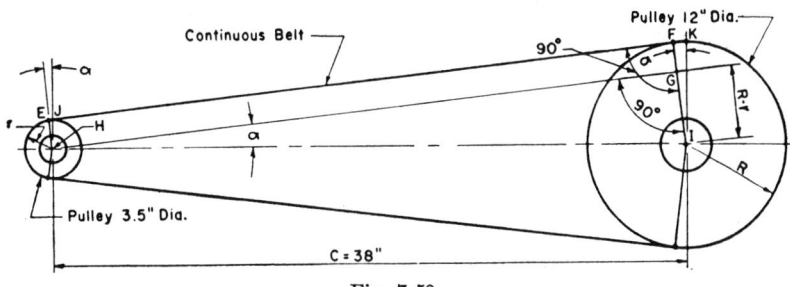

Fig. 7-59.

53. Find the length of the crossed belt in Fig. 7-60, assuming the same pulley diameters and center distance C as in Fig. 7-59.

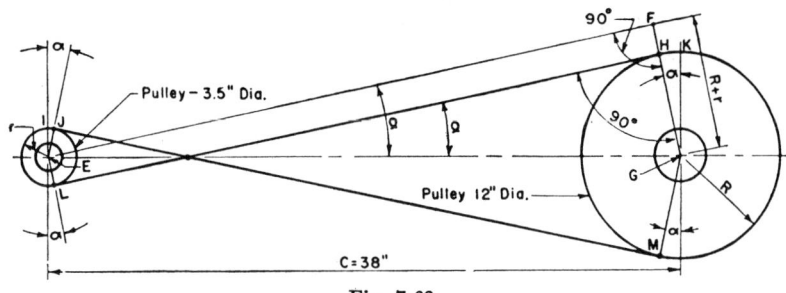

Fig. 7-60.

Hint: Construct right $\triangle EFG$, in which $\sin \alpha = \dfrac{FG}{EG} = \dfrac{R+r}{C}$. In the same triangle EF may be calculated. But $HL = JM = EF$. Lengths of arcs IJ and HK may be calculated by the method suggested in Problem 52.

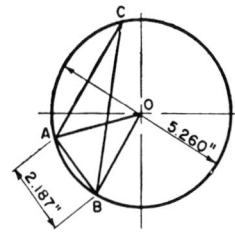

Fig. 7-61.

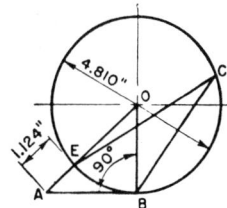

Fig. 7-62.

54. Find the length of the circular arc subtended by chord AB, and the number of degrees and minutes in $\angle AOB$, as shown in Fig. 7-61.
55. Find the number of degrees and minutes in $\angle ACB$, Fig. 7-61.
56. Find the length of line AB in Fig. 7-62.

57. Find the number of degrees and minutes in $\angle AOB$, Fig. 7-62.
58. Find the number of degrees and minutes in $\angle ECB$, Fig. 7-62.

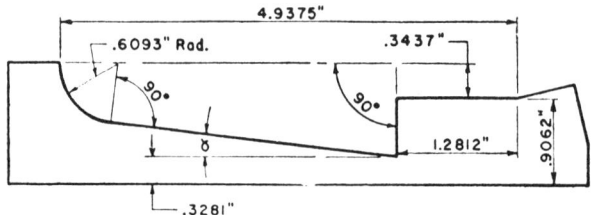

Fig. 7-63.

59. Find $\angle \alpha$ in degrees and minutes, as shown in Fig. 7-63.
60. Two holes are spaced as shown in Fig. 7-64. Find the dimension x and $\angle \alpha$ in degrees and minutes.

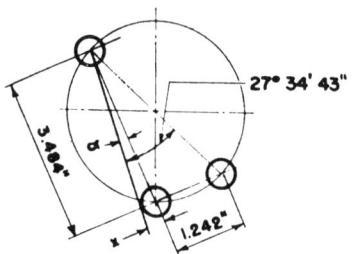

Fig. 7-64.

61. In the oblique $\triangle ABC$ shown in Fig. 7-65, $a = 2.8125''$, $c = 3.7812''$, and $\angle C = 52° 27'$. Find angles A and B and dimension b.
62. In the oblique $\triangle ABC$ shown in Fig. 7-66, $b = 5.987''$, $\angle A = 63° 52'$ and $\angle B = 106° 24'$. Find dimensions a and c and angle C.
63. In the oblique $\triangle ABC$ shown in Fig. 7-67, $a = 3.562''$, $b = 5.562''$ and $\angle C = 63° 30'$. Find dimension c and angles A and B.

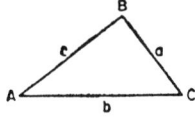 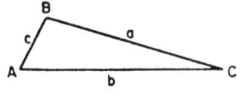 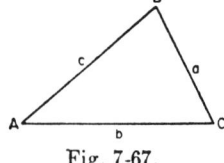

Fig. 7-65.　　　　　　Fig. 7-66.　　　　　　Fig. 7-67.

64. In the oblique $\triangle ABC$ shown in Fig. 7-68, $a = 3.125''$, $c = 3.750''$, and $\angle B = 116° 40'$. Find angles A and C and dimension b.
65. In the oblique $\triangle ABC$ shown in Fig. 7-69, $a = 3.437''$, $b = 5.531''$, and $c = 4.250''$. Find angles A, B, and C.

66. In the oblique $\triangle ABC$ shown in Fig. 7-70, $a = 7.500''$, $b = 4.750''$, and $c = 3.625''$. Find angles A, B, and C.

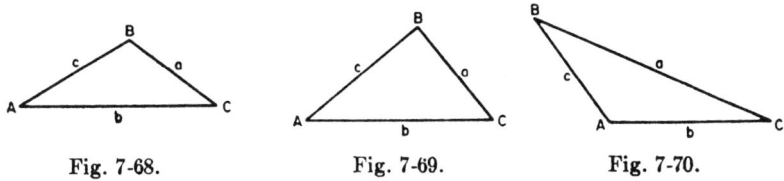

Fig. 7-68. Fig. 7-69. Fig. 7-70.

67. In the oblique $\triangle ABC$ shown in Fig. 7-71, $a = 4.252''$, $b = 5.406''$, and $c = 2.430''$. Find angles A, B, and C. (Use the projection formulas.)

68. In the oblique $\triangle ABC$ shown in Fig. 7-72, $a = 7.000''$, $b = 4.800''$, and $c = 2.790''$. Find angles A, B, and C. (Use the projection formulas.)

69. In the oblique $\triangle ABC$ shown in Fig. 7-73, $b = 5.625''$, $\angle A = 40° 38'$, and $\angle C = 24° 50'$. Find $\angle B$ and dimensions a and c. (Use the cotangent formulas.)

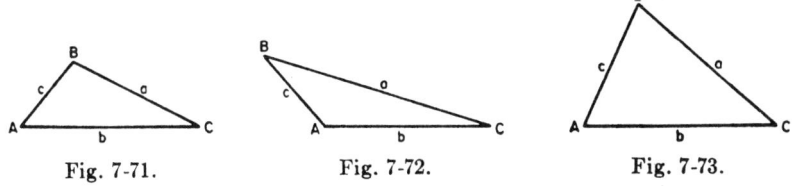

Fig. 7-71. Fig. 7-72. Fig. 7-73.

70. In the oblique $\triangle ABC$ shown in Fig. 7-74, $b = 4.518''$, $\angle A = 18° 24'$, and $\angle C$ $138° 42'$. Find $\angle B$ and dimensions a and c. (Use the cotangent formulas.)

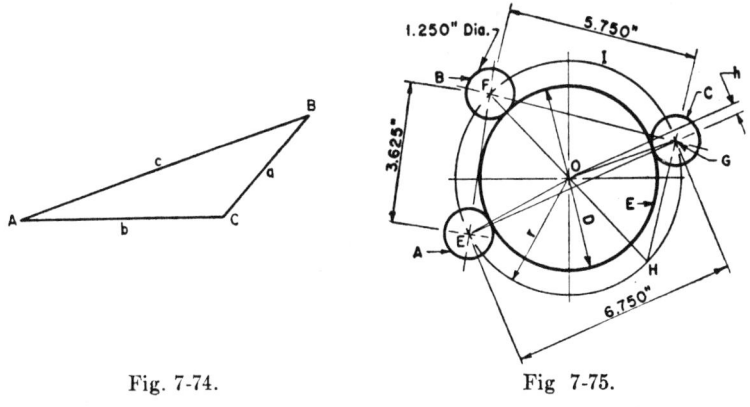

Fig. 7-74. Fig 7-75.

71. Three pinions A, B, and C of the same pitch diameter $(1.250'')$ are in mesh with gear E and are located as shown in Fig. 7-75.

Find the center O of gear E and its pitch diameter, i.e., the dimension h and the pitch diameter D.

Hint: In the oblique $\triangle EFG$, the three sides are known, and $\angle FEG$ may be calculated. $\angle FHG = \angle FEG$, since both angles subtend the same arc FIG.

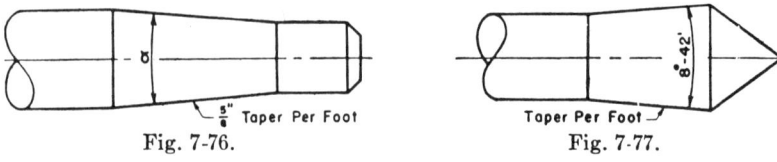

Fig. 7-76. Fig. 7-77.

72. What is the included angle α in Fig. 7-76?
73. What is the taper per foot of part shown in Fig. 7-77?

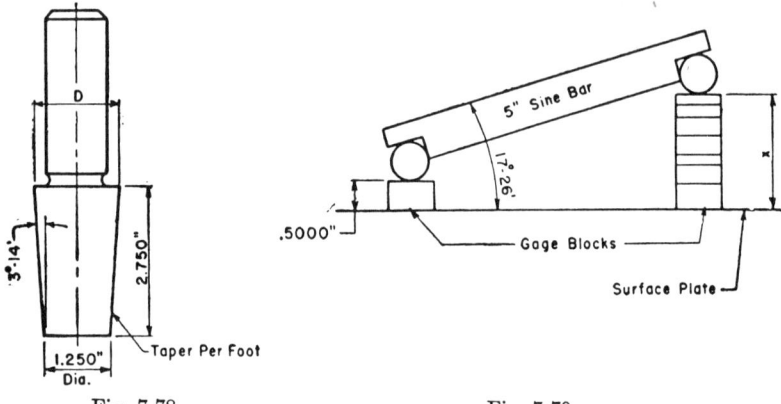

Fig. 7-78. Fig. 7-79.

74. Calculate the diameter D of the taper plug gage shown in Fig. 7-78. Also calculate the taper per foot.
75. Calculate the dimension x in Fig. 7-79.

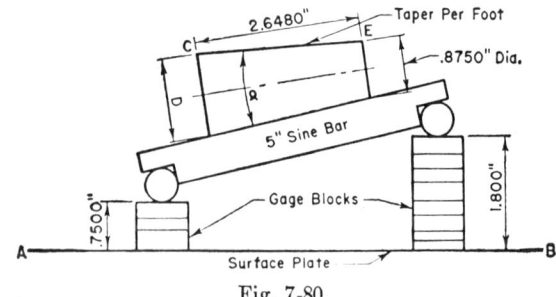

Fig. 7-80.

76. Calculate $\angle \alpha$, the taper per foot, and the diameter D in Fig. 7-80. Assume CE is parallel to AB.

TRIGONOMETRY

77. Find the angle α and calculate x in Fig. 7-81.

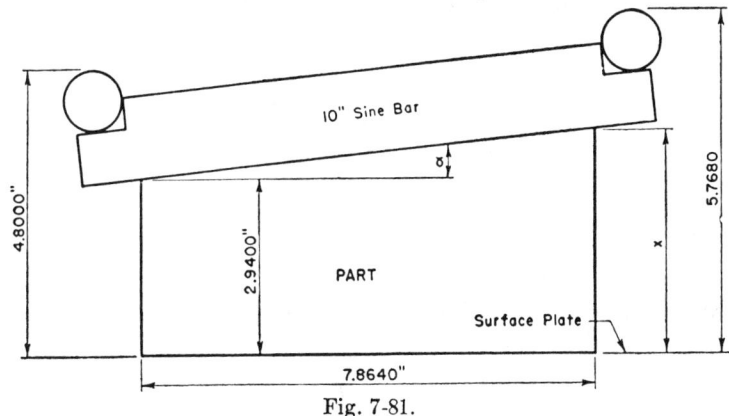

Fig. 7-81.

78. The specification for checking a V-block in Fig. 7-82 calls for a definite measurement (M_1) between the top of the surface plate and the

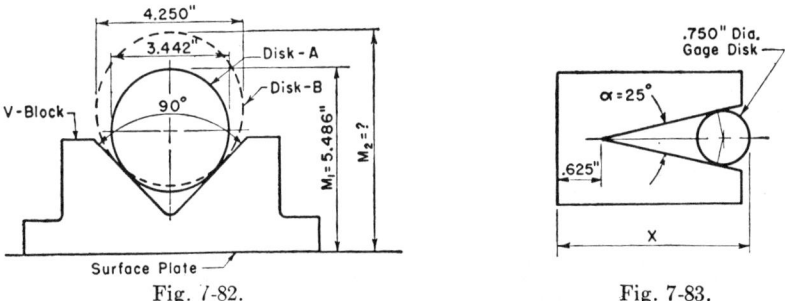

Fig. 7-82.   Fig. 7-83.

top of a recommended gage disk (A) of 3.442″ diameter. Calculate the correct measurement (M_2) from the top of the surface plate to the top of an available gage disk (B) of 4.250″ diameter.

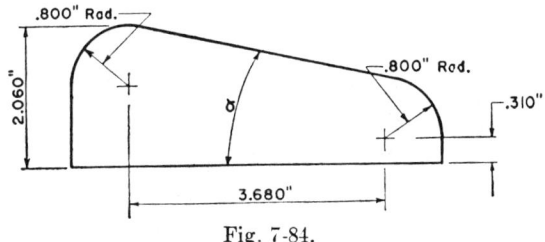

Fig. 7-84.

79. Calculate the dimension x in Fig. 7-83, if the included angle α and the diameter of the gage disk are known.

80. Calculate angle α in Fig. 7-84.

290   MATHEMATICS FOR INDUSTRY

81. Calculate angle α, which is formed by the line AB (tangent to both given arcs) and the horizontal base line, CD, as shown in Fig. 7-85.

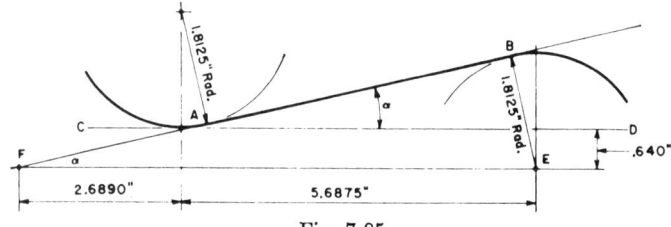

Fig. 7-85.

82. Determine the included angle α and the diameters D_1 and D_2 of the ends of the taper plug gage shown in Fig. 7-86, where two rollers of known diameter and two gage blocks of known thickness are used in the setup.

83. Determine the included angle α and the diameters D_1 and D_2 of the ends of the taper ring gage shown in Fig. 7-87, where two balls of known diameters are used in the setup.

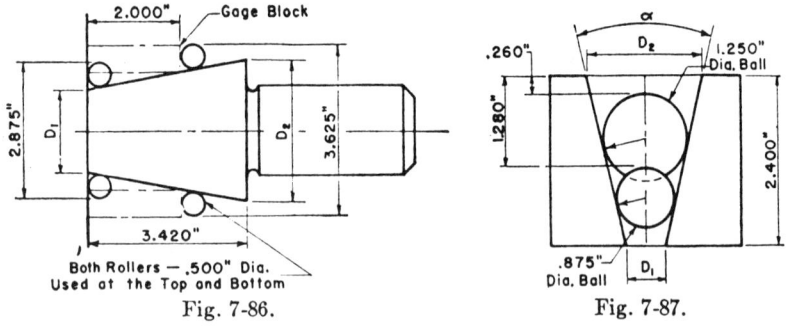

Fig. 7-86.   Fig. 7-87.

84. Calculate the included angle α and the dimension x of the dovetail shown in Fig. 7-88. Two rollers of known diameter are used in the setup.

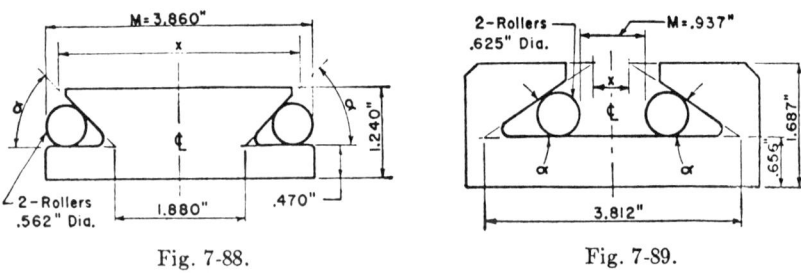

Fig. 7-88.   Fig. 7-89.

85. Calculate the included angle α and the dimension x of the dovetail

TRIGONOMETRY

shown in Fig. 7-89. Two rollers of known diameter are used in the setup.

86. Find the length of line x, which is tangent to the given disks in Fig. 7-90. The disk diameters and the space between disks are known.

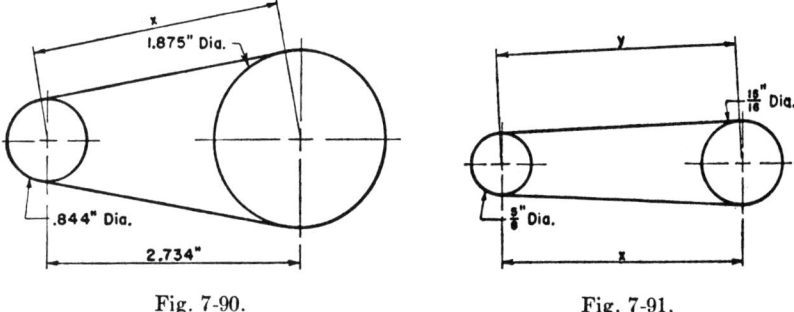

Fig. 7-90.   Fig. 7-91.

87. Construct a taper gage of $\frac{5''}{8}$ taper per foot, if two disks of known diameters are available.

Hint: Calculate dimension x in Fig. 7-91 to determine the correct gage. Also calculate dimension y.

CHAPTER VIII

SCREW THREADS

The screw thread is one of the most fundamental inventions of the human race. The ancients applied the principle of the screw thread in the invention of many useful tools and mechanical devices such as machines for elevating well water for irrigating the fields, machines for pressing cloth and paper, and presses for extracting oil from olives and juice from grapes and berries.

The standardization of various shapes of screw threads was achieved only recently and, at present, the thread forms used in mass production are restricted in number in order that a greater interchangeability of bolts, screws, studs, nuts and other threaded items may be provided in the assembly of industrial machines, household appliances and so on.

Threaded parts are manufactured in large quantities, and since almost all assembled units use them, an attempt was made, for the sake of interchangeability, to standardize the more basic definitions of screw threads. Before any further discussion of screw threads is presented, the reader should become acquainted with the definitions which follow. Fig. 8-1 illustrates some of the important thread elements and the terminology proper to them.

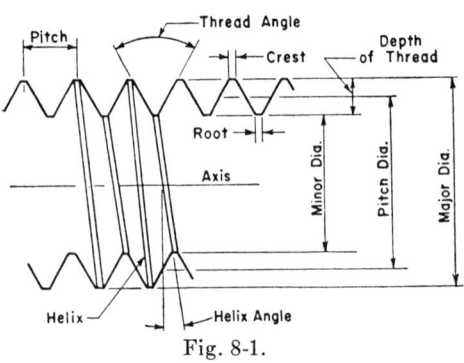
Fig. 8-1.

An *external thread* is a thread on the outside of a member; for example, a threaded plug or bolt.

An *internal thread* is a thread on the inside of a member; for example, a threaded hole.

The *major diameter* (outside diameter) is the largest diameter of the screw thread and applies to both internal and external threads.

The *minor diameter* (core diameter) is the smallest diameter of the screw thread and applies to both internal and external threads.

The *crest* is the top surface joining the two sides of a thread.

The *root* is the bottom surface joining the sides of two adjacent threads.

The *pitch diameter* is the diameter of an imaginary cylinder passing through the threads at points at which the solid thread width and the width of the space between the threads is the same. If the major diameter of the threaded part is *basic*, it is exactly as the nominal size of the bolt; and if the minor diameter, the thread angle, and the thickness of the crest and root are basic—i.e., are exactly as calculated—the pitch diameter bisects the depth of the thread.

The *depth of the thread* is the vertical distance between the crest and the root.

The *thread angle* is the included angle, such as the 60° angle adopted in the American National form of thread.

The *pitch* is the linear distance from a given point on a thread to the corresponding point on the next thread, measured parallel to the axis of the screw.

The *lead* is the axial distance (parallel to the axis) moved by the screw thread in one revolution. In a single-thread screw, the pitch and lead are the same; in a double-thread screw, the lead is twice the pitch, etc.

The *helix angle* is the angle formed by the thread helix at the pitch diameter and a plane perpendicular to the axis.

Various types of screw threads meet the general functions listed above. The American National Screw Thread System, the most generally used in the United States, and some of the commonest other types of screw threads will be discussed.

AMERICAN NATIONAL SCREW THREADS

In 1928 the National Screw Thread Commission was authorized by Congress to establish the American National form of thread, shown in Fig. 8-2. This system was adopted in 1935 as the standard in the United States. The included angle in this system is 60°, and the crest and root width is equal to $\frac{1}{8}$ pitch.

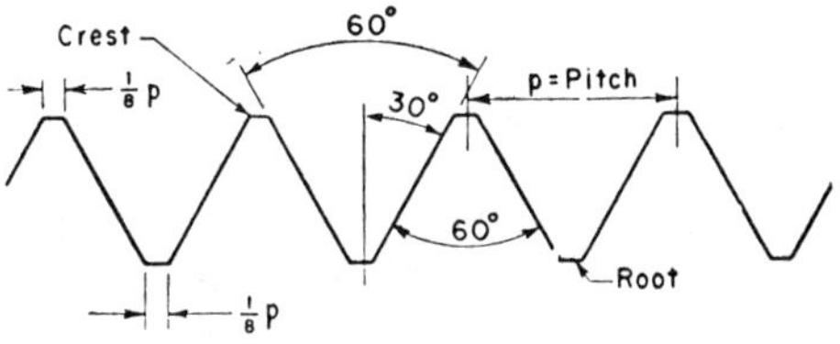

Fig. 8-2.

The American National thread system is divided into five groups called *thread series*. These five groups differ from one another in the diameter-pitch relations of the threads included, but all employ the American National form of thread. These series are: the coarse-thread, the fine-thread, the 8-pitch-thread, the 12-pitch-thread, and the 16-pitch-thread series.

Thread Fits. Since threaded members are used in the assembly of a wide variety of objects, some of which require more precision than others, the accuracy in manufacturing threads proper is also of a varied nature. Some threads are cut or formed with a higher degree of precision than other threads, hence the National Screw Thread Commission established for general use four distinct classes of screw thread fits to be used for particular types of assembly work. These classes were intended to insure a uniform practice for screw threads not included in the American National coarse- or fine-thread series, nor in the 8-, 12-, or 16-pitch series.

Class 1, Loose Fit. This fit is for work that requires easy and fast assembly of threaded parts, such as certain units of agricultural equipment, road-building machinery, and commercial equipment of more or less rough quality, in which a certain amount of play between threaded members does not affect proper functioning of the unit.

Class 2, Free Fit. This fit is for finished and semifinished bolts and screws and for other threaded members used in common grades of work, such as the bulk assembly in automobile and machine manufacture.

Class 3, Medium Fit. Class 3 fit is for work of a more precise nature, where free play between parts is not permissible, as in units of automobile and airplane assembly, for example.

The three classes of screw-thread fits just enumerated are appropriate to the needs of large-scale production in that these fits provide an interchangeability that is greatly desired.

Class 4, Close Fit. Class 4 fit is for work that requires careful fitting of parts that have been manufactured with small tolerances and small allowance for fit. This includes units of automobile and airplane engines composed of parts manufactured with high precision. It is difficult to achieve interchangeability in such work because of the selectivity involved in assembling component threaded parts. Each unit is fitted by careful selection of precisely mated parts.

Thread Formulas. The following formulas are derived for the American National form of thread, as illustrated in Fig. 8-3. All calculations and results are basic, i.e., ideal conditions are assumed, and tolerances or allowances are not considered.

Formula 1. The pitch $= p = \frac{1}{N}$, where $N =$ number of threads per inch.

SCREW THREADS

Formula 2. The basic depth = d = .64952 × pitch. It is derived as follows:

Assume a sharp V-thread as shown at ABC in Fig. 8-3. FC bisects $\triangle ABC$, and $\triangle BFC$ is a right triangle, in which FC = depth of sharp V-thread = h, and $BC = p$, since $\triangle ABC$ is equilateral.

In right $\triangle BFC$, $h = p \times \cos 30° = .86603p$.

The small triangles at the crest and at the root are also equilateral and are similar to $\triangle ABC$.

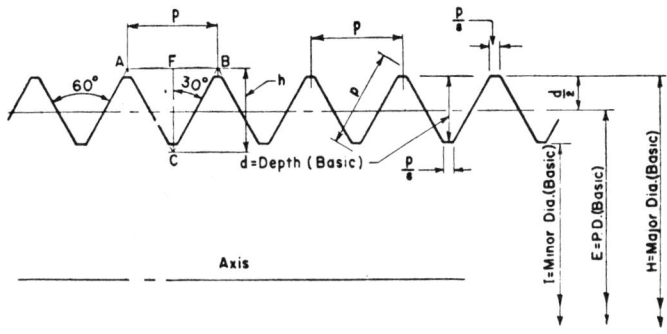

Fig. 8-3.

In similar triangles, corresponding elements are proportional, therefore, the base of each small triangle being $\frac{1}{8}$ of the pitch, the altitude of each small triangle is $\frac{1}{8}$ of the altitude of $\triangle ABC$, i.e., $\frac{1}{8}h$.

$$d = h - \left(2 \times \frac{1}{8}h\right) = h - \frac{h}{4} = .75h$$

$$h = .86603p$$

$$d = .75h = .86603p \times .75 = .64952p$$

$$d = .64952p = \frac{.64952}{N}, \text{ since } p = \frac{1}{N}$$

Formula 3. The basic minor diameter = I = basic major diameter − two-thread depths. $I = H - 2d$.

Formula 4. The basic pitch diameter = E = basic major diameter − one-thread depth. $E = H - d$.

The pitch diameter is an imaginary diameter and is one of the most important elements in thread calculations. The tolerances and allowances of the component parts are given on the pitch diameter for proper fitting of one part into the other. When computing the helix angle, one of the elements of computation is the pitch circumference, which is obtained by multiplying the pitch diameter by π.

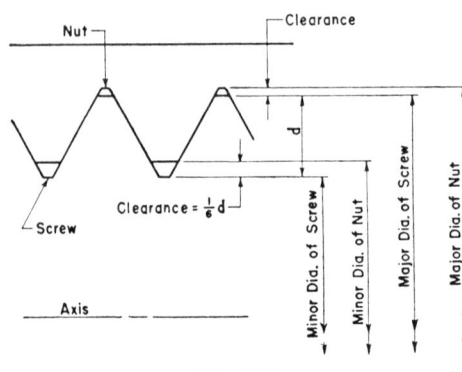

Fig. 8-4.

Formula 5. The tap drill diameter for the American National form of thread is derived and illustrated in Fig. 8-4. Before cutting a thread in a nut with a tap, a preliminary operation of drilling is required. Since there is sufficient clearance between the minor diameter of the nut and the screw, a sufficiently large diameter drill, called a *tap drill*, must be used before the thread is cut. This procedure serves to prevent breakage of the taps as well.

The amount of clearance depends upon the material of the nut. For ordinary materials, such as low-carbon steels and many copper-base alloys, the clearance is $\frac{1}{6}$ of the depth d, as shown in Fig. 8-4.

The tap drill diameter, T, which is somewhat larger than the minor diameter of the nut, is equal to the basic major diameter of the screw, H, less two basic depths of the thread, and plus two clearances.

Expressed algebraically, $T = H - 2d + \left(2 \times \frac{1}{6}d\right)$.

$$T = H - \left[\left(2 \times \frac{.64952}{N}\right) - \left(2 \times \frac{.64952}{6 \times N}\right)\right] =$$
$$H - \left(2 \times \frac{.64952 - .10825}{N}\right).$$

$T = H - 2 \times \frac{.54127}{N} = H - \frac{1.08254}{N}$, since $d = \frac{.64952}{N}$.

Example: Find the tap drill diameter T for a $\frac{1''}{2}$ —13 nut. $T = H -$

$$\frac{1.08254}{N} = .5000 - \frac{1.08254}{N} = .5000 - .0833 = .4167'', \text{ where } N =$$

13 = number of threads per inch. The nearest fractional size drill is $\frac{27''}{64}$ diameter.

Drills from .234'' to .413'' are specified by their letter sizes and those from .0135'' to .2280'' by their wire gage sizes.

Drills above .413'' are specified by their nearest fractional sizes.

When the clearance is $\frac{1}{6}$ of the basic depth of thread, the working depth of thread is equal to $\frac{5}{6}$ of the basic depth, and is referred to as $83\frac{1}{3}$ per cent thread, since $\frac{5}{6} = .833 = 83\frac{1}{3}$ per cent.

In tough and soft materials, such as copper, Norway iron, and drawn aluminum, a 75 per cent thread or less is recommended—i.e., the clearance is equal to $\frac{1}{4}$ of the basic depth. If screws enter more than one and one-half times the diameter, a smaller percentage, say 60 per cent or even 50 per cent, is sufficient.

OTHER TYPES OF SCREW THREADS

Various other forms of screw threads have been introduced to meet special requirements. Some of the most frequently used will be mentioned. The formulas for them will also be given.

Sharp V-Thread. This thread has the same angle of thread as the American National form but the top and bottom of the thread are sharp rather than flat. This thread is useful in some cases—especially where a steam-tight joint is necessary. The sharp V-thread is shown in Fig. 8-5. The included angle of this thread is 60°.

The formulas for the sharp V-thread are as follows:

1. Pitch = $p = \frac{1}{N}$, where N = number of threads per inch.

2. Depth of thread = $d = \frac{.86603}{N}$.

3. H = major diameter of screw.
4. I = minor diameter of screw = $H - 2d$.
5. E = pitch diameter = $H - d$.

Square Thread. The square thread is shown in Fig. 8-6. It is used on lead screws for transmission of power. However, owing to the difficulty of cutting it with dies and other disadvantages, the square thread has been displaced to some extent by the Acme thread.

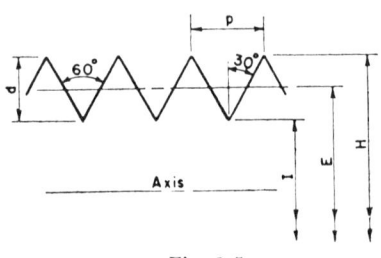

Fig. 8-5.   Fig. 8-6.

The formulas for the square thread follow:

1. Pitch $= p = \dfrac{1}{N}$, where $N =$ number of threads per inch.

2. Depth of thread $= d = \dfrac{p}{2} = \dfrac{1}{2N}$. In practice, the space between the threads is slightly over $\dfrac{p}{2}$, and the thickness of the thread is slightly less than $\dfrac{p}{2}$.

3. $H =$ major diameter of screw.
4. $I =$ minor diameter of screw $= H - 2d$.
5. $E =$ pitch diameter of screw $= H - d$.

Acme Thread. The Acme thread is shown in Fig. 8-7. It is also used on lead screws for transmission of power in preference to square threads because of its smoothness of operation and lack of backlash.

The included angle equals 29°.

The formulas for the Acme thread are:

1. Pitch $= p = \dfrac{1}{N}$, where $N =$ number of threads per inch.

2. Depth of thread $= d = \dfrac{p}{2} + .010'' = \dfrac{1}{2N} + .010''$.

3. $H =$ major diameter of screw.
4. $I =$ minor diameter of screw $= H - 2d$.
5. $E =$ pitch diameter of screw $= H - d$.
6. $w = .3707p - .005'' = \dfrac{.3707}{N} - .005''$.

SCREW THREADS

Worm Thread. The worm thread is shown in Fig. 8-8. It is used in worm gearing, its form is similar to that of an Acme thread, except the thread depth is greater. The thread angle in most cases is 29°, but in some cases it is 40°.

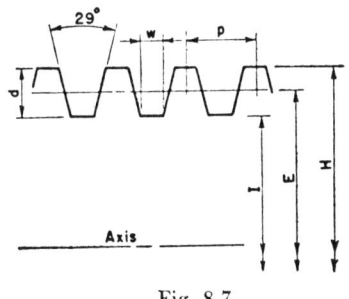

Fig. 8-7.

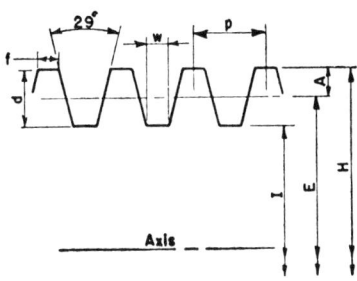

Fig. 8-8.

For the 29° worm the formulas are:

1. Addendum of worm = $A = .3183p = \dfrac{.3183}{N}$.

2. Pitch = $p = \dfrac{1}{N}$, where N = number of threads per inch.

3. Depth of thread = $d = .6866p = \dfrac{.6866}{N}$.

4. H = major diameter of screw.
5. I = minor diameter of screw = $H - 2d$.
6. E = pitch diameter of screw = $H - 2A = H - 2 \times \dfrac{.3183}{N}$.

$E = H - \dfrac{.6366}{N}$.

7. $w = .310p = \dfrac{.310}{N}$.

8. $f = .335p = \dfrac{.335}{N}$.

Buttress Thread. The buttress thread is shown in Fig. 8-9. It is used for transmission of power in mechanisms of the heavy-duty type, such as heavy jackscrews. It takes a bearing

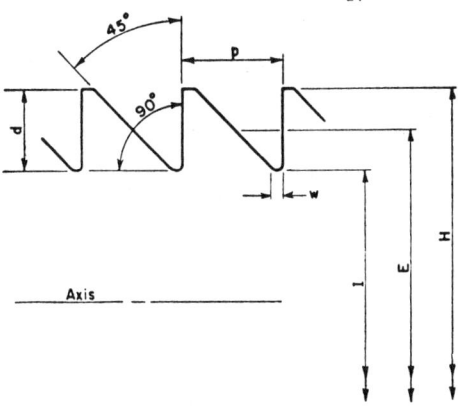

Fig. 8-9.

on the vertical side only, hence the power is transmitted in one direction.

The formulas for the buttress thread are:

1. Pitch = $p = \frac{1}{N}$, where N = number of threads per inch.
2. Depth of thread = $d = .750p + .010'' = \frac{.750}{N} + .010''$.
3. H = major diameter of screw.
4. I = minor diameter of screw = $H - 2d$.
5. E = pitch diameter of screw = $H - d$, usually $H - .750p$.
6. $w = .125p - .010''$.

International (Metric) Thread. The International (Metric) system of threads is the same as the American National form, shown in Fig. 8-3, except that all sizes and pitches are expressed in millimeters. This thread is used on spark plugs and machine parts of airplanes.

American Standard Taper Pipe Thread. The American standard taper pipe thread is shown in Fig. 8-10.

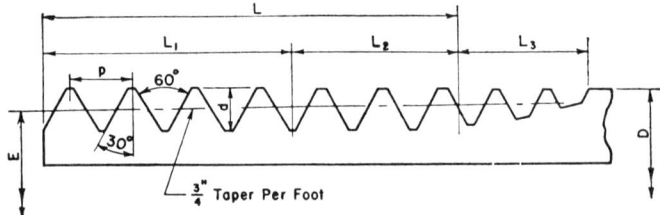

Fig. 8-10.

It is employed in pipe fittings designed for joining pipes, tubes, etc., such as engine manifolds, crankcase and radiator drain plugs, and marine equipment, and in oil refineries and power plants—in short, wherever a tight joint is required.

The front portion of the pipe extending for a distance L_1 includes threads that are perfect at top and bottom.

$L_1 = (.8D + 4.8) \times p$, where D = outside diameter of pipe, and p = pitch of thread. This portion is $\frac{3''}{4}$ taper per foot.

The threads have a 60° included angle.

The pitch diameter E at the end of the pipe = $D - (.050D + 1.1) \times p$. The threads in this portion are only slightly flattened on top and bottom, and the thread depth $d = .800 \times p = \dfrac{.800}{N}$, where N = number of threads per inch.

In the middle portion extending through length L_2, there are two threads whose roots are similar to those in portion L_1 but whose crests are flat.

In the rear portion extending through length L_3, there are several imperfect threads whose origin is due to the chamfer in the thread-cutting die. The crests and roots in this portion are flattened.

The front and middle portions constitute the working portion of the pipe thread, and its length $L = L_1 + L_2 = (.8D + 6.8) \times p$.

MEASURING SCREW THREADS

In precision measurement of screw threads, the outside diameter of the threaded part is not regarded as the basic element determining the accuracy of the entire thread; rather, it is the pitch diameter that determines the accuracy and proper performance of the screw thread, and careful measurement of the pitch diameter constitutes one of the most important operations in thread technology. Three methods for measuring screw threads will be discussed, the thread micrometer method, the three-wire method, and the optical methods.

Thread Micrometer Method. The pitch diameter of an American National form thread and a sharp V-thread can be measured with a special thread micrometer fitted with conical points that bear only on the thread angle, as shown in Fig. 8-11.

The points of the micrometer are rounded off to prevent interference with the threaded part being checked, and the sides of these points fit directly with the sides of the thread.

When the micrometer is closed in the position shown in Fig. 8-11, the pitch lines MN of anvil and spindle coincide, and the micrometer reading is O.

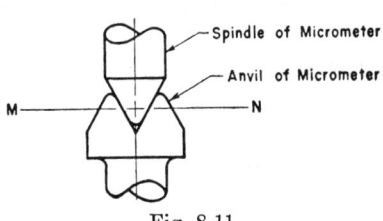

Fig. 8-11.

When the micrometer is open to the extent where the spindle and anvil points fit directly on the sides of the thread being checked, the reading is the actual value of the pitch diameter of the part.

The disadvantages of this method are as follows:
1. A large number of special conical points must be carried in stock to accommodate various pitches of coarse and fine threads.
2. The possible occurrence of slight distortion in the reading of the thread micrometer owing to the fact that its anvil and spindle are in line while the ridge of the thread on one side of the part and the groove on the opposite side may not be in line because of the thread helix.

Three-Wire Method. The three-wire method is recommended by the National Bureau of Standards as one of the most accurate operations in securing uniformity when checking the pitch diameter of screws.

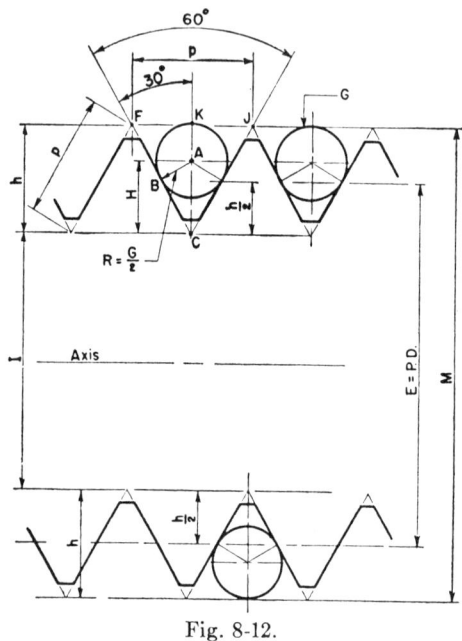

Fig. 8-12.

Three wires of a chosen diameter are placed in the threads, as shown in Fig. 8-12, and a measurement is taken over these wires. By a simple calculation the actual pitch diameter of the screw is determined and compared with the basic pitch diameter for a screw of the size and form indicated. Fig. 8-12 illustrates this method as used for finding the pitch diameter of an American National form thread.

Three steel wires of a chosen diameter are placed in the threads for best results. This diameter G can be calculated as follows:

In the right $\triangle ABC$, AB is equal to the radius of the wire which is tangent to the sides of the thread angle.

The point of tangency B is on the pitch line and it bisects FC, making $BC = \dfrac{FC}{2} = \dfrac{p}{2}$, since $\triangle CFJ$ is equilateral and $FC = FJ = p =$ pitch.

In the right $\triangle ABC$, $\dfrac{AB}{BC} = \dfrac{R}{p/2} = \tan 30° = .57735$. $R = \dfrac{p}{2} \times$

.57735. $G = 2R = 2 \times \dfrac{p}{2} \times .57735 = .57735p = \dfrac{.57735}{N}$, where $N =$ number of threads per inch and $p = \dfrac{1}{N}$.

The formula $G = \dfrac{.57735}{N}$ determines the size of wire which is preferred for measuring the pitch diameter of an American National form thread. It is referred to as the *best* wire diameter.

In the same right $\triangle ABC$, $\dfrac{AB}{AC} = \sin 30° = .5000$, and $AC = \dfrac{AB}{.5000} = \dfrac{R}{.5000} = 2R = G =$ wire diameter.

It has been shown in previous derivations that $h = .86603p = \dfrac{.86603}{N}$, therefore $I = E - \left(2 \times \dfrac{h}{2}\right) = E - h = E - \dfrac{.86603}{N}$.

Combining several components, we finally compute M, which is the measurement over the wires.

$$M = I + (2 \times AC) + (2 \times AK)$$

$$M = I + 2G + G, \text{ since } AC = G \text{ and } AK = \dfrac{G}{2}$$

$$M = E - \dfrac{.86603}{N} + 3G, \text{ since } I = E - \dfrac{.86603}{N}$$

$$E = M + \dfrac{.86603}{N} - 3G, \text{ where } E = \text{pitch diameter}$$

$N =$ number of threads per inch and $G =$ wire diameter.

A best wire diameter is recommended for precise results, although other diameters may be suitable, provided they are not too small or too large, thus preventing the taking of an accurate measurement M over the wires. The formula just derived holds good for variations in the wire diameter.

The recommended minimum diameter of wire is $\dfrac{.56}{N}$.

The recommended maximum diameter of wire is $\dfrac{.90}{N}$.

Note that when measuring a sharp V-thread with best wires placed in the threads, the basic major diameter of the V-thread will be tangent to the wires, and the three-wire formula is reduced to the following form:

$$M = E + h = E + \frac{.86603}{N}, \text{ since } h = \text{depth of V-thread.}$$

Then $E = $ pitch diameter $= M - \dfrac{.86603}{N}$.

When checking a sharp V-thread, wires somewhat larger than the best size should be chosen to make certain that measurement M is over the wires rather than over the sharp crests of the threads.

The three-wire formula just given for computing the pitch diameter of a screw is for the American National form thread only, the thread angle of which is 60°. **The** general formula for computing the pitch diameter of a screw whose thread angle may be any value besides 60° is as follows:

$$E = M + \frac{\cot \alpha}{2N} - G(1 + \csc \alpha),$$

where $E = $ pitch diameter, $M = $ measurement over the three wires, $\alpha = $ one-half of the included angle of the thread, $N = $ number of threads per inch, and $G = $ best wire diameter (preferred).

Since the American National form thread has such wide application in industry, the three-wire formula $\left(E = M + \dfrac{.86603}{N} - 3G\right)$ for checking the pitch diameter of the screw is generally adopted in toolrooms, gage laboratories, inspection departments, and thread manufacturing departments. However, the formula should be used for threads whose thread helix angle does not exceed 6°. If the thread helix angle is greater than 6°, the variation due to the helix angle will be quite pronounced, and another formula for calculating the pitch diameter of the screw should be used.

This formula is $E = M + \dfrac{\cot \alpha}{2N} - G\left(1 + \csc \alpha + \dfrac{m^2}{2} \cos \alpha \times \cot \alpha\right)$,

where $E = $ pitch diameter, $M = $ measurements over the three wires, $\alpha = $ one-half of the included thread angle, $N = $ number of threads per inch, $G = $ best wire diameter (preferred), $m = $ tan of helix angle $= \dfrac{l}{\pi E}$, where $l = $ lead of thread in inches.

An example in the technique of checking the pitch diameter of a precision screw by means of the three-wire formula will be demonstrated.

The given threaded part is specified $\frac{1''}{2}$ —13 threads per inch.

The best wire diameter $= G = \frac{.57735}{N} = \frac{.57735}{13} = .04441$.

Assume the measurement over the wires $= M = .5155''$.
The actual pitch diameter of the part is calculated as follows:

$$E = M + \frac{.86603}{N} - 3G$$

$$E = .5155 + \frac{.86603}{13} - (3 \times .04441) = .44889, \text{ say } .4489''$$

Now the basic pitch diameter of the screw is the basic major diameter of the screw less one basic depth of thread.

Pitch diameter $= .5000 - \frac{.64952}{N} = .5000 - \frac{.64952}{13} = .5000 - .04996 = .45004$, say $.4500''$.

The actual pitch diameter of the screw is less than the basic pitch diameter by the amount $.4500 - .4489 = .0011''$, which may be satisfactory, depending upon the class of fit required in the given part, whether it is Class 1, 2, 3, or 4.

Screws usually have a pitch diameter less than the basic in order to fit in the nut whose pitch diameter is either that of the basic or slightly less.

Pitch diameter gaging limits for various classes of fits may be found in standard engineering handbooks such as *Machinery's* Handbook, *American Machinist's* Handbook, *S.A.E.* Handbook, and reports of the Screw Thread Commission.

The following table includes the essential thread formulas in general used throughout the manufacturing industry.

Note: In the 29° and 40° worm threads the addendum above the pitch diameter is equal to $\frac{.3183}{N}$. The full depth of thread is equal to $\frac{.6866}{N}$. The clearance at the bottom of thread $= \frac{.6866}{N} - \left(\frac{2 \times .3183}{N}\right) = \frac{.6866 - .6366}{N} = \frac{.0500}{N}$.

Optical Methods. Screw threads are also measured by optical methods, which consist of comparing a magnified image of the thread with a standard pattern. These methods are getting a wider application in the checking of precision threads, especially gages and thread-cutting tools.

TABLE 1. THREAD FORMULAS IN GENERAL USE

Kind of Thread	Incl. Thread Angle	Thread Depth	Formulas for Finding Pitch Diameter by Three-Wire Method
Am. Nat. form	60°	$\dfrac{.64952}{N}$	$E = M + \dfrac{.86603}{N} - 3G$
National pipe	60°	$\dfrac{.8}{N}$	$E = 1.00049M + \dfrac{.86603}{N} - 3G$
Sharp V	60°	$\dfrac{.86603}{N}$	$E = M + \dfrac{.86603}{N} - 3G$
International (Metric)	60°	$\dfrac{.64592}{N}$	$E = M + \dfrac{.86603}{N} - 3G$
Whitworth	55°	$\dfrac{.64033}{N}$	$E = M + \dfrac{.96049}{N} - 3.1657G$
British Assoc.	47½°	$\dfrac{.6}{N}$	$E = M + \dfrac{1.13634}{N} - 3.4829G$
Lowenherz- German	53° 8'	$\dfrac{.75}{N}$	$E = M + \dfrac{1}{N} - 3.23594G$
Acme screws	29°	$\dfrac{1}{2N} + .010''$	$E = M + \dfrac{1.93334}{N} - 4.9939G$
Acme taps	29°	$\dfrac{1}{2N} + .020''$	$E = M + \dfrac{1.93334}{N} - 4.9939G$
29° worm	29°	$\dfrac{.6866}{N}$	Use wire $= \dfrac{.5149}{N}$, flush with top of thread
40° worm	40°	$\dfrac{.6866}{N}$	Use wire $= \dfrac{.51234}{N}$, flush with top of thread

The symbols used in the table are:
 E = pitch diameter of screw.
 M = measurement over the three wires.
 N = number of threads per inch.
 G = size of wire diameter.

THREAD HELIX ANGLE

The thread helix angle depends upon the pitch diameter and the lead of the screw. The formula is $m = \dfrac{l}{\pi E}$, where m = tan of helix angle, l = lead of thread = pitch of a single-thread screw, and E = pitch diameter of screw.

For a multiple-thread screw, l = the pitch × number of threads.

SCREW THREADS

Example: The tan of the helix angle of a $\frac{3''}{8}$ — 16 double-thread screw = $m = \dfrac{l}{\pi E} = \dfrac{2 \times p}{\pi E} = \dfrac{2 \times 1/16}{\pi \times .3344} = \dfrac{2 \times .0625}{3.1416 \times .3344} = \dfrac{.1250}{1.05055} =$ .11899, since E = pitch diameter = $\dfrac{3}{8} - \dfrac{.64952}{N} = .375 - \dfrac{.64952}{16} =$ .3750 — .0406 = .3344″. The helix angle (whose tan = .11899) = 6° 47′ 8″.

CHECKING THREAD GAGES

The checking of various types of thread gages by mathematical aid will be described.

Checking Thread Plug Gage. In checking thread gages of various designs and applications, the accuracy of the included thread angle can be ascertained by a simple setup, as shown in Fig. 8-13, where the actual thread angle of a thread plug gage is to be determined in order that a comparison may be made with the required thread angle.

A steel plug of d diameter is inserted in the thread as shown in Fig. 8-13, and a measurement c is taken from the top of the thread to the top of the plug.

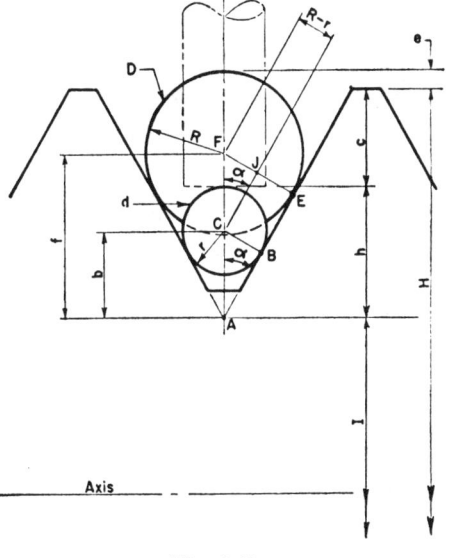

Fig. 8-13.

Measurement H can be obtained by means of a micrometer or any other precision instrument.

The inside diameter I can be calculated by the method used in Fig. 8-12, and the dimension $h = \dfrac{H - I}{2} - c$.

In right $\triangle ABC$, $AC = b = h - r$.

$\therefore \sin \alpha = \dfrac{BC}{AC} = \dfrac{r}{b}$, and α may be determined.

The included thread angle $= 2\alpha$.

A similar procedure is followed by inserting another plug of diameter D in the thread as shown, and a measurement (e) is taken from the top of the thread to the top of the plug.

$$\frac{H-I}{2} + e - R = AF = \text{hypotenuse of right } \triangle AEF.$$

$\sin \alpha = \dfrac{EF}{AF} = \dfrac{R}{f}$, and α may be determined from this calculation, which should equal α of the previous calculation, where the plug of d diameter was inserted in the thread.

If the variation in the values of α is considerable, the calculations should be rechecked for errors. In case the variation is negligible, the average value of α may be taken.

Another check value of α may be obtained by considering the right $\triangle CJF$ in Fig. 8-13. $\angle FCJ = \alpha$, and $\sin \alpha = \dfrac{FJ}{CF} = \dfrac{R-r}{CF}$. CF may be calculated by subtracting b from f.

Note: Combining calculated values, such as I, with actual values, such as c and H, yields a result not strictly accurate, and thus a slight error may be introduced in determining the value of α. The error is slight, however, and the actual value of α so determined is accurate enough for practical purposes.

Checking Thread Ring Gage. The actual thread angle of a thread ring gage can be determined by a setup shown in Fig. 8-14, where a steel ball of a suitable diameter is inserted in the thread.

Dimension I may be obtained by a suitable plug gage or a caliper. Dimension c may be measured by a suitable device. Dimension H may be taken as the calculated major diameter of the sharp thread shown in the figure.

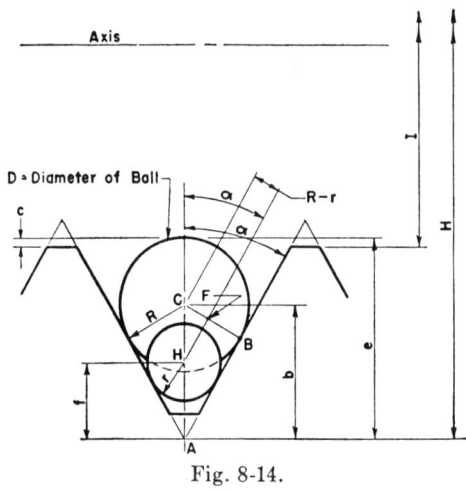

Fig. 8-14.

$$e = \frac{H-I}{2} + c$$

In right $\triangle ABC$, $AC = b = e - R$, and $\sin \alpha = \dfrac{BC}{AC} = \dfrac{R}{b}$.

α may be determined, and the included thread angle is 2α.

In the manner shown in Fig. 8-13, another steel ball of a different diameter may be inserted in the thread of the ring gage. The included thread angle is calculated and compared with that previously calculated. Also α may be calculated from right $\triangle CFH$ in Fig. 8-14, where $\dfrac{CF}{CH} = \sin \alpha = \dfrac{R - r}{b - f}$.

The note on page 308 referring to the setup in Fig. 8-13, with respect to inaccuracies, also applies to the setup in Fig. 8-14.

It should also be noted that the method of checking the thread angles, as illustrated in Figs. 8-13 and 8-14, are more convenient when the thread plug and ring gages are of a larger size, thus enabling the checker to obtain the necessary measurements with greater ease.

Checking Thread-Cutting Tools. Thread-cutting tools such as taps, dies, etc., are checked for accuracy perhaps with greater precision than the manufactured parts themselves, since the accuracy of these tools will definitely reflect upon the parts which they produce.

A method used for checking the pitch diameter of a three-fluted tap by means of one precision wire and a suitable V-block is shown in Fig. 8-15.

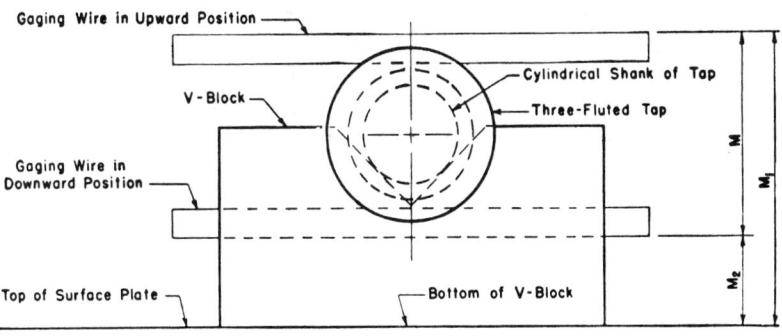

Fig. 8-15.

The V-block supports the cylindrical shank of the tap, the body of the tap proper projects outward, and the precision wire is placed in the thread at the top.

A measurement (M_1) is taken from the top of the wire to the bottom of the V-block.

The shank of the tap is rotated in the V-block until the thread appears at the bottom in such position that the wire can be placed there, and a measurement (M_2) is taken from the bottom of the wire and the bottom of the V-block.

The difference between the two measurements is the M value used in the three-wire formula for finding the pitch diameter of a screw.

$E = M + \dfrac{.86603}{N} - 3G$, where E = pitch diameter of tap, M = $M_1 - M_2$, N = number of threads per inch, and G = wire diameter.

Instead of using a V-block to contact the cylindrical shank of the three-fluted tap, a suitable steel block having at least one finished surface may serve the purpose, if a true hole of the shank diameter is drilled and reamed through this block with a tolerance just permitting the shank of the tap to enter without force. The setup will be similar to that in Fig. 8-15. The shank may be rotated in the reamed hole of the steel block to afford suitable positions of the thread grooves for containing the gaging wire, thus enabling the taking of accurate measurements M_1 and M_2.

Checking Snap Thread Gage. A method of checking the pitch diameter of a snap thread gage by means of the three-wire formula is shown in Fig. 8-16. The snap thread gage shown in the figure is designed for an American National form thread.

According to previous derivations, the dimension h in Fig. 8-16 is equal to $\dfrac{.86603}{N}$, where N = number of threads per inch. $\dfrac{h}{2} = \dfrac{.86603}{2N}$. In right $\triangle ABC, \dfrac{AC}{BC} = \csc 30° = 2.0000$. $\therefore AC = 2.0000 \times BC$, and $H = 2.0000 \times R = 2R = G$ = wire diameter.

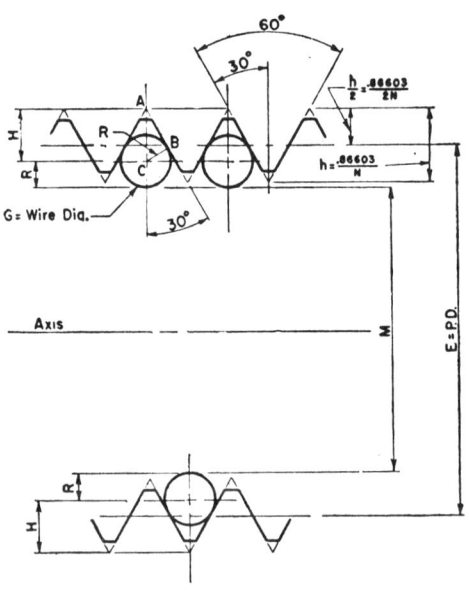

Fig. 8-16.

Combining the components:

$$M + 2R + 2H - \left(2 \times \frac{h}{2}\right) = E, M + G + 2G - \frac{.86603}{N} = E, \text{ or}$$

$$E = M - \frac{.86603}{N} + 3G,$$ which is the formula to be used to calculate the pitch diameter of the snap thread gage, if the measurement M is taken as shown in Fig. 8-16.

In the case of a thread ring gage, three steel balls may be used instead of the wires, and a measurement M is taken in the manner shown in Fig. 8-16. It is rather difficult to keep the balls contained in the thread grooves, but with the aid of grease and a little practice, satisfactory results may be obtained, especially if the diameter of the thread is fairly large.

CUTTING SCREW THREADS ON A LATHE

Lathes of late design are provided with quick-change gear combinations which are rapidly selected as required in proper thread-cutting. Some of the older machines have been rebuilt with the quick-change gear feature incorporated in them, but many of the older machines are still operated and are used to cut threads by changing the stud and lead screw gears.

On the basis of thread-cutting, lathes are grouped as follows: (1) simple-geared lathes; (2) compound-geared lathes; and (3) quick-change gear lathes. The simple-geared lathes and the compound-geared lathes will be discussed.

Simple-Geared Lathe. The gear diagram of a simple-geared lathe is shown in Fig. 8-17. The part on which the threads are to be cut rotates with the lathe spindle at an equal rate of speed. The rotation of the spindle is, of course, caused by the power applied to the machine, and this power is transmitted to the lead screw by means of the gear train shown in Fig. 8-17. The lead screw operates the tool carriage and controls the rate of feed of the thread-cutting tool attached to the carriage provided that the proper size gears are in the train.

Assume that the spindle gear and the inside stud gear have the same number of teeth. Then, if identical

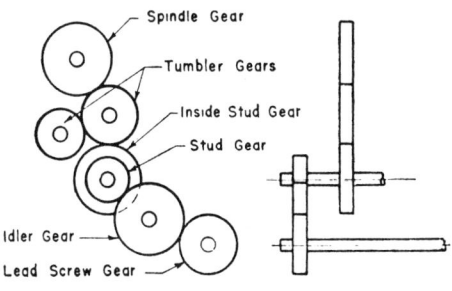

Fig. 8-17.

gears are placed on the stud and the lead screw, the number of threads cut on the part will be the same as the number of threads on the lead screw.

Note that the tumbler gears are idler gears and are in the train for the purpose of reversing the direction of rotation of the lead screw or feed rod in order that both right-hand and left-hand threads may be cut.

For example, if the lead screw has 6 threads per inch, the number of threads cut on the part is also 6. But, in practice, any number of threads may be cut on the part. To provide such accommodation, the gears on the stud and lead screw must be calculated for the required number of threads. In most cases, the spindle gear and the inside stud gear are identical, and the only variable gears are the stud and lead screw gears. Let us demonstrate the procedure of calculation, using a practical example.

It is required to cut 13 threads per inch on a lathe whose spindle gear and inside stud gear are the same, and whose lead screw has 6 threads per inch.

When the cutting tool travels a distance of one inch, the lead screw revolves six times during the interval of travel. In order to cut 13 threads on the part in the same interval, the part and the spindle to which it is attached must revolve 13 times, i.e., the speed of the spindle is to the speed of the lead screw as 13 is to 6, hence the gears on both members must be in a reversed ratio, since the small gear is faster than the large gear.

The formula can be stated as follows: $\frac{N_L}{N_C} = \frac{T_S}{T_L}$, where N_L = number of threads per inch on lead screw, N_C = number of threads per inch to be cut, T_S = number of teeth on stud gear, and T_L = number of teeth on lead screw gear. In our example $\frac{6}{13} = \frac{T_S}{T_L}$.

Usually a convenient and available number of teeth in the gears is selected, as long as the ratio $\frac{N_L}{N_C}$ is maintained.

$\frac{6}{13} = \frac{18}{39}$, where T_S = 18 teeth and T_L = 39 teeth

If the spindle gear and the inside stud gear are not of the same size, their ratio must be taken into account when calculating the number of teeth in the stud gear and the lead screw gear.

Example: It is required to cut 18 threads per inch on a lathe whose lead screw has 5 threads per inch and spindle gear and inside stud gear have 25 and 42 teeth, respectively.

The lead screw revolves five times for every inch of movement of the cutting tool. During this interval the spindle and the work attached to it must revolve 18 times. If the spindle gear and the inside stud gear had a ratio of $\frac{1}{1}$, the previously explained formula, $\frac{N_L}{N_C} = \frac{T_S}{T_L}$, could have been applied. But since the spindle gear revolves faster than the inside stud gear, the rotation of the spindle must be slowed down according to the ratio $\frac{42}{25}$.

The formula for this setup can be stated as follows:

$\frac{G_i}{G_S} \times \frac{N_L}{N_C} = \frac{T_S}{T_L}$, where G_i = number of teeth on inside stud gear, G_S = number of teeth on spindle gear. N_L, N_C, T_S and T_L have values equivalent to identical expressions in the previously given formula.

$$\frac{G_i}{G_S} \times \frac{N_L}{N_C} = \frac{T_S}{T_L}$$

$$\frac{42}{25} \times \frac{5}{18} = \frac{T_S}{T_L}$$

$$\frac{7}{15} = \frac{T_S}{T_L}$$

A convenient number of teeth is selected in the change gears, thus $\frac{7}{15} = \frac{21}{45}$, where T_S = 21 teeth and T_L = 45 teeth.

The necessary change gears required for cutting standard threads are indicated on index plates located on some prominent part of most of the lathes having this type of change gears.

Compound-Geared Lathes. At times it is necessary to cut fine threads on lathes whose lead screws have coarse threads, and a simple geared setup suited to the purpose might call for a pair of gears that are not available; or the distance between the center lines of the stud gear and the lead screw gear on the given lathe will not be great enough to accommodate a gear of large diameter. In this event a compound-geared method, illustrated in Fig. 8-18, is applied.

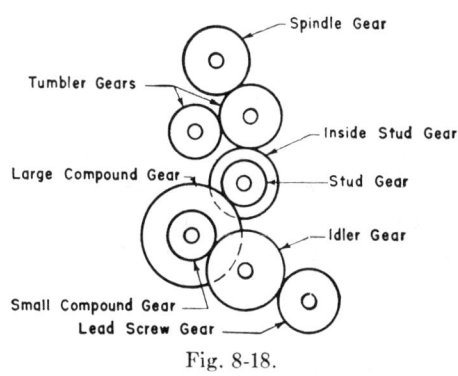

Fig. 8-18.

The method is similar to that of the simple-geared lathe except that two pairs of change gears are selected instead of one. This makes it possible to use the gears that are available.

Let us demonstrate this method by a practical example. It is required to cut 32 threads per inch on a lathe having a lead screw with 6 threads per inch. Assume that the spindle and the inside stud gears are the same. Then, as previously stated, the formula $\dfrac{N_L}{N_C} = \dfrac{T_s}{T_L}$ would apply in this case:

$$N_L = 6$$

$$N_C = 32$$

$$\frac{6}{32} = \frac{T_s}{T_L}$$

Using the minimum number of teeth (24 in the small gear), $T_s = 24$ teeth and $T_L = 128$ teeth.

A gear having 128 teeth may be too large for the machine, or a gear of this specification may not be available for use.

Another pair of compound gears are placed on one shaft, as shown in the figure, thus permitting the use of available gears on the stud and lead screw shafts.

Usually the pair of compound gears are selected in a ratio of 2 to 1, but this is not the rule; the choice, rather, is one of convenience.

The numerator and the denominator of the fraction $\dfrac{6}{32}$ representing the ratio $\dfrac{N_L}{N_C}$ are factored, if possible, in such a way that any one factor in the denominator is twice as large as any one factor in the numerator.

$$\text{Then, } \frac{N_L}{N_C} = \frac{6}{32} = \frac{3 \times 2}{8 \times 4} = \frac{3}{8} \times \frac{2}{4} = \frac{24}{64} \times \frac{20}{40}.$$

We select the pair of compound gears of 20 teeth and 40 teeth, respectively, and the other pair of gears counting 24 teeth on the stud gear and 64 teeth on the lead screw gear; the small compound gear and the stud gear are the drivers, and the large compound gear and the lead screw gear are the driven.

As stated before, the pair of compound gears may be selected of any ratio as long as the deviation from standard procedure makes possible the use of available gears.

Note: If the spindle and the inside stud gears are not the same, use a factor $\frac{G_i}{G_s}$, to be multiplied by $\frac{N_L}{N_C}$, as previously stated.

PRACTICE PROBLEMS APPLICABLE IN THE SHOP AND TOOLROOM

1. Determine the pitch of a $\frac{3''}{8}$—16 N.C. thread screw.
2. What is the pitch of a $\frac{3''}{8}$—16 N.C. thread nut?
3. What is the basic major diameter of the screw in Problem 1?
4. What is the basic depth of the thread in Problem 1?
5. Calculate the basic minor diameter of the screw in Problem 1.
6. Calculate the basic pitch diameter of the screw in Problem 1.
7. Calculate the diameter of the best wire to be used when checking the pitch diameter of the screw in Problem 1 by the three-wire method.
8. Assume that the measurement over the three wires in Problem 1 is .3870″, what is the actual pitch diameter of the screw, if best wires are used in checking the pitch diameter of this screw?
9. Assuming that the screw in Problem 1 is single-threaded, what is its lead?
10. What is the helix angle of the screw in Problem 1?
11. What tap drill diameter is recommended for correct fit of the nut and screw in Problem 1, assuming an $83\frac{1}{3}$ per cent working depth of thread in this nut? Express the value of this drill diameter to the nearest $\frac{1}{64}''$.
12. What are the values of the maximum and minimum wire diameters recommended for checking the pitch diameter of the screw in Problem 1?
13. What is the pitch of a screw with $2\frac{1}{4}$ threads per inch?
14. How many threads per inch are on a screw of .025″ pitch?
15. How many threads are in the threaded portion of the bolt in Fig. 8-19? Express the number of threads as a whole number, omit fractions, if any.
16. What is the lead of the bolt in Fig. 8-19, if it is triple-threaded?
17. What is the basic pitch diameter of the bolt in Fig. 8-19?
18. What is the helix angle of the bolt in Problem 16?
19. What is the basic pitch diameter of the bolt in Fig. 8-19, if the threads are sharp? Assume the major diameter $= \frac{5''}{8}$.

20. What is the depth of the thread in Problem 19?

21. Find the tap drill diameter of a drawn aluminum nut specified as follows: $1\frac{1}{8}''$ —12 N.F. threads. Express this diameter as the nearest common fraction in the decimal equivalent chart.

22. Find the pitch diameter of a $\frac{7''}{8}$ —$4\frac{1}{2}$ square thread.

23. Find the pitch diameter of a $\frac{3''}{4}$ —8 Acme 29° screw thread.

24. Find the pitch diameter of a $1\frac{1}{4}''$ —4, 29° worm thread.

25. Find the pitch diameter of a $1\frac{1}{8}''$ —3 buttress thread.

26. Assume a pipe thread as shown in Fig. 8-10. Its actual outside diameter (D) equals $2\frac{7}{8}''$, and the number of threads per inch is 8. Calculate the pitch diameter (E) at the front end of the thread, the length (L), and the depth (d) of the thread.

27. What should be the measurement over the three wires when checking the pitch diameter of the screw thread in Fig. 8-20? Assume the required pitch diameter of the screw thread is .0020″ below its basic pitch diameter. The available wire diameter is the maximum diameter recommended.

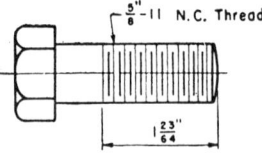

Fig. 8-19.

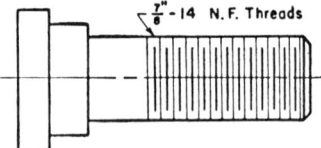

Fig. 8-20.

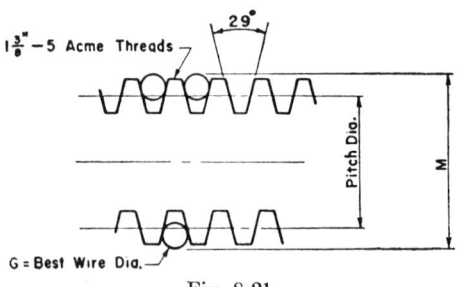

Fig. 8-21.

28. What should be the measurement over the three wires when checking the pitch diameter of the Acme screw thread in Fig. 8-21? Assume the required pitch diameter of the screw thread is equal to its basic pitch diameter. Also use the best wire diameter, i.e., wires touching the thread angle at the pitch line.

29. Assume an American National form thread with a helix angle greater than 6° similar to one shown in Fig. 8-22. Calculate the basic pitch diameter of the screw thread, the best wire diameter recommended for checking the pitch diameter, the measurement over the three wires, the lead and the helix angle.

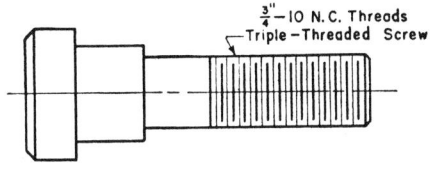

Fig. 8-22.

30. Determine the pitch of a $\frac{9}{16}''$ —18 N.F. thread screw.
31. What is the basic major diameter of the screw in Problem 30?
32. What is the basic depth of the thread in Problem 30?
33. Calculate the basic minor diameter of the screw in Problem 30.
34. Calculate the basic pitch diameter of the screw in Problem 30.
35. Calculate the minimum wire diameter recommended for checking the pitch diameter of the screw in Problem 30.
36. What should the measurement be over the three wires when checking the pitch diameter of the screw in Problem 30, assuming that the pitch diameter of the screw is .0015″ less than its basic pitch diameter? Also assume that the recommended minimum wire diameter is used.
37. What is the lead of the screw in Problem 30, assuming it is double-threaded?
38. Calculate the helix angle of the screw in Problem 37.
39. Calculate the tap drill diameter for accurate fit of the nut and screw in Problem 30. Assume the working depth of thread is 80 per cent. Express the value of the drill diameter to the nearest $\frac{1}{64}''$.
40. Calculate the pitch diameter of the nut to fit the screw in Problem 30. Assume the pitch diameter is .0015″ over the basic pitch diameter.
41. Calculate the actual included thread angle of the screw in Fig. 8-23, assuming the plug diameters inserted in the thread are .112″ and .230″. Also assume that measurement H is 1.997″, $c = .008''$, and $e = .169''$.
42. Calculate the actual included thread angle of the screw in Fig. 8-24, assuming the plug diameters inserted in the thread are .125″ and .175″. Also assume that measurement H is 1.498″, $c = .022''$ and $e = .100''$.
43. Calculate the actual included thread angle of the nut in Fig. 8-25, assuming the ball diameters inserted in the thread are .028″ and .056″. Also assume that measurement $I = 1.172''$, $a = .006''$ and $c = .036''$.

44. Determine the pitch diameter of the three-fluted tap in Fig. 8-15 with the aid of a single wire and a V-block. The tap is designed to

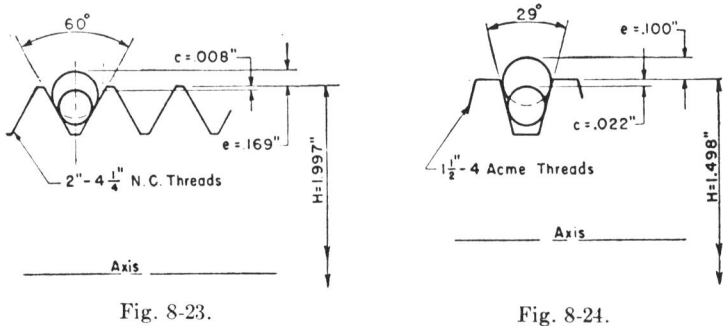

Fig. 8-23.   Fig. 8-24.

cut $\dfrac{7''}{8}$—14 N.F. threads. The wire selected for checking the tap is .0420″. Assume that measurement M_1 in Fig. 8-15 is 1.886″ and measurement M_2 is .992″.

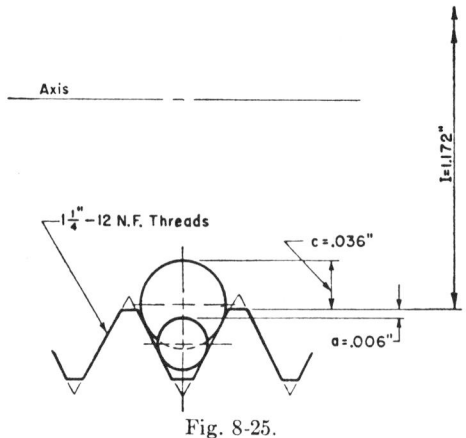

Fig. 8-25.

45. Determine the pitch diameter of the three-fluted tap in Fig. 8-26 with the aid of a single wire and a steel block with reamed hole of the same diameter as the shank of the tap. The tap is designed to cut $\dfrac{9''}{16}$—12 N.C. threads. The wire selected for checking the tap is .0480″. Assume that measurement M_1 in Fig. 8-26 is 1.764″, showing the upward position of the gaging wire. After this measurement is obtained the tap is rotated in the steel block until a convenient position of the thread groove is secured, then measurement $M_2 = 1.183''$ is taken, showing the downward position of the gaging wire.

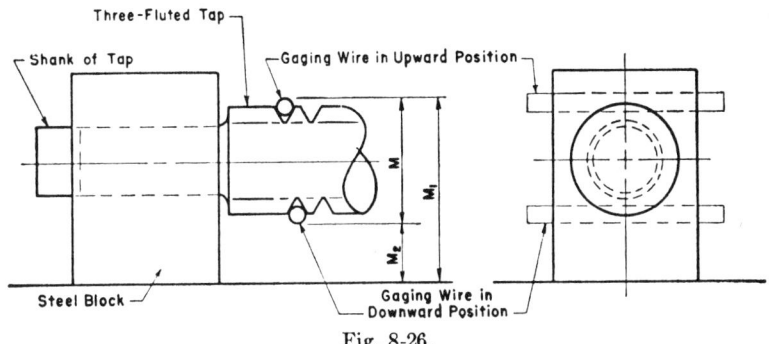

Fig. 8-26.

46. Determine the pitch diameter of the snap-thread gage shown in Fig. 8-27. This gage is designed to check $1\frac{1}{8}''$—12 N.F. screws.

Three wires are used to obtain the pitch diameter of the gage. Assume the wire diameters are all .0500″ and the measurement M is .9750″.

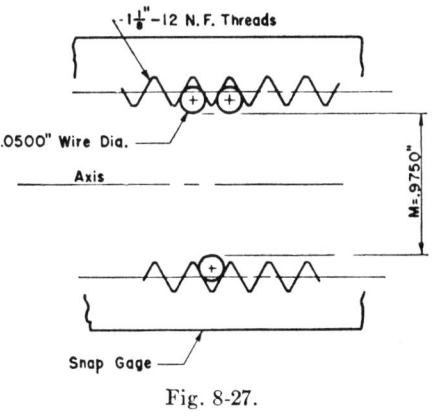

Fig. 8-27.

The following problems (47 to 53, inclusive) deal with thread cutting on a simple geared lathe. Assume that the spindle gear and the inside stud gear have the same number of teeth, and the lead screw has 6 threads per inch. The available gears that constitute standard equipment of this lathe are: 18, 20, 24, 28, 30, 32, 36, 40, 44, 48, 52, 54, 56, 60, 64, 68, 72, 76, and 80 teeth.

47. Select the proper pair of gears for the stud and lead screw shafts in order to cut 7 threads per inch.
48. Select gears for 8 threads per inch.
49. Select gears for 9 threads per inch.
50. Select gears for 12 threads per inch.
51. Select gears for 13 threads per inch.
52. Select gears for 14 threads per inch.
53. Select gears for 18 threads per inch.

The following problems (54 to 60, inclusive) deal with thread cutting on a simple-geared lathe whose spindle gear has 36 teeth and inside stud gear 48 teeth. The lead screw has 8 threads per inch, and the available gears are the same as those used in the preceding problems. Also assume that the above gears are furnished in pairs of each kind.

54. Select the proper pair of gears for the stud and lead screw shafts in order to cut 6 threads per inch.
55. Select gears for 9 threads per inch.
56. Select gears for 12 threads per inch.
57. Select gears for 8 threads per inch.
58. Select gears for 4 threads per inch.
59. Select gears for 5 threads per inch.
60. Select gears for 10 threads per inch.

The following problems (61 to 65, inclusive) deal with thread cutting on a compound geared lathe. Assume that the spindle gear and the inside stud gear have the same number of teeth, and the lead screw has 6 threads per inch. The available gears are similar to those used in the previous problems.

61. Select the proper set of gears for both compounds and for the stud and lead screw shafts in order to cut 30 threads per inch.
62. Select gears for 32 threads per inch.
63. Select gears for 36 threads per inch.
64. Select gears for 40 threads per inch.
65. Select gears for 48 threads per inch.

The following problems (66 to 71, inclusive) deal with thread cutting on a compound geared lathe whose spindle gear has 24 teeth and inside stud gear 36 teeth. The lead screw has 6 threads per inch, and the available gears are the same as those used in the preceding problems. Also assume that the above gears are furnished in pairs of each kind.

66. Select the proper set of gears for both compounds and for the stud and lead screw shafts in order to cut 28 threads per inch.
67. Select gears for 32 threads per inch.
68. Select gears for 36 threads per inch.
69. Select gears for 40 threads per inch.
70. Select gears for 44 threads per inch.
71. Select gears for 48 threads per inch.

CHAPTER IX

GEARS

Gears are used for uniform transmission of rotary motion between two shafts. This motion may also be transmitted between the shafts in several other ways. For example, if the shafts are parallel and are a fairly good distance apart, they may be attached to pulleys of proper diameter, depending upon the velocity ratio, i.e., the ratio of the speeds of the driving and the driven shafts. Open belts are then used to connect the pulleys if the direction of motion of both shafts is the same, and crossed belts are used if the direction of motion is reversed.

Another method of transmission of rotary motion is to attach rolls of proper diameter to the shafts, provided the rolls are not far apart and frictional contact is established between their surfaces. The objection to this method of transmission of motion is the difficulty of obtaining rolls of perfect diameter and of maintaining an exact center distance between them; hence the impossibility of maintaining a desired velocity ratio between the driving and the driven members.

When gears are used to transmit motion between the shafts, the velocity ratio of the driving and the driven members is maintained with fairly good accuracy owing to the fact that the gear-tooth profiles are shaped to insure the desired accuracy, although the center distance between the shafts and the other elements may vary a slight amount.

The chief problem in gearing, therefore, is the correct development of the gear-tooth shape for smooth transmission of motion. For best results the involute-shape curve is adopted for the gear-tooth profile. The cycloid curve is one of the oldest shapes of gear teeth, but it is seldom used at the present time.

The transmission of motion by means of gears can be accomplished for positions of driving and driven shafts other than the one described, which is between parallel shafts. Of course, different types of gears are required for various positions of the shaft.

Before discussing the different types of gears and their formulas, the following general definitions will be given.

Fig. 9-1 shows two spur gears—i.e., gears in mesh between two parallel shafts—whose elements are parallel to each other and to the axes of their shafts.

The *pinion* is the smaller of the two gears; usually it is the driver, though in some cases it may be the driven gear.

The *gear* is the larger of the two gears.

The *pitch circles* of a pair of gears are imaginary circles that are constantly in contact with each other. Most gear calculations are based upon these circles, hence their importance.

The *pitch diameter* (D) is the diameter of the imaginary pitch circle and is designated D_g for the gear and D_p for the pinion.

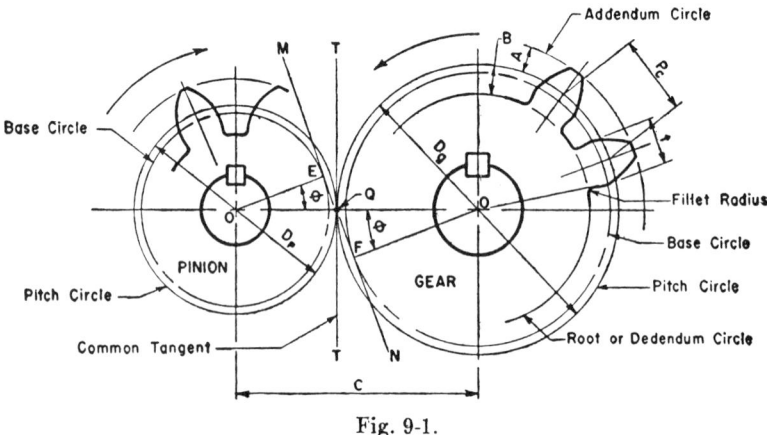

Fig. 9-1.

The *center distance* is the distance between the axes of the shafts and is designated C in Fig. 9-1.

The *circular pitch* is the distance in inches between two adjacent teeth taken on the pitch circle at corresponding points such as the centers of the teeth (see Fig. 9-1). It is designated P_c.

The *number of teeth* in any gear is expressed as N. In the gear it is designated by N_g, and in the pinion by N_p.

The *diametral pitch* is the ratio between the number of teeth in the gear and the pitch diameter. Since a ratio is also an indicated division, the diametral pitch expresses the number of teeth in the gear for every inch of the pitch diameter. It is designated P_d, and is often referred to as the *pitch*.

Both the circular pitch and the diametral pitch are measures of the spacing of the teeth, and the diametral pitch is also a measure of the size of the teeth.

Certain relationships exist between the pitch diameter, the number of teeth, the circular pitch, and the diametral pitch of a gear. The relationships are expressed as follows:

$$P_c = \frac{\pi D}{N} \text{ or } NP_c = \pi D$$

$$P_d = \frac{N}{D} \text{ or } N = P_d D$$

$$P_c \times P_d = \frac{\pi \cancel{D}}{\cancel{N}} \times \frac{\cancel{N}}{\cancel{D}} = \pi$$

$$P_c = \frac{\pi}{P_d} \text{ and } P_d = \frac{\pi}{P_c}$$

The involute shape of the gear tooth is the most practical curvature and it is developed as shown in Fig. 9-2. This curve can be imagined as traced by a point at the end of a string being unwound from a cylindrical body or a circular disk. The circumference of this disk is called the *base circle* of the gear.

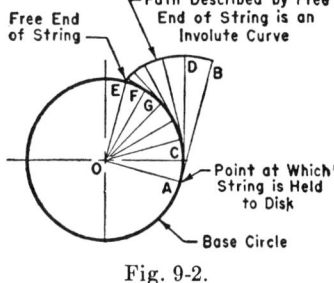

Fig. 9-2.

The origin of the involute curve is at the base circle, and the tooth profile in properly shaped gears is an involute curve extending from the base circle to the outside circle of the gear.

The method of developing an involute curve consists of dividing the base circle or a part of it, such as the circular arc AE in Fig. 9-2 into a number of equal parts, namely, 6. The division points, F, G, etc., are connected to the center O, forming radii OE, OF, OG, etc. Lines AB, DC, etc., are drawn perpendicular to the corresponding radii, such as OA, OC, etc. These lines are made equal in length to their corresponding circular arcs. Thus, line AB is equal in length to circular arc AE, line CD is equal to circular arc CE, etc.

A curve is traced through the ends of these lines, such as B, D, etc., and part of this curve is used for the profile of the teeth on a gear, whose base circle is drawn with radius OA, as shown in Fig. 9-2.

The pressure line MN and pressure angle ϕ of Fig. 9-1 is illustrated and explained in more detail in Fig. 9-3. In this illustration, an enlarged tooth of the pinion from Fig. 9-1 drives an enlarged tooth of its mating gear, and the direction of rotation of the pinion tooth is clockwise, while that of the gear tooth is counterclockwise.

The force at the point of tangency (*Q*) created by rotation of the pinion tooth is in the direction *QT*; but the force exerted by the pinion tooth (by which force the gear tooth rotates) is in the direction *QN* perpendicular to line *RS*, which is the line of contact of the meshing teeth.

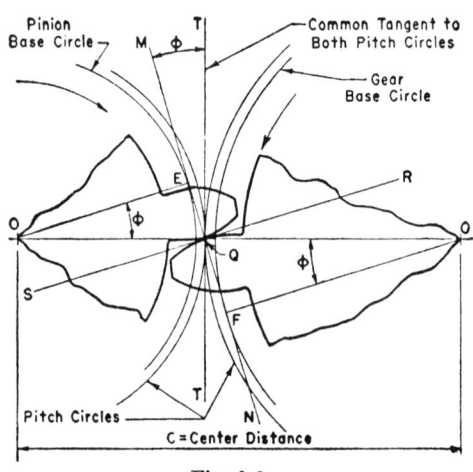

Fig. 9-3.

The angle ϕ between the lines of forces *TT* and *MN* is the *pressure angle*, or the angle of action.

The line *MN* is the *pressure line*, the generating line, or the line of action, since a pair of meshing teeth of the pinion and the gear are in contact along this line from the start of engagement to the end.

The pressure angle is usually $14\frac{1}{2}°$, this angle having been found by experiment to afford the best efficiency, since the force or thrust that widens the distance between the gear shafts is at this angle counteracted by the friction between the teeth. Also, at this angle the tooth thickness at the base is strong enough to resist breakage that could cause failure.

At times a greater pressure angle, such as 20°, or more, is used to design a stronger gear tooth; yet the tooth is not longer than one having a pressure angle of $14\frac{1}{2}°$, in fact, it is sometimes even shorter, in which case it is called a *stub tooth*. Of course, better grade bearings must be provided to resist the action of greater pressure angles.

The base circles of the pinion and gear are tangent to the pressure line *MN* in Fig. 9-3, and the tooth profile above the base circle is an involute curve. The shape of the tooth below the base circle is a radial

GEARS

line that is rounded off at the root circle by a *fillet radius*, as shown in Fig. 9-1.

It is evident from study of Figs. 9-1 and 9-3, that for the same gear the base circle diameter decreases with increase of the pressure angle. Therefore, the involute shape of the gear tooth extends farther toward the center in a gear which has an increased pressure angle.

It is also evident from the drawings that a gear of large diameter has teeth whose involute profiles extend farther toward the center than do those of a gear of smaller diameter. Hence, in a pair of meshing gears, the pinion teeth have a smaller range of involute profile than have those of the gear, and the gear teeth are likely to interfere with the pinion teeth unless the latter are relieved or the pinion has a sufficient number of teeth to prevent interference.

The *addendum* of the gear tooth is the portion above the pitch circle, or distance from pitch circle to top of tooth, and is designated by the letter A in Fig. 9-1.

The *dedendum* is the portion of the tooth below the pitch circle, or distance from pitch circle to the bottom of the tooth space, and is designated by the letter B.

The *whole depth* of the tooth is an amount equal to the sum of the addendum and the dedendum or the total height of the tooth.

The *clearance* is the difference between the dedendum and the addendum.

The *fillet* is the joint between the tooth and the root circle, and the fillet radius is large enough to increase the strength of the tooth at the base.

The *working depth* of the tooth is equal to twice the addendum.

The *thickness* of the tooth t is measured along the pitch circle.

The *width of space* between two adjacent teeth is measured along the pitch circle, and, theoretically, it is equal to the thickness of the tooth, but for practical purposes it is somewhat greater to provide *backlash* or clearance.

TYPES OF GEARS

Since gears are employed for various positions of shafts, their shapes are designed to meet different conditions, hence the variety of gears used in industry. In the paragraphs that follow, the principal types of gears are described. Later in the chapter the formulas for some of these gears will be given.

Spur Gears. *Spur gears* operate on parallel shafts. The contact between the teeth is in a straight line parallel to the shafts. The pitch surfaces of both meshing gears are of cylindrical shape and are tangent to each other at all times. Spur gears are the most commonly used type of gears.

Helical and Herringbone Gears. For better performance gears other than spur gears are employed to operate on parallel shafts, namely, *helical* and *herringbone* gears. The contact between teeth is in a straight line, forming an angle with the shaft axis. The herringbone gear is really a pair of helical gears (right- and left-hand), either assembled together or cut integral from the same blank. Helical gears create an end thrust on the shafts that must be counteracted by suitable thrust bearings, whereas each side of a herringbone gear neutralizes the opposite side, thus canceling the end thrust and eliminating the use of thrust bearings. The pitch surfaces of helical and herringbone gears are of cylindrical shape and are tangent to each other. An advantage of these gears is that at least two or more teeth are constantly in mesh, hence a greater load can be transmitted through the gears. Moreover, engagement of the gear teeth is smoother, and there is less noise in the operation of gears of this type.

Bevel Gears. For transmission of rotary motion between two intersecting shafts that lie in one plane, *bevel gears* are employed. Usually these shafts intersect at 90°, although the angle of intersection may be less or more than 90°. The pitch surfaces of these gears are cones that roll on each other.

The shape of the gear teeth in bevel gears is of two designs: (1) Straight teeth which contact along straight lines directed toward the apex of the rolling cones. (2) Curved teeth which contact along curved lines. The shape of the curves is approximately spiral.

Worm Gears. For nonparallel and nonintersecting shafts *worm gears* may be employed. The angle between the shafts is usually 90°. In the combination, one member is the worm gear and the other the worm, the latter being threaded. The pitch line of this worm thread is a helix. In properly designed worm gears there is a line contact between the mating members. This type of gearing is employed where a large velocity ratio is desired, i.e., where the ratio between the speed of the driver member and that of the driven is large. Usually the worm helix is complete for one or more revolutions of the worm, depending upon whether a single-thread or multiple-thread worm is employed. The worm in almost all cases drives the worm wheel.

Spiral Gears. These gears are similar to helical gears except for the positions of the shafts with which they operate. The shafts are nonparallel and nonintersecting. There is a great amount of sliding action between the teeth. Spiral gears are not recommended for heavy loads, but are very useful in conjunction with shafts that operate in different directions and where velocity ratios between the driver and the driven members differ at various positions of the shaft.

Internal Gears. Where a compact design is required, internal gears are employed, since the center distance between the internal gear and its mating pinion is much shorter than that provided by regular spur gears. The teeth of an internal gear point inward toward the center of the gear and are stronger than those of an external spur gear. The tooth action lasts a longer period of time in an internal gear than in a corresponding external spur gear, and there is less slippage. Internal gears are used in speed reducers, rear-axle drives for heavy-duty vehicles, and in planetary gear trains.

A *rack* is a straight gear that can be regarded as a gear of infinitely long radius. Its motion is in a straight line and its pitch line is a straight line that is in contact with the pitch circle of the mating gear.

Skew Bevel and Hypoid Gears. For nonparallel and nonintersecting shafts other types of gears, called *skew bevel* and *hypoid* gears, may be employed. The pitch surfaces of these gears are hyperboloids, and they contact along straight lines.

Hypoid gears have been adopted extensively in the automobile industry for use in the differential because they permit lower build of the body than do standard bevel gears. When hypoid gears are used the drive shaft can be dropped, since the pinion shaft of the differential is lower than that of the gear.

The advantage which skew bevel gears provide in industrial applications is that a continuous shaft which carries several pinions that, in turn, drive several gears, may be employed. The use of skew bevel gears also makes it possible to place bearings on each side of the gear, as the gear shaft can be extended outside the gear proper without intersecting the mating gear shaft.

Other Gears. Other kinds of gears are made in small quantities to meet special conditions. Some of these special gears are *eccentric gears, planetary bevel gears, cone gears,* and *pin gears.*

FORMULAS FOR SPUR GEARS

In order to be interchangeable, involute gears must have the same diametral pitch, the same pressure angle, and the same depth of tooth.

It is obvious that gears having the same diametral pitch also have the same circular pitch. Also, for interchangeability, the addendum in each mating gear must be equal to the dedendum less the clearance.

Interchangeable involute gears are divided into two main classifications: *standard involute* gears and *stub-tooth* gears.

The standard involute tooth gears are gears having full-depth teeth. Gears of this type are divided into three systems: (1) Brown and Sharpe system; (2) $14\frac{1}{2}°$ full-depth involute system; (3) $20°$ full-depth involute system.

The stub-tooth gears are divided into two systems, namely, the American Standards Association system and the Fellows Gear Shaper Company system.

Formulas for Full-Depth Tooth Gears. The first two systems mentioned have a $14\frac{1}{2}°$ pressure angle, and the third system a $20°$ pressure angle.

The Brown and Sharpe system is also known as the *composite* system, and the shape of the tooth is made only partially involute so as to avoid interference. The upper and lower thirds of the tooth depth are approximately cycloidal in shape, and the central portion near the pitch line is of involute shape. These gears are produced with form milling cutters.

The gears of the $14\frac{1}{2}°$ and $20°$ full-depth involute systems are produced in large lots on generating and hobbing machines, and the shape of the teeth is an involute curve from the base circle to the addendum circle.

The formulas for full-depth tooth gears follow. Refer to Fig. 9-1.

1. $A = \dfrac{1}{P_d}$, where A = addendum and P_d = diametral pitch.

2. $B = \dfrac{1.157}{P_d}$, where B = dedendum.

3. Full depth of tooth $= \dfrac{2.157}{P_d}$.

4. Working depth of tooth $= \dfrac{2}{P_d}$.

5. Clearance $= \dfrac{.157}{P_d}$.

6. Fillet radius $= \dfrac{4}{3} \times$ clearance $= \dfrac{4}{3} \times \dfrac{.157}{P_d}$, except for the $20°$ full-

depth system, in which the fillet radius $= \frac{3}{2} \times$ clearance $= \frac{3}{2} \times \frac{.157}{P_d}$.

7. $P_c = \frac{\pi D}{N}$, where $P_c =$ circular pitch, $D =$ pitch diameter, and $N =$ number of teeth.

8. $P_d = \frac{N}{D}$, where $N =$ number of teeth and $D =$ pitch diameter.

    Note that the number designating the diametral pitch varies inversely with the tooth size—the smaller the tooth, the greater the number designating the pitch.

9. Tooth space $=$ tooth thickness on pitch circle $= \frac{P_c}{2} = \frac{\pi D}{2N}$ (theoretical).

10. Backlash is not standardized by all manufacturers, and it varies for different types of gears, depending upon the material of which the gears are made, the service, the speed, the precision of workmanship, etc. However, for ordinary industrial use the backlash is generally expressed in terms of the diametral pitch.

$$\text{Maximum backlash} = \frac{.040}{P_d}$$

$$\text{Minimum backlash} = \frac{.020}{P_d}$$

$$\text{Average backlash} = \frac{.030}{P_d}$$

For example, the average backlash of an 8-pitch gear is $\frac{.030}{P_d} = \frac{.030}{8} = .0038''$.

The difference between the tooth space and the tooth thickness on the pitch circle is $.0038''$. Half of this amount is subtracted from the theoretical tooth thickness and the other half is added to the theoretical tooth space.

11. $C = \frac{D_g + D_p}{2}$, where $C =$ center distance, $D_g =$ pitch diameter of gear, and $D_p =$ pitch diameter of pinion.

12. $D = \dfrac{N}{P_d}$, or $D = \dfrac{P_c \times N}{\pi}$, where D = pitch diameter of any gear and N = the number of teeth in any gear.

13. Outside diameter = $\dfrac{N + 2}{P_d}$.

14. The base circle radius of a gear = its pitch circle radius × cos of the pressure angle. In the case of the $14\frac{1}{2}°$ pressure angle, the cos = .96815. In the case of the 20° pressure angle the cos = .93969.

    OF is the base circle radius of the gear and OE is the base circle radius of the pinion in Fig. 9-3. ϕ = pressure angle. OQ = pitch circle radius of gear.

Formulas for the Stub-Tooth Gears. The formulas for the American Standards Association stub-tooth system are as follows:

1. The pressure angle is 20°.

2. The addendum = $\dfrac{.8}{P_d}$, where P_d = diametral pitch.

3. The clearance = $\dfrac{.2}{P_d}$.

4. The dedendum = $\dfrac{1}{P_d}$.

5. The whole depth of tooth = $\dfrac{1.8}{P_d}$.

6. The working depth of tooth = $\dfrac{1.6}{P_d}$.

7. The radius of the fillet = $\dfrac{3}{2} \times$ clearance = $\dfrac{3}{2} \times \dfrac{.2}{P_d}$.

The remaining formulas are the same as those outlined for the standard involute tooth gears.

The formulas for the Fellows Gear Shaper Company system follow:
1. The size of the tooth is usually expressed by two diametral pitches indicated in the form of a fraction. For example, a $\dfrac{6}{8}$ pitch in this system means that the thickness of the tooth on the pitch circle is that of a tooth of the standard system whose diametral pitch is 6.

GEARS

The addendum of this tooth is the same, however, as that of a tooth of the standard systems, whose diametral pitch is 8, therefore the addendum is $\frac{1}{8}''$. The clearance of this tooth is $\frac{.25}{8}$, making the whole depth of the tooth equal to $\left(2 \times \frac{1}{8}\right) + \frac{.25}{8} = \frac{2.25}{8} = .281''$, and the working depth equal to $2 \times \frac{1}{8} = \frac{1}{4} = .250''$.

The stub tooth is stronger than the standard involute full-depth tooth, for though it is shorter than the corresponding standard tooth it is just as wide at the pitch circle, and is wider at the base.

2. The pressure angle is 20°.

3. The radius of the fillet $= \frac{3}{2} \times$ clearance $= \frac{3}{2} \times \frac{.25}{P_d}$, where P_d is the diametral pitch, which serves as the basis for tooth depth calculation.

The diametral pitches for the Fellows system are in the following order: $\frac{4}{5}, \frac{5}{7}, \frac{6}{8}, \frac{7}{9}, \frac{8}{10}, \frac{9}{11}, \frac{10}{12}$, and $\frac{12}{14}$.

The remaining formulas are the same as those outlined for the standard involute tooth gears. The backlash may be calculated as previously explained.

Formulas Used in Solution of Gear Problems. The Problems 1 to 15, inclusive, pertain to standard involute-tooth gears.

1. Find the diametral pitch of a gear that has 44 teeth and a pitch diameter equal to 5.500''.

    Solution: $P_d = \frac{N}{D} = \frac{44}{5.5} = 8$.

2. Find the number of teeth in a gear if its pitch diameter equals 6.200'' and its diametral pitch equals 10.

    Solution: $N = P_d \times D = 10 \times 6.2 = 62$ teeth.

3. Find the pitch diameter of a gear if the number of teeth equals 36 and the diametral pitch equals 8.

    Solution: $D = \frac{N}{P_d} = \frac{36}{8} = 4.500''$.

4. Find the pitch diameter of a gear if the number of teeth equals 58 and the circular pitch equals .3927''.

Solution: $D = \dfrac{N \times P_c}{\pi} = \dfrac{58 \times .3927}{3.1416} = \dfrac{58}{8} = 7\dfrac{1}{4} = 7.250''$.

5. Find the center distance between gear and pinion in mesh if the pitch diameter of the gear equals 7.250" and the pitch diameter of the pinion equals 4.500".

Solution: $C = \dfrac{D_g + D_p}{2} = \dfrac{7.250 + 4.500}{2} = \dfrac{11.750}{2} = 5.875''$.

6. Find the outside diameter of a gear if its pitch diameter equals 6.666" and its diametral pitch equals 6".

Solution: $N = P_d \times D$.

Outside diameter $= \dfrac{N + 2}{P_d} = \dfrac{(P_d \times D) + 2}{P_d} = \dfrac{(6 \times 6.666) + 2}{6} = \dfrac{42}{6} = 7.000''$.

7. Find the outside diameter of a gear if its number of teeth equals 64 and its circular pitch equals .5236".

Solution: Outside diameter $= \dfrac{(N + 2) \times P_c}{\pi} = \dfrac{(64 + 2) \times .5236}{3.1416} = \dfrac{66}{6} = 11.000''$.

8. Find the circular pitch if the center distance between gears equals 5.875", the number of teeth in gear equals 62, and that of the pinion equals 32.

Solution: $C = \dfrac{D_g + D_p}{2}$; $P_c = \dfrac{\pi \times D_g}{N_g}$; $P_c = \dfrac{\pi \times D_p}{N_p}$; $P_c = \dfrac{\pi(D_g + D_p)}{N_g + N_p}$; $P_c = \dfrac{2\pi \left(\dfrac{D_g + D_p}{2}\right)}{N_g + N_p}$; but $\dfrac{D_g + D_p}{2} = C$;

$\therefore P_c = \dfrac{2\pi \times C}{N_g + N_p}$.

$P_c = \dfrac{2 \times 3.1416 \times 5.875}{62 + 32} = \dfrac{36.9138}{94} = .3927''$.

GEARS

9. Find the whole depth of tooth if the circular pitch equals .3927".

   Solution: $P_d = \dfrac{\pi}{.3927}$. Whole depth $= \dfrac{2.157}{P_d} = \dfrac{2.157 \times .3927}{3.1416} = .2696''.$

10. Find the clearance if the circular pitch equals .3927".

    Solution: Clearance $= \dfrac{.157}{P_d} = \dfrac{.157}{\dfrac{\pi}{.3927}} = \dfrac{.157 \times .3927}{3.1416} = .0196''.$

11. Find the average backlash between the teeth if $D_g = 20.500''$, $D_p = 10.500''$, $N_g = 123$ teeth and $N_p = 63$ teeth.

    Solution: Average backlash $= \dfrac{.030}{P_d}$, $P_d = \dfrac{N_g}{D_g} = \dfrac{123}{20.5} = 6.$

    Average backlash $= \dfrac{.030}{P_d} = \dfrac{.030}{6} = .005''.$ P_d may also be obtained as follows: $P_d = \dfrac{N_p}{D_p} = \dfrac{63}{10.5} = 6.$

12. Find the theoretical tooth thickness and the width of space on the pitch circle if the number of teeth in gear equals 62, and the diametral pitch equals 10.

    Solution: $D = \dfrac{N}{P_d} = \dfrac{62}{10} = 6.200''.$

    $t = \dfrac{\pi D}{2N} = \dfrac{3.1416 \times 6.2}{2 \times 62} = .1571.$

13. Assuming an average backlash, what is the thickness of the tooth and width of space in Problem 12?

    Solution: Total backlash $= \dfrac{.030}{P_d} = \dfrac{.030}{10} = .003''.$

    Half of the total backlash is added to t in Problem 12 for the width of space, and the other half is subtracted from t for the thickness of the tooth.

Thickness of tooth $= .1571 - \dfrac{.003}{2} = .1556''$.

Width of space $= .1571 + \dfrac{.003}{2} = .1586''$.

14. Find the base circle diameter of a gear having 64 teeth and a diametral pitch of 12. The pressure angle is $14\tfrac{1}{2}°$.

    Solution: $D = \dfrac{N}{P_d} = \dfrac{64}{12} = 5.3333''$.

    Pitch circle radius $\dfrac{D}{2} = \dfrac{5.3333}{2} = 2.6666''$.

    Base circle radius $= \dfrac{D}{2} \times \cos 14\tfrac{1}{2}° = 2.6666 \times .96815 = 2.5817''$.

    Base circle diameter $= 2.5817 \times 2 = 5.1634''$.

15. Find the base circle diameter of the gear in Problem 14 if the pressure angle equals $20°$.

    Solution: Base circle radius $= \dfrac{D}{2} \times \cos 20° = 2.6666 \times .93969 = 2.5058''$.

    Base circle diameter $= 2.5058 \times 2 = 5.0116''$.

Problems 16 to 18, inclusive, refer to an American Standards Association stub-tooth gear.

16. Find the addendum of an American Standards Association stub-tooth gear having a diametral pitch of 8.

    Solution: Addendum $= \dfrac{.8}{P_d} = \dfrac{.8}{8} = .100''$.

17. Find the whole depth of tooth in Problem 16.

    Solution: Whole depth $= \dfrac{1.8}{P_d} = \dfrac{1.8}{8} = .225''$.

18. Find the radius of the fillet in Problem 16.

    Solution: Clearance $= \dfrac{.2}{P_d} = \dfrac{.2}{8} = .025''$.

GEARS

Radius of fillet $= \frac{3}{2} \times$ clearance $= \frac{3}{2} \times .025 = .0375''$.

Problems 19 to 23, inclusive, refer to a Fellows Gear Shaper Company gear.

19. Find the theoretical thickness of the tooth on the pitch circle of a Fellows system gear having a diametral pitch of $\frac{9}{11}$ and a pitch diameter of 8.3333".

Solution: $t = \dfrac{\pi \times D}{2N} = \dfrac{3.1416 \times 8.3333}{2 \times 9 \times 8.3333} = \dfrac{3.1416}{18} = .1745''$.

t may also be found as follows: $t = \dfrac{\pi}{2P_d} = \dfrac{3.1416}{2 \times 9} = .1745''$.

20. Find the addendum of the gear in Problem 19.

Solution: Addendum $= \dfrac{1}{11} = .0909''$.

21. Find the clearance of the tooth in Problem 19.

Solution: Clearance $= \dfrac{.25}{11} = .0227''$.

22. Find the whole depth of tooth in Problem 19.
Solution: Whole depth $= (2 \times .0909) + .0227 = .1818 + .0227 = .2045''$.

23. Find the radius of the fillet in Problem 19.

Solution: Radius of fillet $= \dfrac{3}{2} \times .0227 = .034''$.

CHECKING THE THICKNESS OF INVOLUTE TEETH

Since gears are used for transmission of motion from one shaft to another, and this motion must be positive and the velocity ratio maintained throughout the revolution, various methods have been adopted to check the accuracy of the gear teeth.

One of the methods consists of checking the thickness of the gear tooth on a straight line going through the points at which the pitch circle touches the gear tooth, which is to say, the chordal thickness t is checked, as shown in Fig. 9-4. Various instruments are used for this purpose, and one of them is the *gear tooth vernier* consisting of two scales, one horizontal and the other vertical.

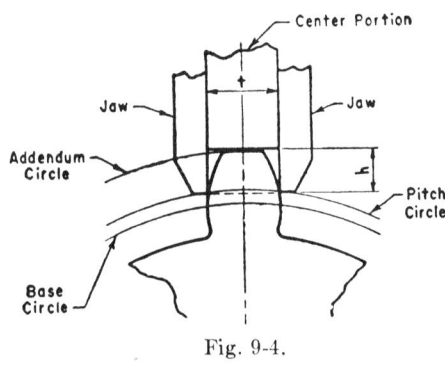

Fig. 9-4.

The horizontal scale, aided by a vernier attachment, indicates the chordal thickness of the tooth, provided the jaws contact the tooth at points on the pitch circle. This can be accomplished if the center portion of the instrument touches the addendum circle, as shown in the drawing, and the jaws are moved down a distance, h, which value represents the corrected addendum. The distance h is read on the vertical scale of the instrument.

Tables of chordal thicknesses and corrected addendums are shown in standard engineering handbooks for a range of gears from 10 teeth to 140 and over, based on a diametral pitch of 1. For other pitches, the values from the tables are divided by the particular diametral pitch.

The chordal thickness (t) for standard involute-tooth gears can also be calculated as follows: $t = D \times \sin \frac{90°}{N}$, where D = pitch diameter of gear, and N = number of teeth in gear.

The corrected addendum (h) for the same type of gears is calculated as follows: $h = \frac{D}{2}\left(1 - \cos \frac{90°}{N}\right)$ + regular addendum.

Note that when checking the chordal thickness of the tooth by means of the gear-tooth vernier, the actual outside diameter of the gear is measured and compared with the calculated outside diameter of a similar gear.

For example, if the calculated outside diameter of a gear equals 9.250″ and the actual outside diameter equals 9.246″, the corrected addendum (h) is reduced by the amount $\frac{9.250 - 9.246}{2} = \frac{.004}{2} = .002″$. The new corrected addendum is then set on the vertical scale of the instrument.

If the calculated outside diameter of the gear is less than the actual diameter, half of the difference is added to the corrected addendum (h),

GEARS

and the new corrected addendum is set on the vertical scale of the instrument.

Other instruments and gages are employed for checking the thickness of the tooth. Some of the gages are designed to extend a distance spanning two or more teeth, thus checking not only the accuracy of one tooth but also the spacing of the teeth, their concentricity, etc.

Some gages are designed to check the pressure angle of the tooth below and above the pitch circle, as there is a relation between the pressure angle and the included angle for the involute curve.

For detailed information regarding comprehensive gear checking, the reader is advised to consult texts on gear development, design, and inspection.

CHECKING THE PITCH DIAMETER OF A SPUR GEAR

For best results the pitch diameter of the spur gear is checked by means of two precision wires placed in two opposite tooth spaces. The wire diameters are calculated to contact the pitch circle of the gear. A measurement is taken across the wires, and the pitch diameter is computed with the aid of the following formulas. (The method of checking is illustrated in Fig. 9-5.) Since not all gears have an even number of teeth, a formula is presented for a gear having an odd number of teeth.

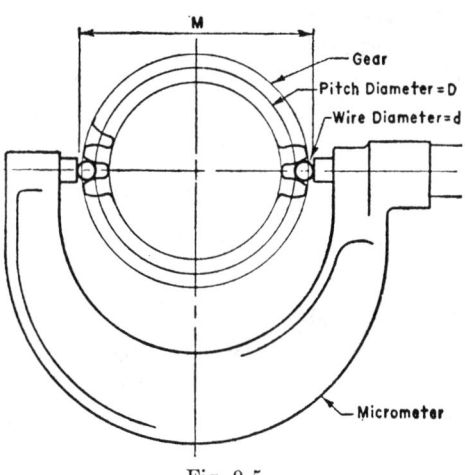

Fig. 9-5.

1. $d = D \left[\cos \phi \tan \left(\phi + \dfrac{90°}{N} \right) - \sin \phi \right]$, where d = wire diameter which is tangent to the teeth at the pitch circle, D = pitch diameter of gear, ϕ = pressure angle, N = number of teeth in gear.

2. For an even number of teeth, $M = D \cos \phi \sec \left(\phi + \dfrac{90°}{N} \right) + d$, where M = measurement across the wires.

3. For an odd number of teeth, $M_o = (M - d) \cos \dfrac{90°}{N} + d$, where M_o = measurement across the wires, M = value calculated in accordance with Formula 2. The other symbols are similar to those in Formulas 1 and 2.

Note: One of the wires is placed in the first space above or below the center line in Fig. 9-5.

The operation consists of taking a measurement across the wires and comparing it with the values calculated in accordance with the formulas. Any deviation indicates the amount of error in the pitch diameter of the gear. Of course, there are allowable tolerances in the pitch diameter of the gear, and, when comparing the values obtained by measurement with those indicated in the formulas, these tolerances must be taken into consideration.

It is difficult at times to obtain wires that contact the gear teeth exactly at the pitch circle. Other wires may be used, but the preceding formulas cannot be employed.

In engineering practice, the pitch diameter of the gear shown on the drawing is usually checked with wires the diameter of which is specified on the drawing. If the gear is within allowable limits, the reading over the wires is also specified.

FORMULAS FOR INTERNAL GEARS

The following formulas are especially for internal gears.

1. $C = \dfrac{N_g - N_p}{2P_d}$, where C = center distance, N_g = number of teeth in gear, N_p = number of teeth in pinion, and P_d = diametral pitch.

2. $P_d = \dfrac{N_g - N_p}{2C}$.

3. Inside diameter of internal gear = $\dfrac{N_g - 2}{P_d}$.

All other formulas are similar to those of standard gears.

FORMULAS FOR BEVEL GEARS

Bevel gears transmit rotary motion between intersecting shafts. In most cases straight bevel gears are employed, i.e., gears whose elements are straight lines that converge to a point. Also, in most cases, gears are employed whose shafts intersect at 90°, and these gears are referred to as right-angle drives. A pair of meshing bevel gears is shown in Fig. 9-6.

Bevel gears, whose shafts intersect at angles other than 90°, are referred to as angular drives.

Other types besides the straight-tooth bevel gears shown in Fig. 9-6 are spiral bevel gears, skew bevel gears, and hypoid gears.

The following formulas are common to straight-tooth bevel gears and have reference to elements shown in Fig. 9-6.

1. The pitch diameters of the gear and pinion are taken at the outer ends of the teeth and are designated D_g for the gear and D_p for the pinion.

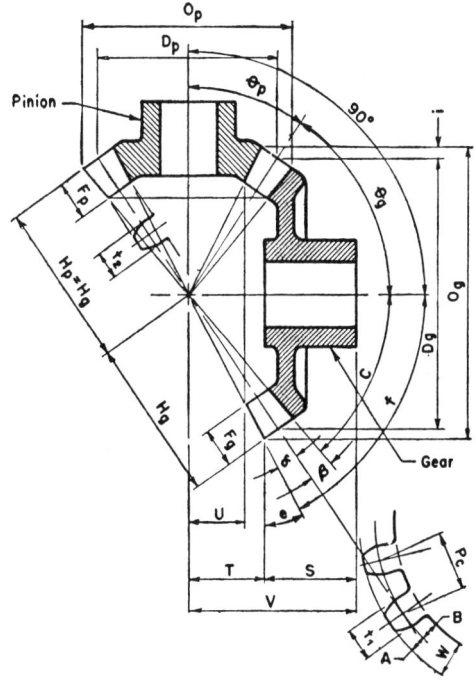

Fig. 9-6.

$$D_g = \frac{N_g}{P_d}, \text{ where } N_g =$$

number of teeth in gear and P_d = diametral pitch.

$$D_p = \frac{N_p}{P_d}, \text{ where } N_p = \text{ number of teeth in pinion.}$$

Note: The meaning of the diametral pitch P_d in bevel gears is the same as that in standard involute spur gears.

2. $P_c = \frac{\pi D}{N}$, where P_c = circular pitch = distance between two adjacent teeth on the pitch circle at corresponding points, taken at the outer ends of teeth.

D = pitch diameter of any gear, and N = number of teeth in any gear.

3. $A = \dfrac{1}{P_d}$, where A = addendum, taken at outer ends of teeth.

4. $B = \dfrac{1.157}{P_d}$, where B = dedendum, taken at outer ends of teeth.

5. $W = \dfrac{2.157}{P_d}$, where W = whole depth of tooth at outer ends.

6. $\text{Tan } \theta_g = \dfrac{N_g}{N_p}$, where θ_g = pitch-cone angle of gear.

   $\text{Tan } \theta_p = \dfrac{N_p}{N_g}$, where θ_p = pitch-cone angle of pinion.

   $\theta_g + \theta_p = 90°$ in straight-tooth bevel gears.

7. $H_g = \dfrac{D_g}{2 \times \sin \theta_g}$, where H_g = cone distance of gear.

   $H_p = H_g = \dfrac{D_p}{2 \times \sin \theta_p}$, where H_p = cone distance of pinion.

8. $F_g = \dfrac{H_g}{3}$, where F_g = maximum width of gear face.

   $F_p = \dfrac{H_p}{3}$, where F_p = maximum width of pinion face.

9. $\text{Tan } \alpha = \dfrac{A}{H_g}$ or $\dfrac{A}{H_p}$, where α = addendum angle.

10. $\text{Tan } \beta = \dfrac{B}{H_g}$ or $\dfrac{B}{H_p}$, where β = dedendum angle.

11. $c = \theta - \beta$, where c = bottom or cutting angle and θ = pitch-cone angle.

12. $f = \theta + \alpha$, where f = face angle of bevel gear.
13. $e = 90° - f$, where e is sometimes called the face angle.
14. $i = A \times \cos \theta$, where i = increment.
15. $O_g = D_g + 2i$, where O_g = outside diameter of gear.
    $O_p = D_p + (2 \times \text{pinion increment})$.

    Note: The pinion increment = $A \times \cos \theta_p$.

16. $T = \dfrac{O_g}{2} \times \tan(90° - f)$.

   Note: To calculate a similar dimension for the pinion, O_p and f_p are placed in the above equation, where f_p = face angle of pinion.

17. $U = \dfrac{H_g - F_g}{H_g} \times T$ (for gear). When calculating U for the pinion, the corresponding dimensions for the pinion are used in the formula given.

18. $S = V - T$, where V = mounting distance.

19. The arc thickness t_1 is taken on the pitch circle at the outer ends of the teeth. The arc thickness t_2 is taken on the pitch circle at the inner ends of the teeth.

20. The chordal thicknesses of the teeth and the corrected addenda are calculated in the same manner as those of standard involute-tooth spur gears. These dimensions may be calculated at either the outer or inner ends of the teeth.

21. The velocity ratio $= \dfrac{\text{speed of driver}}{\text{speed of driven}} = \dfrac{\text{number of teeth in driven}}{\text{number of teeth in driver}} =$

   $\dfrac{\text{pitch diameter of driven}}{\text{pitch diameter of driver}} = \dfrac{\sin \theta \text{ (driven)}}{\sin \theta \text{ (driver)}}$.

FORMULAS FOR HELICAL AND SPIRAL GEARS

Helical gears operate on parallel shafts. When a pair of helical gears are in mesh, the contact between the engaging teeth is on a straight line that makes an angle with the gear shaft. The pitch surfaces of helical gears are cylinders.

Spiral gears operate on nonparallel and nonintersecting shafts. They are similar to helical gears.

The following formulas are common to helical and spiral gears, and are referred to elements shown in Figs. 9-7 and 9-8:

1. Referring to Figs. 9-7 and 9-8, it may be seen

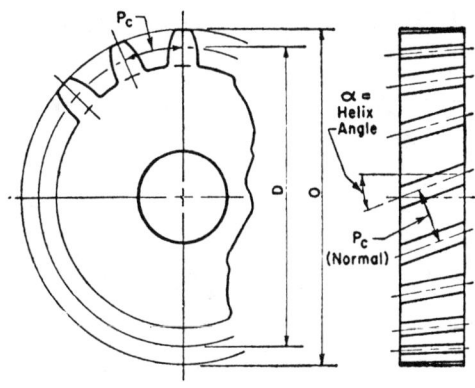

Fig. 9-7.

that the tooth slopes across the face of the pitch cylinder. This slope is measured by the angle between the helix and a line parallel to the center line of the shaft. The angle is the helix angle and is designated by α.

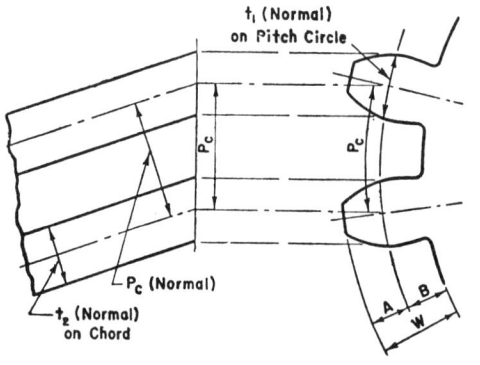

Fig. 9-8.

2. There are two diametral pitches and two circular pitches in helical and spiral gears: one of each pitch is in a diametral plane, and another is in a plane normal to the pitch helix. The circular pitch is designated P_c and the diametral pitch P_d when reference is made to the diametral plane. When referring to the plane normal to the pitch helix, the circular pitch will be designated P_c (normal), and the diametral pitch P_d (normal).

P_c (normal) $= P_c \times \cos \alpha$.
P_d (normal) $= P_d \times \cos \alpha$.

3. $D = \dfrac{N}{P_d} = \dfrac{N}{P_d \text{ (normal)} \times \cos \alpha} = \dfrac{P_c \text{ (normal)} \times N}{\pi \times \cos \alpha}$, where $D =$ pitch diameter, and $N =$ number of teeth in gear.

4. $A = \dfrac{1}{P_d \text{ (normal)}}$, where $A =$ addendum of tooth.

5. $B = \dfrac{1.157}{P_d \text{ (normal)}}$, where $B =$ dedendum of tooth.

6. $W = \dfrac{2.157}{P_d \text{ (normal)}}$, where $W =$ whole depth of tooth.

7. $O = D + \dfrac{2}{P_d \text{ (normal)}}$, where $O =$ outside diameter of gear.

8. $t_1 = \dfrac{P_c \times \cos \alpha}{2} = \dfrac{\pi}{2 P_d \text{ (normal)}}$, where $t_1 =$ normal thickness of tooth on pitch circle.

GEARS 343

9. $t_2 = \dfrac{N}{P_d \text{ (normal)}} \times \sin \dfrac{90°}{N}$, where t_2 = normal chordal thickness of tooth.

10. $L = \pi \times D \times \cot \alpha$, where L = lead.

Note: The reader will recall the definition of *lead* from the chapter on threads.

11. The center distance between two spiral gears in mesh $= \dfrac{D_g + D_p}{2}$, where D_g = pitch diameter of gear, and D_p = pitch diameter of pinion.

The center distance between two spiral gears in mesh $= \dfrac{D_g + D_p}{2}$, which is nearly correct for most practical purposes. If a fixed center distance is required, an adjustment of helix angles and diameters is necessary.

For most comprehensive study of spiral gears, the reader is referred to the series *Manuals of Gear Design* (Sections I, II, and III), by Earle Buckingham.

Herringbone gears are similar to right- and left-hand helical gears joined together, though in modern manufacturing practice they are made from an integral gear blank.

FORMULAS FOR WORM GEARS

Worm gears are similar to helical and spiral gears, which transmit motion between nonparallel and nonintersecting shafts.

The worm is similar to a screw thread, and its pitch line is a helix.

The shaft angle between the gear and the worm is usually 90°. The velocity ratio in worm gearing is usually high.

The following formulas are common to worm gears and are referred to elements shown in Fig. 9-9.

1. $L = P_c \times N_t$, where L = lead of worm thread, P_c = circular pitch of worm gear or linear pitch of worm thread, and N_t = number of separate threads in worm, such as a single-threaded worm, double-threaded worm, etc.

Also $L = \pi \times d \times \tan \alpha$, where d = pitch diameter of worm thread, and α = lead angle of worm, measured from a plane at right angles with the worm shaft.

2. $\operatorname{Tan} \alpha = \dfrac{L}{\pi \times d}$, or $\cot \alpha = \dfrac{\pi \times d}{L}$.

3. $A = \dfrac{P_c}{\pi} = \dfrac{P_c}{3.1416} = .3183 \times P_c$, where A = addendum of tooth in worm gear and addendum of thread in worm. If lead angle is greater than $20°$, $A = .3183 \times P_c \times \cos \alpha$.

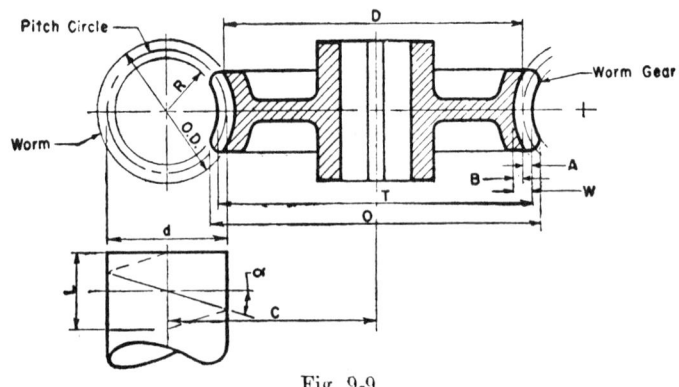

Fig. 9-9.

4. $B = \dfrac{P_c \times 1.157}{\pi} = \dfrac{1.157 \times P_c}{3.1416} = .3683 \times P_c$, where B = dedendum.

5. $W = A + B = (.3183 + .3683) \times P_c = .6866 \times P_c$, where W = whole depth of tooth in gear and whole depth of thread in worm.

6. As with helical and spiral gears, there is a circular pitch P_c in a diametral plane and a normal circular pitch P_c (normal) in a normal plane.

$P_c = \dfrac{\pi \times D}{N_g}$, where D = pitch diameter of gear, and N_g = number of teeth in gear. Also $P_c = \dfrac{L}{N_t}$. P_c (normal) = $P_c \times \cos \alpha$.

7. $d = \dfrac{L \times \cot \alpha}{\pi}$, where d = pitch diameter of worm.

$D = \dfrac{P_c \times N_g}{\pi}$, where D = pitch diameter of gear.

8. $C = \dfrac{D + d}{2}$, where C = center distance between center lines of gear and worm shafts.
9. $T = D + (2 \times A) = D + 2A$, where T = throat diameter of worm gear.
10. $O = D + 3A$, where O = outside diameter of worm gear.
11. $t = .5 \times P_c \times \cos \alpha$, where t = normal thickness of thread on pitch line. This dimension is taken at right angles with the thread sides.
12. $O.D. = d + 2A$, where $O.D.$ = outside diameter of worm.
13. Included angle of worm thread is usually 29°, as explained in Chapter VIII, although some worm threads have an included angle of 40°.
14. The length of the threaded portion on the worm is usually $6 \times P_c$.
15. $R = \dfrac{O.D.}{2} - 2A$, where R = radius of worm gear throat.

PROCESSES FOR SHAPING GEARS

One of the oldest methods of shaping gear teeth is by casting. This process is still used when the shape of the gear teeth does not require high precision. Such gears are made use of in road-making machines, certain types of agricultural machines, and other machines in which component parts are slow-moving.

The various processes for shaping gears follow.

Casting. A casting is made by pouring molten metal into a cavity. The metal retains the shape of the cavity after it cools. There are three types of castings: (1) sand; (2) permanent mold; and (3) die casting.

Molding. Molding is the process of shaping in or on a mold. Gears for light loads are often shaped by plastic molding.

Forming on Power Presses. Sheet-metal stamped gears used in instruments, clocks, watches, and meters.

Rolling. A master gear of a hard material is rolled against a heated gear blank, thus forming teeth in the hot blank.

Machining. The machining of gears is the process of widest use in the industry for the reason that other processes are not accurate enough to meet conditions of heavy load transmitted through the gears at the high speeds at which modern machines and engines are operated, nor are they accurate enough to withstand continuous performance or to meet demands of present-day machinery in other respects.

The machining of gears is classified according to the equipment used

and the procedure adopted in producing the required shape of gear. Some of the common processes are:
1. Machining on milling machines, where cutters of circular form are used to remove enough material from the gear blanks to give the teeth the required shape. The process is slow and is used in cases where volume production of certain types of gears is not warranted.
2. Broaching of gear teeth. This process is recommended for producing internal gears.
3. Machining gear teeth on reciprocating machines, such as gear shapers, using form cutters that are fed into the blank the full depth of the gear tooth.
4. Machining with the aid of a template that is shaped like the tooth it is desired to produce. The template guides the reciprocating cutter. The process is recommended for finishing rough-cast teeth, or to complete the shaping of gear teeth that have been roughed-out by a preliminary machining operation.
5. Generating processes consist of producing gear teeth by means of cutters that not only shape the teeth, but also mesh with the gears, thus assuring the correct shape and spacing of the teeth. This process is adopted in volume production of gears, such as the gears used in automobile transmissions.

The processes here described are performed with the aid of rack cutters, pinion cutters, single-point cutters, rotary side-cutters, and hobs.

GEAR RATIOS

When a pair of gears are in mesh, the driving gear transmits rotary motion to the driven gear, and the speeds of the shafts on which the gears are mounted depend upon the number of teeth on each gear, or on their pitch diameters.

Usually the gears are of different diameters, since in most cases the speed of the rotary motion transmitted is slowed down, although the condition may be reversed, depending upon the type of transmission.

The ratios between the number of teeth or pitch diameters of the gears are called gear ratios, and if these ratios are between the speeds of the driver and the driven members they are velocity ratios.

Spur-Gear and Rack Ratios. For example, a pair of spur gears in mesh is shown in Fig. 9-10.

Assume the pinion is the driver, then the velocity ratio = $\frac{\text{speed of driver}}{\text{speed of driven}} = \frac{1750 \text{ r.p.m.}}{750 \text{ r.p.m.}} = 2\frac{1}{3}$.

This ratio can also be obtained by dividing the number of teeth of the driven by that of the driver, or by dividing the pitch diameter of the driven by that of the driver. Hence $V.R. = \dfrac{S \text{ driver}}{S \text{ driven}} = \dfrac{N_g}{N_p} = \dfrac{D_g}{D_p}$, where $V.R.$ = velocity ratio, S = speeds in r.p.m., N_g = number of teeth in the driven, N_p = number of teeth in the driver, D_g = pitch diameter of the driven, and D_p = pitch diameter of the driver.

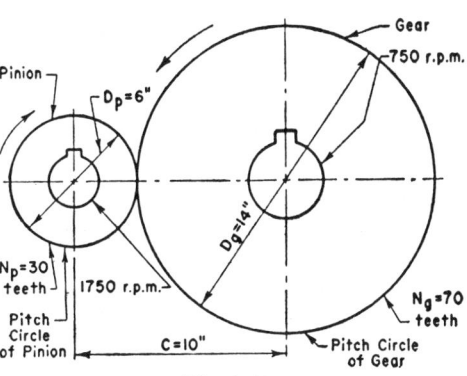

Fig. 9-10.

Any unknown value in this combination may be solved if the remaining values are known. The center distance $C = \dfrac{D_g + D_p}{2} = \dfrac{14 + 6}{2} = 10''$.

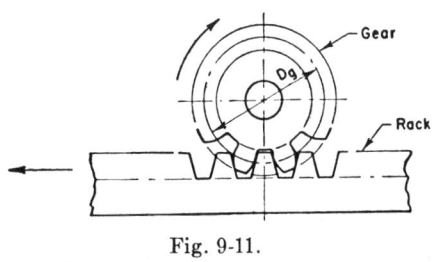

Fig. 9-11.

The ratio between a gear and a rack is calculated in the following manner:

The axial movement of the rack in Fig. 9-11 for one revolution of the gear is equal to the circular pitch of the rack × the number of teeth in the gear, or it is equal to π × pitch diameter of gear.

For example, if the pitch diameter (D_g) of the gear = 12.000'', and the number of teeth (N_g) in the gear = 72, the axial movement of the rack in the direction for one revolution of the gear = $P_c \times N = \dfrac{\pi D_g \times N_g}{N_g} = \pi D_g = 3.1416 \times 12.000 = 37.699''$.

If it is desired to calculate the axial movement of the rack for one sixth of a revolution of the gear, the circular pitch is multiplied by $\dfrac{N_g}{6}$, thus the axial movement of the rack = $\dfrac{P_c \times N_g}{6} = \dfrac{\pi D_g \times N_g}{N_g \times 6} =$

$$\frac{37.699}{6} = 6.283.$$ If, on the other hand, it is desired to calculate the number of revolutions of the gear for a certain axial movement of the rack, say 5″, the number of revolutions of gear =

$$\frac{\text{axial movement of rack}}{P_c \times N_g} = \frac{\text{axial movement} \times N_g}{\pi \times D_g \times N_g}$$

$$= \frac{5}{3.1416 \times 12} = \frac{5}{37.699} = .133.$$

The gear will rotate .133 of a revolution.

Ratios for Other Gears. The gear ratios for helical, spiral, and bevel gears are calculated in the same manner as for spur gears, i.e., the

$$V.R. = \frac{\text{speed of driver}}{\text{speed of driven}} = \frac{\text{number of teeth in driven}}{\text{number of teeth in driver}} = \frac{\text{pitch diameter of driven}}{\text{pitch diameter of driver}}.$$

In worm gears, the worm is usually the driver, and the $V.R. =$

$$\frac{\text{speed of driver}}{\text{speed of driven}} = \frac{\text{number of teeth in worm gear}}{\text{number of separate threads in worm}}$$

Note: Do not consider the pitch diameter of the worm in calculation of gear ratios.

Gear Trains. For calculation of gear trains, the reader is referred to the section on compound ratios and proportions in Chapter IV, and to the numerous problems pertinent thereto in the section of practice problems at the end of that chapter.

INDEXING

The indexing head or dividing head is a device that can be attached to a milling machine for the purpose of rotating the work in the machine through a definite portion of a revolution.

For example, if it is desired to cut 36 teeth in a given gear blank on a milling machine, the indexing head is attached to the machine for the purpose of revolving the gear blank one thirty-sixth of a revolution between two successive cuts.

Another example is the machining of a reamer to produce a certain number of flutes or grooves. The reamer is rotated a portion of a revolution between two successive cuts in the machine, the amount of rotation depending upon the number of grooves to be cut.

Indexing consists of dividing the work into a number of equal parts. There are various methods by which this is accomplished, the most common of which are the direct and simple methods. Other methods of indexing are used under specific conditions, which we shall discuss briefly.

Direct Indexing. Direct indexing is accomplished without the aid of the gear mechanism by means of a plate attached to the spindle of the indexing head. The plate is about 8″ in diameter and is drilled and reamed usually with 24 equally spaced holes which are concentric with the spindle. A pin inserted in the frame of the device holds the indexing plate securely.

The indexing consists of turning the spindle and the work the required amount while the pin is released, then the pin is inserted in the proper hole of the plate.

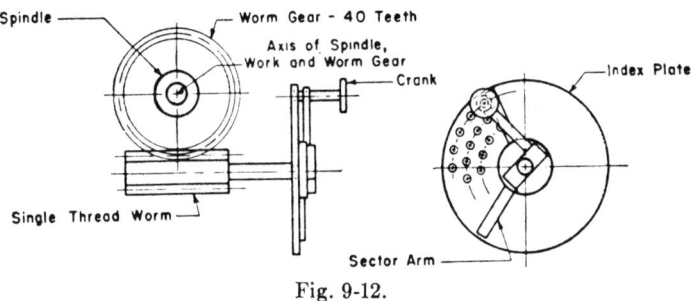

Fig. 9-12.

For example, if six grooves are to be cut in a reamer, the spindle and the work are rotated one sixth of a revolution between successive cuts, a distance spanning four holes on the indexing plate.

The direct method of indexing is limited to cases where the numbers of holes through which the plate rotates are factors of 24, such as 2, 3, 4, 6, 8, 12, and 24.

Simple Indexing. Simple indexing is performed by means of the device illustrated in Fig. 9-12. The spindle is rotated by means of a hand crank which turns a single-threaded worm in mesh with a 40-tooth worm gear. The spindle, the work, and the worm gear have the same axis. Since the worm gear has 40 teeth, one revolution of the single-threaded worm turns the gear, the spindle, and the work one fortieth of a revolution.

In order to turn the work any amount, an indexing plate containing a number of concentric circles of equally spaced holes is used. Three indexing plates are provided as standard equipment to accommodate a good variety of work, and, for a wide range of work, additional plates containing various combinations of concentric circles of equally spaced holes are available.

The three standard indexing plates provide the following combinations of holes.

#1 plate: 15, 16, 17, 18, 19, and 20 holes.
#2 plate: 21, 24, 27, 29, 31, and 33 holes.
#3 plate: 37, 39, 41, 43, 47, and 49 holes.

The method of simple indexing can best be demonstrated by a practical example.

Assume that a 27-tooth gear is to be cut on a milling machine. The gear blank is attached to the spindle of the indexing head, and a milling cutter of proper tooth form shapes a space between two teeth, as shown in Fig. 9-13.

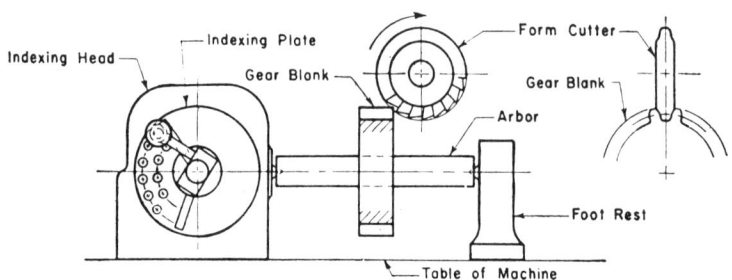

Fig. 9-13.

One turn of the crank rotates the threaded worm once, and the 40-tooth worm gear turns one fortieth of a revolution.

There are 27 teeth to be cut in the gear blank, therefore, after completing one cut as shown in Fig. 9-13, the gear blank must be turned one twenty-seventh of a revolution, and so must the worm gear.

For one complete turn of the worm gear, the crank must be turned 40 times, but for a turn of the worm gear measuring one twenty-seventh of a revolution, the crank must turn $\frac{40}{27}$ revolutions, or $1\frac{13}{27}$ revolutions.

It is convenient to turn the crank once, but the fraction of a turn of thirteen twenty-sevenths can be accomplished by selecting an indexing plate containing the proper number of holes, such as the #2 plate, which has a 27-hole circle. By turning the crank once and then through 13 holes in the plate, which is made stationary, the worm gear and the work will be turned through the required fraction of a revolution, namely, one twenty-seventh.

The crank is furnished with a pin which is inserted in the proper hole of the indexing plate at the proper time, thus making the crank and the mechanism operated by it stationary.

The calculation is performed as follows: The number of turns of the crank $= \frac{40}{N}$, where 40 is a constant, and $N =$ number of equal divisions of the work.

An indexing plate containing a suitable number of holes is selected by inspection of the denominator N, which may or may not be the required number of divisions of work, this depending upon whether the numerator and the denominator of the fraction $\frac{40}{N}$ are prime to each other or can be reduced.

For example, if the number of division of work is 25, the turns of the crank $= \frac{40}{N} = \frac{40}{25} = 1\frac{15}{25} = 1\frac{3}{5}$.

The crank is turned completely around once, and for the fraction $\frac{3}{5}$, the #1 plate containing a 15-hole circle may be used. Since $\frac{3}{5}$ equals $\frac{9}{15}$, the crank is turned additionally through nine holes in the chosen indexing plate.

For convenience, a sector with two arms is furnished with the device, as shown in Fig. 9-12. This sector can be adjusted to span the required number of holes, and once adjusted, it is turned in such a way that one of its arms previously in contact with the first hole of the required span will be in contact with the last hole, thus bringing the other arm automatically in contact with the new position on the indexing plate.

If the range of standard indexing plates is exhausted, special plates can be used, which are of larger diameters and contain circles that provide a larger number of holes, thereby offering a greater range of indexing.

However, if the standard indexing plates are the only ones available, other methods must be resorted to for achieving the required results. Among these methods are the differential indexing and the compound indexing.

Differential Indexing. Differential indexing is similar to simple indexing, differing from it only by the fact that the indexing plate also revolves when the crank turns through an available number of holes in the plate approximate to the number desired. The speed of rotation of the plate is automatically controlled by a set of change gears that connects the plate to the dividing head spindle holding the work to be indexed. The plate can be rotated in the same direction as the crank, or in the opposite direction, depending upon the gears. The purpose of rotating the indexing plate while the crank turns is to compensate for choosing a plate of approximate range among the plates available. Compensation is effected through the calculated gear ratio.

The reader is advised to refer to manufacturers' manuals for more detailed information of the subject of indexing.

Compound Indexing. Compound indexing is cumbersome and not much used in actual practice. For study of this method of indexing, the reader is referred to texts on machine shop practice, wherein milling machine operations are described in detail.

Angular Indexing. Angular indexing consists of getting even divisions by taking angular measure. One complete circle is equal to 360°. One turn of the indexing crank will turn the spindle one fortieth of 360°, or 9°.

If the angle desired is a multiple of 9°, it can be obtained by a calculated number of turns of the crank. For example, an angle of 63° may be indexed by turning the crank seven times.

For indexing angles that are not multiples of 9°, proportional parts of a turn of the crank can be made. For example, to index 7°, the crank is turned seven ninths of a revolution and, by choosing a plate of suitable range, such as the #1 plate, the crank may be turned through 14 holes of the 18-hole circle to index fourteen eighteenths, or seven ninths, of a revolution of the work, as required.

An exact angular indexing, consisting of degrees, minutes, and seconds, may be made for any angle so long as enough plates of ample range are available.

For comprehensive study of the subject of gears, the reader is referred to the series, *Manuals of Gear Design* (Sections I, II, and III), by Earle Buckingham, published by Industrial Press, New York. A knowledge of involute trigonometry, which is treated thoroughly in Section I of these manuals, is essential to thorough understanding of the subject matter.

Also, simplified formulas for checking the pitch diameters of spur gears (external and internal), helical gears, and worm gears are given extensively in standard engineering handbooks, such as *Machinery's* handbook, *American Machinist's* handbook, and *Kent's* handbook.

The reader is also referred to pamphlets issued by manufacturers of precision instruments and tools, a few of which are the Van Keuren Company of Watertown, Massachusetts, Brown & Sharpe of Providence, Rhode Island, Fellows Gear Shaper Company of Springfield, Vermont, Illinois Tool Works of Chicago, Foote Gear and Machine Corporation of Chicago, and Gleason Works of Rochester, New York.

GEARS

PRACTICE PROBLEMS APPLICABLE IN THE SHOP AND TOOLROOM

1. Given two spur gears in mesh with each other. The number of teeth in the pinion is 28, and its pitch diameter is 3.500″. The pinion is connected to a 1650 r.p.m. motor and transmits rotary motion to a machine shaft through the gear. The speed of the machine shaft is 600 r.p.m. Find the number of teeth in the gear.
2. Find the pitch diameter of the gear in Problem 1.
3. Find the center distance between the pinion and gear in Problem 1.
4. Find the diametral pitch in Problem 1.
5. Find the circular pitch in Problem 1.
6. Assuming the pinion and gear in Problem 1 have full-depth $14\frac{1}{2}°$ involute teeth, calculate the addendum of the teeth.
7. Calculate the dedendum of the teeth in Problem 6.
8. Calculate the clearance and the fillet radius of the teeth in Problem 6.
9. Calculate the working depth of the teeth in Problem 6.
10. Calculate the whole depth of the teeth in Problem 6.
11. Calculate the theoretical thickness of the teeth on the pitch circle in Problem 6.
12. Calculate the average backlash between teeth in Problem 6.
13. Calculate the actual thickness of the teeth on the pitch circle in Problem 6.
14. Calculate the actual width of space between teeth on the pitch circle in Problem 6.
15. Find the outside diameters of the gear and pinion in Problem 6.
16. Find the inside diameters of the gear and pinion in Problem 6.
17. Find the base circle diameters of the gear and pinion in Problem 6.
18. Given two spur gears having a center distance of 12.750″. Their velocity ratio is 2. Find the pitch diameters of the pinion and gear.
19. Assuming the gears in Problem 18 have a 6 diametral pitch, find the number of teeth in gear and pinion.
20. Find the circular pitch in Problem 18.
21. Assuming the gears in Problem 18 have full-depth 20° involute teeth, calculate the full depth of the teeth.
22. Calculate the fillet radius of the teeth in Problem 21.
23. Calculate the base circle diameters of the gears in Problem 21.
24. Assuming a rack is driven by a 10-pitch spur gear, calculate the axial distance traveled by the rack, if 13 teeth of the gear actively engage the rack throughout an operational cycle.
25. Calculate the number of degrees through which a 7-pitch, 45-tooth spur gear must turn to move a rack an axial distance of π inches.
26. Given two spur gears in mesh with each other. The center distance between the gears is 8.500″ and their velocity ratio is twelve fifths. The gears are of the American Association Stub-tooth system, and their diametral pitch is 5. Find the pitch diameters of both gears.
27. Find the number of teeth in each gear in Problem 26.
28. Find the whole depth of tooth in Problem 26.
29. Find the clearance of the teeth in Problem 26.

354   MATHEMATICS FOR INDUSTRY

30. Find the fillet radii of the teeth in Problem 26.
31. Find the base circle diameters of the gears in Problem 26.
32. Find the outside diameters of the gears in Problem 26.
33. Given a gear of the Fellows Gear Shaper Company system. Its diametral pitch is seven ninths, and its pitch diameter = 9.000″. Find the number of teeth in the gear.
34. Calculate the theoretical thickness of the tooth on the pitch circle of the gear in Problem 33.
35. Calculate the whole depth of tooth of the gear in Problem 33.
36. Calculate the clearance of the tooth of the gear in Problem 33.
37. Calculate the outside diameter of the gear in Problem 33.
38. Given an internal gear meshing with a pinion whose center distance is 2.4167″ and diametral pitch is 12. The velocity ratio of pinion and gear $= \frac{43}{14}$. Calculate the number of teeth in pinion and gear.
39. Calculate the inside diameter of the internal gear in Problem 38.
40. Calculate the outside diameter of the pinion in Problem 38.
41. Calculate the pitch diameters of the pinion and gear in Problem 38.
42. Calculate the whole depth of tooth in Problem 38.
43. Calculate the circular pitch of gears in Problem 38.
44. Calculate the chordal thickness of tooth of gear having a pitch diameter of 6.833″ and a diametral pitch of 12.
45. Calculate the corrected addendum of the tooth of gear in Problem 44.
46. Assume that the outside diameter of the gear in Problem 44 is oversized by .004″. What should the corrected addendum be?
47. Assume that the outside diameter of the gear in Problem 44 is undersized by .003″. What should the corrected addendum be?
48. Assume that a $14\frac{1}{2}°$ standard involute full-depth tooth gear has 64 teeth and a 6-diametral pitch. The pitch diameter of this gear is to be checked with the aid of two precision wires. Calculate the proper wire diameter to be used and the measurement over the wires.
49. Assume that the gear in Problem 48 has a 20° full-depth tooth, and that its pitch diameter is checked with the aid of precision wires. Calculate the proper wire diameter and the measurement over the wires.
50. Assume a $14\frac{1}{2}°$ standard involute full-depth tooth gear having a pitch diameter of 17.400″ and a 5-diametral pitch. Calculate the proper wire diameter in order to check the pitch diameter of this gear with the aid of wires. Also calculate the measurement over the wires.
51. Assume a pair of straight-tooth bevel gears whose shafts intersect at 90°. Their velocity ratio $= \frac{7}{3}$. The pitch diameter of the driver

= 5.500″, and the diametral pitch = 6. Calculate the number of teeth in the pinion and gear. Also calculate the pitch diameter of the driven member.
52. Calculate the pitch-cone angles of gear and pinion in Problem 51.

Hint: $\dfrac{\sin \theta_g}{\sin \theta_p} = V.R. = \dfrac{7}{3}$

$\dfrac{\sin \theta_g}{\sin (90° - \theta_g)} = \dfrac{7}{3}$, but $\sin (90° - \theta_g) = \cos \theta_g$

∴ $\dfrac{\sin \theta_g}{\cos \theta_g} = \dfrac{7}{3} = \tan \theta_g$

Using another method: $\tan \theta_g = \dfrac{N_g}{N_p} = V.R. = \dfrac{7}{3}$

53. Calculate the cone distance of each gear in Problem 51.
54. Calculate the maximum width of gear and pinion face in Problem 51.
55. Calculate the addendum angle, dedendum angle, and the cutting angle of the gear and pinion in Problem 51.
56. Calculate the face angles f and e of both gears in Problem 51.
57. Calculate the outside diameters of the gear and pinion in Problem 51.
58. Calculate dimensions T and U for gear and pinion in problem 51. See Fig. 9-6, bevel gears.
59. Assume a pair of helical gears between parallel shafts. Their velocity ratio = 3. The pitch diameter of the driver = 3.750″, and its number of teeth = 30. The helix angle = 6°. Find the normal diametral pitch and the normal circular pitch of the gears. Find the number of teeth in the gear.
60. Find the whole depth of tooth in Problem 59.
61. Find the outside diameters of both gears in Problem 59.
62. Calculate the lead of the driven gear in Problem 59.
63. Calculate the center distance between the gears in Problem 59.
64. Assume a worm gear driven by a single-threaded worm. The worm gear has 36 teeth, and the speed of the worm is 1800 r.p.m. The worm has a linear pitch = .3927″. Calculate the lead of the worm and the speed of the worm gear.
65. Calculate the diametral pitch and the circular pitch of the worm gear in Problem 64.
66. Assume the pitch diameter of the threaded worm in Problem 64 is .875″. Calculate the lead angle of the worm and the whole depth of the worm thread in Problem 64.
67. Calculate the whole depth of the gear tooth in Problem 64 and also the pitch diameter of the worm gear.
68. Calculate the center distance between the gear and worm shafts in Problem 64.

69. Calculate the throat diameter of the worm gear in Problem 64.
70. Calculate the outside diameters of the worm and worm gear in Problem 64.
71. Calculate the radius of the worm gear throat in Problem 64, and also the length of the threaded portion of worm.
72. Assume two spur gears in mesh with each other. The pinion is the driver and is keyed to a shaft whose speed = 1550 r.p.m. The velocity ratio = $2\frac{1}{2}$. How many r.p.m. will be transmitted by the driven shaft, which is keyed to the gear?
73. Assume the gear in Problem 72 has 125 teeth, how many teeth are in the pinion?
74. If the gears in Problem 72 have a 12-diametral pitch, what is the center distance between the gears?

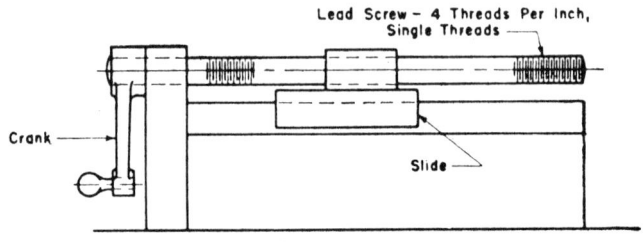

Fig. 9-14.

75. Assume the pinion in Problem 72 is the driven. What will the velocity ratio be between gears, if the size of the gears and their speeds remain the same?
76. Assume a pair of bevel gears with a velocity ratio of 5:3. The pinion is the driver and rotates 900 r.p.m. Its pitch diameter = 3.600″ and the diametral pitch = 5. How many teeth are in the gear and pinion?
77. What is the speed of the gear in Problem 76?
78. Assume a worm drive. The worm gear has 56 teeth and the worm is double-threaded. How many revolutions will the worm turn per minute in order to rotate the worm gear 47 r.p.m.?
79. A 14-diametral pitch gear drives a rack an axial distance of 7.854″. How many revolutions must the gear rotate in order to move the rack the required distance, if the pitch diameter of the gear = 2.000″?
80. How many turns must the crank make in order to move the slide $5\frac{11''}{16}$, as shown in Fig. 9-14?
81. Calculate the distance traveled in one minute by the slide, as shown in Fig. 9-15. This figure illustrates a gear train driving a lead screw that is connected to the slide, the axial motion of which is to be determined.

GEARS

82. A gear train is shown in Fig. 9-16. The power is transmitted from the motor through the worm and the gears to shaft A. Calculate the r.p.m. of shaft A.

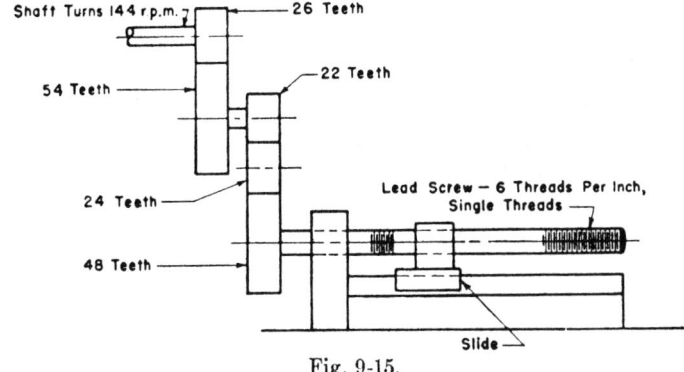

Fig. 9-15.

83. Assume a plate is to be indexed on a circle for 35 holes equally spaced. The dividing head of a milling machine is employed for simple indexing, and the three-standard indexing plates are available. Choose the proper plate and calculate the number of holes on the proper hole-circle of the plate.

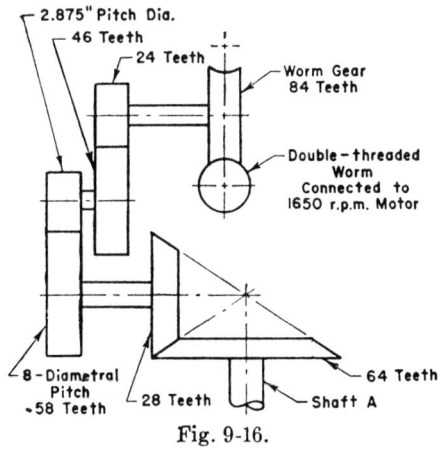

Fig. 9-16.

84. Index a 120-tooth gear by the equipment used in Problem 83.
85. Index a 66-tooth gear by the equipment used in Problem 83.
86. Index a 34-tooth gear by the equipment used in Problem 83.
87. Assume that a circular saw is to be indexed for 45 teeth by the equipment used in Problem 83.

88. Design a special indexing plate with the required circle of holes for simple indexing of a 122-tooth gear. Select the least number of holes.

89. Design a special indexing plate having the required circle of holes for simple indexing of 96 equally spaced slots on a circular plate. Select the least number of holes in the indexing plate.

90. How many times must the indexing crank turn in order to index an angle of 72°?

91. How many times must the indexing crank turn in order to index an angle of 60°?

92. How many complete turns must the indexing crank make, and through how many holes of the proper indexing plate must the crank turn, in order to index an angle of 94°?

Chapter X
LOGARITHMS

In order to perform rapid calculations involving multiplication, division, raising numbers to various powers, and extracting roots of numbers, logarithms are employed. By means of logarithms, the cumbersome operations of multiplication and division of large numbers and decimal fractions can be changed to simple operations of addition and subtraction. Instead of raising a number to a power by established procedure, the simpler process of multiplication is employed to obtain the same result. Similarly, the process of division is simpler than extracting a root by the customary procedure.

Many kinds of calculations are performed with difficulty when conventional processes are employed; and some calculations are impossible without recourse to logarithms. For instance, if a number must be raised to a very high power, or if extraction of a very small root of a number is desired, the operations of arithmetic are found unduly long and cumbersome. The same is true when a number is to be raised to a power whose exponent is a fractional or a mixed number. In a computation of this kind the arithmetical operation becomes impossible, and one is compelled to use logarithms.

A logarithm (designated *log*) of any number is the exponent of the power to which a predetermined base is raised to create that number. For example, the expression $2^3 = 8$ means that the exponent 3 is the logarithm of 8 when the base is 2. Any convenient base may be chosen, and a system of logarithms can be devised for all the positive numbers to the chosen base.

There are only two important systems of logarithms. The system of *common logarithms* (also called the *Briggs* system), uses the base 10 and in practice this is preferred for the convenience it affords when performing the various operations. The *natural* or *Napierian* system of logarithms employs the base $e = 2.71828$ and is used in theoretical computations and in higher mathematics.

REVIEW OF LAWS OF EXPONENTS

For a better understanding of logarithms, a brief review of the law of exponents from the chapters on algebra is presented.

1. $a^n = a \times a \times a \cdots \times a$, where a is a factor, and n is the exponent, and it is equal to the number of factors of a.

2. $a^{-n} = \dfrac{1}{a^n}$, where $-n$ is the negative exponent.

3. $a^{m/n} = \sqrt[n]{a^m}$.

4. $a^m \times a^n = a^{m+n}$.

5. $\dfrac{a^m}{a^n} = a^{m-n}$.

6. $(a \times b \times c)^n = a^n b^n c^n$.

7. $\left(\dfrac{a}{b}\right)^n = \dfrac{a^n}{b^n}$.

8. $(a^m)^n = a^{mn}$.

9. $a^o = 1$.

THE CHARACTERISTIC AND MANTISSA

The common system of logarithms with a base of 10 has wide use in practical engineering and shop computations, therefore it is presented in this chapter.

Since the base 10 raised to a certain power is equal to the given number, and the exponent of the power is the logarithm of the number, the following tabulation may be written:

$10^0 = 1$, therefore the log of $1 = 0$.
$10^1 = 10$, therefore the log of $10 = 1$.
$10^2 = 100$, therefore the log of $100 = 2$.
$10^3 = 1000$, therefore the log of $1000 = 3$.

$10^{-1} = \dfrac{1}{10} = .100$, therefore the log of $.100 = -1$

$10^{-2} = \dfrac{1}{100} = .010$, therefore the log of $.010 = -2$.

$10^{-3} = \dfrac{1}{1000} = .001$, therefore the log of $.001 = -3$.

The logarithm of a number between 1 and 10 is greater than 0 and smaller than 1; the logarithm of a number between 10 and 100 is greater than 1 and smaller than 2; the logarithm of a number between .100 and 1 is greater than -1 and smaller than 0, etc.

It is evident that the logarithms of numbers greater than 1 are positive, and those of numbers less than 1 are negative. However, the logarithms of numbers less than 1, if they are whole numbers, are written in the negative form only, i.e., -1, -2, -3, etc. If the above logarithms

LOGARITHMS

are mixed numbers, they are written thus: $\bar{1}.3242$, $\bar{2}.0467$, etc., indicating that only the whole numbers are negative, while the decimal fractions are always positive.

In most cases, the logarithm is a mixed number and its integral part is called the *characteristic* and the decimal fraction is the *mantissa*.

Determining the Mantissa. Tables of logarithms have been computed for finding the mantissa of any number desired. Condensed tables to five places may be found in the Appendix of this text, and, for extreme accuracy in computations, the reader is referred to standard handbooks and specially prepared manuals containing tables of logarithms to six or seven places.

Determining the Characteristic. The characteristic must be decided upon by mere inspection of the given number, which is governed by the following rules.

Rule 1. If the given number is greater than 1, the characteristic is positive and is equal to *one less* than the number of digits to the left of the decimal point in the given number. For example, if the given number is 6,473, the characteristic of its logarithm is equal to 4, which is the number of digits in 6,473, less 1, i.e., $4 - 1 = 3$.

Rule 2. If the given number is less than 1 and expressed as a decimal fraction, the characteristic of its logarithm is negative and is equal to *one more* than the number of zeros between the decimal point and the first nonzero figure in the given number. For example, if the given number is .0052, the characteristic of its logarithm is negative and is equal to 2, which is the number of zeros in .0052, plus 1, i.e., $2 + 1 = 3$, and since the characteristic is negative, it is equal to -3.

Rule 3. The mantissa is the same for all numbers that consist of identical significant figures. For example, the mantissa for 374.60 and .0037460 is the same, namely, .57357, the difference between the logarithms of both numbers being their characteristics, namely, 2 for the first number and -3 for the latter. The complete logarithms of the given numbers are: log 374.60 = 2.57357, and log .0037460 = $\bar{3}.57357$.

FINDING THE LOGARITHM OF A NUMBER

Finding the logarithm of a number consists of writing down the characteristic in accordance with the rules indicated in the previous paragraphs and finding the proper mantissa from the tables.

The tables in the Appendix show the first three significant figures of the given number in the extreme left-hand vertical column designated by "N." The fourth significant figure of the given number is located in the upper horizontal line to the right of "N."

If the given number consists of four or less significant figures, the mantissa of the logarithm of the given number is found directly in the tables by locating the first three significant figures of the number in the left-hand column, and then tracing a horizontal line to the right of the three significant figures until a place is reached on the line below the fourth significant figure located in the upper horizontal line. The mantissa is found at the intersection of the two traced lines.

Example 1: Find log 4,763. The 476 is located in the left-hand column, and at the intersection of a horizontal line traced to the right and a vertical line traced downward from 3 (at the top) is the mantissa of the logarithm of the given number, namely, .67788. Note that the decimal point is omitted in the table to prevent confusion in printing. The characteristic of the logarithm (3) is determined by the procedure previously indicated. Finally, the log 4,763 = 3.67788.

Example 2: Find log .0004763. The mantissa of this logarithm is the same as that of 4,763, namely, .67788. The characteristic, however, according to the rule, is -4, so the log .0004763 = $\bar{4}$.67788.

If the given number consists of more than four significant figures, its logarithm may be obtained directly from tables in handbooks and manuals giving values of the mantissa of any number up to six or seven significant figures. But if the reader wishes to use the tables in the Appendix of this text, the very nearly correct mantissa for the given number can be obtained by the simple process of *interpolation*.

Example: If the logarithm of 67,438 is desired, we first determine the mantissa of 6,744, which is .82892, and then the mantissa of 6,743, which is .82885. The mantissa of 67,438 lies between .82892 and .82885. To find the increment that must be added to .82885, or the increment that must be subtracted from .82892, we obtain the difference between the two mantissas .82892 − .82885 = .00007.

The next step is to estimate by proportion the increment which must be added to .82885. Since the difference between the numbers is 10, the given number 67,438 is between 67,440 and 67,430. Note that 67,438 is 8 larger than 67,430, thus 67,438 is 67,430 plus $\dfrac{8}{10}$ of the difference between 67,430 and 67,440. It is assumed that the mantissas behave proportionately; so take $\dfrac{8}{10}$ of the difference, .00007, and add it to the mantissa of 67,430 to get the mantissa of 67,438. This gives

$.00007 \times \dfrac{8}{10} = .000056$. The correct mantissa $= .82885 + .000056 = .828906$.

Note: When the estimated proportional increment is added to the mantissa of the smaller number, the resulting mantissa may not be as accurate as that in tables of special manuals and handbooks, since the mantissas in the Appendix of this text are given to only 5 places and frequently the fifth place of the mantissa is rounded off to the nearest number.

FINDING THE ANTILOGARITHM OF A LOGARITHM

When numbers or antilogarithms (designated *antilog*) which correspond to given logarithms are to be determined, the operation is reversed. The correct mantissa of the given logarithm is located in the tables, and a corresponding number is determined in the left-hand vertical column, and at the top.

Example: What number corresponds to the logarithm 3.81097? The mantissa .81097 is opposite 647 in the "N" column and in line with 1 at the top, therefore the significant figures of the required number are 6,471. Since the characteristic is equal to 3, the number of significant figures in the given number must be equal to 4. Hence, the given number $= 6,471$.

If the mantissa of the given logarithm is not in the tables, the corresponding number may be determined with close approximation by means of interpolation.

Example: What number corresponds to the logarithm $\bar{2}.95300$? The nearest mantissas in the tables are: .95303 and .95299. The number which corresponds to the mantissa .95303 contains the significant figures of 8,975. The number which corresponds to the mantissa .95299 contains the significant figures of 8,974. The number which corresponds to the mantissa .95300 is greater than 8,974 by the proportional increment of
$$\dfrac{.95300 - .95299}{.95303 - .95299} = \dfrac{1}{4} = .25.$$

It is evident that the required number, so far as the significant figures are concerned, lies between 8975 and 8974. Hence it is $8974 +$ the increment $.25 = 897,425$.

Since the characteristic is a negative 2, the required number is a decimal fraction, and its value is .0897425.

COLOGARITHMS

When simplifying various operations, such as multiplication, division, etc., the use of cologarithms (designated *colog*) instead of logarithms is more convenient and less cumbersome.

The cologarithm of a number is obtained by subtracting the logarithm of the number from 10 and then attaching -10 to the result. For example, the logarithm of 8,465 = 3.92763, and the cologarithm of the same number $= (10 - 3.92763) - 10 = 6.07237 - 10$.

In the simplification of the various operations that follow, it will be shown how the use of cologarithms and logarithms combined is more convenient at times than use of logarithms alone. The simplification is illustrated by the fact that it is more convenient at times to add a log of one number to the colog of the other than to subtract the log of one number from the log of the other.

MULTIPLICATION OF NUMBERS BY MEANS OF LOGARITHMS

Multiplication of numbers can be simplified by adding the logarithms of the given numbers and then finding a number or antilog in the tables whose logarithm is equal to the sum of the logarithms obtained.

Example 1: 642.7 × 835.4.

$$
\begin{array}{ll}
\log 642.7 & = 2.80801 \\
\log 835.4 & = 2.92189 \\
\hline
\text{sum of logs} & = 5.72990 = \log \text{ of product}
\end{array}
$$

The number or antilog which corresponds to log 5.72990 is found from the tables with the aid of interpolation.

The number is estimated to be very nearly 536,912, since the characteristic of its logarithm is 5, and therefore the number must have six significant figures.

Example 2: 24.072 × .000485.

$$
\begin{array}{ll}
\log 24.072 & = 1.381516 \\
\log\ \ \ .000485 & = \bar{4}.68574 \\
\hline
\text{sum of logs} & = \bar{2}.067256 = \log \text{ of product}
\end{array}
$$

The number which corresponds to log $\bar{2}.067256$ is found from the tables with the aid of interpolation.

The number is estimated to be very nearly .011675 since the characteristic of its logarithm is $\bar{2}$, and therefore the number is a decimal fraction containing a single zero between the decimal point and the first significant figure.

Example 3: 572.6 × .00873 × 49.06.

$$
\begin{array}{ll}
\log 572.6 & = 2.75785 \\
\log\ \ \ .00873 & = \bar{3}.94101 \\
\log\ \ 49.06 & = 1.69073 \\
\hline
\text{sum of logs} & = 2.38959 = \log \text{ of product}
\end{array}
$$

LOGARITHMS

The number which corresponds to log 2.38959 is found from the tables with the aid of interpolation. It is estimated to be very nearly 245.239.

DIVISION OF NUMBERS BY MEANS OF LOGARITHMS

Division of numbers can be simplified by subtracting the logarithm of the divisor from the logarithm of the dividend. The difference is the logarithm of the quotient and the quotient is the number that corresponds to its logarithm.

Example 1: 418.50 ÷ 76.08.

$$\begin{array}{ll} \log 418.50 & = 2.62170 \\ \log\ \ 76.08 & = 1.88127 \\ \hline \text{difference of logs} = & .74043 = \log \text{ of quotient} \end{array}$$

The number or antilog which corresponds to log .74043 is found from the tables with the aid of interpolation. The number is estimated to be very nearly 5.501.

This example can be worked with the aid of logarithms and cologarithms as follows:

$$\begin{array}{ll} \log\ \ \ 418.50 & = 2.62170 \\ \text{colog}\ \ 76.08 & = 8.11873 - 10 \\ \hline \text{sum of log and colog} = 10.74043 - 10 = .74043 = \log \text{ of quotient} \end{array}$$

By adding the log of the dividend and the colog of the divisor, we obtain directly the log of the quotient, which at times is more convenient than subtracting the log of the divisor from the log of the dividend.

Example 2: 7.008 ÷ 832.04.

$$\begin{array}{ll} \log\ \ \ 7.008 & = .84559 \\ \log 832.04 & = 2.92014 \\ \hline \text{difference of logs} = \bar{3}.92545 \end{array}$$

The number which corresponds to log $\bar{3}.92545$ is found from the tables to be very nearly .0084226.

It is evident that this example can be worked with more ease by adding the log of the dividend and the colog of the divisor, thus obtaining the log of the quotient.

$$\begin{array}{ll} \log\ \ \ \ 7.008 & = .84559 \\ \text{colog } 832.04 & = 7.07986 - 10 \\ \hline \log \text{ of quotient} = 7.92545 - 10 = \bar{3}.92545 \end{array}$$

This checks with the log obtained by the first method. The suggestion is offered that both methods be used for checking purposes.

MULTIPLICATION AND DIVISION COMBINED

The following examples illustrate the use of logarithms and cologarithms in multiplication and division combined.

Example 1: $\dfrac{7.048 \times 6482.5}{.0473 \times 598.25}$.

This example furnishes ample evidence of the saving of time by the use of logs and cologs. The logs of both factors in the numerator or dividend and the cologs of the factors in the denominator or divisor are added together, and the sum is the log of the number required.

$$
\begin{array}{lll}
\log & 7.048 & = .84807 \\
\log & 6482.5 & = 3.81174 \\
\text{colog} & .0473 & = 11.32514 - 10 \\
\text{colog} & 598.25 & = 7.22312 - 10 \\
\hline
\text{log of result} & & = 23.20807 - 20 = 3.20807
\end{array}
$$

The number which corresponds to log 3.20807 is found from the tables to be very nearly 1614.62.

Example 2: $\dfrac{23.080 \times 42.58}{263.7 \times 92.07 \times 74.36}$.

$$
\begin{array}{lll}
\log & 23.080 & = 1.36324 \\
\log & 42.58 & = 1.62921 \\
\text{colog} & 263.7 & = 7.57889 - 10 \\
\text{colog} & 92.07 & = 8.03588 - 10 \\
\text{colog} & 74.36 & = 8.12866 - 10 \\
\hline
\text{log of results} & & = 26.73588 - 30 = \bar{4}.73588
\end{array}
$$

The number which corresponds to log $\bar{4}.73588$ is a decimal fraction having three zeros between the decimal point and the first significant figure. It is very nearly .00054435.

EXTRACTION OF A ROOT BY MEANS OF LOGARITHMS

This operation is simplified considerably by the use of logarithms. It consists of finding the logarithm of the given number, then dividing it by the index of the root. The quotient is the log of the root.

Example 1: $\sqrt[3]{724.85}$.

The log of the given number 724.85 = 2.86025. Dividing 2.86025 by the index of the root = $\dfrac{2.86025}{3}$ = .953416. The root is equal to the number whose log = .953416. The root = 8.983.

LOGARITHMS

Example 2: $\sqrt[3]{.08635}$.

The log of .08635 = $\bar{2}$.93626. To eliminate the negative characteristic $\bar{2}$, it is more convenient to add 30 and subtract 30 from $\bar{2}$.93626. The operation is as follows: $30 + \bar{2}.93626 = 28.93626 - 30$. Then $\frac{1}{3}$ of the log of the given number $= \dfrac{28.93626 - 30}{3} = 9.64542 - 10 = \bar{1}.64542$. The number which corresponds to log $\bar{1}.64542$ is the root. It is .4420.

Example 3: $\sqrt{.5738}$.

The log of .5738 = $\bar{1}$.75876. To eliminate the negative characteristic $\bar{1}$ before dividing by the index of the root, which is 2, it is more convenient to add 10 and subtract 10 from $\bar{1}$.75876. Thus $10 + \bar{1}.75876 - 10 = 9.75876 - 10$. Then, dividing by the index of the root, we obtain the log of the root, which is $\dfrac{9.75876 - 10}{2} = 4.87938 - 5 = \bar{1}.87938$. The root = .7575.

RAISING A NUMBER TO ANY POWER BY MEANS OF LOGARITHMS

The simplified operation consists of finding the logarithm of the given number and multiplying it by the exponent of the power. The product is the logarithm of the power.

Example 1: $(43.08)^3$.

The log of $(43.08)^3 = 3 \times \log 43.08 = 3 \times 1.63428 = 4.90284$, $(43.08)^3 = 79,954$.

Example 2: $(.0684)^4$.

The log of $(.0684)^4 = 4 \times \log .0684 = 4 \times \bar{2}.83506 = \bar{5}.34024$, $4 \times \bar{2} = \bar{8}$ and $.83506 \times 4 = 3.34024$. Combining $\bar{8}$ and 3.34024, we obtain $\bar{5}.34024$, $(.0684)^4 = .00002189$.

When a number is raised to a fractional power, the procedure is the same, i.e., the exponent of the power is multiplied by the logarithm of the given number, and this product is the logarithm of the result.

Example 1: $(6457)^{\frac{3}{5}}$.

The log of $(6457)^{\frac{3}{5}} = \dfrac{3}{5} \times \log 6457 = \dfrac{3}{5} \times 3.81003 = .6 \times 3.81003 = 2.286018$. The number which corresponds to log 2.286018 is 193.204.

Example 2: $(.0748)^{\frac{4}{7}}$.

The log of $(.0748)^{\frac{4}{7}} = \dfrac{4}{7} \times \log (.0748) = .5714 \times \bar{2}.87390 =$

$(-1.1428) + .49935 = \bar{1}.35655$, since $.49935 - .1428 = .35655$. The number which corresponds to log $\bar{1}.35655$ is .22727.

Another way to work this example is to multiply $\frac{4}{7}$ or its decimal equivalent .5714 by $(8.87390 - 10)$. Thus $.5714 \times (8.87390 - 10) = 5.07055 - 5.714 = 5.07055 + (\bar{6}.286) = \bar{1}.35655$, which is the logarithm of the required number.

Note that extracting a root of a number is the same as obtaining a fractional power of the number, the exponent of the power being equal to the reciprocal of the root.

Examples: 1. $\sqrt[n]{a} = a^{1/n}$.

2. $\sqrt[n]{a^m} = a^{m \times 1/n} = a^{m/n}$.

Therefore, it is often convenient to consider the fractional power of the number instead of its root when dealing with logarithms.

The preceding example, $\sqrt[n]{a^m}$, is more convenient to calculate in the form $a^{m/n}$.

Then log $a^{m/n} = \frac{m}{n} \times \log a$.

Finally a number is found whose log $= \frac{m}{n} \times \log a$.

Computations performed with the aid of logarithms give results with as many correct figures as there are places in the logarithmic tables available. The results will not always match those obtained through operations of the conventional arithmetical processes, but a fairly close approximation is possible.

Many engineering problems that are stated in the form of complex algebraic expressions involving multiplication, division, and powers and roots of symbols and numbers can be conveniently solved with the aid of logarithms. It is important in the operations involving logarithms that the methods of solution be clearly understood and correctly followed.

PRACTICE PROBLEMS APPLICABLE IN THE SHOP AND TOOLROOM

Find the logarithms of the numbers in Problems 1 to 44, inclusive:

1. 34	**2.** 163	**3.** 472
4. 823	**5.** 746	**6.** 838
7. 64.6	**8.** 37.2	**9.** 8.06
10. 7.32	**11.** .865	**12.** .0742
13. .037	**14.** .0926	**15.** .0834
16. .00368	**17.** .00792	**18.** .00108

LOGARITHMS

19. .00834
20. .00384
21. .00072
22. .00608
23. .000704
24. .00207
25. 3628
26. 7391
27. 8406
28. 5004
29. 8006
30. 84.076
31. 7.0082
32. 26.782
33. 104.06
34. 7000.2
35. .002763
36. .008467
37. 13.00027
38. $\frac{2}{7}$
39. $13\frac{6}{11}$
40. $103\frac{3}{5}$
41. $430\frac{7}{13}$
42. 126.08
43. $74\frac{3}{10}$
44. 3786.06

Find the numbers of the logarithms in Problems 45 to 68, inclusive:

45. .4639
46. .5378
47. 2.8102
48. 1.9201
49. 3.9974
50. 2.9415
51. $\bar{1}.7356$
52. $\bar{1}.6170$
53. $\bar{2}.9154$
54. $\bar{4}.9795$
55. $\bar{3}.7528$
56. $\bar{2}.8021$
57. .7360
58. .4556
59. 2.7304
60. 1.9476
61. 2.9848
62. 3.7116
63. 4.9660
64. $\bar{2}.39858$
65. $\bar{3}.46774$
66. 2.72005
67. .03608
68. $\bar{4}.00868$

Solve Problems 69 to 108, inclusive, by means of logarithms.

69. 636 × 23
70. 28.2 × 54.6
71. 76.04 × 3.7
72. 4.006 × 36
73. .0087 × 64.3
74. 4872 × 23.78
75. $36\frac{3}{5} \times 472\frac{5}{7}$
76. $84\frac{6}{11} \times \frac{13}{18}$
77. $62.084 \times 23\frac{11}{16}$
78. $.00372 \times \frac{57}{64}$
79. 392 × 712 × 93
80. 64.8 × 38.4 × 72
81. 57.6 × 7.05 × 6.7
82. 634 × .0083 × 57
83. 7.0083 × 38.02 × 7.08
84. $92\frac{3}{7} \times 62.8 \times 17.006$
85. $37.04 \times 17\frac{3}{5} \times 68.47$
86. 87.016 × 92.78 × 214.03
87. 2714 × 6.0082 × 3782
88. 63.007 × 18.008 × .00486
89. 3478 ÷ 82.6
90. 1782 ÷ .0073
91. 6276 ÷ 3.005
92. 8314 ÷ 27.072
93. 6758 ÷ 82.008
94. 8347 ÷ .000974
95. $\dfrac{6382.0086}{743.8052}$
96. $\dfrac{743.756}{38.072}$
97. 827.846 ÷ .086
98. $\dfrac{.00832}{238.064}$

99. $\dfrac{3476.37}{834.009}$

100. $3417 \div 17\dfrac{3}{64}$

101. $8317 \div \dfrac{51}{64}$

102. $\dfrac{64\dfrac{11}{13}}{36\dfrac{5}{17}}$

103. $917\dfrac{7}{19} \div \dfrac{5}{13}$

104. $.00827 \div \dfrac{47}{64}$

105. $.000734 \div .00942$

106. $.00268 \div 37.0087$

107. $745.864 \div 18\dfrac{11}{23}$

108. $\dfrac{7}{134} \div .004762$

Solve Problems 109 to 134, inclusive, with the aid of logarithms:

109. $\dfrac{72.82 \times 78.06}{232.04 \times 7.64}$

110. $\dfrac{692.04 \times 17.009}{472.008 \times 73.08}$

111. $\dfrac{2392.6 \times 74.08}{487.4 \times 792.6}$

112. $\dfrac{4784.7 \times 782.016}{867.08 \times 64.0085}$

113. $\dfrac{.0078 \times 23.142}{868 \times .00004926}$

114. $\dfrac{28.6 \times 14.02 \times 734}{672 \times .0034 \times 812.7}$

115. $\dfrac{748.005 \times 92.6 \times .00476}{346.5 \times 692.6 \times .00007928}$

116. $\dfrac{7912 \times 8763 \times 2918.08}{3914.8 \times .0008928}$

117. $\dfrac{347 \times .00179 \times 62.004}{743 \times 971.007 \times 40026}$

118. $\dfrac{2.345 \times 3.456 \times 4.567}{54.32 \times .006543 \times 7.00654}$

119. $\dfrac{7200 \times 89.040 \times 19000}{2314 \times .00084123}$

120. $\dfrac{.000347 \times 82000}{3.1416 \times .00007432}$

121. $\dfrac{.0253 \times .00364 \times .000475}{.00586 \times .0473 \times 62.0072}$

122. $\dfrac{\dfrac{3}{8} \times 6\dfrac{11}{16} \times 14\dfrac{1}{64}}{2312 \times 6982}$

123. $\dfrac{7\dfrac{9}{64} \times 12\dfrac{4}{11} \times 46\dfrac{5}{13}}{.00723 \times .0438}$

124. $\dfrac{550 \times \pi \times 32.16}{1680 \times .7854 \times 83}$

125. $\dfrac{\dfrac{\pi}{4} \times 382.00416}{4726 \times .0000283}$

126. $\dfrac{3\dfrac{11}{16} \times 52\dfrac{3}{5} \times 437\dfrac{13}{17}}{\dfrac{53}{64} \times \dfrac{13}{16} \times \dfrac{27}{32}}$

127. $\dfrac{\pi \times 682.71 \times .0045}{32800 \times 630000}$

128. $\dfrac{234.56 \times 78.654}{\pi \times 1200 \times 37000}$

129. $\dfrac{382 \times 417 \times 89 \times 364}{283 \times 714 \times 98 \times 463}$

130. $\dfrac{3.82 \times 4.17 \times 3.64}{4.63 \times 98 \times 7143}$

LOGARITHMS 371

131. $\dfrac{\dfrac{5}{13} \times \dfrac{7}{11} \times \dfrac{9}{17} \times \dfrac{11}{19}}{\dfrac{23}{17} \times \dfrac{58}{17} \times \dfrac{34}{18}}$

132. $\dfrac{32382 \times \dfrac{13}{64}}{37800 \times 63120}$

133. $\dfrac{7264 \times 5820 \times .32}{\pi \times 3882000 \times \dfrac{3}{64}}$

134. $\dfrac{\pi \times 823\dfrac{17}{35} \times 328}{288300 \times 11\dfrac{5}{16} \times \dfrac{17}{64}}$

Simplify the solution of Problems 135 to 168, inclusive, with the aid of logarithms:

135. $(572)^3$
136. $(364)^2$
137. $(846)^4$
138. $(7.064)^3$
139. $(.4672)^2$
140. $(.02764)^3$
141. $(.00465)^4$
142. $(372.87)^3$
143. $(4203.08)^2$
144. $(30706.3)^3$
145. $\sqrt{673.3}$
146. $\sqrt[3]{3.60703}$
147. $\sqrt[3]{.00456}$
148. $\sqrt{.04276}$
149. $\sqrt[4]{.00027846}$
150. $\sqrt[3]{\dfrac{374.86 \times 67.843}{68.734 \times 87.643}}$
151. $\sqrt[5]{\dfrac{72.008 \times 2700.4}{3.472 \times .0004975}}$
152. $(694.3)^3 \div (493.06)^2$
153. $(459.6)^2 \times (53.4608)^3$
154. $(.9458)^4 \times .000375$
155. $(72.0052)^3 \times (27.0284)^2$
156. $\sqrt[3]{.002928} \div \sqrt[4]{13\dfrac{17}{64}}$
157. $\sqrt[5]{\dfrac{384.26 \times 62.836}{.00945 \times 23.017 \times 728}}$
158. $\sqrt[7]{\dfrac{.00845 \times \pi \times 394.6}{276.4 \times 87.273 \times .7854}}$
159. $(3846)^{\frac{5}{8}}$
160. $(72.005)^{\frac{5}{8}}$
161. $(28720)^{\frac{17}{7}}$
162. $(.0009824)^{\frac{3}{5}}$
163. $(546.8)^{\frac{1}{3}} \times (548)^{\frac{1}{5}}$
164. $(.004836)^{\frac{3}{5}} \times (842.6)^{\frac{5}{8}}$
165. $(2346.8)^{\frac{1}{3}} \times (8432.7)^{\frac{1}{4}}$
166. $(528.006)^{\frac{3}{5}} \div (347.8)^{\frac{5}{8}}$
167. $\dfrac{(.00729)^{\frac{1}{4}}}{(57300)^{\frac{1}{3}}}$
168. $\dfrac{\left(571\dfrac{17}{64}\right)^3}{(.04367)^{\frac{1}{4}}}$

Chapter XI

THE SLIDE RULE

The slide rule is an instrument used for rapid calculations involving multiplication, division, raising a number to a power, extracting the root of a number, proportions, trigonometric functions, logarithms, and other mathematical processes. Addition and subtraction cannot be performed with the slide rule.

The results of calculations made by the slide rule are, of course, not entirely precise as are those obtained by standard arithmetical operations performed in longhand, but the accuracy of calculations made on the slide rule is substantially the same as that found in tables of logarithms to four places, and the factor of error is about one fourth of one per cent.

When exact calculations are demanded, such as for various types of accounting work, precision and tool inspection work, and scientific work demanding a high degree of accuracy, the slide rule serves only as a check. However, for engineering computations involving strength of materials, stress analysis, design of structural members, estimating amounts of materials for engineering and construction projects, etc., the slide rule is indispensable, since the time saved by its use is great in comparison to that consumed in making longhand computations. Furthermore, in design of machine structures and elements, several assumptions of sizes of details are sometimes made before a suitable size and proportion is decided upon, thus a number of computations can be made with an acceptable degree of accuracy when a slide rule is employed.

PARTS OF THE RULE

Most slide rules consist of three essential parts: the body or rule called the *stock*, the *slide*, and the glass *runner* or indicator with a hairline scratched on its surface.

There are many designs and types of slide rules to suit individual requirements, such as special instruments for chemical computations, electrical, steam, and power computations, hydraulic computations, etc. The simplest type is the 10-inch Mannheim slide rule, which is the most common type of instrument used and the one we shall describe.

SCALES OF THE RULE

There are four principal scales on the slide rule: scales A and D on the rule proper (stock), and scales B and C on the slide. When the rule and the slide are lined up from end to end, it can be observed that scales A and B are identical and so are scales C and D. The extreme ends of these scales are marked 1, and are known as the *indexes*. The index at the beginning end of any scale is the initial or left index and the index at the far end is the final or right index.

There are scales on the Mannheim and other slide rules besides those we have designated, but these, being the most important, will be studied first in this chapter inasmuch as the ability to read them accurately is essential to successful manipulation of the rule. The C and D scales will be explained in detail in this section. A reasonable amount of practice in reading these scales is necessary before actual operations can be performed on them. The A and B scales, together with other scales, will be discussed in detail later in the chapter.

C and D Scales. Most of the operations of multiplication and division are performed with the aid of the C and D scales. The C and D scales are also used with other scales to perform operations such as finding the square, square root, cube, or cube root of a number. A detailed description of these scales follows.

The C and D scales are identical. Their divisions are unequal, and, unlike other scales, such as measuring scales, they have no beginning and no end, but contain any number from zero to infinity.

The divisioning of the C and D scales is unequal because the spaces marked off represent distances equal to the logarithms of 1, 2, 3, 4 and so on up to the logarithm of 10; therefore the slide rule actually is a logarithmic scale. Since the mantissa of any number from 1 to 10—say 4—is the same as that of numbers 40, 400, and 4,000, or of .4, .04, .004, etc., the particular location on either the C or D scale for the number 4 is also the location for numbers 40, 400, and 4,000, or for .4, .04, .004, etc.

To differentiate between the various numbers that occupy the same location on the scales, a careful placement of the decimal point is essential, and this will be discussed later on.

To enable an easy explanation of the scales, they have been divided into three sections. Between the left index and the right index there are numbered the *main divisions* 2 (about three-tenths of the total length from the left index), 3, 4, 5, 6, 7, 8, and 9. The first section includes from the left index to the main division 2, the second section includes from main division 2 to main division 4, and the third section includes from main division 4 to the right index 1 or main division 10.

Section 1. The first section of the scales from left index 1 to main division 2 is divided into 10 *secondary divisions*, which are numbered 1, 2, 3, etc. (see Fig. 11-1).

374 MATHEMATICS FOR INDUSTRY

Each of the secondary divisions is divided into 10 smaller divisions (usually called *tertiary divisions*) that are not numbered, and these smaller divisions constitute tenths of the secondary divisions. To read this section of the scale, we must note that the first main division line at the left, marked *1* (called the left or initial index), represents 1, 10, 100, 1,000, and .1, .01, .001, etc. Let us assume that the left index represents 100, then the first secondary division line to the right, marked *1*, represents 110, and the next secondary division line to the right, marked *2*, represents 120, etc., up to the main division line marked *2*, which represents 200 (see extreme right of Fig. 11-1).

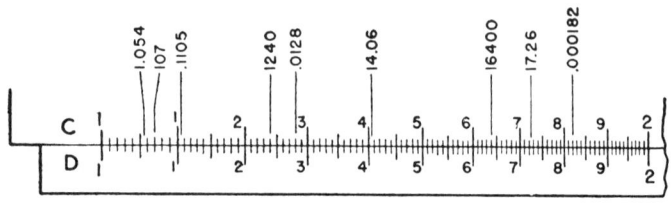

Fig. 11-1.

If we assume that the first main division line at the left represents 100, the next unmarked division line (tertiary division line) to the right is 101, and the division lines following are 102, 103, 104, 105, etc., up to the extreme right main division line, marked *2*, which is 200. The unmarked division lines immediately to the left of the main division *2* are 199, 198, 197, etc.

It is evident from the description of this section of the C and D scales that the first three digits of any number whose first digit is 1 can be located accurately in this area. Therefore, only the first four digits of such numbers are considered in slide rule calculation, the first three being accurate and the fourth being estimated by eye.

In order to practice reading the various division lines, the C and D scales are matched as shown in Fig. 11-1, and the glass runner is used to indicate the numbers desired. Note, for practice, the following readings in Fig. 11-1: 107, 1,240, 17.26, .1105, .0128, 14.06, 1.054, 16,400, and .000182.

Section 2. The next section of the C and D scales is from main division 2 to main division 4, and the spaces from 2 to 3 and 3 to 4 are each divided into 10 secondary divisions which are not marked but are extended for the sake of contrast with the smaller divisions (tertiary divisions), as shown in Fig. 11-2. Each of these secondary divisions is divided into five smaller unmarked tertiary divisions, and these divisions constitute fifths of the main divisions.

To read this section of the scale, we must note that the first main division line at the left, marked *2*, represents 2, 20, 200, and 2,000 or .2, .02, .002, etc.

We must also note that the main division line marked *3*, represents 3, 30, 300, and 3,000 or .3, .03, .003, etc. This applies as well to the main division line *4*.

If we assume that the main division line, marked *2*, represents 200, then the next secondary division line to the right, which is unmarked,

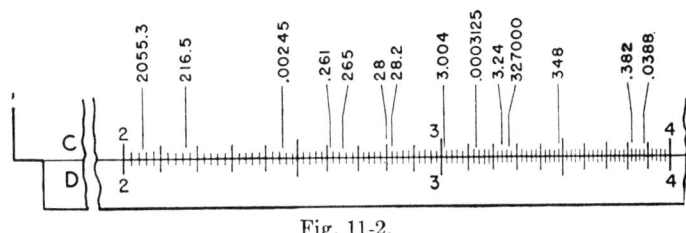

Fig. 11-2.

represents 210, and the following unmarked secondary division line represents 220, etc., up to the main division line marked *3*, which represents 300, and from there on up to the main division line marked *4* which represents 400.

A similar progression occurs where the main division line marked *2* represents .200, the next unmarked secondary division line then representing .210, and that following representing .220, etc., on up to the main division line marked *3*, which represents .300, and from there to the main division line marked *4*, which represents .400.

If we assume the main division line marked *2* represents 200, then the very next division line (tertiary division line) to the right represents 202, the following 204, etc.

If a number such as 201 is to be located on the scale, we must divide the space between 200 and 202 equally, and estimate the position of the missing division line by eye.

It is evident from the description of this section of the scales that the first three digits of any number whose first digit is 2 or 3 can be located accurately on the scale if the third digit is even. Otherwise the number can be accurately located on the scales up to the third digit, and this third digit must be estimated by eye.

Therefore, to locate a number consisting of many digits, only the required number of digits plus one digit are taken into account, the remaining digits being discarded for the purpose of setting up the required number on the scale. Of course, the proper position of the decimal point and the actual number of digits are considered in the final result, and the method followed will be described later.

In order to practice reading the various division lines in this section, note the following readings in Fig. 11-2; 28, 348, 265, 3.24, 28.2, 216.5, .382, .261, .0388, .00245, .0003125, 327,000, 2055.3, and 3,004.

Section 3. The next section of the C and D scales is from main division 4 to right index 1 or main division 10, and the spaces from 4 to 5, 5 to 6, etc., up to 10, are each divided into 10 secondary divisions, which, though not marked, owing to a lack of space, are extended to make a contrast with the smaller divisions (tertiary), as shown in Fig. 11-3.

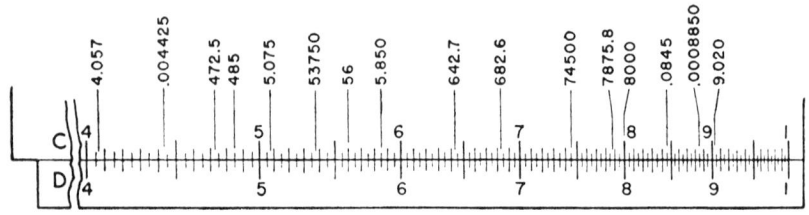

Fig. 11-3.

Each of these secondary divisions is divided into two smaller unmarked divisions, and these divisions constitute halves of the secondary divisions.

To read this section of the scale, we must note that the first main division line at the left, marked *4*, represents 4, 40, 400, and 4,000 or .4, .04, .004, etc. We must also note that the main division marked *5* represents 5, 50, 500, and 5,000 or .5, .05, .005, etc. The same applies to the main division lines *6*, *7*, etc., up to *1* (or 10) at the extreme right of the rule.

Let us assume that the main division line marked *4* represents 400, then the first secondary division line to the right, which is unmarked, represents 410, and the following unmarked secondary division line represents 420, etc., up to the main division line marked *5*, which represents 500, and then on up to the main division line at the extreme right, marked *1*, which represents 1,000.

A similar progression occurs where the main division line marked *4* represents .004, the first unmarked secondary division line then representing .0041, and that following representing .0042, etc., on up to the main division line marked *5*, which represents .005, and from there to the main division line marked *1*, at the extreme right of the rule, which represents .010.

If we assume the main division line marked *4* represents 400, then the very next division line (a tertiary division line) to the right represents 405 and the lines following represent 410, 415, 420, etc.

If a number, say 4,025, is to be located on the scale, we must divide the space separating 400 and 405 equally and fix the position of the missing division line by eye. If any other number to be located on the scale requires that a space be divided into more equal parts than two, then estimation of the missing division line is no more than approximate. For example, to locate numbers such as 401, 402, 403, and 404, would demand considerable experience for estimation of $\frac{1}{5}, \frac{2}{5}, \frac{3}{5}$, and $\frac{4}{5}$ of the space indicated. The difficulty would increase as the readings advanced to the right, the spaces between the division lines narrowing steadily.

It is evident from the description of this section of the scales that any number whose first digit is 4 or greater, up to and including 9, can be located accurately on the scales for three digits, if the third digit is 5 or 0. Otherwise the number can be located accurately on the scales up to the third digit, and this third digit must be estimated by eye. Therefore, to locate a number consisting of many digits, only the required number of digits plus one digit are taken into account, the remaining digits being discarded for the purpose of setting up the required number on the scale. The actual number of digits and the proper position of the decimal point are considered in the final result.

In order to practice reading the various division lines in this section, note the following readings in Fig. 11-3: 56, 485, 8,000, 74,500, 5.850, 472.5, 9.020, 642.7, .0845, 4.057, 5.075, 7875.8, 682.6, 53,750, .0008850, and .004425.

The Decimal Point. The location of the decimal point in a particular number cannot be determined by the slide rule, since various scales of the rule show only the possible significant figures of the number in the order in which they are related, as previously explained. The real evaluation of a group of significant figures representing a given number must be made by processes that will be explained later when various mathematical operations, such as multiplication, division, etc., are described in detail with the aid of the rule.

MULTIPLICATION AND DIVISION

The C and D scales are most easily and commonly used in multiplication and division. The A and B scales, however, can also be used. In this section multiplication of two numbers, multiplication of several numbers, division, and multiplication and division combined will be explained with the use of the C and D scales of the slide rule.

Multiplication of Two Numbers. Since the slide rule is a logarithmic rule, the operation of multiplication of two numbers consists first of finding the sum of logarithms of both numbers and then the number which corresponds to the sum of the logarithms. However, instead of performing the double operation, the numbers themselves are marked in the respective places of the rule, and the operation of multiplication of the two numbers merely becomes one of addition on the slide rule.

Usually the C and D scales are employed for multiplication and division, although the A and B scales may also be employed for these operations, as will be explained later.

Since the C and D scales are identical, each one of the scales is employed for setting one of the given factors, and, with the aid of the glass runner, the product of the two factors is determined. A rule for multiplication will be first stated and then examples given illustrating the rule.

The rule for multiplication is as follows: *To find the product of two numbers, set the proper index of the C scale opposite either of the numbers on the D scale. Next set the hairline of the runner to the second number on the C scale and read the answer on the D scale under the hairline.*

In order to determine which index (right or left) of the C scale is used to locate the first factor on the D scale, a simple rule is suggested: Multiply the first digits of both factors; if the product is less than 10, use the left index of the C scale, otherwise use the right index.

Example 1: 34×23.

Procedure: Since the product of the first digits of both factors is less than 10, the left index is used. Set the left index of the C scale over 34 on the D scale and move the glass runner until its hairline indicates 23 on the C scale. The hairline also indicates 782 on the D scale, which amount is the product of the two given numbers. The third digit in the answer must be estimated by the eye.

To determine the number of digits in the product, one of two methods may be used. By the first method, the number of digits is determined by inspection or simple estimation. Say the approximate value of the first factor is 30, that of the second is 20, and the product of both is 600, a sum consisting of three digits. It is seen, then, that the product of the given factors also consists of three digits, and the correct answer so far as slide rule calculation is concerned is 782.

The second method is to count the number of the digits in both factors and, if the slide rule is moved out to the right, 1 is subtracted from the sum of the digits, producing the same result as that obtained by the first method. If the slide is moved out to the left, the number of digits in both factors is counted and the sum determines the number of digits in the product.

Example 2: 56 × 45.

Procedure: In this case the right index is used (product of first digits of both factors is more than 9). Set the right index of the C scale over 56 on the D scale, and move the glass runner until its hairline indicates 45 on the C scale. This hairline also indicates 2,520 on the D scale, which is the product of the two given numbers. The number of digits in the product is equal to four, in accordance with the procedure just indicated, since the slide is moved out to the left.

The approximate method of inspection or simple estimation would show one factor to be 50 and the other 40, the product being 2,000. which figure contains four digits.

It is left to individual preference which method to follow in determining the number of digits in the product.

In multiplying decimals the problem arises as to where to place the decimal point. The general rule previously stated regarding the counting of the digits and subtracting 1 only if the slide is moved to the right applies to decimals also. However, in a mixed decimal the digits to the left of the decimal point only are counted and they are considered positive digits. The digits to the right of the decimal point are ignored. For example, in the number 465.71 there are three positive digits, hence the digit count is three (+3); the numerals to the right of the decimal point are ignored. When the number is a decimal fraction there are never digits to the left of the decimal point. If the first digit to the right of the decimal is not a zero but a significant figure, then the digit count is zero; for example, the digit count of 0.325 is zero. However, the digit count for any number less than 0.1 is a negative number—the negative number being equal to the number of zeros between the decimal point and the first significant figure. For example, the digit count of the number 0.046 is one negative digit (−1). The following examples illustrate the multiplication of numbers involving decimals.

Example 1: 52.6 × 2.42.

Procedure: Since the product of the first digits of both factors is greater than 9, the right index of the C scale is set over 52.6 on the D scale, and the glass runner is moved to 2.42 on the C scale. The product, which is 127.3, is indicated by the glass runner on the D scale.

In accordance with the preceding rule, the number of digits in the product is equal to the sum of the number of the digits in both factors, since the slide is moved out to the left. The number of digits may also be determined by simple inspection. Since the factors are mixed numbers and the digits to the left of the decimal point only are counted, there will be three digits to the left of the decimal point in the answer.

Example 2: 146.3 × .000237.

Procedure: In setting the factors on the respective scales, the decimal points and the zeros are disregarded. Only the possible numbers of significant figures are located on the C and D scales. Since the product of the first significant digits of both factors is 1 × 2 = 2, the left index of the C scale is employed to be set over 1463 on the D scale, and the glass runner is moved to indicate 237 on the C scale. The product, .0347, is indicated by the glass runner on the D scale.

In the decimal fraction .000237 there are three negative digits, and in the number 146.3 there are three positive digits. Therefore, the sum of the digits = 3 − 3 = 0. But the slide is moved out to the right, hence the final total of digits = 0 − 1 = −1, and the product in the answer is .0347, containing one negative digit.

The inspection method in the case of decimal fractions is not as simple as that applying to whole numbers, but with some practice it can be mastered. In the present example, the first factor is approximately 100 and the other factor is .0002, therefore the approximate product is 100 × .0002 = .02, proving that there should be only one zero in the product.

When dealing with common fractions, they are converted into decimals and treated in accordance with the rules previously stated.

Multiplication of Several Numbers. The operation is similar to the multiplication of two numbers, except that it is simplified when successive products are determined. The procedure is illustrated by the following example.

Example: 384.3 × 642 × 4.075.

Procedure: Consider the first two numbers. The right index of the C scale is set over 384.3 on the D scale, and the glass runner is moved to indicate 642 on the C scale. The runner also indicates the product of the two numbers on the D scale, but this product is not read. The right index of the C scale is moved over to the hairline of the glass runner, and then the runner is moved over to indicate 4.075 on the C scale. The final product, which is 1,005,000, is indicated by the hairline of the runner on the D scale. The number of digits in all factors is equal to $3 + 3 + 1 = 7$, and the slide was moved out to the left in both instances. Therefore, all the digits are added.

To use the inspection method or the method of estimation, the first factor is approximately 400, the second factor 600, and the third factor 4. The product of $400 \times 600 \times 4 = 960{,}000$, approximately 1,000,000. Therefore, the number of digits in the answer apparently is all right.

Division. Since the process of division is the reverse of that of multiplication, the operation of the slide rule in the case of division is reversed from that shown for multiplication.

The following rule is given for division with the slide rule: *To find the quotient of two numbers, set the hairline of the runner to the dividend on the D scale. Then slide the divisor on the C scale under the hairline. Finally, read the answer on the D scale under one index of the C scale.*

Only one index can give a reading on the D scale at one time. For this reason there is never a question as to which index of the C scale locates the answer.

Example: $825 \div 26.4$.

Procedure: Set the glass runner to indicate 825 on the D scale and move the C scale so that 26.4 on it coincides with the hairline on the glass runner. In other words, both given numbers are in line. Read the quotient on the D scale, under the left index of the C scale, which is 31.25.

The number of digits in the quotient can be estimated easily in this manner: The dividend is approximately 800 and the divisor is 25. Hence the quotient is approximately 30, consisting of two digits.

Another method is to subtract the number of digits in the divisor from that in the dividend, and if the slide is moved out to the right, one digit is added to the result, which is the case in the present example: 3 digits in the dividend, less 2 digits in the divisor, and 1 added because the slide is moved out to the right, make a total of 2 digits.

When the slide is moved out to the left, the number of digits in the divisor is subtracted from that in the dividend, and the difference is the number of digits in the quotient. For example: $3{,}372 \div 64.02 = 52.7$, in which the slide is moved out to the left, and the number of digits in the quotient is equal to the difference between the digits in the dividend and the divisor.

An example involving decimals follows.
Example: 872.05 ÷ .00463.

Procedure: The significant figures of the dividend are 872, and that of the divisor are 463. The operation is performed as described in the previous example: The divisor (463) is set in line with the dividend (872) with the aid of the hairline of the glass runner. The quotient, which is 188,300, is read under the left index of the C scale. The number of digits in the quotient is determined as follows: The number of digits in the dividend = 3, that in the divisor = -2, and the slide is moved out to the right, which makes a total of $3 - (-2) + 1 = 3 + 2 + 1 = 6$. The approximation method is as follows: $800 \div .004 = 800 \times 250 = 200,000$, which consists of six digits, hence the number of digits in the quotient is all right.

Multiplication and Division Combined. General rules for performing multiplication and division combined are apt to become complicated and long. For this reason, the procedure can best be described by the use of examples.

Example 1: $\dfrac{624 \times 36}{78}$.

Procedure: Let us perform the operation of multiplication first by setting the right index of C over 624 on the D scale and moving the glass runner over 36 on the C scale. The reading on the D scale under the hairline is the product of the two numbers. The setting is not disturbed, except that the C scale is now moved so that the divisor 78 on this scale comes under the hairline. The quotient, 288, is read on the D scale under the right index of the C scale.

The number of digits in the final result can be determined by noting the position of the slide at each operation and adding or subtracting the digits properly, or by using the simple method of estimation, which method, for the present example, is as follows: The factors in the numerator are approximately 600 and 40, and the denominator is approximately 80, so the operation is reduced to the simple form $\dfrac{600 \times 40}{80} = 300$, which proves that the number of digits of the result is equal to 3.

The preceding example can also be solved by first dividing 624 by 78 and then multiplying the quotient by 36. The operation consists of setting 624 on the D scale in line with 78 on the C scale and then moving the glass runner to 36 on the C scale. The result is read on the D scale under the hairline.

Example 2: $\dfrac{172.84 \times 62.3 \times .0406 \times 4.075}{.002567 \times 83750 \times 2.070}$.

THE SLIDE RULE 383

Procedure: This example can be solved by operations such as those performed in the previous example, except that the number of steps is naturally increased, depending upon the number of factors in the numerator and the denominator.

The procedure will be given in six steps:

Step 1. Move the glass runner with its hairline over 172.84 on the D scale and slide the C scale until .002567 is under the hairline.

Step 2. Move the hairline over 62.3 on the C scale and keep it there.

Step 3. Move the C scale until 83,750 is under the hairline.

Step 4. Move the hairline over .0406 on the C scale.

Step 5. Move the C scale until 2.070 is under the hairline.

Step 6. Move the hairline over 4.075 on the C scale and read the result, which is 4.01, on the D scale.

From the sequence of steps of operations the following is observed: The first, third, and fifth steps consist of moving the slide. The second, fourth, and sixth steps consist of moving the hairline. After a certain amount of practice the operations become automatic no matter how many factors are in the numerator and the denominator of the given arithmetical expression.

The number of digits and the position of the decimal point in the result are determined by previously explained methods. The method of approximate estimation seems conducive of satisfactory results in operations of this type.

The approximate factors in the numerator and the denominator are:
$$\frac{2\emptyset\emptyset \times 5\emptyset \times .04 \times 4}{.002 \times 100,\emptyset\emptyset\emptyset \times 2} = \frac{.8}{.2} = 4,$$ which proves the correct number of digits and the placement of the decimal point in the result.

Note: When making an approximate estimate, the given numbers are rounded off to the numbers nearest those which can be handled rapidly, such as tens, hundreds, etc., since the values of the given numbers must be altered by a considerable amount before the position of the decimal point is affected. Bear in mind that moving the decimal point just one place to right or left alters the value of the number in the ratio of 10 to 1.

SQUARES AND SQUARE ROOTS, CUBES AND CUBE ROOTS

This section explains how to find the square of a number, the square root of a number, the cube of a number, and the cube root of a number with the aid of the slide rule. In this connection, the A and B scales will be explained and also the K scale.

Squaring a Number and Extracting Square Roots. Obtaining the square or square root of a number is one of the easiest calculations made with the slide rule. For these operations a new scale is employed in addition to the D scale, and that is the A scale.

The A and B scales on the slide rule are identical in every respect, and each one of these scales consists of two complete scales—1 (left index) to 10 (center index) and 10 (center index) to 1 (right index).

The graduations on these scales are much smaller than those on the C and D scales, since the space allowed for them is only half that allowed for the C and D scales. For example, the space between main division 1 and main division 2 on the A scale, as on the D scale, contains ten secondary divisions, but each of these secondary divisions contains only five tertiary divisions, whereas each of the secondary divisions on the D scale spans ten tertiary divisions.

Note that the A and B scales may also be employed for operations involving multiplication and division, except that the accuracy of results obtained on these scales is only half that provided by the C and D scales.

For the operation of extracting the square root of a number and also for that of squaring a number, only the A and D scales are employed. The only moving member of the slide rule is the glass runner, as the A and D scales are stationary.

The square of a number is the product of the number multiplied by itself. The A scale is constructed so that when the hairline is set to a number on the D scale, the square of that number is found under the hairline on the A scale. For example, 2 on the D scale is opposite 4 on the A scale, 3 on the D scale is opposite 9 on the A scale, and 12 on the D scale is opposite 144 on the A scale. Thus we have the following rule for finding the square of a number: *Set the hairline of the runner to the number on the D scale. Read the square of the number on the A scale under the hairline.*

The number of digits in the square of a number found on the A scale is determined by inspection or by a method which will be described later.

The square root of a number is a second number which when multiplied by itself will give the original number.

The rule for extracting the square root is as follows: *Set the hairline of the runner to the number on the A scale. Read the square root of the number on the D scale under the hairline.*

The A scale is a double scale and the problem arises as to where a given number should be located. The proper location of the given number on the A scale depends upon the number of digits in the number. A number consisting of an odd number of digits is placed in the left section of the A scale, and one consisting of an even number of digits is placed in the right section. For a decimal fraction, if the number of zeros is odd, the fraction is placed in the left section of the A scale, and if the number of zeros is even, or if there are no zeros, the fraction is placed in the right section of the A scale.

The number of digits and the location of the decimal point in the square root are determined by a simple procedure: Divide the given number into groups of two digits each, beginning with the decimal point and moving to the left of it when dealing with whole numbers. If the number is a mixed number, the division into groups of two digits each begins at the decimal point and is moved to the left of it—thus the whole numbers only are counted. If the number is a decimal fraction, the division into groups of two digits each begins at the decimal point and is moved to the right of it. Each group in the given number accounts for one digit of the square root.

A few examples of extracting the square root of a number are given.

Example 1: $\sqrt{344}$.

Procedure: This number consists of 3 digits, and therefore it is indicated in the left section of the A scale by means of the hairline. The square root is read on the D scale below the hairline, and it is 18.55. The number of groups of 2 digits each to the left of the decimal point is 2. The group at the extreme left of the number consists of only one digit, still it constitutes a group. Hence the number of digits in the square root is 2, since only the number of digits in the whole number is taken into consideration.

Example 2: $\sqrt{2862.815}$.

Procedure: The given number, consisting of 4 digits, is placed in the right section of the A scale, and the square root is found on the D scale to be 53.5. The number of groups in the given number is equal to 2, therefore the square root must contain 2 digits.

Example 3: $\sqrt{.007832}$.

Procedure: The given decimal, consisting of 2 zeros, is placed in the right section of the A scale, and the square root is found on the D scale to be .0885. The number of zeros to the right of the decimal point in the given decimal is 2. Therefore, there should be one zero in the square root, as the first group in the given decimal is .00, the second group is 78, and so on.

Example 4: $\sqrt{.0002648}$.

Procedure: The given decimal, consisting of 3 zeros, is placed in the left section of the A scale, and the square root is found on the D scale to be .01626. The first group in the given decimal is .00, and the first digit in the square root should be .0. The second group in the given decimal is 02, and the second digit in the square root should be 1, and so on.

When squaring a number or a decimal, the method of finding the number of digits and the decimal point in the square of the given number or decimal is the reverse of that used in the previous examples.

Cubing a Number and Extracting Cube Roots. Most slide rules today have a special K scale, which consists of three separate sections or smaller scales ranging from 1 to 10.

The cube of a number is the product of the number multiplied by itself twice. The cubes of the numbers are on the K scale, and the cube roots are on the D scale.

As with squares and square roots, the glass runner is used to aid the location of the numbers.

The following rule enables one to locate the cube of a number: *Set the hairline of the runner to the number on the D scale. Read the cube of the number on the K scale under the hairline.*

The cube root of a number is a second number whose cube is the given number. The rule for extracting the cube root is as follows: *Set the hairline of the runner to the number on the K scale. Read the cube root of the number on the D scale under the hairline.*

Since the K scale is divided into three sections, the given number must be located in the proper section. The location of the number depends on the number of digits it contains. If the given number contains 1, 4, 7, 10, etc., digits, it is placed in the left section of the K scale. If the given number contains 2, 5, 8, 11, etc., digits, it is placed in the middle section of the K scale. If the given number contains 3, 6, 9, 12, etc., digits, it is placed in the right section of the K scale. In the case of decimals, the zeros are considered as negative digits, and the decimals are treated according to rule.

Another method to determine the proper location of the given number on the K scale follows: To find the cube root of a number between 1 and 10, use the left section of the K scale. To find the cube root of a number between 10 and 100, use the middle section of the K scale. To find the cube root of a number between 100 and 1,000, use the right section of the K scale.

The method for determining the number of digits is similar to that used when dealing with square roots, only instead of counting digits in multiples of two they are counted in multiples of three.

Example 1: $\sqrt[3]{6{,}275}$.

Procedure: The given number contains 4 digits, and therefore it is indicated with the aid of the hairline in the left section of the K scale. The cube root is read on the D scale under the hairline, and it is 18.44.

The method of finding the number of digits in the cube root consists of dividing the given number into groups of 3 digits each, beginning at the decimal point and moving to the left of it. In this example, the given number consists of 2 groups and therefore the cube root must have 2 digits to the left of the decimal point.

Example 2: $\sqrt[3]{34{,}728{,}063}$.

Procedure: The given number consists of 8 digits and it is indicated in the middle section of the K scale. The cube root, which is 326.4, is found under the hairline on the D scale. There are 3 groups in the given number. Therefore, the cube root must have 3 digits to the left of the decimal point.

Example 3: $\sqrt[3]{.003946}$.

Procedure: In the case of decimals, the given decimal is divided into groups of 3 digits each to the right of the decimal point, and since there are 2 groups in this example, there will be 2 digits to the right of the decimal point in the cube root. If more digits are wanted to the right of the decimal point in the cube root, more zeros are added to the right of the given decimal. Three zeros must be added to the right of the given decimal for each additional digit in the cube root. Since the given decimal contains two zeros to the right of the decimal point, it is indicated in the left section of the K scale by the hairline, and the cube root (.1581) is found on the D scale under the hairline. The first digit in the cube root is accounted for by the first group in the given decimal, the second digit by the second group, and so on.

Square and Cube Roots of Decimals. The operations of extracting the square and cube roots of decimals can be simplified by magnifying the given decimals, i.e., by multiplying the given decimals by 100, 10,000, etc., in the operation of extracting the square root; and by multiplying the given decimals by 1,000, 1,000,000, etc., in the operation of extracting the cube root. Then the final results are demagnified in accordance with the degree of magnification of the given decimals.

For example, if the given decimal is multiplied by 100, the square root is divided by 10; if the given decimal is multiplied by 1,000, the cube root is divided by 10, and so on.

In other words for a square magnify the value an even number of places to obtain a number between 1 and 100 and demagnify by moving the decimal point one-half as many places as it was moved in the original number but in the opposite direction. For a cube magnify the value by moving the decimal point over three places at a time until a number between 1 and 1,000 is obtained and demagnify by moving the decimal point one-third as many places as it was moved in the original number but in the opposite direction.

A few examples will illustrate the simplicity of operations.

Example 1: $\sqrt{.0002463}$.

Procedure: Magnify the given decimal by multiplying it by 10,000 = $\sqrt{2.463}$. The new number under the radical is indicated in the left section of the A scale by the hairline, and the square root is found on the D scale under the hairline to be 1.57. The next step is dividing the square root 1.57 by 100, obtaining the final square root, .0157.

Example 2: $\sqrt{.0000725}$.

Procedure: Magnify the given decimal by multiplying it by 1,000,000 = $\sqrt{72.5}$. The new number under the radical is indicated in the right section of the A scale by the hairline, and the square root is found on the D scale under the hairline to be 8.52. The next step is dividing the square root 8.52 by 1,000, obtaining the final square root, .00852.

Example 3: $\sqrt[3]{.0000428}$.

Procedure: Magnify the given decimal by multiplying it by 1,000,000 = $\sqrt[3]{42.8}$. The new number under the radical is indicated in the middle section of the K scale by the hairline, and the cube root is found on the D scale under the hairline to be 3.5. The next step is demagnifying or dividing the cube root 3.5 by 100, obtaining the final cube root, .035.

Example 4: $\sqrt[3]{.00000847}$.

Procedure: Magnify the given decimal by multiplying it by 1,000,000 = $\sqrt[3]{8.47}$. The new number under the radical is indicated in the left section of the K scale by the hairline, and the cube root is found on the D scale under the hairline to be 2.04. The next step is demagnifying or dividing the cube root 2.04 by 100, obtaining the final cube root, .0204.

Extracting Cube Root Without K Scale. A common 10-inch Mannheim slide rule does not have a K scale for cubes and cube roots, and these operations must be performed by means of the A, B, C, and D scales.

The cube of a given number is found on the common Mannheim rule by setting the proper index of the C scale over the given number on the D scale, then moving the glass runner to indicate the given number on the B scale. The cube of the given number is read on the A scale under the hairline.

Example: Find the cube of 26.4.

Procedure: The right index of the C scale is set over 26.4 on the D scale, and the glass runner is moved to indicate 26.4 on the B scale. The cube of the given number, which is 18,400, is read on the A scale under the hairline. By inspection, the given number is approximately 30, and the cube of 30 is 27,000, consisting of 5 digits.

To extract the cube root of a number, the procedure is as follows: The given number is divided into groups of three digits each, beginning at the decimal point and either moving to the left if it is a whole number or to the right if it is a decimal fraction. If the first group has one significant figure, the glass runner is moved to indicate the given number in the left section of the A scale, and if more than one, the glass runner is moved to indicate the given number in the right section of the A scale. Then the slide is moved until the number on the D scale under the index of the C scale is identical with the number on the B scale under the hairline. The number is the cube root. The number of digits in the cube root is equal to the number of groups in the given number.

Example 1: $\sqrt[3]{7,825,000}$.

Procedure: The given number consists of 3 groups, and the first group contains one significant figure. Therefore, the glass runner is moved to indicate 7,825,000 in the left section of the A scale. The next step is moving the slide until the number on the D scale under the left index of the C scale is the same as the number on the B scale under the hairline. This number is 198.5, and it is the cube root of the given number. It consists of 3 digits, since the given number contains 3 groups.

Example 2: $\sqrt[3]{23,654}$.

Procedure: The given number consists of two groups and the first group has two significant figures, therefore, the runner is moved to indicate 23,654 in the right section of the A scale. The slide is then moved until the number on the D scale under the left index of the C scale is the same as the number on the B scale under the hairline. This number is 28.71, and it is the cube root of the given number.

DIAMETERS, CIRCUMFERENCES, AND AREAS OF CIRCLES

The A and B scales can be used conveniently for finding the circumference of a circle if the diameter is known, and vice versa. The circumference of a circle is equal to $\pi \times$ diameter, and the A and B scales have a special division mark π, therefore the operation of finding the circumference of a circle consists of setting the left index of the B scale below the diameter of the circle on the A scale, then moving the hairline to π on the B scale. The circumference can now be read on the A scale under the hairline. The operation can also be performed by setting the left index of the B scale below π on the A scale and moving the hairline to the figure designating the diameter on the B scale. The circumference is then read on the A scale under the hairline. Since π is indicated on the left section of both A and B scales, the left section is used.

Example: Find the circumference of a circle whose diameter is equal to 3.85.

Procedure: Set left index of B below 3.85 on A. Move hairline to π on B. Read circumference 12.1 on A under the hairline.

The reversed operation, i.e., finding the diameter of a circle, if its circumference is known, can best be illustrated by an example.

Example: Find the diameter of a circle if its circumference is equal to 8.46.

Procedure: Move hairline to 8.46 on A. Set π on B under the hairline. Read diameter 2.7 on A above the left index of B.

Note that on some slide rules the foregoing operations can be performed conveniently by means of the DF and D scales. By setting the hairline over the known diameter of the circle on the D scale, the circumference can be read directly on the DF scale under the hairline. When the hairline is set over the known circumference of the circle on the DF scale, the diameter of the circle is read directly on the D scale below the hairline.

The CF and the DF scales, which are on some slide rules, are called *folded scales*. Their purpose is to facilitate certain operations that become limited if the C and D scales are the only scales available. For example, if for a certain setting there is a definite relationship between the C and D scales, a similar relationship is found to exist between the CF and DF scales throughout the length of the scales.

Let us set the left index of C over 1.8 on D. It can be observed that the relation of 1.8 to 1 exists throughout the range of the C and D scales, and also the CF and DF scales. For examples, observe 7.2 on the D scale opposite 4 on the C scale, 1.6 on the DF scale opposite .89 on the CF scale, and so on. At times it is desired to obtain a reading on the D scale opposite a certain number on the C scale, but because of the positions of the scales, it is impossible to do so. In such event the desired reading may be obtained on the DF scale opposite the known number on the CF scale, without changing the direction of the slide.

The area of a circle can be determined conveniently by a simple operation of the slide rule if the diameter is known. The procedure consists of setting the right or left index of the C scale over the known diameter on the D scale and reading the area on the A scale over .7854 on the

B scale. The division mark .7854, which is $\frac{\pi}{4}$, is plainly engraved on either the A or B scales in their right-hand sections.

Example 1: Find the area of a circle whose diameter is 2.84.

Procedure: Set the right index of C over 2.84 on D, and read area 6.33 on A over .7854 on B.

Example 2: Find the area of a circle whose diameter is .0845.

Procedure: Set the right index of C over .0845 on D and read area .00561 on A over .7854 on B.

If the area of a circle is known, the diameter can be obtained by reversing the process.

Example 1: Find the diameter of a circle if its area is 4.075.

Note: If the number representing the area consists of an odd number of digits, it is set in the left section of the A scale, otherwise it is set in the right section.

Procedure: Set the given number in the left section of A with the glass runner, and move .7854 on B under the hairline. Read the diameter 2.278 on D under the right index of C.

Example 2: Find the diameter of a circle if its area is 86.50.

Procedure: Set the given number in the right section of A with the glass runner, and move .7854 on B under the hairline. Read the diameter 10.49 on D under the left index of C.

RATIO, PROPORTION, AND PERCENTAGE

According to previous discussion of the C and D scales used in combination with the CF and DF scales, many problems involving ratios, proportions, and percentages may be solved by simple manipulation of the slide rule. Whenever the slide is moved to any position, the ratio of any number on the D scale to the number opposite it on the C scale is constant throughout the entire range of the scales. (Furthermore, the same ratio exists between any number on the DF scale to its opposite on the CF scale.)

Example 1: $\dfrac{x}{2.47} = \dfrac{87.50}{4.375}$

Procedure: The constant ratio is known: it is $\dfrac{87.50}{4.375}$; therefore, set 87.50 on C over 4.375 on D, and read $x = 49.4$ on C over 2.47 on D.

Example 2: $\dfrac{x}{.0287} = \dfrac{165.2}{54.30}$

Procedure: Set 165.2 on the C scale over 54.30 on the D scale. Since it is impossible to set .0287 on the D scale, it may be set on the DF scale and the required number x (which is .0873) read on the CF scale.

Example 3: Find 34.6% of 1,765.

Procedure: Since 100 per cent equals 1,765, set the left index of C over 1,765 on D, and move the glass runner to 34.6 on C. Read the required number, which is 611, on D under the hairline. The operation is like that performed in multiplying the given number by the percentage, then dividing the product by 100.

RECIPROCAL SCALES

The CI scale, which is on some slide rules, is a reciprocal scale, and it is the same as the C scale inverted. When the hairline is set on a number on the C scale, the reciprocal of this number is found on the CI scale under the hairline.

For example, if the hairline is set on 4 on the C scale, the reciprocal of 4, which is .25, is found on the CI scale below the hairline. The CI scale is convenient if a number instead of being divided by another number is multiplied by the reciprocal of the latter number

Example 1: $632 \div 25$ is the same as $632 \times \frac{1}{25}$.

Procedure: The result can be obtained by setting the index of C to 632 on D, and, opposite 25 on CI, we read the result 25.28. The CI scale can be used conveniently in calculating inverse proportions.

Example 2: A gear of 72 teeth revolves 800 r.p.m. How many r.p.m. will a gear of 27 teeth revolve, if both gears are in mesh?

Procedure: Set 800 on D opposite 72 on CI, and since most of the slide is moved to the right, the right index of CI is placed in the same position occupied by the left index of CI, which operation is not always necessary. The next step is to move the glass runner to 27 on the CI scale. The answer, which is 2,133, is on the D scale under the hairline.

Some rules also have DI and CIF scales, which together with the CI scale are similar to the D, CF, and C scales, respectively, except that they are inverted.

A valuable use of the CI scale will be shown later in the discussion pertaining to trigonometric functions.

TRIGONOMETRIC SCALES

To determine sines use the S and A scales after the ends of the scales are matched together. The angles are on the S scale, and their sines are found above them on the A scale.

If the reading of the sine is in the left section of the A scale, a zero must be included between the decimal point and the first significant place.

Example 1: Sin $23°30'$ = .399.

Procedure: Set the hairline over $23°30'$ on the S scale (each small division = $30'$). Read .399 on the A scale below the hairline.

Example 2: Sin $4°20'$ = .0756.

Procedure: Each small division in this section of the S scale = $5'$. Also note that the reading of the sine is in the left section of the A scale, therefore a zero is included between the decimal point and the first significant place.

For cosines, the sine of the complement is found on the A scale opposite the complement on the S scale.

If the sine of a large angle is required, note that it is almost impossible to obtain a fairly accurate reading because of the very small space for range of angles in that section on the S scale. The procedure then is as follows:

Assume the given large angle = α; $\sin \alpha = 1 - \dfrac{\sin^2(90° - \alpha)}{2}$

Step 1. Obtain the sin of the complement of the given angle = $\sin(90° - \alpha)$.

Step 2. Square the result in Step 1, and obtain $\sin^2(90° - \alpha)$.

Step 3. Divide the result in Step 2 by 2, and obtain $\dfrac{\sin^2(90° - \alpha)}{2}$.

Step 4. $\sin \alpha = 1 - \dfrac{\sin^2(90° - \alpha)}{2}$.

Example: Find sin $78°40'$.

Procedure: $\sin(90° - \alpha) = \sin(90° - 78°40') = \sin 11°20' = .1962$. $\sin^2(11°20') = (.1962)^2 = .0386$. $\sin \alpha = 1 - \dfrac{.0386}{2} = 1 - .0193 = .9807$.

For tangents use the T and D scales after their ends are matched together. The angles are on the T scale and the tangents are found below them on the D scale.

If the angle is less than $6°$, the sin of the angle is taken, as in that range the sin and the tan of the angle are nearly the same. The cot of an angle equals the reciprocal of its tan. The CI scale is employed for this purpose.

If an angle is greater than $45°$, the cot of the complement of the angle is found instead of the tan of the angle proper, since there is no angle greater than 45° on the T scale. This value of the cot is also that of the tan required.

Example 1: Find the tan of 36°40′.

Procedure: Set hairline over 36°40′ on the T scale and read its tan .744 on the D scale.

Example 2: Find the cot of 18°20′.

Procedure: Tan 18°20′ = .3315 is found on the D scale opposite 18°20′ on the T scale. The reciprocal of .3315 is found to be 3.018 on the CI scale, and this value is the cot of 18°20′.

Example 3: Find the tan of 68°50′.

Procedure: The complement of the given angle = 90° − 68°50′ = 21°10′. The tan of 21°10′ = .3872. The cot of 21°10′ = the reciprocal of .3872 = 2.583. The tan of 68°50′ also equals 2.583.

LOGARITHMIC SCALES

The L scale is used together with the D scale. The given number is located on the D scale, and the mantissa of its logarithm is opposite it on the L scale. Both of these scales are stationary, and the hairline of the glass runner is used to locate the mantissa corresponding to a certain number, and vice versa.

Example 1: Find the log of 328.64.

Procedure: Set 328.64 on the D scale with the hairline and read the mantissa, which is .5167, on the L scale under the hairline. The log of 328.64 = 2.5167, since the characteristic is 3.

Example 2: Find the number whose logarithm = $\bar{2}.3482$.

Procedure: Set 3482 on the L scale and read 2225 on the D scale under the hairline. The correct number is .02225, since the characteristic is $\bar{2}$.

APPLICATION OF SLIDE RULE TO SHOP PROBLEMS

Many engineering problems may be solved rapidly with the aid of the slide rule. These problems involve strength of materials, power required to run engines and machine tools, estimates of the amount of production for given conditions, estimates of weights of materials and possible costs of materials and labor, speeds and feeds of machine tools, industrial engineering work, such as time and motion study, etc.

The preceding enumeration of problems deals with only a portion of mechanical engineering. One could continue enumerating problems involving many branches of mechanical engineering, electricity, physics, chemistry, navigation, and other theoretical and applied sciences.

THE SLIDE RULE

PRACTICE PROBLEMS APPLICABLE IN THE SHOP AND TOOLROOM

Locate the numbers in Problems 1 to 9, inclusive, on the D scale with the aid of the hairline of the glass runner:

1. 135, 184, 168, 14,700, 12.600, 10.80, 1.565, .01955, .1005, 1,002, .000174.
2. 2,345, 26.03, 2.425, .000288, .0222, 216.50, 2,975.
3. 3.020, 38,600, .000345, .00395, 3,526, 3.105, 37.27, 34.64.
4. 4,250, 4.125, 48.75, 40.25, 463, 47.01, 4,444, 49,800, .0004525.
5. 58.30, 5,264, 5.825, .0005625, .05146, 5.0082, 59,800, 5,763.
6. 6.0020, 6,345, 682.8, 60.85, .0006425, .06752, 6.958, 6,736.
7. 73,450, 7.962, 78.05, 748.02, 7.6005, 758.2, 78,220, .007225.
8. .00824, 8,888, 8.054, 8.0026, .000842, 82.56, 815.4, 899.02.
9. 9.035, 946,000, .00924, .0926, 9,353, 9,191, 94.04, 9.333, 9.0602.

Multiply the numbers in Problems 10 to 17, inclusive, with the aid of the C and D scales:

10. 4×1.8
    3×63
    2×3.7
    8×64
    3.28×4.08
11. 7.06×928
    $.784 \times .0408$
    $8,430 \times 6,668$
    7.208×3.065
    $2,314.3 \times 72.016$
12. $.00376 \times 2,384$
    $80.26 \times .0902$
    $3.008 \times .0102$
    $4,265 \times .00906$
    7.0825×63.086
13. $.000246 \times 8,110$
    $.00308 \times .0702$
    $9,246 \times .00078$
    $7.0005 \times 4,500$
    $8,725 \times .00266$
14. $7\frac{1}{4} \times 6\frac{1}{32}$

    $2\frac{11}{64} \times 3\frac{7}{16}$

    $8\frac{5}{8} \times 40\frac{11}{32}$

    $16\frac{3}{4} \times .00827$

    $220\frac{1}{64} \times 3.026$
15. $23.016 \times 184.00 \times .0372$

    $18.06 \times 6\frac{23}{64} \times 4,824$

    $282 \times 363.8 \times .00826$

    $3.064 \times 92.02 \times 74.05$

    $23\frac{1}{4} \times 16\frac{1}{16} \times 82\frac{7}{64}$
16. $823 \times 654 \times 526$

    $328 \times 456 \times 625$

    $.00927 \times .0746 \times .000536$
    $.0216 \times 4.052 \times 346,000$
    $.852 \times 4\frac{47}{64} \times .00426$
17. $4,168 \times 2,328 \times 9,460 \times 23.280$

    $8.004 \times 26.54 \times 3\frac{27}{64} \times .00326$

    $682 \times 7.048 \times 420.8 \times .000562$
    $.0005102 \times 482,000 \times .0426 \times 245$
    $1234 \times 4.567 \times .08923 \times .0000876 \times 542$

    $482 \times \pi \times .0435 \times \frac{\pi}{4} \times .000258$

Divide the numbers in Problems 18 to 23, inclusive, with the aid of the C and D scales:

18. $17 \div 3$
$760 \div 23$
$4.26 \div 32$
$62.4 \div 2.3$
$807 \div 40.2$

19. $482 \div \pi$
$2.84 \div 825$
$.000264 \div \dfrac{\pi}{4}$
$3.25 \div .0725$
$523.6 \div .00527$

20. $9{,}216 \div 562$
$61.29 \div 2.65$
$34.002 \div .0026$
$62\dfrac{17}{64} \div 5\dfrac{29}{32}$
$32\dfrac{17}{32} \div .0574$

21. $2{,}694 \div 4.025$
$4.962 \div .0962$
$.00654 \div 7.024$
$15.02 \div 80.08$
$9.008 \div 3\dfrac{19}{64}$

22. $826{,}032 \div .0296$
$6.2802 \div 28.05$
$.00426 \div 78{,}750$
$3\dfrac{63}{64} \div 847.52$
$23\dfrac{17}{32} \div 3{,}842\dfrac{15}{16}$

23. $.2904 \div 385\dfrac{5}{8}$
$17\pi \div 63\dfrac{\pi}{4}$
$\dfrac{\pi}{4} \div 302.8$
$32{,}000 \div .00017$
$20{,}408.2 \div .08072$

Calculate the results in Problems 24 and 25 with the aid of the C and D scales:

24. $\dfrac{26.02 \times 1{,}405}{368}$

$\dfrac{89.06 \times 23\dfrac{25}{64}}{642.5 \times 82.42}$

$\dfrac{2.3904 \times 62.06}{864.3 \times 36.05 \times 72.004}$

$\dfrac{47 \times 82 \times .0032 \times 8.0402}{.0026 \times 2.048 \times 62.32 \times 82.05}$

$\dfrac{.0027 \times 327 \times 53.6 \times 29.8}{82{,}000 \times 5.820 \times 37.025 \times 42.05 \times 114\dfrac{11}{16}}$

25. $64.3 \times \dfrac{78.5}{1{,}140} \times \dfrac{\pi}{4} \times 26{,}580$

$$\frac{114.2 \times 346.8 \times 7.982}{76.2 \times .00284}$$

$$\frac{328.02 \times 842.605 \times 408 \times \pi}{92.008}$$

$$\frac{8.0602 \times 52.7 \times 18\frac{5}{16} \times 2{,}328 \times 86.405}{236.8 \times \frac{\pi}{4} \times 100.52}$$

$$62.84 \times 18\frac{5}{8} \times 32\frac{17}{64} \times .000832 \times \frac{57}{64} \times \pi$$

Find the squares of the numbers in Problems 26 to 30, inclusive:

26. 3.2	**27.** 32.8	**28.** 82.017	**29.** .00429	**30.** 572
23	823	264.2	.000265	.035
4.06	6.04	246.2	.0000424	$23\frac{29}{64}$
6.42	43.02	6.008	.000292	$4\frac{21}{32}$
7.28	27.028	7.025	.00856	$\frac{55}{64}$

Find the square roots of the numbers in Problems 31 to 34, inclusive:

31. 62.4	**32.** 26.58	**33.** .000926	**34.** $56\frac{19}{32}$
388.05	85.62	.02745	$\frac{61}{64}$
8426	74.98	.0000268	.000849
23.008	38.004	384,000	30.8520
.00482	72.512	2170.68	346,531

Find the cubes of the numbers in Problems 35 to 38, inclusive:

35. 3.06	**36.** .000426	**37.** 42.5	**38.** .0284
842	$17\frac{3}{16}$	50.18	.000743
.036	28.084	31.72	9.008
64.2	34.208	20.08	4.123
2.008	$\frac{57}{64}$	1.602	60.06

Find the cube roots of the numbers in Problems 39 to 42, inclusive:

39. 368	**40.** 6.038	**41.** 2.0048	**42.** .000426
29.04	18.84	6.032	$\frac{13}{16}$

52.82	2826.02	$28\frac{47}{64}$	.0000278
64.02	870,000	62.418	.00000546
80.34	472,642	846.06	389,281,000
900.26	862,000	23.008	78,261,120
726.83	24.0682	.00426	$682\frac{19}{64}$

Compute the results in Problems 43 to 52, inclusive, with the slide rule:

43. $\sqrt{46.06 \times 14.342}$

44. $\sqrt[3]{642.22 \times .624}$

45. $\dfrac{\sqrt{248\frac{17}{64} \times 32.5}}{\pi \times 43.27}$

46. $\dfrac{\sqrt[3]{.00482} \times .0437}{372.8 \times 2.34}$

47. $\dfrac{2356.2 \times \frac{\pi}{4}}{\sqrt{482.5} \times \sqrt[3]{.0718}}$

48. $\dfrac{\sqrt{721.9} \times .00921}{3.175 \times \sqrt[3]{24.968}}$

49. $\dfrac{(23.62)^3 \times .4265}{345.6 \times (7.12)^2}$

50. $\dfrac{(23.46)^{\frac{1}{2}} \times \sqrt[3]{.02749}}{\sqrt{61.0925} \times (34.17)^3}$

51. $\dfrac{512.003 \times \sqrt[3]{928.5}}{(.00573)^2 \times (31.416)^{\frac{1}{2}}}$

52. $\dfrac{\sqrt{343.5} - (.08264)^3}{46.03 + \sqrt[3]{.00982}}$

Find the circumferences of the circles whose diameters are given in Problems 53 to 56, inclusive:

53.	54.	55.	56.
1.24	3.008	$2\frac{17}{64}$	$\frac{51}{64}$
2.86	27.12	$3\frac{19}{32}$	$5\frac{15}{16}$
.375	8.005	$5\frac{7}{16}$	$23\frac{1}{64}$
21.04	17.59	$4\frac{29}{64}$	$8\frac{5}{8}$
18.002	23.06	$5\frac{5}{16}$	11.008

Find the diameters of the circles whose circumferences are given in Problems 57 to 60, inclusive:

57.	58.	59.	60.
326	682	312.37	81.068
58.6	286.5	218.02	$34\frac{5}{16}$
72.05	842.02	$5\frac{17}{64}$	$83\frac{17}{32}$
80.02	23.007	$21\frac{19}{32}$	314.02

THE SLIDE RULE

5.006	18.002	$300\frac{5}{8}$	38.008
218.2	20.005	$108\frac{11}{16}$	34.062

Find the areas of the circles whose diameters are given in Problems 61 to 65, inclusive:

61. $3\frac{1}{2}$ **62.** 4.06   **63.** $5\frac{9}{16}$ **64.** $4\frac{5}{8}$ **65.** $2\frac{3}{4}$

4.65	6.12	4.002	$3\frac{11}{32}$	$4\frac{5}{8}$
12.03	5.003	6.23	2.005	12.085
8.005	7.108	3.08	8.924	10.804
$13\frac{5}{16}$	$8\frac{17}{32}$	4.085	7.098	9.255

Find the diameters of the circles whose areas are given in Problems 66 to 70, inclusive:

66. 256   **67.** 654   **68.** 452.8   **69.** $100\frac{11}{16}$ **70.** $8\frac{13}{16}$

743.2	$112\frac{3}{4}$	240.02	$343\frac{1}{4}$	.0812
80.05	$87\frac{11}{16}$	312.68	252.2	.000275
72.56	$63\frac{57}{64}$	218.72	600.86	.00846
61.08	$42\frac{9}{16}$	27.005	512.05	.0000758

Find the value of the unknown in the expressions given in Problems 71 to 75, inclusive:

71. $x \div 3.45 = 4 \div 7$

$58.08 \div 3.04 = x \div 8.5$

$28.52 \div x = 11 \div 5$

$.345 \div x = 11.5 \div 16.28$

$875.03 \div x = 3\frac{5}{8} \div 5\frac{11}{16}$

72. $\dfrac{354}{x} = \dfrac{.72}{12.8}$

$\dfrac{35.8}{23.06} = \dfrac{x}{12.7}$

$\dfrac{.0285}{46.23} = \dfrac{.0053}{x}$

73. $\dfrac{52\frac{51}{64}}{2.172} = \dfrac{x}{34}$

$\dfrac{x}{23.6} = \dfrac{3.385}{3,580}$

74. $x = \dfrac{35\frac{3}{8} \times 16.35}{317}$

$\dfrac{38.6 \times 45.05}{x} = 265$

$$\frac{62.7}{18.9} = \frac{2.152}{x}$$

$$846.5 = \frac{79.8 \times .0058}{x}$$

$$34.9x = 927.03 \times 37$$

75. $572 = \dfrac{354.9}{x}$

$\dfrac{1}{86.4} = \dfrac{514x}{35.28}$

$9.82 = \dfrac{514x}{3.26}$

$2.082 = \dfrac{.0748x}{6.872}$

76. Find 37.5% of 384.80.
77. Find 68.2% of 57,840.
78. Find 42.7% of 37.68.
79. Find 18% of 675.6.
80. Find .83% of 82.05.
81. Find 1.08% of 328.16.
82. What per cent of 534 is 26.5?
83. What per cent of 826 is 40.8?
84. What per cent of 8.062 is .058?
85. Find the reciprocals of the following numbers: 528.6, 87.03, 25.72, .052, .00768.
86. A 6″ pinion drives a 21″ gear. Find the speed of the pinion if the gear revolves 170 r.p.m. Compute on the slide rule.
87. A 24-tooth pinion drives a gear. The speed of the pinion is 750 r.p.m., and that of the gear 225 r.p.m. Compute the number of teeth in the gear.
88. A 10″ diameter pulley rotating 800 r.p.m. drives a pulley whose speed is 240 r.p.m. Calculate the diameter of the driven pulley.
89. Four machine operators can complete a certain project in 3.375 days. How many days will it take 3 operators to complete the same project?
90. A machine operator completes a certain task in 7 days if he works 8 hours per day. How many days will it take the same operator to complete the same task if he works 10 hours per day?
91. Find the sines of the following angles by means of the slide rule: 31°, 20°10′, 41°40′, 8°30′, 26°50′, 4°20′, 68°18′, 74°53′, 69°50′.
92. Find the cosines of the following angles by means of the slide rule: 24°, 18°40′, 34°18′, 62°43′, 6°28′, 78°31′, 40°50′, 28°10′, 58°35′.
93. Find the tangents of the following angles by means of the slide rule: 3°20′, 5°45′, 22°36′, 18°22′, 34°30′, 42°10′, 58°12′, 63°20′, 72°40′.
94. Find the cotangents of the following angles by means of the slide rule: 4°18′, 12°20′, 23°50′, 32°30′, 48°50′, 65°20′, 72°10′, 28°37′, 34°50′, 56°10′.

THE SLIDE RULE

95. Find the secants of the following angles by means of the slide rule:
28°40', 32°25', 41°18', 52°42', 40°14', 18°24', 16°18', 38°28', 48°49'.

Note: Secant = the reciprocal of cosine.

96. Find the cosecants of the following angles by means of the slide rule: 6°48', 13°28', 28°48', 34°50', 42°40', 51°20', 58°49', 64°50', 72°20', 82°50', 76°41'.

Note: Cosecant = the reciprocal of sine.

97. Find the angles whose sines are as follows: .083, .097, .121, .142, .223, .261, .324, .410, .551, .660, .870, .895, .995.
98. Find the angles whose tangents are as follows: .167, .220, .296, .373, .433, .541, .635, .821, .849, .910, 1.327, 1.732, 2.106.
99. Find the logarithms of the following numbers by means of the slide rule: 32, 46, 318, 54.6, 685, 1,872, 34.082, .0387, .000576, 468,000.
100. Find the numbers whose logarithms are as follows: .622, 2.750, 1.025, 3.439, $\bar{2}$.890, .790, $\bar{3}$.665, 2.509.

CHAPTER XII
ENGINEERING COMPUTATIONS

This chapter discusses methods which are followed in computing engineering problems frequently encountered in shop work.

Realizing the space limitations of a text such as this, an attempt was made to cover only those features which would aid the shop man in making necessary computations, thus saving the valuable time of the professional engineer, whose abilities are directed toward comprehensive analysis of the project undertaken.

PROBLEMS DEALING WITH STRENGTH OF MATERIALS

A few fundamental principles involved in strength of materials are briefly explained before proceeding with actual problems.

Assume that a weight of 50 lb. is attached to a metal rod that is suspended from a hook in the ceiling. This weight, designated the *load* or *force*, in the terminology of engineering, stretches the rod a slight amount and, by so doing, alters its shape and size. The alteration which takes place is termed *deformation* or *strain*.

The metal rod, due to its strength (which depends on its size, shape, and composition), resists action of the given load to distort it by counteraction of an internal force set up within it. This resisting internal force and the applied external force are called the *stresses*.

If the external force acting upon the rod is greater than the resisting internal force, the rod will fail by breaking.

The stress developed in the rod which tends to pull it apart is known as the *tensile stress*.

If the load is applied in such a manner that it tends to crush the object, as might a load acting on a column, the stress developed within the column and tending to crush it is called the *compressive stress*.

If two plates are riveted together, each of the rivets holding the plates may fail by *shearing*, i.e., the plates will exert enough pressure to cause one-half of the rivet to slide past the other half at the joint.

There are also bending stresses to be considered. These are tension and compressive stresses combined, such as stresses developed in structural beams, loaded machine shafts, etc.

Torsion or *twisting stresses* are stresses developed in rotating shafts keyed to pulleys, gears, flywheels, and so on. These stresses are a form of shear stresses.

In computing problems involving strength of materials, all conditions which enter the problem as factors to be considered must be known and understood. Some of these factors are unit stress, unit strain, elastic limit, modulus of elasticity, ultimate strength, factor of safety, and safe working unit stress. A brief explanation is given of these factors.

Unit stress is the stress per unit of area of the object upon which the load applied acts. It is usually expressed in pounds per square inch and abbreviated *p.s.i.*

Unit strain is the unit deformation occurring in the object as a result of action of the load applied. For example, in the case of a tensile force acting upon an object, the strain created is the elongation per unit length of the object, usually expressed in inches per inch of length.

In elastic bodies—i.e., bodies which revert to their original size and shape upon removal of the external forces acting upon them—the deformation which takes place is directly proportional to the stresses imposed. This proportionality is effective only to a definite limit, which is called the *elastic limit*.

The elastic limit is the unit stress denoting the degree of elasticity of a material and determining the limit to which that material can be restored to its original size and shape upon removal of the load acting upon it.

The law of proportionality which elastic bodies obey is called Hooke's Law. This law states that in elastic materials the unit stress is proportional to the unit strain up to the elastic limit. The relation of the stress to the strain up to the elastic limit is expressed by a ratio (a constant) and it is called the *modulus of elasticity*.

The mathematical expression of this ratio is $E = \dfrac{S}{e}$, where E = modulus of elasticity, S = unit stress in pounds per square inch, and e = unit strain in inches per inch length.

The modulus of elasticity of a material plays an important part in strength computations, and a table of moduli of elasticity of common materials is to be found in the Appendix of this book. Tables covering a more comprehensive range of materials may be found in standard engineering handbooks and shop manuals.

Other expressions designating the ratio of unit stress to unit strain are:

1. $S = \dfrac{P}{A}$, where S = unit stress in pounds per square inch, P =

external load or force acting on the body in pounds, and A = cross-sectional area upon which the external force acts, in square inches.

2. $e = \dfrac{\Delta}{L}$, where e = unit strain in inches per inch length, Δ = total deformation in inches, and L = total length of body in inches.

The expression for modulus of elasticity can now be expanded to the form $E = \dfrac{S}{e} = \dfrac{P}{A} \div \dfrac{\Delta}{L} = \dfrac{PL}{A\Delta}$.

The *ultimate strength* of a material is that unit stress at which the material ruptures. (A table containing ultimate strengths of common engineering materials appears in the Appendix of this book.)

The *factor of safety* is the ratio of the ultimate strength of a material to the actual stress imposed upon it; i.e., the object in question is designed to be twice or, perhaps, three or more times as strong as is required to resist breakage when load is applied. (A table containing factors of safety appears in the Appendix of this book.)

The *safe working unit stress* is the stress obtained by dividing the ultimate strength by the factor of safety.

A few examples are presented to demonstrate the application of the preceding formulas.

Example 1: Find the diameter of a round steel rod that is to carry safely a suspended load of 70,000 pounds if the allowable tensile stress of the rod = 12,500 p.s.i.

Procedure: $S = \dfrac{P}{A}$, $A = \dfrac{P}{S}$, $\dfrac{\pi}{4} d^2 = \dfrac{P}{S}$, where d = diameter of rod.

Solving for the diameter, $d = \sqrt{\dfrac{4P}{\pi S}} = \sqrt{\dfrac{4 \times 70{,}000}{3.1416 \times 12{,}500}} = \sqrt{7.13} = 2.67''$.

The nearest commercial size obtainable is $2\tfrac{11}{16}''$ diameter.

Problems of this type can be solved quickly with the aid of the slide rule, the operation of which is described at length in Chapter XI. The reader is strongly advised to use the slide rule in all computations encountered in this chapter, since the accuracy of the results will not be affected appreciably.

Example 2: What is the size of a square steel column supporting a load of 180,000 pounds if the ultimate strength of the material is 65,000 p.s.i. and the recommended factor of safety is 5?

Procedure: $S = \dfrac{65,000}{5} = 13,000$ p.s.i., where S = safe working stress. $A = \dfrac{P}{S} = \dfrac{180,000}{13,000} = 13.85$ sq. in. The side of the square column = $\sqrt{13.85} = 3.727''$. The nearest commercial size is $3\tfrac{3}{4}'' \times 3\tfrac{3}{4}''$.

Example 3: How many pounds can a $\tfrac{3}{4}''$-10 N.C. steel bolt support safely in tension if its ultimate strength in tension = 60,000 p.s.i. and the recommended factor of safety = 8?

Procedure: The net cross-sectional area of this bolt is calculated on the basis of its root diameter.

The depth of thread $= \dfrac{.64952}{N}$. (See the chapter on threads, where .64952 = a constant and N = number of threads per inch = 10.)

The net diameter in tension $= d_r = d - \left(\dfrac{2 \times .64952}{10}\right) =$ .750 − .130 = .620″, where d = major diameter of bolt = $\tfrac{3}{4}''$.

The net area in tension $= A = \dfrac{\pi}{4} \times (d_r)^2 = .7854 \times (.620)^2 =$ .302 sq. in. The safe stress $= \dfrac{\text{ultimate stress}}{\text{factor of safety}} = \dfrac{60,000}{8} = 7,500$ p.s.i. $= S$.

The safe load $= P = S \times A = 7,500 \times .302 = 2,265$ lb.

Example 4: What is the elongation of a steel rod measuring $\tfrac{5}{8}''$ diameter $\times 52''$ long and supporting a load of 3,250 lb.?

Procedure: $E = \dfrac{S}{e} = \dfrac{PL}{A\triangle}$, where E for steel in accordance with the table of moduli of elasticity = 30,000,000 p.s.i., and $A = \dfrac{\pi}{4} d^2$.

$\triangle = \dfrac{PL}{EA} = \dfrac{3,250 \times 52}{30,000,000 \times .7854 \times (.625)^2} = .0184''$.

Example 5: Find the diameter of a steel rivet that is to resist shear if the applied load = 3,250 lb., the ultimate shearing strength = 42,000 p.s.i., and the factor of safety = 4. (See Fig. 12-1.)

Procedure: The force P, which is equal to 3,250 lb., acting on the rivet tends to shear it along the line AB, and the area of the rivet

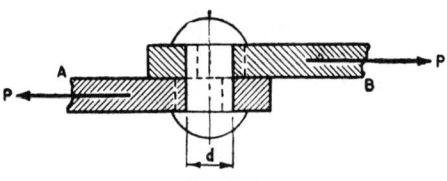

Fig. 12-1.

resisting this shearing action is $A = \frac{\pi}{4}d^2$. The safe working stress $S = \frac{42,000}{4} = 10,500$ p.s.i. $A = \frac{P}{S}$. $\frac{\pi}{4}d^2 = \frac{P}{S}$. $d = \sqrt{\frac{4P}{\pi S}} = \sqrt{\frac{4 \times 3,250}{3.1416 \times 10,500}} = \sqrt{.3938} = .627''$. The nearest commercial size rivet is $\frac{5}{8}''$ diameter.

Example 6: Find the diameter of a steel rivet that will safely resist shear if the applied load = 5,840 lb. The ultimate shearing strength and the factor of safety are the same as those in the preceding example. (See Fig. 12-2.)

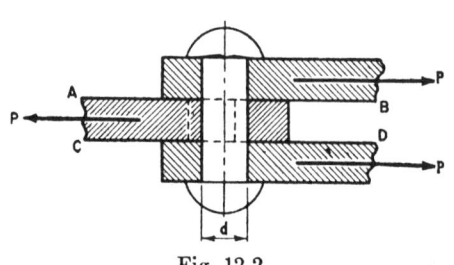

Fig. 12-2.

Procedure: The force P, which is equal to 5,840 lb., acting on the rivet tends to shear it along lines AB and CD simultaneously, and there are two cross-sectional areas of the rivet along these lines that resist the shearing action of the force. Therefore, the total area resisting this force is $A = 2 \times \frac{\pi}{4}d^2 = \frac{\pi}{2}d^2$. $A = \frac{P}{S}$. $\frac{\pi}{2}d^2 = \frac{P}{S}$. $d = \sqrt{\frac{2P}{\pi S}} = \sqrt{\frac{2 \times 5,840}{3.1416 \times 10,500}} = \sqrt{.3536} = .595''$. The nearest commercial size rivet is $\frac{19}{32}''$ diameter.

The rivet in this example is smaller than is that of the preceding example, yet it will safely resist the shear of a much greater applied load. This is because of the double shear encountered in this example as compared to the single shear of the other.

Example 7: What force will strip the threads on a $1''$-8 N.C. bolt, considering that the ultimate shearing strength of the material of the bolt is 38,000 p.s.i.? (See Fig. 12-3.)

Procedure: When tightening the nut as shown in the drawing, a shearing force (P) is developed, which tends to

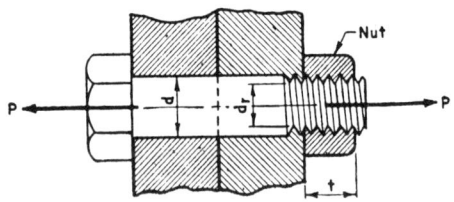

Fig. 12-3.

strip the threads from the bolt, thus leaving a bare cylindrical rod whose diameter is the root diameter, d_r. An area strong enough to resist this shearing action depends upon ample circumference of the root section of the bolt and adequate thickness of the nut.

This area can be expressed: $A = \pi \times d_r \times t$, where t = thickness of nut = $\frac{7}{8}''$, and $d_r = d - \dfrac{2 \times .64952}{N} = 1 - \dfrac{1.299}{8} = .8376$.

The force that will strip the threads is $P = S \times A$, where S = ultimate shearing strength of the bolt = 38,000 p.s.i. P = 38,000 $\times \pi \times$ .8376 $\times$ .875 = 87,500 lb.

Example 8: What force can be safely applied in tightening the nut in Fig. 12-3 if we assume a factor of safety equal to 5?

Procedure: Safe load = $\dfrac{\text{ultimate load}}{\text{factor of safety}} = \dfrac{87,500}{5} = 17,500$ lb.

Example 9: An interesting example involving calculation of the shearing strength of a material occurs in determining the unit of force which must be applied to punch holes through a material of known composition and thickness—What unit of force is required to punch 5 holes $\frac{9}{16}''$ in diameter through a plate $\frac{17}{32}''$ thick?

Procedure: In this example, the area resisting the shearing action of one punch is equal to πdt, where d = diameter of hole punched through the plate and t = the thickness of the plate. The total resisting area is equal to $5\pi dt$. Since, in punching through, the material will fail in shearing, the total force required to overcome the shearing resistance of the material is equal to its shearing strength of 32,000 p.s.i. multiplied by the total area of $5\pi dt$ resisting the shearing action. $P = SA = S \times 5\pi dt = 32,000 \times 5 \times \pi \times .562 \times .531 = 150,300$ lb. = 75.15 tons.

Example 10: The strength of pipes to resist the pressure of gas and liquids is calculated as follows: Pressure in the pipe tends to burst it along its length, therefore the pipe material is under tension.

The area of the diametral plane subjected to the pressure of the gas or liquids is equal to the inside diameter of the pipe (d) multiplied by the length of pipe (l). See Fig. 12-4 for the cross section of the pipe.

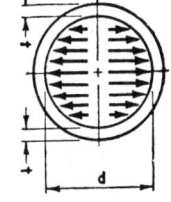

Fig. 12-4.

$P = pdl$, where P = total pressure in pounds on the diametral plane of the pipe and p = pressure in pounds per square inch. The bursting tendency of the pipe contents is resisted by the tension in the pipe walls, and this resistance depends upon the wall thickness (t) and the length of the pipe (l).

Assume that the total resisting tensile stress in the pipe is S, and the unit tensile stress is s. Then $S = 2tsl$.

To withstand bursting, the pipe must be so designed that its total resisting tensile stress is the same as the total pressure on the diametral plane. $S = P$; $2tsl = pdl$. Therefore, $t = \dfrac{pdl}{2sl} = \dfrac{pd}{2s}$.

Assume that the inside diameter of a pipe $= 6''$, and that it is required to calculate the thickness of the pipe wall for safe resistance to pressure of its contents, which is 250 p.s.i. The pipe material is known, and its allowable tensile stress $= 6,500$ p.s.i. Then $t = \dfrac{pd}{2s} = \dfrac{250 \times 6}{2 \times 6,500} = .115''$. The nearest practical, then, is $\tfrac{1}{8}''$ thickness.

Example 11: What is the strength of the riveted joint in Fig. 12-5?

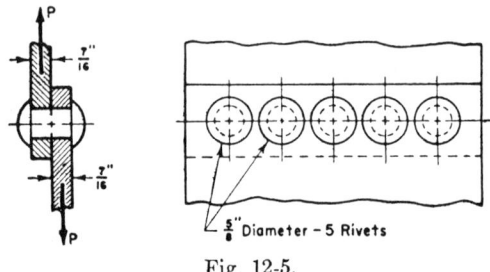

Fig. 12-5.

Procedure: The safe shearing load on each rivet $= S_s$, the safe shearing strength per square inch, times A, the cross-sectional area of the rivet.

Assume that $S_s = 12,000$ p.s.i. and that $A = \dfrac{\pi}{4} d^2$, where $d =$ diameter of the rivet. Then, the safe shearing load on each rivet $= S_s \times A = S_s \times \dfrac{\pi}{4} d^2 = 12,000 \times .7854 \times (.625)^2 = 3,690$ lb.

The *safe shearing load* on five rivets is $3,690 \times 5 = 18,450$ lb. The safe bearing or crushing load on each rivet is determined by S_b, the safe crushing strength per square inch, and A, the area of the plate, which is equal to the rivet diameter (d) times the plate thickness (t). The safe bearing load on each rivet $=$

$S_b \times A = S_b \times d \times t = 18,000 \times .625 \times .437 = 4,920$ lb.

The *safe bearing load* on five rivets $= 4,920 \times 5 = 24,600$ lb.

The smaller of the two values of safe loads, which is 18,450 lb., is taken as the final strength of the riveted joint in Fig. 12-5.

FORCE, WORK, POWER, AND VELOCITY

In this section force, work, power, and velocity will be explained and formulas involving them will be given.

Force. Force can be defined as that cause which produces motion of a body at rest, changes the motion of a moving body, or brings its motion to a stop. Force is identified by its *direction*, its *point of application*, and its *magnitude*. Usually it is measured in pounds.

A force can be represented by a straight line or vector. The direction of the force can be indicated by an arrow, the point of application by one end of the line, and the magnitude by the length of the line. (See Fig. 12-6.)

When several forces are acting on a body, they may be replaced by one resultant force producing a cumulated effect. We refer to this unification as *composition of forces*.

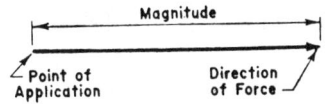

Fig. 12-6.

On the other hand, we may replace one given resultant force by several component forces, and this is referred to as *resolution of forces*.

If two known forces acting at one point can be drawn as adjacent sides of the parallelogram, the resultant force can be found by completing the parallelogram and obtaining its diagonal. (See Fig. 12-7.)

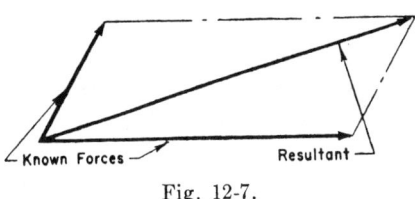

Fig. 12-7.

If two given forces act in the same direction, their resultant is found by adding the two forces.

If the two given forces take opposite directions, their resultant is obtained by subtracting the smaller force from the greater one. The direction of the resultant is that of the greater force.

Work. Work is the result obtained by multiplying a force by the distance through which it travels.

Example 1: If a weight of 10 pounds is raised a distance of 2 feet, 20 foot-pounds of work is done.

Example 2: If a weight of 100 pounds is pushed a distance of 3 feet by a force of suitable magnitude, the amount of work done is 300 foot-pounds.

A *foot-pound* (ft.-lb.), then, is the amount of work done by exerting a force of one pound through a distance of one foot. It is a value establishing a definite unit or work. Another unit of work is the *inch-pound* (in.-lb.), and there are other units appropriately designated, such as those of the metric system.

Power. Power is the rate of work done in a given time. Usually it is measured in foot-pounds per unit of time, as by minutes and seconds.

If work which is the product of force and distance is divided by the time consumed in accomplishment of the given task, we obtain power. It should be noted, therefore, that whenever power is computed, the time element is essential. In the second of the preceding examples given in the discussion on work, the amount of work done in pushing a weight of 100 pounds through a distance of 3 feet is 300 ft.-lb., and if it takes 2 minutes to accomplish the task, the power will be $\frac{100 \times 3}{2} = 150$ ft.-lb. per minute, or $\frac{150}{60} = 2\frac{1}{2}$ ft.-lb. per second. The power in this example can also be expressed as $2.5 \times 12 = 30$ in.-lb. per second.

The most common unit of power is the *horsepower* (h.p.). It is equal to 550 ft.-lb. of work done in one second, or 33,000 ft.-lb. per minute. To find the equivalent horsepower, reduce the foot-pounds of work done in a given time to foot-pounds of work done per minute and divide this by 33,000. For example, if a machine does 53,000 ft.-lb. of work in 1.4 minutes, the horsepower is $\frac{53{,}000}{1.4 \times 33{,}000} = 1.14$.

Velocity. The rate of motion is measured by velocity. It is obtained by dividing the distance traveled by the time consumed. Velocity usually is expressed in feet per minute or in feet per second. It is used in computations independent of the force.

Moment of Force. If we assume a fixed point and a force acting upon a body in such a way that the perpendicular distance from the fixed point to the direction of force is known, we can compute the moment of force by multiplying the force by the perpendicular distance.

The fixed point is referred to as the *center of moment*, and the perpendicular distance is the *lever arm* of the force.

The moment of force is measured in inch-pounds or foot-pounds.

Moment of force is a factor involved in many problems of mechanics. Good examples of the operation of moment of force are furnished by lever arms, which we shall employ in Figs. 12-8, 12-9, and 12-10. In lever computations the moment of force is calculated by multiplying the force or weight by its respective distance or "reach" of the lever arm from the fulcrum.

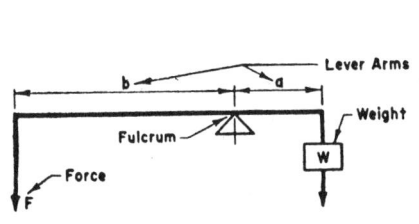

Fig. 12-8.

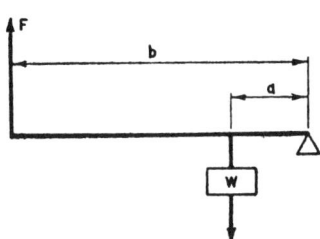

Fig. 12-9.

There are three forms of levers in common use. These are:

Form 1: In levers of this form, the fulcrum is located between the weight (W) and its balancing force (F). Both the weight and the force must act in the same direction. (See Fig. 12-8.)

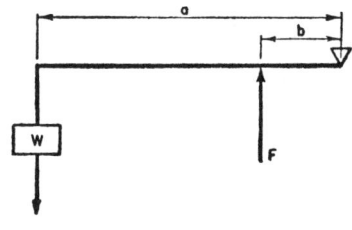

Fig. 12-10.

The mathematical relations for this form of lever are $\frac{W}{F} = \frac{b}{a}$, $Wa = Fb$; $W = F\frac{b}{a}$; $F = W\frac{a}{b}$.

Form 2: In levers of this form, the fulcrum is located outside the weight and its balancing force, and the weight is between the fulcrum and its balancing force. (See Fig. 12-9.)

The mathematical relations for these levers are the same as for those of Form 1, since $Wa = Fb$.

Form 3: Levers of this form are similar to the lever shown in Fig. 12-9 except that the force is between the fulcrum and the weight. (See Fig. 12-10.)

The mathematical relations for levers of this form are the same as those indicated for the preceding forms, since $Wa = Fb$.

Note that the weight of the lever has not been considered in any of the three forms. If the weight of the lever must be considered in the computation, an additional moment is determined by the product of the weight of the lever arm and the distance of its center of gravity from the fulcrum.

The principle of moment of force is utilized in many industrial applications such as the windlass, shown in Fig. 12-11 and the pulleys in Fig. 12-12, where considerably smaller forces are employed to lift much greater weights. We refer to this as the *mechanical advantage*.

In Fig. 12-11, the weight (W) is attached to a rope wound on the drum. The force (F) is calculated as follows: $Fb = Wa$, and $F = \dfrac{Wa}{b}$.

The mechanical advantage obtained by the use of pulleys for lifting weights by considerably reduced force is illustrated in Fig. 12-12. There

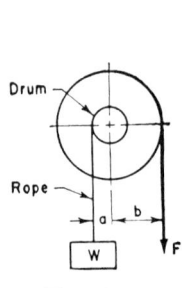

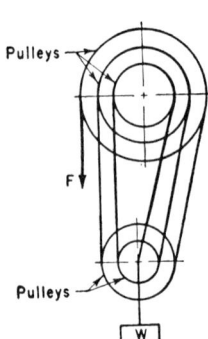

Fig. 12-11.   Fig. 12-12.

are five pulleys that are employed when the weight (W) is raised by application of the force (F), therefore there are five ropes, and the magnitude of the force is obtained by dividing the weight by the number of ropes. Then, $F = \dfrac{W}{5}$. If the weight in Fig. 12-12 is 800 pounds, it will take a force of $\dfrac{800}{5} = 160$ pounds to raise it.

The mechanical advantage in the case illustrated is 5 to 1, but the velocity or speed at which the weight is raised is only $\dfrac{1}{5}$ of the velocity of the force. Also the distance traveled by the weight is $\dfrac{1}{5}$ of the distance traveled by the force.

Bending Moments. Beams are subjected to bending actions of loads imposed on them, and these actions are referred to as bending moments. A beam and its bending moment are illustrated in Fig. 12-13. The beam is supported at one end, and the load (P) acts downward upon the free end of the beam. This is known as a *cantilever beam*, and the bending moment (M) is equal to the load (P) times the distance (L), which is referred to as the lever arm. $M = P \times L$.

The beam resists the bending action of the load, and this resistance is called the *moment of resistance* of the beam. The moment of resistance is the product of the unit stress occurring in the farthest fiber of the beam

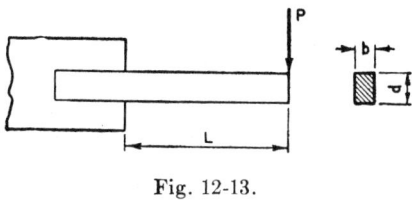

Fig. 12-13.

and the section modulus. The section modulus in turn depends upon the size and shape of the cross section of the beam, which information is available in tables found in standard engineering handbooks. In addition to the section moduli, moments of inertia for various cross sections are given in these handbooks. The moment of inertia is needed to compute the section modulus, and the relation between the two is $Z = \dfrac{I}{c}$, where Z = section modulus, I = moment of inertia, and c = distance from the neutral surface to the farthest fiber.

The moment of resistance must be equal to the bending moment, and according to the formula $M = P \times L$, we have $M = PL = \dfrac{SI}{c} = SZ$, where S = the unit stress in the farthest fiber in pounds per square inch, and the bending moment is expressed in inch-pounds.

A beam may be supported and loaded in many different ways. Therefore the bending moment produced by the load imposed varies according to the manner of support and loading. The point at which the unit stress is maximum is called the *critical point*, and the stress at this point is equal to the bending moment divided by the section modulus. Expressed mathematically, $S = M \div Z = M \div \dfrac{I}{c} = \dfrac{Mc}{I}$.

There are various formulas for calculating the stresses that occur at the critical point, and these may be found in standard engineering handbooks. For an illustrative example, assume that the beam in Fig. 12-13 is of rectangular cross section, the load (P) = 1,200 lb., and the distance (L) = 22″. Assume that the ultimate strength = 65,000 p.s.i., and a factor of safety = 6. It is required to find the cross section of the beam that is to support the given load.

According to the foregoing, $M = PL = 1{,}200 \times 22 = 26{,}400$ in.-lb.

S = safe fiber stress = $\dfrac{65{,}000}{6}$ = 10,830 p.s.i. $M = SZ$. 26,400 = 10,830 × Z. $Z = \dfrac{26{,}400}{10{,}830} = 2.438$.

According to the handbook, $Z = \dfrac{bd^2}{6}$ for rectangular cross section, where b = the length of the shorter side and d = the length of the longer side. Assuming that the dimension $b = 1''$, then $2.438 = \dfrac{1 \times d^2}{6}$, and $d^2 = 6 \times 2.438 = 14.63$. $d = \sqrt{14.63} = 3.83''$, say $3\tfrac{7}{8}''$. The cross section of the beam is $1'' \times 3\tfrac{7}{8}''$.

For commercial reasons, this section may not be standard and is difficult to obtain, so the dimensions b and d are varied until a standard section is obtained. Assuming that the dimension $b = 1\tfrac{1}{2}''$, then $2.438 = \dfrac{1.5 \times d^2}{6}$, and $d = \sqrt{\dfrac{2.438 \times 6}{1.5}} = 3.126$, say $3\tfrac{1}{8}''$. In this case a section $1\tfrac{1}{2}'' \times 3''$ will be satisfactory, as by using $3''$ instead of $3\tfrac{1}{8}''$ the factor of safety is reduced only a slight amount.

If the section of the beam in the preceding example is cylindrical $Z = \dfrac{\pi d^3}{32} = .098 d^3$ (according to standard engineering handbooks), and the suitable diameter d may be calculated.

The beam may be supported at both ends, and the load imposed upon it causes a deflection, or sagging, in the beam. In that case it is desired to calculate the maximum deflection, which occurs at the center of the beam length, if the load is concentrated there.

The deflection = $\dfrac{PL^3}{48EI}$, where P = load in pounds, L = length of unsupported beam in inches, E = modulus of elasticity of the material used, and I = moment of inertia of the cross section about the center of gravity, this last factor depending upon the shape of the beam as characteristically designated: **I** beam, **H** beam, channel beam and so on. Calculation of the deflection is very important, as the amount of deflection may be excessive and a stiffer beam is designed to reduce the maximum deflection.

TORQUE, HORSEPOWER, AND MECHANICAL EFFICIENCY OF MACHINES

In this section computations of the torque and horsepower of various engines and motors, and of the mechanical efficiency of machines are given.

ENGINEERING COMPUTATIONS

Torque of Various Engines and Motors. When a shaft or any other structural member is subjected to twisting or turning, torsional (shearing) stresses develop in the member, and consequently torque is transmitted. (See Fig. 12-14.)

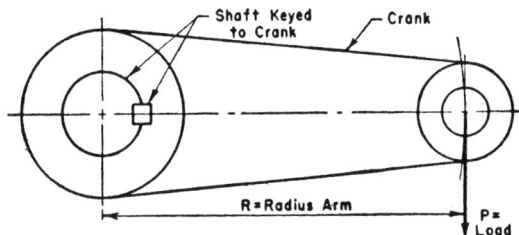

Fig. 12-14.

The torque transmitted by the shaft as a result of the load (P) imposed on the end of the crank (which is keyed to the shaft) is equal to the *turning moment*, and the mathematical expression of the torque is $T = P \times R$, where T = torque in inch-pounds, P = load in pounds, and R = radius arm of the crank in inches.

To balance the effect of the turning moment, the resisting moment of the shaft or any other member must be equal to the turning moment This resisting moment = $T = \dfrac{S \times J}{c}$ in.-lb., where J = polar moment of inertia for the particular size and shape of the object, c = distance in inches from the neutral axis to the farthest fiber, $\dfrac{J}{c}$ = polar section modulus (which value may be found in standard engineering handbooks), and S = maximum unit shearing stress occurring in the farthest fiber.

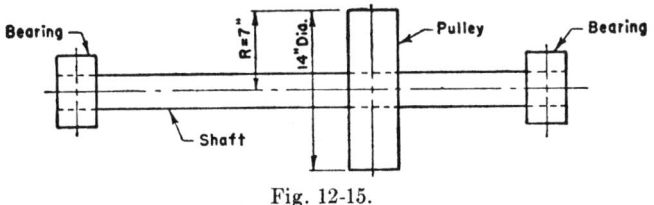

Fig. 12-15.

Example: Assume that the shaft in Fig. 12-15 is subjected to a twisting moment produced by a tangential load (P) of 6,000 pounds applied at the circumference of the pulley of known diameter, which is keyed to the shaft. Also assume that the safe maximum shearing stress (S) of the shaft material is 8,000 p.s.i. Calculate the shaft diameter so that it can safely transmit the developed torque.

Procedure: $T = P \times R$ = turning moment.

$$T = \frac{SJ}{c} = \text{resisting moment.}$$

$$PR = \frac{SJ}{c}, \text{ where } \frac{J}{c} = \frac{\pi d^3}{16} = .196d^3, \text{ found in handbooks.}$$

$6{,}000 \times 7 = 8{,}000 \times .196d^3.$

$42{,}000 = 1{,}568d^3.$

$$d^3 = \frac{42{,}000}{1{,}568}, \text{ and } d = \sqrt[3]{\frac{42{,}000}{1{,}568}} = \sqrt[3]{27} = 3''.$$

Horsepower of Various Engines and Motors. If we assume a load P applied continuously at the end of a rotating crank in Fig. 12-14, a certain amount of work is done in one revolution of the crank, and this work is equal to $P \times 2\pi R$ in.-lb.

If the crank revolves N revolutions, the work done is equal to $P \times 2\pi R \times N$ in.-lb. If it takes 1 minute for the crank to revolve N revolutions, the work above is done in 1 minute.

Since power is the rate at which work is done, and the unit of power is the horsepower, which factor is equal to 33,000 ft.-lb. per minute, or $33{,}000 \times 12$ in.-lb. per minute, the formula is:

$$\text{Horsepower} = \frac{P \times 2\pi RN}{12 \times 33{,}000} = \frac{P \times R \times N}{63{,}000}$$

Also, $P \times R = T$ = torque, therefore the preceding formula is:

$$\text{Horsepower} = \frac{TN}{63{,}000}, \text{ and } T = \frac{63{,}000 \times \text{horsepower}}{N}$$

Example: A shaft rotates at 350 r.p.m. and transmits a torque of 18,000 lb. What is the horsepower transmitted by the shaft?

$$\text{Horsepower} = \frac{TN}{63{,}000} = \frac{18{,}000 \times 350}{63{,}000} = 100$$

When calculating the horsepower of a steam engine, the indicated horsepower is usually determined — i.e., the power is computed on the basis of the diameter of the cylinder in inches, the length of the stroke in feet, the mean effective pressure, and the number of revolutions per minute.

The formula for the horsepower is $\frac{PLAN}{33{,}000}$, where P = mean effective pressure on piston in pounds per square inch, L = length of stroke in feet, A = area of piston in square inches, and N = number of strokes per minute. N also equals 2 × the r.p.m. for a double-acting steam engine.

When calculating the horsepower of a gas engine, the standard formula for a four-cycle engine is: Horsepower = $\frac{D^2 N}{2.5}$, where D = diameter of cylinder in inches and N = number of cylinders.

According to the Society of Automotive Engineers (S.A.E.), this formula is correct for the piston speed of 1,000 feet per minute, with an average, or mean effective, pressure of 90 p.s.i.

At times it is desired to compute the actual horsepower delivered by the engine at the flywheel. This value is referred to as the *brake horsepower*, which is equal to $\frac{2\pi L N W}{33{,}000}$, where W = net pressure obtained by reading the scale of the Prony brake, a device used for determining the brake horsepower. The weight of the lever is subtracted from the scale reading. L = lever arm length expressed in feet. N = r.p.m.

In electrical machinery, such as generators and motors, the unit of power is the watt and the kilowatt (kw.). One (1) kilowatt = 1,000 watts; 746 watts = 1 h.p. = ¾ kw., approximately. The formula is:

$$\text{Horsepower} = \frac{\text{amperes} \times \text{volts}}{750}$$

Mechanical Efficiency of Machines. *Energy*, which is the capacity for performing work, exists in different forms, as mechanical energy, electrical energy and so on.

Mechanical energy is of two kinds, *potential* and *kinetic*. Potential energy is energy of position or configuration. An example of potential energy can be demonstrated by a weight of 100 pounds, suspended at a height of 15 feet, and possessing 100 × 15 = 1,500 ft.-lb. of potential energy. If the weight is allowed to fall by action of gravity, the weight receives energy which can do 1,500 ft.-lb. of work. The formula is: Potential energy = $W \times h$ = 100 × 15 = 1,500 ft.-lb., where W = weight of the body in pounds and h = height of the body in feet.

Kinetic energy is the energy possessed by a body in motion, the body having calculable capacity to perform work. The formula is: Kinetic energy = $\frac{WV^2}{2g}$ ft.-lb., where W = weight of moving body in pounds,

V = velocity, or speed, of moving body in feet per second, and g = the acceleration of gravity in feet per second, in this case, 32.2.

A machine delivers less power than the amount it receives because of friction which occurs in the moving parts of the machine, and also because of losses encountered in the system of power transmission.

The power received by the machine is the *input*, and that delivered by the machine is the *output*. The ratio of output to input is computed in the form of a percentage, and is designated by the term *efficiency*.

Thus, efficiency = $\dfrac{\text{output}}{\text{input}}$.

For example, if 15 h.p. is received by a motor and it delivers 13 h.p. to other machines, the efficiency = $\dfrac{13}{15} \times 100 = 86.7$ per cent.

Another example of efficiency is demonstrated by the following problem.

Example: How many amperes (current) does a motor operating on 220 volts take if the capacity of the motor is $7\frac{1}{2}$ h.p. and its efficiency is 75 per cent?

Procedure: Efficiency = $\dfrac{\text{output}}{\text{input}} = \dfrac{7.5}{x} \times 100 = 75$ per cent.

$\dfrac{7.5 \times 100}{x} = 75.$

$x = \dfrac{7.5 \times 100}{75} = 10$ h.p. = input.

To convert 10 h.p. to watts: $10 \times 750 = 7{,}500$ watts
Watts = volts × amperes.

Amperes = $\dfrac{\text{watts}}{\text{volts}} = \dfrac{7{,}500}{220} = 34.1$, say 34 amperes.

LENGTH OF BELTS AND POWER TRANSMITTED BY BELTS

Power is carried from one line of shafting to another by means of belts which pass over pulleys fastened to the shafts. If these belts are open the pulleys turn in the same direction, if the belts are crossed the pulleys turn in opposite direction.

A close approximation of the length of an open belt is expressed in the following formula: $L = \pi \dfrac{(D + d)}{2} + 2C$, where L = length of open belt, D = diameter of large pulley, d = diameter of small pulley, and C = center distance between the pulleys. (See Fig. 12-16.)

The length of a crossed belt is computed by means of the following formula: $L = \frac{\pi}{2}(D + d) + 2\sqrt{C^2 + \left(\frac{D+d}{2}\right)^2}$, where L = length

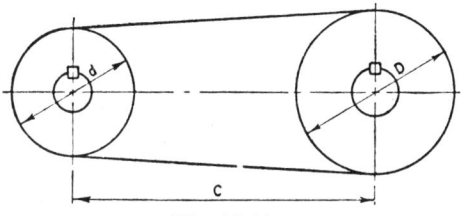

Fig. 12-16.

of crossed belt, D = diameter of large pulley, d = diameter of small pulley, and C = center distance between the pulleys. (See Fig. 12-17.)

The speed of the belt is the same as the rim speed of the pulleys, and the rim speed of both pulleys is the same. It is calculated as follows:

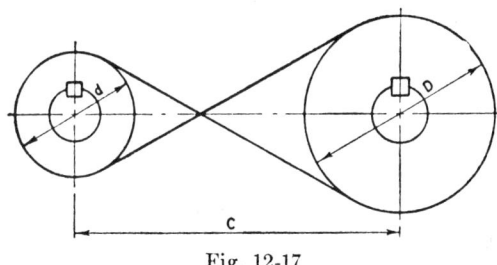

Fig. 12-17.

Speed of belt = rim speed of pulley = $\frac{\pi D \times N}{12}$ = feet per minute, where D = diameter of any pulley in inches and N = speed of the respective pulley in r.p.m.

The horsepower transmitted by the belts equals: Speed of belt in feet per minute $\times w \times T$, where w = width of belt in inches and T = the difference in tension between the tight and slack sides of the belt, expressed in pounds per inch width of belt.

The values of T are usually given in standard engineering handbooks for various belts and conditions of use.

An example will demonstrate the application of the horsepower formula.

Example: Assume that the belt in Fig. 12-16 is 6″ wide and that the diameters of the pulleys are 8" and 26", respectively. The small pulley revolves 340 r.p.m., and the difference in tension per inch width of belt = 50 lb. What horsepower is transmitted by the belt?

Procedure: Speed of belt = rim speed of pulley = $\dfrac{\pi d N}{12}$ = $\dfrac{\pi \times 8 \times 340}{12}$ = 712 feet per minute.

Horsepower = $\dfrac{712 \times w \times T}{33{,}000} = \dfrac{712 \times 6 \times 50}{33{,}000} = 6.47.$

The width of the belt required to transmit a given horsepower = $\dfrac{33{,}000 \text{ h.p.}}{\text{speed of belt in feet per minute} \times T}.$

CUTTING SPEEDS AND FEEDS FOR VARIOUS MACHINE TOOLS

In order to obtain the highest efficiency and economy in machine shop production, the machines should operate at the highest permissible speeds and feeds, these conditions depending, of course, upon the type of material of the product, the type of material of the cutting tool, the size (depth and width) of the cut, whether it is a roughing or finishing cut, the shape of the cutting tool, and the type and strength of the machine.

By the cutting speed we mean the velocity or surface speed with which the cutting tool engages the work. It is usually expressed in feet per minute, and the mathematical expression of this value depends upon the type of machine in use.

For example, the cutting speed in feet per minute on a lathe is measured on the surface of the rotating product, which passes the cutting edge of the tool in an interval of one minute.

The cutting speed on a milling machine is the speed, measured in feet per minute, of a point on the periphery of the milling cutter engaging the work.

On a drilling machine it is the speed, measured in feet per minute, of a point on the surface of the drill.

On a shaper it is the speed, measured in feet per minute, at which the cutting edge of the tool passes the work.

On a planer it is the reverse of conditions indicated for the shaper, the cutting speed in feet per minute being the speed at which the work passes the cutter.

The cutting speed in feet per minute on a lathe, milling machine, and

ENGINEERING COMPUTATIONS

drill press can be expressed by the following formula: $\frac{\pi d N}{12}$, where d = diameter of the work, diameter of the milling cutter, or diameter of the drill in inches, and N = r.p.m.

Tables appearing in standard engineering handbooks give appropriate values of cutting speeds in feet per minute for various metals shaped on designated machines.

Example: A cutting speed of 115 feet is recommended for a cylindrical piece of work to be turned on a lathe. The average diameter of the piece is $3\frac{1}{4}$ inches. What is the safe speed of the lathe spindle?

Procedure: The cutting speed in feet per minute = $\frac{\pi d N}{12}$ = 115.

$$N = \frac{115 \times 12}{\pi d} = \frac{115 \times 12}{3.1416 \times 3.25} = 135 \text{ r.p.m.}$$

This is the calculated r.p.m. of the spindle, and since there is a series of selective speeds of the spindle on every lathe used for production, a speed can be chosen that will equal or be close to the computed r.p.m.

ESTIMATING TIME REQUIRED FOR VARIOUS CUTTING OPERATIONS

To estimate the machine time required to complete a cut over a piece of work turned on a lathe, the following formula may be used: $T = \frac{L}{f \times N}$, where T = machine time in minutes, L = length of cut in inches, f = feed of cutter along work in inches per revolution, and N = r.p.m. of the spindle. (See Fig. 12-18.)

Example: A cylindrical bar in Fig. 12-18 is turned on a lathe. The average diameter is $2\frac{3}{4}''$, and the length of cut $L = 14\frac{1}{2}''$. Assume that the cutting speed in feet per minute recommended for the type of material of work and cutter is 165, and that the suitable feed f = .006″ per revolution.

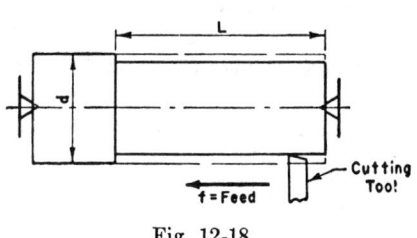

Fig. 12-18.

Procedure: Estimate the machine time required to complete one cut of the given length.

Cutting speed in feet per minute = $\dfrac{\pi d N}{12}$ = 165.

$N = \dfrac{165 \times 12}{\pi \times d} = \dfrac{165 \times 12}{3.1416 \times 2.75} = 229$ r.p.m. (computed).

The actual r.p.m. on the machine may be a value slightly different from the computed value. Assume that the nearest value on the machine is 220 r.p.m. To estimate the machine time necessary to make a complete cut, use the formula $T = \dfrac{L}{f \times N}$. $T = \dfrac{14.5}{.006 \times 220} = 11$ minutes.

The suitable cutting speed in feet per minute on a shaper or planer can be found in the proper table if the materials of the work and the cutter and other necessary data are known. The number of strokes the ram of the shaper makes in one minute can be computed as follows:

Assume that the recommended cutting or surface speed is 135 feet per minute. Then by modifying the formula given for the lathe we can compute the number of cutting strokes per minute on the shaper. $\dfrac{l \times N}{12} = 135$, where l = length of stroke on the shaper in inches, say 17″, and N = number of cutting strokes per minute. $N = \dfrac{12 \times 135}{17} =$ 95.3 strokes per minute.

Assume the return, or idle, stroke on the shaper is twice as fast as the cutting stroke, then one cutting stroke takes $\dfrac{60}{95.3}$ seconds, and one return stroke takes $\dfrac{60}{95.3 \times 2} = \dfrac{30}{95.3}$ seconds. One complete stroke on the shaper (cutting and return, that is) takes $\dfrac{60}{95.3} + \dfrac{30}{95.3} = \dfrac{90}{95.3}$ seconds. Number of complete strokes per minute = $60 \div \dfrac{90}{95.3} = \dfrac{60 \times 95.3}{90} = 63.5.$

Assume that the feed per stroke on the shaper = $\tfrac{3}{64}$″ = .047″, and that the width of work to be shaped = $16\tfrac{3}{4}$″.

The machine time required to shape the entire width is $T = \dfrac{W}{f \times N_c}$, where W = width of work = $16\tfrac{3}{4}$″, f = .047″, and N_c = number of complete strokes per minute = 63.5. $T = \dfrac{16.75}{.047 \times 63.5} = 5.62$ minutes.

ENGINEERING COMPUTATIONS

POWER REQUIRED FOR CUTTING GIVEN AMOUNT OF METAL

The formula for calculating the horsepower developed at the point at which the cutting tool engages the work is: Horsepower $= \dfrac{F \times C}{33,000}$, where $F =$ force required to cut the metal, acting in a direction tangent to the path of the cutting tool, and expressed in pounds. (This force depends upon the feed and depth of cut.) $C =$ velocity or cutting speed in feet per minute.

The power of the motor will be greater than the net power required to make a certain cut in a proportion determined by the machine's efficiency.

For example, if the net h.p. to cut the metal $= 2.75$, and the machine efficiency $= 60$ per cent, the h.p. of the motor $= \dfrac{2.75}{.6} = 4.6$.

In some cases it is required to compute the horsepower per cubic inch of metal cut per minute.

This value can be obtained if the foregoing horsepower required to make a cut is divided by the number of cubic inches of metal removed per minute.

The number of cubic inches of metal removed per minute $= 12 \times C \times f \times h$, where $C =$ cutting speed in feet per minute, $f =$ feed in inches, and $h =$ depth of cut in inches.

The horsepower per cubic inch of metal per minute $=$

$$\dfrac{F \times \cancel{C}}{12 \times 33,000 \times f \times h \times \cancel{C}} = \dfrac{F}{396,000 \times f \times h}.$$

These values of horsepower are given for different metals cut under various conditions designated in engineering handbooks and pamphlets published by the machine tool builders.

For press operations such as blanking, shearing, and punching holes it is frequently desired to calculate the pressure required for the particular operation in order that a press capable of withstanding the pressure exerted may be used.

The maximum pressures are calculated as follows:

P maximum $= L \times t \times S_s$, where P maximum $=$ maximum pressure in pounds, $L =$ length of cut in inches, which is the perimeter of the outline of the cut. $t =$ thickness of material in inches. $S_s =$ ultimate shearing strength in p.s.i. The pressure for round work is P maximum $= \pi \times d \times t \times S_s$, where $d =$ diameter of hole or outline of object to be cut.

The average pressure (P av.) required to make a cut is equal to the maximum pressure times the per cent of penetration, which varies from 35 to 65 per cent of the maximum pressure, depending upon the clearance between the cutting dies, the amount of shear on the dies, the ductility of the metal and so on.

The energy required to perform the preceding operation = W inch-pounds per stroke. $W = P$ av. $\times t \times 1.16$, where the factor 1.16 allows 16 per cent for friction in the machine.

The motor horsepower required for these operations is computed as follows: Horsepower $= \dfrac{W \times N}{12 \times 33{,}000}$, where $N =$ the number of strokes per minute.

The preceding operations are illustrated in Fig. 12-19.

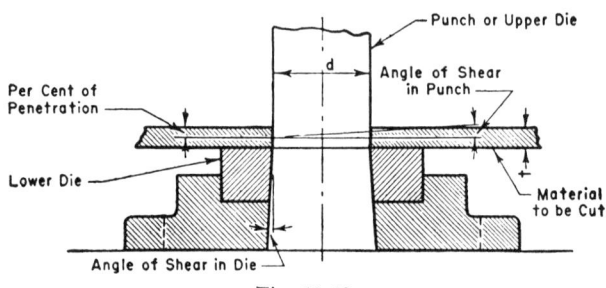

Fig. 12-19.

The pressure required for drawing operations such as those by which automobile, airplane, and various other stampings are produced is computed as follows:

$P = L \times t \times S_t$, where $P =$ pressure in pounds, $L =$ perimeter of outline of formed object in inches, $t =$ thickness of metal, and $S_t =$ ultimate tensile strength of material in pounds per square inch.

The pressure required for drawing deep shells and similar objects is $P = \pi \times d \times t \times S_t$, where $d =$ mean diameter of shell wall.

The power required for blanking, shearing, and punching a number of holes or parts per minute in h.p. $= \dfrac{t^2 \times d \times N}{3.78}$, where $t =$ thickness of metal in inches, $d =$ diameter of holes punched, and $N =$ number of holes or parts.

This formula is according to Pomeroy[*] and allows 80 per cent of motor efficiency.

[*] *American Machinists' Handbook*, pp. 997-998; published by McGraw-Hill Book Co., 1940.

ENGINEERING COMPUTATIONS

Another formula originated by Pomeroy for calculating the horsepower required in machining operations is:

Horsepower = $f \times t \times C \times 12 \times N \times K$, where f = feed per revolution or per stroke in inches, t = depth of cut in inches, C = cutting speed in feet per minute, N = number of cutting tools on machine, and K = constant, depending upon the material to be machined, such as .35 to .50 for cast iron, .45 to .70 for soft steel, .7 to 1.0 for medium hard steels, 1.0 to 1.10 for hard steels.

PRACTICE PROBLEMS APPLICABLE IN THE SHOP AND TOOLROOM

1. A square steel rod supports a steady load of 7,500 lb. in tension. Calculate the side of the square cross section of the rod, assuming the ultimate tensile strength of the material is 65,000 p.s.i., and a factor of safety of 5.
2. A cylindrical aluminum alloy rod, $3\frac{1}{4}''$ diameter, is subjected to a tensile load of 41,500 lb. Find the factor of safety, if the ultimate tensile strength of the material is 40,000 p.s.i.
3. A cast-iron column of rectangular cross section is subjected to a live compressive load of 53 tons. Assuming that the ultimate compressive strength of the material is 80,000 p.s.i., and that the factor of safety is 16, what are the dimensions of the cross section of the column if the width is twice the thickness?
4. Assume a riveted joint, as shown in Fig. 12-20. The rivet is $\frac{5''}{8}$ diameter and of low-carbon steel. What is the safe load to which the rivet can be subjected, if its ultimate shearing strength is 42,000 p.s.i., and the recommended factor of safety is 6?

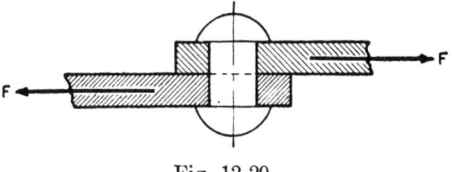

Fig. 12-20.

5. A brass tube whose outside diameter is $1\frac{3}{4}''$ and wall thickness is $\frac{3''}{16}$ is subjected to a tensile load of 4,500 lb. in the direction of the tube axis. Find the induced stress in the tube material.
6. If the brass tube in Problem 5 is 30'' long and is stretched .0128'' when subjected to the 4,500 lb. load, what is the modulus of elasticity of the material?
7. In the knuckle joint shown in Fig. 12-21, the force (F), which is equal to 5,500 lb. and acts along the axis, tends to elongate the yoke and the rod. It also tends to shear the clevis pin at ab and cd. Assume that the safe load (F) is computed from the ultimate tensile strength of the material, which is 65,000 p.s.i., and that the factor of safety is 8. Compute the diameter D_1 of the rod. Also compute the diameter D_2 of the clevis pin, if the ultimate shearing strength of the material is 42,000 p.s.i., and the factor of safety is 8.

8. What force P will strip the threads on a $\frac{3}{4}''$-16 N.F. bolt, considering that the ultimate shearing strength of the material of the bolt is 42,000 p.s.i.? See Fig. 12-22.

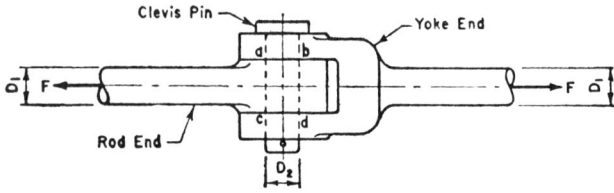

Fig. 12-21.

9. What force can be safely applied in tightening the nut in Fig. 12-22 if we assume a factor of safety equal to 6?

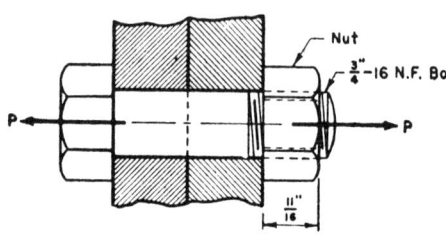

Fig. 12-22.

10. What maximum force is necessary to punch 7 holes of $\frac{7}{16}''$ diameter through a steel plate $\frac{13}{32}''$ thick, assuming that the ultimate shearing strength of the material is 32,000 p.s.i.?

11. Calculate the wall thickness of a pipe having an inside diameter of 5 inches for the purpose of providing safe resistance to an inside pressure of 175 p.s.i. Assume that the safe tensile strength of the material of the pipe is 6,250 p.s.i.

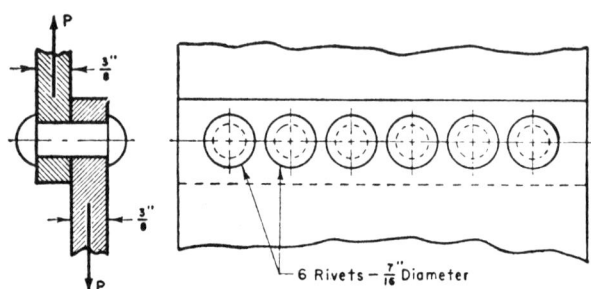

Fig. 12-23.

12. What is the strength of the riveted joint in Fig. 12-23? Assume that the safe shearing strength of each rivet is 11,000 p.s.i., and the safe crushing strength is 17,000 p.s.i.

13. Calculate the amount of work done in foot-pounds when a weight of 225 pounds is pushed a distance of 5 ft. 9 in.
14. Calculate the power in foot-pounds per minute in Problem 13 if it took $3\frac{1}{4}$ minutes to accomplish the task.
15. Calculate the power in inch-pounds per second in Problem 14.
16. Calculate the horsepower of a machine that does 82,400 ft.-lb. of work in $2\frac{1}{4}$ minutes.
17. Compute the resultant force in Fig. 12-24, if the directions and magnitudes of the known forces are as indicated in the drawing.

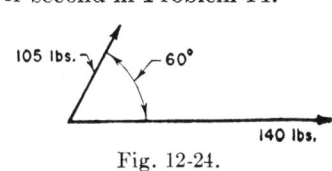

Fig. 12-24.

18. Compute the force (F) in Fig. 12-8 if $W = 165$ lb., $a = 6.5''$, and $b = 11''$.
19. Compute the weight (W) in Fig. 12-9 if $F = 325$ lb., $a = 7''$, and $b = 2'\text{-}3''$.
20. Compute the distance b in Fig. 12-9 if $W = 242$ lb., $F = 93$ lb., and $a = 5''$.
21. Compute the distance a in Fig. 12-10 if $W = 510$ lb., $F = 774$ lb., and $b = 8''$.
22. Calculate the weight (W), which is attached to the rope wound on the drum, as shown in Fig. 12-11, assuming that the force F is equal to 735 lb. and the distances a and b are $4\frac{1}{4}''$ and $7\frac{5}{8}''$, respectively.
23. Calculate the force (F) in Fig. 12-11 if the weight $(W) = 1{,}200$ lb., $a = 5\frac{1}{8}''$ and $b = 8\frac{7}{16}''$.
24. Calculate the force (F) required to raise the weight (W) in Fig. 12-25.
25. Calculate the distance through which the force (F) is acting downward in order to raise the weight (W) a distance of $3''$ in Fig. 12-25.

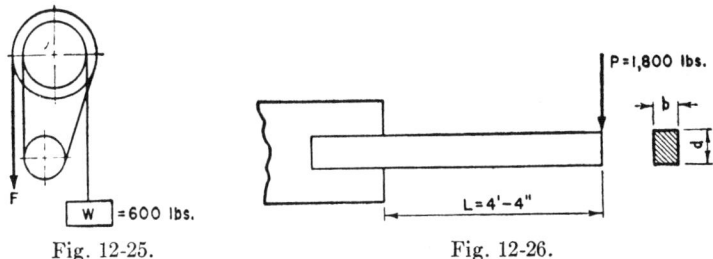

Fig. 12-25.   Fig. 12-26.

26. Calculate the bending moment (M) of the beam in Fig. 12-26, if the load (P), acting downward, is equal to 1,800 lb., and the distance $(L) = 4$ ft. 4 in.
27. Assuming that the unit stress in the farthest fiber of the beam in Fig. 12-26 is 12,000 p.s.i., calculate the section modulus of the beam.
28. Assume that the cross section of the beam in Fig. 12-26 is rectangular, and that $b = \dfrac{d}{2}$. Calculate the cross section of the beam in

Problem 27. Note that the section modulus of a beam of rectangular cross section, as shown in the drawing, is $\dfrac{bd^2}{6}$.

29. Assuming that the beam in Fig. 12-26 is of cylindrical cross section, calculate its diameter on the basis of $P = 1,800$ lb. and $L = 4$ ft. 4 in. Note that the section modulus of a beam of cylindrical cross section is $\dfrac{\pi d^3}{32}$, where $d =$ diameter of beam in inches.

30. Assume that the cross section of the beam in Problem 29 is a hollow circle whose outside diameter (d_o) is twice that of the inside diameter (d_i). Calculate the outside and inside diameters of the cross section of the beam. Note that the section modulus of a hollow circle is $\dfrac{\pi(d_o^4 - d_i^4)}{32 d_o}$.

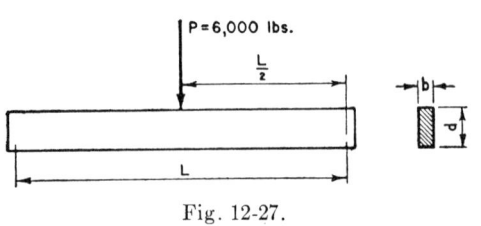

Fig. 12-27.

31. Calculate the maximum deflection in the steel beam loaded as shown in Fig. 12-27. Assume that $L = 3$ ft. 7 in.; modulus of elasticity of the beam material $= 30,000,000$ p.s.i.; moment of inertia of the beam cross section about the center of gravity $= \dfrac{bd^3}{12}$, where $b = 2''$ and $d = 5''$.

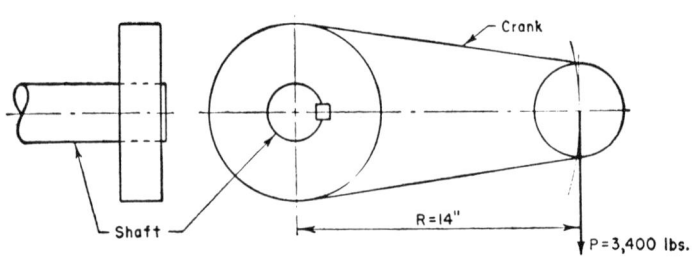

Fig. 12-28.

32. Calculate the torque transmitted by the shaft in Fig. 12-28, if the load (P) imposed on the end of the crank $= 3,400$ pounds and the radius arm $(R) = 14''$.

33. Calculate, in Fig. 12-28, the shaft diameter (d) that will safely transmit the torque computed in Problem 32 if the safe maximum unit shearing stress of the shaft material is 8,000 lb. Note that the polar section modulus of the shaft is $\dfrac{\pi d^3}{16}$.

ENGINEERING COMPUTATIONS

34. Assume the force P is applied continuously at the end of the crank in Fig. 12-28, rotating at 450 r.p.m. Calculate the developed horsepower.
35. Calculate the horsepower transmitted by a shaft rotating at 210 r.p.m. and transmitting a torque of 26,500 in. lb.
36. Compute the indicated horsepower of a double-acting stress engine if the piston diameter = 10″, the stroke is 8″ long, the mean effective pressure is 60 p.s.i., and the r.p.m. = 300.
37. Compute the horsepower of a 4-cycle gas engine if the diameter of the cylinder = 4″ and the number of cylinders in the engine = 6. Assume that the piston speed does not exceed 1,000 ft. per min., and that the mean effective pressure is 90 p.s.i.
38. Compute the horsepower of a generator if the current is 60 amperes and the voltage is 660.
39. Compute the actual horsepower delivered by an engine at the flywheel, if the reading of the net pressure on the scale of the Prony brake is 250 lb., the length of the lever arm is 3 ft. 9 in., and the r.p.m. is 300.
40. Compute the potential energy of a body weighing 230 lb. suspended at a height of 22 ft.
41. Compute the kinetic energy of a body weighing 75 lb. moving with a speed of 75 ft. per second.
42. Compute the kinetic energy of a body weighing 12 lb. and rotating about a center 9″ distant, assuming that the body rotates at 200 r.p.m.

Note: The given distance of 9″ is assumed to be between the center of rotation and the center of gravity of the material.

43. What is the efficiency of a motor receiving 25 h.p. and delivering 22 h.p. to other machines?
44. If a motor delivers 18 h.p. to a planer in a machine shop, and its efficiency is 80 per cent, what is the power received by the motor?

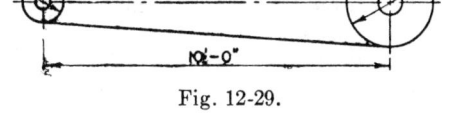

Fig. 12-29.

45. How much current does a motor operating on 220 volts take if its capacity is 10 h.p. and its efficiency is 80 per cent?

46. What is the length of the belt in Fig. 12-29?
47. What is the length of the belt in Fig. 12-30?
48. What is the speed of the belt in Problem 46, if the large pulley rotates at 160 r.p.m.?

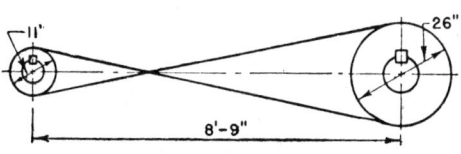

Fig. 12-30.

49. What is the horsepower transmitted by the belt in Problem 48, if the width of the belt is 4″ and the difference in tension between the tight and slack sides of the belt (i.e., the effective belt pull per inch width of belt) is 30 lb.?

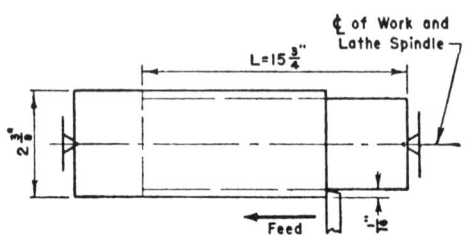

Fig. 12-31.

50. What is the width of belt required to transmit 10 h.p., if one of the pulleys is 12″ in diameter and it rotates at 210 r.p.m.? The effective belt pull per inch width of belt, which is the difference in tension between the tight and slack sides of the belt, is 40 lb.

51. What is the safe speed of the lathe spindle in Fig. 12-31 if a cylindrical rod of 2⅜″ diameter is turned to a smaller diameter, as shown in the drawing? Assume that a recommended cutting speed in feet per minute = 85 and that the feed per revolution is .008″.

52. What is the machine time required for turning the part in Fig. 12-31 and Problem 51? The length of cut = 15¾″.

53. Assume that the speed of the spindle in Fig. 12-31 is 160 r.p.m. Compute the cutting speed in feet per minute.

54. Assume that the time required to complete the machine operation in Fig. 12-31 is 16 minutes and the spindle rotates at 200 r.p.m. Compute the speed per revolution.

55. Compute the machine time to complete the operation of shaping the top of a casting of rectangular shape, as shown in Fig. 12-32. Assume that the stroke on the shaper is along the long side of the casting and that it begins ¼″ before reaching the casting and ends ¼″ past the length of it. The feed is along the shorter side of the casting, and its value is .040″ per stroke. The recommended cutting speed is 90 feet per minute, and the return stroke takes 60 per cent of the time required for the forward cutting stroke.

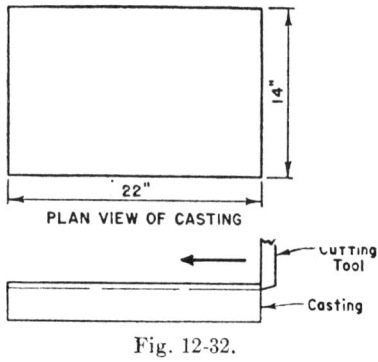

Fig. 12-32.

56. Compute the horsepower developed at the point at which the cutting tool engages the work, assuming that the force required to cut the metal is 1,100 lb. and that the recommended cutting speed in feet per minute is 115.

ENGINEERING COMPUTATIONS

57. Compute the power of the motor driving the machine tool in Problem 56, assuming that the efficiency of the motor is 70 per cent.

58. Compute the horsepower required to cut a cubic inch of metal per minute in a machining operation if the force required to cut the metal is 1,350 lb., the cutting speed in feet per minute is 95, the feed is .015", and the depth of cut is .062".

59. Compute the maximum pressure required to punch 6 holes of $\frac{9}{16}$" diameter through a steel plate on a press similar to that shown in Fig. 12-19. The thickness of the plate is $\frac{3}{16}$" and the ultimate shearing strength of the material is 42,000 p.s.i.

60. Compute the average pressure required for the operation in Problem 59, if the per cent of penetration is 50.

61. Compute the energy required to produce the operation in Problem 60.

62. Compute the motor horsepower in Problem 61, if the number of strokes on the press is 20 per minute.

63. Compute the pressure required for drawing a sheet metal panel if the perimeter of the panel outline is $16\frac{5}{8}"$, the thickness of metal is .062", and the ultimate tensile strength of the material is 65,000 p.s.i.

64. Compute the pressure required for drawing a steel cylindrical shell whose outside diameter $= 1\frac{1}{8}"$ and thickness of metal $= .040"$. The ultimate tensile strength of the material $= 65,000$ p.s.i.

65. Calculate the horsepower required for punching 40 holes per minute if the diameter of each hole $= \frac{5}{8}"$, and the thickness of metal $= .030"$. Use the formula taken from Pomeroy.

66. Calculate the horsepower required for machining operations according to Pomeroy. Assume the feed $= .010"$, the depth of cut $= .050"$, the recommended cutting speed in feet per minute $= 140$, the number of cutting tools $= 7$, and the constant depending upon the material to be machined $= .50$.

Chapter XIII

USE OF GRAPHS IN SOLUTION OF ENGINEERING PROBLEMS

Engineers, accountants, and executives are frequently called on to present mathematical or numerical information to persons who are unskilled in reading or understanding figures. One of the best ways to show such information is by means of graphs or charts. The idea or conclusion that the graph shows is readily understood although the actual figures are not. The terms graph and chart are often used synonomously although some people make a distinction between them using graph to mean the presentation of numbers or figures, while chart means the presentation of other types of facts.

There are many ways of showing information in graph form. The commonest are by use of lines, bars, areas, pictures, flows and nomographs. There are several variations of each.

LINE GRAPHS

The simplest is the line graph which consists of a line connecting points plotted on graph paper. This is paper that is ruled in two direc-

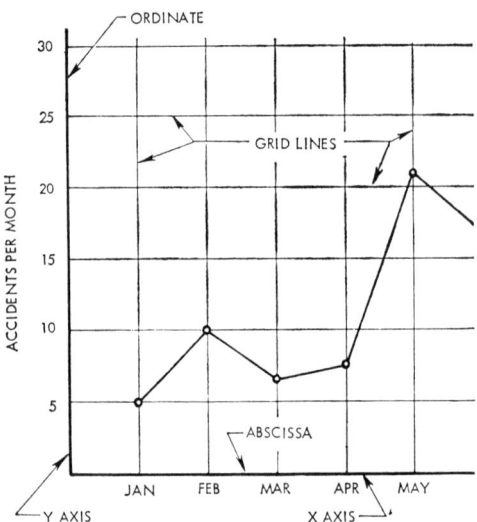

Fig. 13-1. Line graph showing terms used.

GRAPHS IN ENGINEERING

tions so that one set of lines can be numbered or named in a vertical direction and the other in a horizontal direction. The vertical line at the extreme left of the graph is called the Y-axis or *ordinate* and is marked with one set of figures or facts while the line at the bottom of the graph running horizontally is termed the X-axis or *abscissa*. This line contains the second set of numbers or facts. See Fig. 13–1.

In this figure the number of accidents in a plant is recorded on the vertical lines while the months are recorded on the horizontal lines. The number of accidents for each month is marked as a point where the two lines cross. These points are then joined to give a line or curve.

A variation is shown in Fig. 13–2 where two sets of figures are plotted on the same graph. Note that one is a broken line which is used to prevent confusion. Colored lines, dashes, dots, and combinations can be used to show various kinds of information. Generally only related facts are used on the same graph.

BAR GRAPHS

This is a variation where the information is plotted as in the line graph but a solid bar is drawn from one line to the proper place on the other. The bars may be drawn either from the abscissa or from the ordinate. In Fig. 13–3 team standings are shown in relation to their percentages of wins. This graph can be used to show several sets of information if bars with different markings are used.

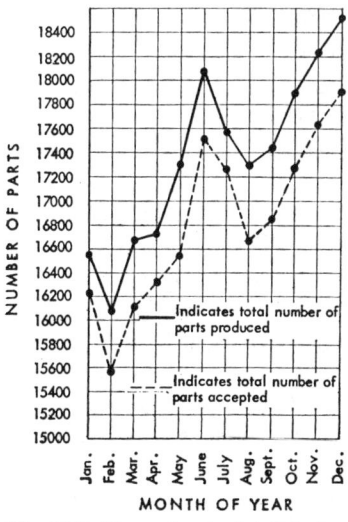

Fig. 13-2. Two sets of related facts.

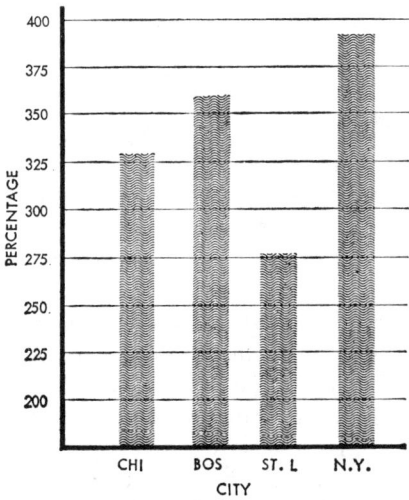

Fig. 13-3. A bar graph.

AREA GRAPHS

These may be made as bars, circles or pies, or rectangles and are most often used to show percentages. The area of the bar or pie is figured as being 100%. The various percentages to be shown are calculated as percentages of the total length of the bar or the circumference of the circle. In Fig. 13-4 it can be seen that Asia has half the world population in the bar at the right side, while it has only a fifth of the land area in the other bar.

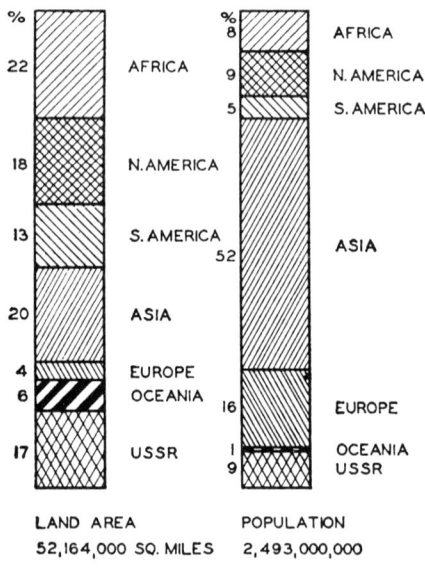

Fig. 13-4. Percentage chart.

In Fig. 13-5 a pie shape is used to express the relationship between the numbers of workers in several categories. The pie shows that half the department store employees are salespeople, while a fourth are engaged in stock work, delivery, and store maintenance. The remaining fourth is almost equally divided between executives such as managers, buyers, and adjusters, and office personnel.

The pie shape is used quite often to show dollar percentages with the circle being drawn to give a resemblance to a silver dollar. Note the use of rules, shadings and stippling to differentiate between the various segments of the bar and the pie. To convert percentages to the degrees in a circle multiply by 3.6. Lay these off with a protractor.

HALF OF ALL DEPARTMENT STORE EMPLOYEES ARE SALESPEOPLE

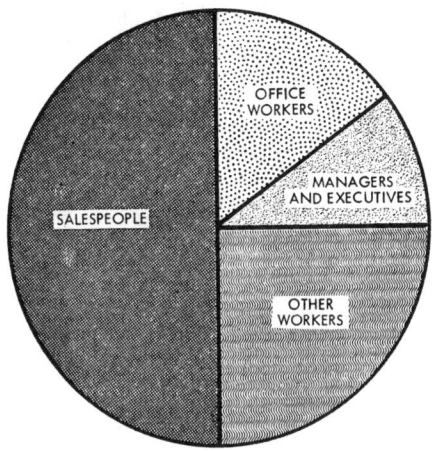

Fig. 13-5. Pie graph is usually based on percentages.

PICTORIAL CHARTS

Pictures can sometimes be used to present figures graphically. This has been done in Fig. 13-6 where the size of the people in the front has been increased one-fifth to show the growth of population and jobs by 20% each. The value of the national product has doubled in size in the same period. The relative sizes bring out the desired information.

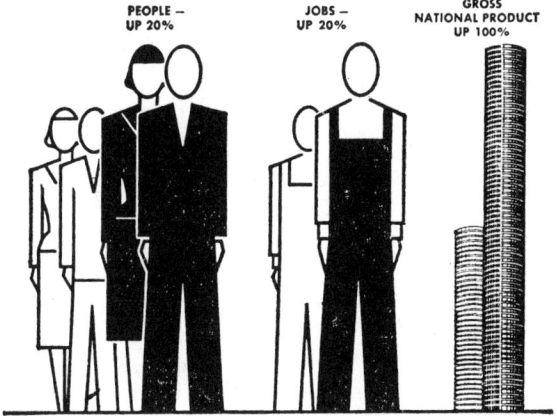

Fig. 13-6. Pictorial chart uses drawings.

FLOW CHARTS

This chart may be used to show the flow of material through a plant or process or to show the flow of authority or duties. In the case of materials flow small sketches of trucks, buildings, machinery, and products may be used to give a better understanding of the process. In showing lines of authority or duties rectangular boxes or circles are used to show how one is related to the others. Fig. 13-7 is a flow chart.

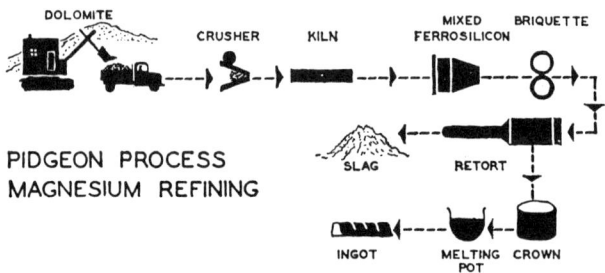

Fig. 13-7. A flow chart may use pictures or boxes.

NOMOGRAPHS

A nomograph is a calculating graph. It is used where a number of problems of a similar nature are to be worked. The nomograph in its simplest form consists of three parallel scales that are an equal distance apart. The two outside scales are the same while the middle one is made with the divisions just half of the other two. Most scales are constructed on logarithmic values. See Fig. 13-8.

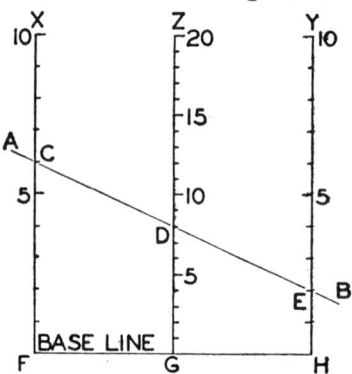

Fig. 13-8. Parallel scale nomograph.

The nomograph is used by connecting the left or X scale with the right or Y scale and reading the answer on the center or Z scale. The nomograph utilizes the same basic elements of the slide rule except that the scales are manipulated geometrically instead of physically. Some nomographs used in engineering may have several scales on them cali-

GRAPHS IN ENGINEERING

brated for the variable factors that enter into a problem. The major difficulty with nomographs is that the scales must be accurately prepared. This may be a time-consuming task and unless the nomograph is to be used for solving many equations using the same scales the time and trouble may not be justified, as other methods, although taking longer for each problem, may effect greater total savings in time.

PRACTICE PROBLEMS APPLICABLE IN THE SHOP AND TOOLROOM

1. Plot a graph showing the relationship between the resistance of the wire and its diameter. Use cross-section paper and spread out the figures to give as large a curve as possible. Plot the diameters on the X-*axis* and the resistance on the Y-*axis*.

Dia. of Wire in Mills	Resistance in Ohms per 1,000 ft. of Wire
50	4.0
75	1.8
100	1.0
150	0.50
200	0.30
250	0.15

2. Use the following figures showing the number of rejected pieces for two weeks to construct a bar graph. You may run the bars horizontally or vertically. Make one week open bars and the other week shaded.

Day of Week	No. of Rejects First Week	No. of Rejects Second Week
Monday	11	9
Tuesday	14	12
Wednesday	10	8
Thursday	12	10
Friday	15	11
Saturday	13	7

3. Construct a flow chart of the organization of your school or some plant that you are familiar with. (Note: Each school has a slightly different organization than any other.)

4. Construct a pie chart showing percentages of students enrolled in different courses in your school. (Note: You can use some other statistics if they are available.)

5. Construct a curve or line graph showing number of students entering your school and the ones that graduate. Are the number that drop out increasing or decreasing?

6. Construct a pictorial graph showing flow or percentages.

Chapter XIV

COMPUTATION OF AUTOMATIC CONTROLS FOR AUTOMATION

Any deviation from prescribed conditions must be detected and accurately measured in order to automatically control a planned production process, such as an operation on a milling machine. After detecting and measuring the error, which can arise from the product specifications or from the production process, a control mechanism should be brought into action to induce corrections. This control mechanism is usually designed so that corrective action is taken only if the deviation is in excess of the permissible range of error, known in the metal working industries as the *commercial tolerance*.

PRODUCT AND PROCESS CONTROL

The ideal or desired specifications for a product or a process are known as *set points*. An automatic control mechanism must be able to detect and measure any deviation from a set point, and to perform the corrective action necessary to restore the product or process to required conditions. Deviations from a set point may occur not only in the fabricating processes, but also in the materials handling processes. Improper delivery, erratic loading and unloading, and erratic transfer of work pieces may occur. In a fully automated productive system, any such deviations are detected and corrected by automatic devices.

In certain situations a control device may consist of a simple switch, such as the common switch for an electrical appliance, except that it would be actuated automatically by responding to certain conditions. A common example of such an automatic switch is the thermostat which turns a furnace on or off, depending on the difference between the *desired temperature* (the set point) and the *actual temperature* measured by the thermostat.

In controlling the accuracy and sequence of machining operations, more complex devices than a simple switch are often required. For example, the maintaining of the proper position of a cutting tool in a machining operation is important, because even very small errors may cause deviations exceeding the commercial tolerance. An automatic control device for such an operation would most probably require a high order of sensitivity for detecting and measuring deviations, and high speed of response for corrective action. Such control systems require more complicated mechanisms, and may employ hydraulic, pneumatic, or electronic systems, either singly or in combination.

The use of hydraulic, pneumatic, and electronic equipment in designing automated systems permits a wide variety of control devices for specific conditions and products. Automated systems may be designed either to permit the automatic transfer of workpieces from tool to tool or from machine to machine. On the other hand, the workpieces may be kept stationary while cutting tools are automatically brought to the piece, positioned, and controlled throughout the process. The means employed depends, of course, on many factors, the chief one being the nature of the product and the process, and the economic problems involved in designing and setting up automatic equipment.

Regardless of the system employed, it should be remembered that *an automatic, self regulating system must be based on closed loop control employing feedback.* In this system two characteristics are outstanding: one being that there is a continuous connection or sequence of events between the input and output ends; and second, there is a means of detecting and measuring the output and transmitting this information back to the input. More specifically, the system must incorporate the means by which any deviations in the system from desired conditions can be detected, measured, reported, and corrected.

Open Loop Control. Many automatic devices which are not self-regulating employ *open loop control.* The elements in this system are not connected in a continuous closed sequence, but in *a linear sequence which is open at both ends.*

A simple example of an open loop system would be a heating system in which a heat-sensitive control unit is exposed to the open air, so that the outside temperature regulates the operation of the furnace, and hence the inside temperature of the room. Fig. 14-1 illustrates the basic elements of such a system. Since the amount of heat entering the room is entirely influenced by the outside temperature, the system is not directly affected by the inside temperature of the room. In other words, the loop is open. In spite of the fact that the regulation of the furnace is automatic, there is no feedback from the output end of the system to the input end.

A relatively simple example of open loop control in industry would be the use of a cam to control the movement of a machining tool. In this type of automatic control, the cam is mounted on the machine. As the machine goes through its cycle, a cam follower transmits the "informa-

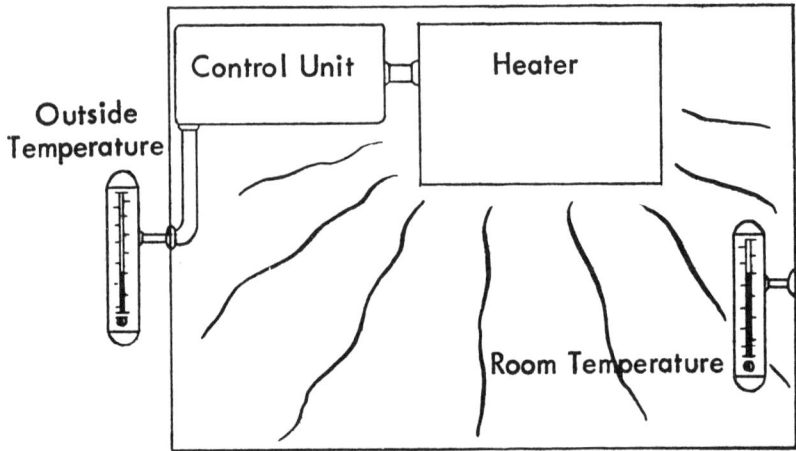

Fig. 14-1. A room heating system is an example of a basic open loop control.

tion" contained in the shape of the cam through a system of linkages to the machining tool. The motion of the machining tool is thus predetermined by the shape of the cam. In this system, there is no means of knowing, during the cycle, whether there are any errors in the machining process, since there is no feedback of information from the workpiece to the controlling end of the machine.

Open loop control has been widely used in industry and will continue to have wide usage in situations where full automation is not feasible or not justified economically. But its basic difference from closed loop control should be understood. It should also be remembered that fully automated systems, to be truly self-regulating, must depend on the use of closed loop control and feedback.

CONTROL DEVICES

In selecting and applying devices for use in automatic control, the precision and care of their manufacture must be given careful consideration. The success of an automated system depends greatly on the reliability of these control devices, along with other components, such as valves, power units, amplifiers, etc.

In many situations a complex array of detecting, measuring, and controlling devices must be coordinated and organized into one self-regulating system—a system necessarily dependent on precise, effective, and diversified instrumentation. Instrumentation, therefore, is the heart of automation, and its continued development has deep implications for the evolution of industry.

There are two groups of controls which are used to accomplish automation through instrumentation. *Mechanical control devices* comprise the first group, which has the primary purpose of actuating tools, valves, material handlers, etc. in a production cycle. This group, known as *effector mechanisms*, is based on hydraulic, pneumatic, and electronic principles, employed singly or in combination.

The other group of automation devices consists of the *data processing or information handling devices*. Basically they include the *digital and analogue computers*, and the various devices for storing and recording information, such as the punched card and tape, and the magnetic wire or tape. From the point of view of automation, the common function of all the devices in this group is to handle information about products or production processes to make production more automatic, or self-regulating. For instance, data processing or information handling devices are used to complete feedback circuits, to rapidly solve production problems involving complex mathematical operations, and to store information relevant to production. In complex automated systems, data processing devices are extremely useful for achieving feedback; that is, for detecting, measuring, interpreting output conditions, and determining what corrective actions, if any, should be put into effect at the input end of a system. These corrective actions are finally carried out at the output end by the automatic effector mechanisms.

It should be clear, then, that data processing and automatic effector mechanisms are closely dependent on each other in a fully automated production system.

ELECTRONIC COMPUTERS

The electronic computer was developed to meet needs in the military, engineering, and scientific fields. Essentially, a computer is a machine which performs a series of individual calculations as steps in the solution of a given problem which is fed or *programed* into it.

Analogue and Digital Computers. The two main types of automatic computers are the *analogue computer* and the *digital computer*, both valuable for theoretical as well as practical applications. The analogue computer is so called because it solves problems essentially by setting up physical analogues or "models" of the problem to be solved. An example is provided by the ordinary slide rule in which the values of the problem, two numbers, are set up as physical analogues; that is, as distances along the scales. The slide rule, in other words, is a relatively simple and non-automatic analogue computer. In the electronic analogue computer, the physical analogues of the problem are set up in the computer, usually in terms of voltages. This process of setting up a problem in a computer of any kind is known as *programing*.

The digital computer differs in that it solves problems by directly handling numbers, or digits. The machine processes these numbers according to the rules of arithmetic or logic, with the answer being expressed as a numerical value.

Comparison of Analogue and Digital Computers. One of the chief advantages of the analogue computer is that it continuously and instantaneously produces solutions to the problems fed into it. This characteristic makes it valuable for control systems requiring immediate response to changing conditions. On the other hand, the digital machine, once programed, has to go through the required steps for the solution of the problem, thus creating a time lag between the presenting of the problem and its solution. With continued development, the speed of the digital computer should increase.

A second important difference is that the analogue computer, because of its principle of operation, need only be as complicated as the analogue or model of the problem it deals with. Hence, for many purposes, fairly simple analogue computers are adequate. On the other hand, digital computers are inherently complicated machines. Where the problems requiring solution are all fairly simple, it would not be feasible to use an extremely complicated and expensive digital computer.

Conversely, the digital computer has a decided advantage in solving complicated problems which can only be processed by an analogue computer of extraordinary size and complexity. Beyond a certain point of complexity, an analogue computer would be too inefficient to be reliable.

The final important difference is that for situations requiring extremely accurate solutions, the digital computer is superior to the analogue computer. This is the result of the analogue computer depending for its accuracy on the accuracy of the physical analogue or model. In other words, the answer is accurate only to the extent that the voltages, or other physical analogues, accurately represent the programed data. Furthermore, electronic analogue computers are subject to errors inherent in electronic circuits, namely, electronic "noise." On the other hand, digital computers have a greater accuracy because they deal directly with numerical values of the problem rather than analogues or models of the problem. As automation progresses, and automatic production systems become more complicated, there will probably be an increasing dependence on the digital computer as the "brains" of an automated system.

One way in which the analogue computer is very useful in industry is in instantaneously determining the values or responses of specific factors in manufacturing processes under different operating conditions. This makes it particularly suitable for production systems involving continuous processes with different factors, and for subjecting the model of the process to conditions not possible or feasible in actuality. It is also very useful in control situations that are not too complex, and which do not require a very high degree of accuracy. Changes or adjustments in the control factors are immediately accounted for and interpreted.

The future of automation, however, will probably be more dependent on the digital computer because of its greater flexibility. The analogue computer is designed so that it solves all problems by going through the same process. This is somewhat as if a human mathematician could solve differential equations, but no other kind of problem. The digital machine, on the other hand, can be designed to perform any number of logical operations; in other words, it can solve a much greater variety of problems than the analogue computer. It can be built with certain standards or criteria with which it can compare its own performance and improve on that performance. It can be incorporated with "memory" storage units—devices such as punched tape or cards, or magnetic tape—in which it can retain its "experience" in solving specific problems. Furthermore, it can coordinate the functions of individual analogue computers in a complex industrial process.

Basic Elements of Digital Computers. Three basic elements are incorporated in all digital computers despite their design. These are:

1. The *computing element*, which performs the arithmetic operations, makes logical choices between given alternatives, and compares different sets of data, according to the instructions contained in its program.

2. The *control element*, which contains the program and schedules the sequence of operations which the computing element must go through to arrive at the solutions.

3. The *memory unit*, which stores information relevant to the operation of the machine in the form of magnetic tape, punched cards, electronic tubes, magnetic drums, or other suitable mediums from which the computer can "read" instructions.

Due to its electronic principle of operation, the digital computer avoids most of the mechanical wear of non-electronic computers. This results in a longer operating life and a minimum of down time for repair and maintenance. The rapid development of miniature components, such as transistors, have considerably reduced size, power, and ventilating requirements.

Secondary Elements of Digital Computers. In addition to the three basic elements, there are two secondary elements necessary for the operation of the digital computer. These are:

1. The *input element*, which provides the computer with the information which it will process. This can be any appropriate device which can introduce pertinent information into the computer in a logical sequence determined by the problem. The computer is a device for "reading" punched cards or magnetic tape, or even a typewriter.

2. The *output element*, which receives and records the solution to a problem. This output element may consist of a device for punching the solution out on a card, or a panel of lights which would indicate the numerical value of the solution.

A significant trend is in evidence towards the designing and production of comparatively simple and low-cost electronic digital computers. This promises the eventual use of digital computers in business and industry by even moderate-sized manufacturers and businesses. Fig. 14-2 shows the desk-control unit of a low-cost, medium-speed, high-capacity digital computer, capable of great flexibility and easy programming.

A feature of this computer is its ability to pick up two sets of instruc-

Logistics Research Inc.

Fig. 14-2. Control units of a low cost digital computer system.

tions simultaneously and carry them out successively without a second reference to the memory unit, and without requiring a resetting of the instructions for repetitive sequences of numbers. The same register acts as an automatic tally for repetitive sequences of operations in the computer. The machine operates with magnetic tape, punched cards, and an automatic typewriter.

Uses of Digital Computers. The extensive development of automatic digital computers with large information-handling capacities has made them suitable instruments for solving complex problems affecting production, inventory, and numerous other industrial operations. Because of their high reading and writing speeds, they can be employed to describe manufacturing processes, record production, publish schedules, and to maintain piecework rate tables and manufacturing specifications. Computers are in operation that can read and write at better than a million characters a minute and can scan and publish enough information to adequately describe even the largest production or supply system.

Digital computers can also perform arithmetic operations at high speeds. Depending on their design, these data processing systems can perform between 250,000 and 1,000,000 additions or subtractions per minute. Multiplications can be performed on these machines from 50,000 to 240,000 per minute. Logical operations can be carried out from 500,000 to 1,000,000 times per minute. Each computer incorporates in its design the logical requirements and procedures that make up its primary set of operations. Thus, in one large centralized inventory control system, it may be necessary to specify as many as 60,000 separate instructions. Some of these will be used for each accounting entry describing a particular commodity or transaction.

Coordinated Data Processing Systems. The principles of high-speed computer design have been applied to various devices for coordinating data processing systems. One of these, called the *Datamatic 100*, speeds up all phases of records-keeping and accounting. This data processing system, shown as a scale-model in Fig. 14-3, consists of several coordinated individual types of computers.

Although suitable for use in industrial processes, this system was designed primarily for office automation, and handles a wide range of clerical operations such as billing, accounting, sales analysis, inventory, and production control. It also provides management with continuous compiling and processing of reports upon which business and administrative decisions are based.

Data is fed in the form of punched cards into the system's input

Datamatic Corporation

Fig. 14-3. A scale model of a data processing system for business.

converter, a device which translates, edits, and transcribes the data on 3-inch wide magnetic tapes at the rate of 900 cards per minute. A 2,700-foot reel of tape can store 37,200,000 digits of information, or the equivalent of information contained on 465,000 cards. The control unit of this system, called the *brain*, can read and write at the rate of 60,000 digits per second, simultaneously handling 1,000 multiplications, 4,000 additions, or 5,000 comparisons.

One magnetic tape file unit of this system can read, write, or file as much information as some of the older types can store in their entire system. For a company doing a large volume of clerical work, a system like this would well justify its cost. The building-block principle, which it employs, permits the incorporation of as many as 100 tape units, any of which can be referred to or questioned without disturbing the rest of the system. Final reports can be turned out at the rate of 6,000 punched cards per hour or 900 printed lines of reports per minute.

In an experimental project, the National Bureau of Standards, in cooperation with the Navy Bureau of Supplies and Accounts, combined two digital computers into a coordinated data processing system to study the application of electronics to the problems of supply and accounts management. After experimentation, it was revealed that two digital computers could coordinate with each other without having identical characteristics, assuming each of the computers has sufficient control flexibility. This experiment showed that a coordinated digital computer system for data processing could replace many small specialized machines.

USE OF TAPE AND PUNCHED CARD CONTROLS

Punched tape or numerically controlled machine tools play an important role in machining and metal forming processes. Magnetic contactless controls for relaying, amplifying, and switching electrical signals are welcomed by builders as well as users of machine tools since they have no working parts and require no complicated circuits. Corrective gaging during and after the production cycle, and efficient work handling systems further reduce production unit costs while maintaining quality.

Numerical control of machine tools is not restricted necessarily to machines producing large volumes of identical parts. The principle can be applied to machine tools, such as turret lathes, for limited volume production such as 50 to 100 identical pieces. Since a large proportion of the metal working industries (about 70%) is based on limited volume production, numerical and punched tape controls are being widely adopted. Furthermore, the vast majority of machining operations consist of turning out pieces with simple geometric forms, thus simplifying the design and arrangement of controls.

Numerical controls are also applicable to other types of production lathes and milling machines. In the case of the milling machine, the International Business Corporation developed a digital control system for milling master cams. This system produces cams quickly, accurately, and automatically from analytical data. Once the workpiece is loaded in the fixture and the punched tape is threaded in the reader unit, the milling operations can be completed without any attention from the machine operator. The machine not only reduces setup and machining time, but produces more uniform pieces as well. Such installations have operated steadily for several months without repair down time. Maintenance problems should not be difficult after production personnel have been properly coached in the understanding and maintenance of the control mechanisms.

AN AUTOMATIC PRODUCTION SYSTEM

Fig. 14-4 is a simplified schematic diagram of an automated production line consisting of three production stages, each of which is tied in by a feedback circuit with the master computer unit. Information concerning the process stages is received by the control units and fed back to the computer unit, which acts on the information and signals the control units to effect any required changes in input.

In the system shown in this schematic, the inspection and quality control functions are also brought under automatic control. The quality control unit detects and feeds back information concerning the products or process to the computer. Any deviations from the set points are noted

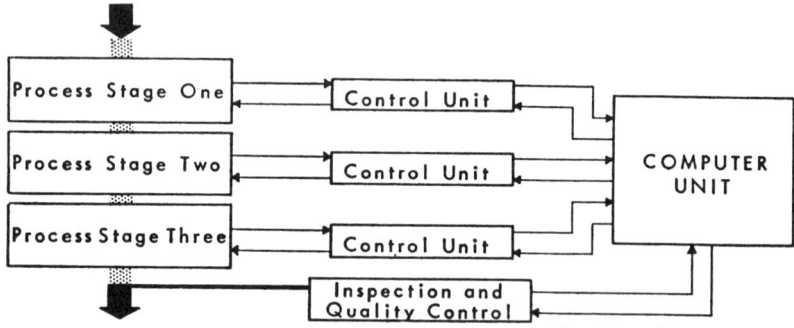

Adapted from Scientific American

Fig. 14-4. Automation in industry is depicted by this schematic of an automated production line.

and measured by the computer, which orders either a re-working of the products to conform to set points, or a change in the process to produce pieces conforming to set points.

Although the schematic shown in Fig. 14-4 is a relatively simple automatic production circuit, it does show the main elements of such a system and the relationships between these elements.

PNEUMATIC CONTROL DEVICES

After the data processing devices have computed and prepared the necessary data for a production process, effector mechanisms actuate the necessary devices to perform that process. These effector mechanisms may operate on any one of several principles; one, which is in common use, being pneumatic.

The pneumatic equipment involved may be in the form of air power cylinders, air motors, control valves, and various safety accessories for the operation of the machine tools. Compressed air, the source of power for various kinds of pneumatic equipment and controls, is suitable for machine tools using the following auxiliary devices and attachments:
1. Fixtures for clamping the workpieces prior to machining
2. Devices for feeding the workpieces during the machining cycle
3. Devices for ejecting the workpieces upon completion of the machining cycle
4. Devices for shifting control levers

AUTOMATIC CONTROLS

5. Devices for tripping valve controls for automatic sequence
6. Devices for driving machine spindles by an air motor
7. Devices for providing an air cushion on a drawing or forming press
8. Devices for moving machine elements for retracting and positioning work, and other similar applications

AIR POWER CYLINDERS

In air cylinders, as in hydraulic cylinders, the aim is to have a unit that will develop the maximum amount of power from a given piston area which is acted upon by a given pressure. This would make the greatest use of the supply of compressed air. Combined with this, the cylinder should be capable of standing up under the required working conditions as long as reasonably possible, with packing replacements for extra work under ordinary conditions. These cylinders are in most cases the non-rotating type and are available either as single- or double-acting cylinders. The requirements of the operation determine the type to be used. If it is desired to have the pressure exerted upon only one side of the piston, single-acting cylinders are chosen. A cut-away view of a non-rotating, single-acting air cylinder is shown in Fig. 14-5. If pressure is to be exerted alternately on both sides of the piston, double-acting cylinders are recommended.

Improved design and the development of better materials have resulted in a wider use of air cylinders. For example, oil-resisting,

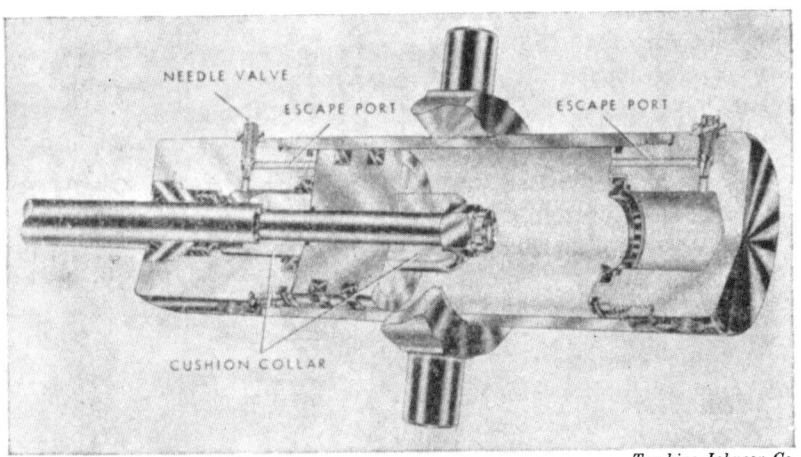

Tomkins-Johnson Co.

Fig. 14-5. A single-acting pneumatic cylinder with cushioning device.

synthetic materials used in piston and piston rod packings have added longer life and have reduced friction.

Air cylinders can be used in automatic operations where machine tools, conveyors, gaging devices, etc., are engaged to process metals and non-metals into finished component parts. Large parts, such as automotive engine blocks, and small parts, such as small gears, can be automatically and continuously processed on machine tools. Air cylinders have considerably reduced the time required for loading, feeding, sizing, gaging, and segregating.

FORCES IN AIR CYLINDER OPERATIONS

In order to determine the forces involved in air cylinder operation a very important factor must not be overlooked. Most available methods for figuring the work that an air cylinder is to do, that is, the required force a cylinder must develop to do the job, are only approximations. Sometimes it is impractical to definitely ascertain the exact pressure that will be required. It is advisable, to cover the unknowns usually present in most applications, that a cylinder be specified which is capable of approximating $1\frac{1}{2}$ times the force which is considered sufficient for the job.

Example: Assume that the air pressure available is 80 p.s.i. (pounds per square inch) and the estimated force to do the job is 150 pounds.

According to the previous recommendation the cylinder size selected should exert $1\frac{1}{2}$ times the estimated force, i.e., $150 \times 1.5 = 225$ pounds.

Thus:

$$1.5F = p\frac{\pi D^2}{4}$$

Where:

F (estimated force) = 150 pounds
p (pressure) = 80 p.s.i.
D (Diameter of piston or cylinder bore) = x

Substituting:

$$1.5 \times 150 = 80\frac{\pi D^2}{4}$$

Where:

$$D^2 = \frac{1.5 \times 150 \times 4}{80\pi} = \frac{22.5}{2\pi} = 3.58$$

$$D = \sqrt{3.58} = 1.9, \text{ or approximately } 2''.$$

AUTOMATIC CONTROLS

Air Operated Boosters. There are numerous cases where certain operations, such as those in metal-cutting, require pressures much higher than are economically obtainable from the compressed air in the shop. In these cases an air-operated device called a booster can be used to provide an intensified hydraulic output pressure that meets the power and control demands of the application, and also eliminates the large hydraulic reservoirs and pumps usually required for hydraulic operation.

The booster is a device which might have, for example, a 5-inch diameter air piston driving a $2\frac{1}{2}$ inch diameter hydraulic ram through a 3-inch stoke to provide a booster ratio of 4 to 1, calculated on the basis of the squares of the respective diameters:

$$(5)^2 : (2\tfrac{1}{2})^2 = 25 : 6.25 = 4 : 1$$

Thus, from 60 p.s.i. shop air input, the booster provides 240 p.s.i. hydraulic output.

Fig. 14-6 shows the application of a booster in precision honing of pinion gears. To properly actuate the feed-out tool in this operation a

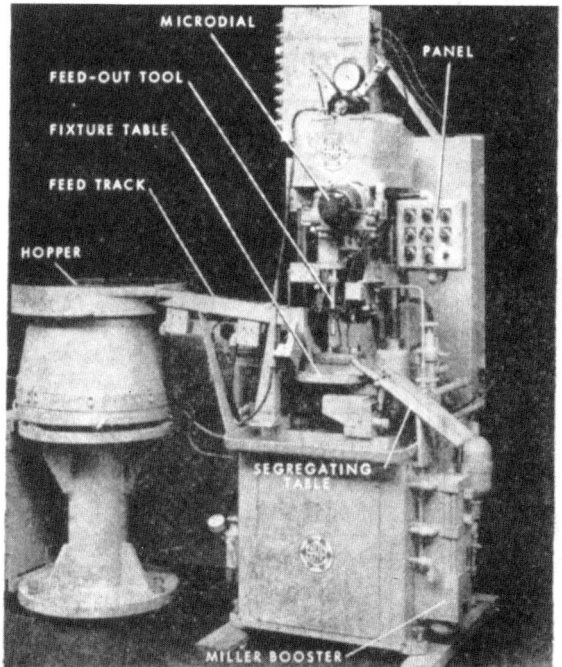

Micromatic Hone Corp.

Fig. 14-6. A pneumatic booster on an automatic honing machine.

pressure much higher than that ordinarily obtainable from shop air would be required. At the same time a smooth feed control is essential for the successful performance of the tool.

PNEUMATIC CONTROL VALVES

For operational purposes, pneumatic valves can be divided into two categories. The first is the *direct operated valves* in which the air supply operates directly on the valve mechanism. Fig. 14-7 is a schematic of such a circuit. The second group of operational valves is the pilot operated valves. This group can be further subdivided into the valves which have an *integral pilot* and the valves which are *remote controlled*.

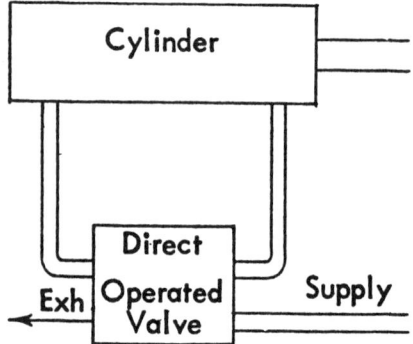

Ross Operating Valve Co.

Fig. 14-7. An example of a direct operated valve installation.

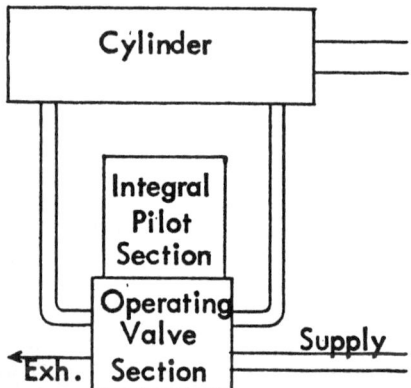

Ross Operating Valve Co.

Fig. 14-8. This direct operated valve installation uses a valve with integral pilot section.

Fig. 14-8 is a schematic diagram of an integral pilot valve. The control of the valve is accomplished by a pilot mechanism which is an integral part of the valve.

HYDRAULIC CONTROL DEVICES

Hydraulic systems utilize another principle to perform as effector mechanisms. Their chief purpose in this respect is to transmit power and to control production processes. However, in this treatment we are primarily concerned with the hydraulic control devices in automated processes where cutting and forming tools, materials handling equipment, etc. have to be controlled according to predetermined cycles and sequences of operations. The treatment of the devices for transmitting power will be restricted to their function as a part of control devices.

Hydraulic equipment and control devices such as pumps, power cylinders and pistons, valves, and pressure boosters are applied in automated processes where high-production machine tools, presses, and materials-handling equipment are extensively employed. They may be employed in machine tools as auxiliary devices and attachments such as work-clamping fixtures, work-feeding mechanisms, mechanisms for ejecting the work, and other automatic installations.

The power-transmitting fluid in hydraulic systems is usually mineral oil under pressure. Either rotary or linear motion may be obtained, depending on special conditions and the devices used. The desired motion may be controlled in accordance with the following conditions and requirements:

1. Direction of the actuating fluid, which depends on the operating requirements of either the particular tool or work-holding fixture, or the mechanism for moving the work through the cycle
2. Velocity or rate of movement
3. Acceleration or deceleration rate
4. Sequence of motions through the processing period
5. Duration of cycle
6. Duration of motions with respect to distance, or time involved in the particular operation
7. Forces involved in the motion, particularly the magnitude of thrust or torque

The use of hydraulic control devices in automation permits a wide range of speeds, flexibility of cycle control, automatic overload protection, constant or variable output, and a wide range of thrust.

Hydraulic systems are assembled by a selection of units which may be simple or complex, depending on the problems in a particular process,

and may consist of a few or a large number of the following components:
1. Pressure pumps
2. Cylinders, pistons, boosters, and motors for converting fluid pressure to mechanical motion
3. Pressure controls, such as the various types of control valves. These may be sequence valves, loading and unloading valves, switch valves, relief valves, shut-off valves, check valves, time-delay valves, or numerous other special valves whose use depends on the nature of the automated operations.

HYDRAULIC PUMPS

Positive displacement pumps are generally used for delivering fluid power to machine tools, and these are manufactured in several classifications, of which the most commonly used are the *constant displacement* and the *variable displacement* pumps. The three principal pump designs are the gear, vane, and piston-type pumps.

Because pump design follows the basic principle that power losses through the operating cycle of any machine must be eliminated, variable displacement (or variable volume) pumps of the vane type are preferred by many users of fluid power for machine tools. In a hydraulic circuit, power consumption is measured by the volume of oil delivered by the pump and the pressure at the pump. Seldom is there an application where maximum volume and pressure from the pumping unit are needed throughout the entire operating cycle. Therefore, with variable volume pumps, there is no excess delivery of pressure oil. The pumps automatically deliver only the amount of oil needed to operate the circuit. Since no excess oil is pumped, relief and by-pass valves with extra piping are unnecessary. Heat losses developed through bypassing of oil are eliminated. Smaller reservoirs, less oil, and lower horsepower motors cut the cost of the initial installation.

The hydraulic horsepower can be computed by the following general power formula:

$$\text{H.P.} = \frac{\text{lb.} \times \text{f.p.m.}}{33000}$$

Where:

lb. = load in pounds
f.p.m. = velocity in feet per minute

AUTOMATIC CONTROLS

The hydraulic horsepower is derived from the above general formula

$$H.P. = \text{pressure (p.s.i.)} \times \text{volume (g.p.m.)} \times .000583$$

Where:

$$\text{p.s.i.} = \text{pounds per square inch}$$
$$\text{g.p.m.} = \text{gallons per minute.}$$

One U.S. gallon contains 231 cubic inches. Moving one gallon or 231 cu. in. one foot in one minute may be expressed as $\frac{231}{12}$ f.p.m. Then, if the movement is at the rate of 1 g.p.m. at a pressure of 1 p.s.i., the horsepower is:

$$H.P. = \frac{1 \times 231}{12 \times 33000} = .000583$$

As stated above, the final formula becomes:

$$H.P. = \text{pressure} \times \text{volume} \times .000583$$

Fig. 14-9 shows two section views of a variable-volume pump with a compensating governor.

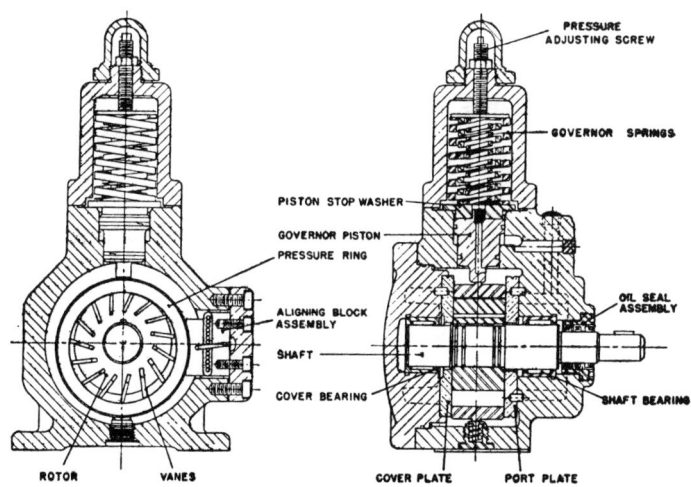

Fig. 14-9. The variable volume pump is used for adjusting oil output to various power requirements.

HYDRAULIC PRESSURE CYLINDERS

In hydraulic cylinders, as in air cylinders, the important thing is to have a unit that will exert, as nearly as possible, the theoretical amount of power that a given piston area, acted upon by a given pressure, can develop. To insure top efficiency for a long period of time, hydraulic cylinder manufacturers have incorporated special features of construction in their models. One of these features, an adjustable cushioned hydraulic cylinder, is illustrated in Fig. 14-10.

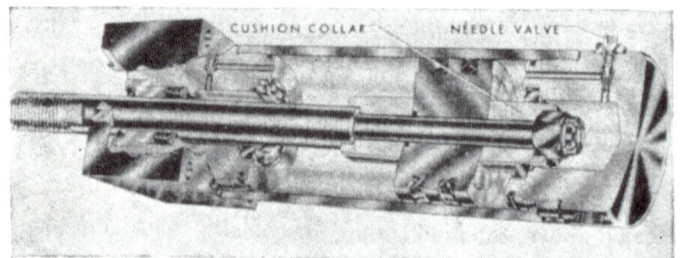

Tomkins-Johnson Co.

Fig. 14-10. An adjustable cushioned, variable stroke hydraulic cylinder.

This type of oil-cushioned cylinder is recommended for retarding the cylinder stroke either to prevent the piston from hitting the cylinder head or to effect slow travel at the end of a stroke.

PRACTICE PROBLEMS APPLICABLE IN THE SHOP AND TOOLROOM

1. Assume that the air pressure available in a shop is 85 pounds per square inch (p.s.i.) and the estimated force to do a certain job is 120 pounds. An air cylinder is employed for this task. Calculate the diameter of the piston or cylinder bore on the basis of it being able to exert a force $1\frac{1}{2}$ times the estimated force.
2. What actual force would an air cylinder with a $2\frac{1}{2}''$ diameter bore be capable of exerting if the available air pressure in a shop is 80 p.s.i.?
3. What would be the safe estimated force to do the job in problem 2 on the basis of the recommendation by manufacturers of air cylinders?
4. Compute the horsepower from the general power formula, if the load is equal to 720 pounds and the velocity in feet per minute is equal to 70.
5. Compute the hydraulic horsepower delivered by a hydraulic pump if the pressure of the pump is equal to 200 pounds per square inch (p.s.i.), and the volume in gallons of oil delivered by the pump per minute is equal to 24 gallons per minute (g.p.m.).
6. Explain how the constant of .000583 in computation of hydraulic pumps is derived. Use the conversion value of 1 U.S. gallon containing 231 cubic inches and 1 horsepower equal to 33,000 foot pounds (ft. lbs.) per minute.

CHAPTER XV

PROBLEMS IN INSPECTION AND QUALITY CONTROL

Parts manufactured on a mass production basis must conform to rigid engineering specifications to facilitate assembly and to allow convenient servicing or component replacement. The chief function of production inspection is to insure that the specifications and standards designated are maintained within tolerance.

Besides checking manufactured parts for specific size, which is the prime function of inspection, this work is important to other aspects of production. Among these are constant maintenance and improvement of the quality of the product, a function known as *quality control*. There are several phases of inspection carried out in a modern manufacturing plant, depending on the products manufactured. These may include dimensional inspection and testing of the physical properties of the parts, surface inspection, and other types of inspection, sometimes unique to the product.

DIMENSIONAL CONTROL

This is one of the most important phases of inspection and is essential to maintaining the pace of modern industry and to economy of production. Close dimensional control is necessary to precision manufacturing, without which rapid assembly of products would be impossible and the metallurgical developments affecting the life of manufactured products would be valueless.

Only recently a manufactured tolerance of .001 inch, even on piston pins and piston pin holes in automobile engines, was considered satisfactory for all practical purposes. This tolerance may even now be acceptable in the manufacture of many products, but for production of automobile and aircraft engines, closer tolerances are demanded. Numerous critical automotive and particularly aircraft engine parts must be produced with an accuracy of .0001 inch or better. If tolerances of .001 were permitted, a mileage of 10,000 miles would call for replacement of engine parts because of wear. The same is true with respect to the breaking-in period of new automobiles. The closer tolerance, precision manufacture, and dimensional control of modern manufacturing processes permit the breaking in of new vehicles at much higher speeds than those permitted by standards of the past.

DIMENSIONAL CONTROL THROUGH GAGING

Measuring instruments, such as the micrometer and vernier caliper, are designed for a wide range of work. These are adjustable measuring devices and have certain disadvantages for such work as testing large numbers of duplicate parts for size. This is especially the case when such tests must be made a larger number of times. In manufacturing plants where many identical parts must be checked for size, the fixed gage is commonly used.

A gage is designed for testing a fixed dimension and is not graduated and, in most cases, has no adjustable members for measuring various lengths as well as various angles. Fixed gages are in constant use in modern inspection and manufacturing divisions. There is small chance of inaccuracy when using a gage; furthermore, it is far more convenient and more rapid to use than an instrument or tool which requires adjustment suited to the individual workpiece or to each measurement taken. Considerable care must be exercised when handling precision gages since serious and permanent damage can be done to these costly instruments when little heed is paid to their proper use.

Modern gaging makes possible substantial savings in both manufacturing and assembly costs. When component parts of manufactured products which must have close tolerances can be classified dimensionally, they may be produced at a faster rate, and by workers with only moderate skill and experience. Parts which have been properly inspected by use of suitable gages and then classified, assemble quickly and easily without requiring costly and cumbersome fitting operations.

DIAL INDICATORS AND GAGES

Dial indicators are used for work of the finest precision and for satisfying the most exacting requirements. Their use tends constantly to increase, since they speed up the work of inspection and eliminate human variations in the sense of touch. One reason for the increasing use of these instruments is the speed of multiple and simultaneous inspection attainable. Several diameters may be quickly and accurately inspected at one time.

The dial indicator is a complete unit and can be mounted singly, or several instruments may be located so as to contact the workpiece at the points to be gaged, and thus permit inspection of all dimensions at the same time. Both setups are shown in Fig. 15-1.

Some of these dimensions bear a definite relationship to each other,

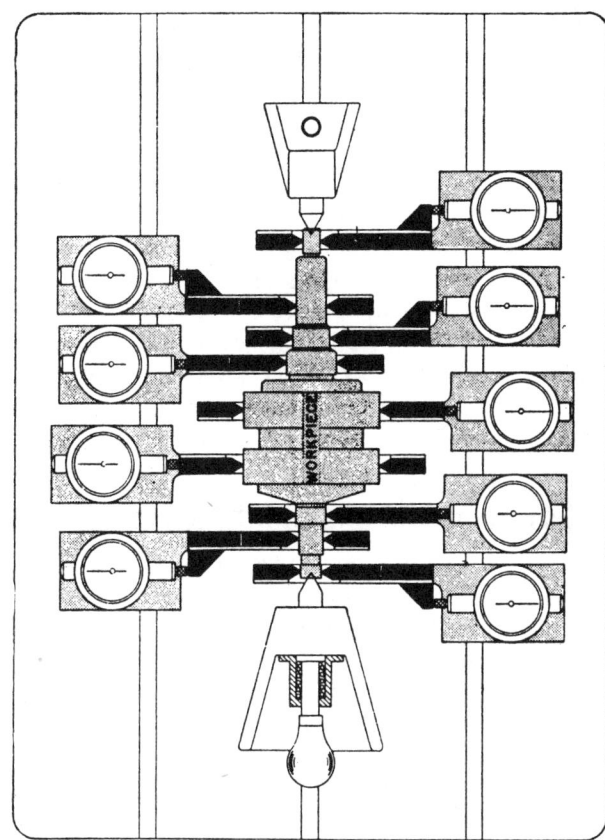

Fig. 15-1. Checking diameters is accomplished by a dial indicator gage (left). Nine dial indicators set up for simultaneous checking of rotor shaft diameters (right).

therefore simultaneous inspection offers a ready solution of the inspection requirements. The gage shown in Fig. 15-1 is designed to simultaneously check the various diameters of a rotor shaft. It is adjustable to any length of shaft up to 12 inches, and can be arranged for various diameters.

Other types of gaging devices for checking a number of measurements are comparators, air gages, electric gages, and multiple or combination gages. The electric gage assembled in a head may be mounted on a production machine, and arranged to stop the machine automatically when the work has been brought to size. The light unit of the gage or indicator may be located directly on the machine, such as a production grinder, or when desired, remotely. When grinding has proceeded far enough to bring the work within tolerance, the grinding machine is stopped automatically. More complex cycles also may be accomplished with this machine.

AN EXPLANATION OF DIMENSIONAL QUALITY CONTROL *

It is practically impossible to make two things alike. Where quantities are involved only the exceptions, relatively, attain the absolute perfection desired. The multitude differs by varying degrees.

Quality control by statistical methods visualizes and can be made to designate the practical and economic limits to which these variations may be permitted to differ from normal before the process should be corrected.

It recognizes that the majority of pieces usually fall, dimensionally, into an average group or cluster close to it, but that "fliers," or pieces digressing to an unusual degree from the uniform, may exist. It shows graphically the tendency or trend of this average or majority group in relation to specifications or tolerances and the effect, in general, that fliers may have on the overall quality of the machine's output.

Quality control indicates the collective quality of a group of units. It supplies the basis for judgment as to whether action should be taken and—equally significant—whether action need not be taken. In other words, it supplies a timely criterion for deciding whether a process is being disturbed by a cause worth while identifying—information which comes too late from 100% final inspection.

Quality control applies essentially the actuarial mathematics of

* "An Explanation of Dimensional Quality Control" is reprinted through the courtesy of Federal Products Corporation, Providence, R.I.; material adapted from *Federal Dimensional Quality Control Primer*, 10th Printing, June 1957.

INSPECTION AND QUALITY CONTROL

probabilities that are used in figuring life expectancy or that are used in the Gallup Poll, in certain financial control plans or even in the well known card hands which form the basis of poker. It therefore estimates scientifically the probable variations and warns of digressions reaching a dangerous degree of percentage.

It is systematic, as compared to haphazard process inspection; it takes the guess out of guess work. Its mathematic-statistical approach neutralizes personal bias, verifies sound reasoning, and refutes poor judgment.

In order to furnish a mental background for the routine, the following birdseye view of the foundation the quality control engineer works from is offered.

If, as an example, fifty pieces are taken from the work of a machine where the O. D. (outside diameter) has been turned and if the pieces are measured individually with an *indicating gage* for this outside diameter and then classified by actual dimension (a sort of selective assembly operation), laid out in rows by dimension in other words, a result considerably like that shown in Fig. 15-2 would be obtained.

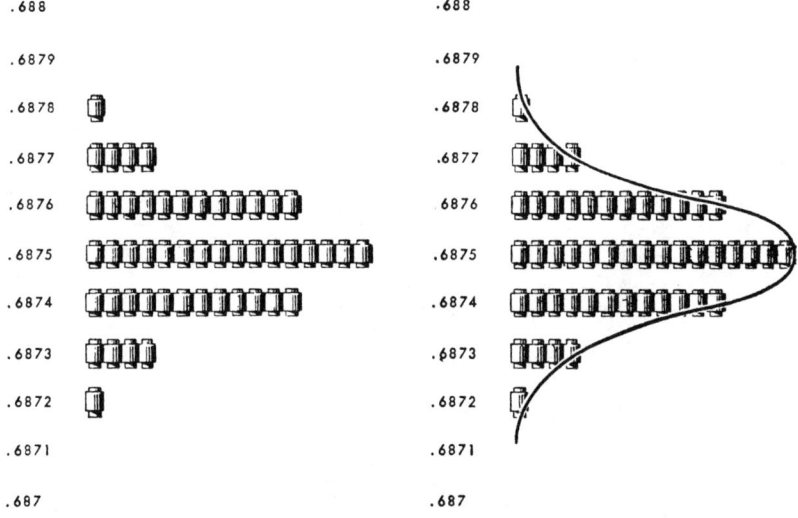

Fig. 15-2. Fig. 15-3.

The outline of the group of work suggests a curve as illustrated in Fig. 15-3.

A group of pieces dimensionally classified in this manner makes what is known as a *frequency distribution*, which illustrates the frequency of occurrence of certain dimensions and their distribution among the whole. The curve itself is called a *frequency distribution curve*.

In actual practice, a frequency distribution and the resulting curve is plotted on graph paper, or in an equivalent manner, as Fig. 15-4 indicates.

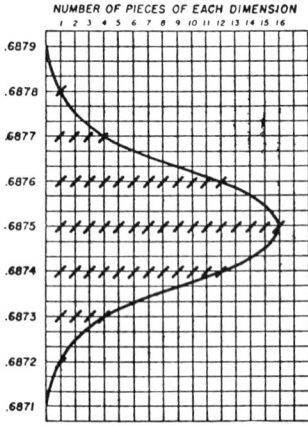

Fig. 15-4.

It is characteristic of pieces classified and distributed according to their dimensions that the largest groups would fall close to the mathematical average of the entire assembly as illustrated by the "Average" line in Fig. 15-5.

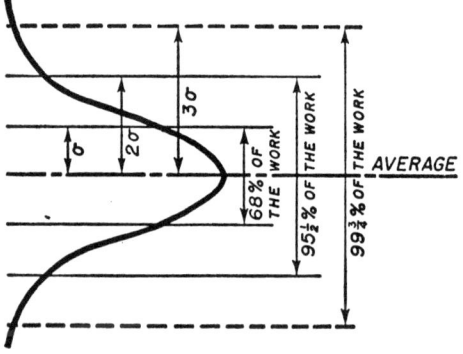

Fig. 15-5.

Furthermore, it has been found that a frequency distribution can be divided into six zones mathematically equal in width, as shown also in Fig. 15-5, three on one—the plus—side of the Average line and three on the minus side. The statistician uses the Greek letter "sigma", or σ, to name or define the individual width of any one of these strips.

Practical use of the frequency distribution becomes immediately available because it has been determined that the number of pieces ordinarily lodging within each of the strips represent percentages of the total, as illustrated in Fig. 15-5.

Following from this, experience has also demonstrated that it is normally unprofitable to attempt a process or method so thorough that no marginal work at all appears from the operations and, since better than 99% of the pieces will appear within the 3 sigma (3 σ) lines, the 3 sigma limits are usually adopted as the economic and practical basis for dimensional quality control work.

Where frequency distributions are made covering a variety of operations and products, the curves seldom if ever come out in shape so symmetric or "normal" as Fig. 15-5 would indicate. Fig. 15-6, therefore, portrays a few examples of conditions taken from practical distributions, where the majority of pieces cluster pretty well together but where, too, a percentage of the work varies in dimension sufficiently to present a distribution with a more or less distorted curve.

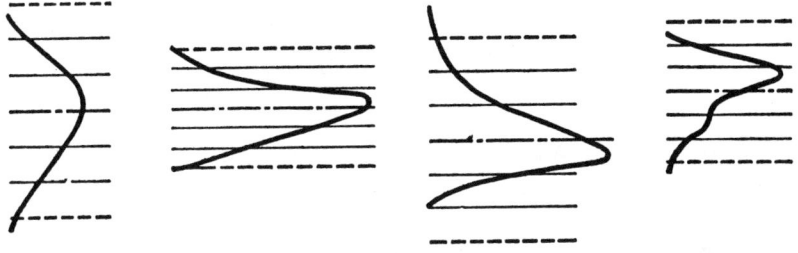

Fig. 15-6.

In this same connection, a most practical and interesting use of the frequency distribution is to compare its view of the dimensional distribution in relation to the specifications or tolerances used. This conception is pictured in several examples in Fig. 15-7.

In the shop application of quality control, however, another method is used in place of the frequency distribution, a system which is simpler,

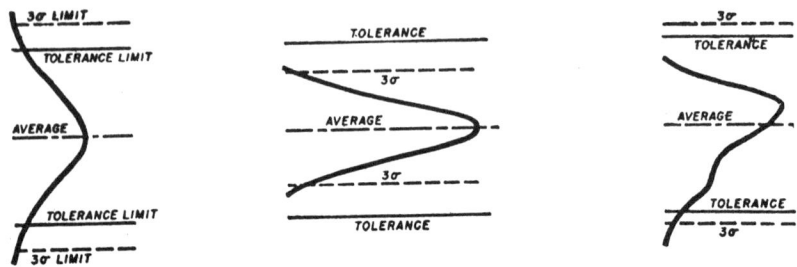

Fig. 15-7.

yet for its purpose even more effective. Carried to an extreme the frequency distribution procedure could resemble or equal 100% inspection in effort and cost.

In the shop application a sampling procedure is adopted, and a chart system replaces the frequency distribution curve.

Rather than sort over fifty pieces, small samples such as 5 pieces at a time are taken more or less regularly from the work as it progresses and certain resulting observations are plotted on a *control chart*, as illustrated in Fig. 15-8.

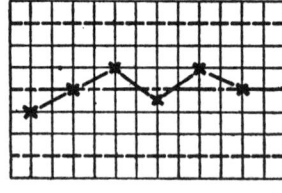

Fig. 15-8.

Three sigma (3 σ) limits also form practical boundaries on a chart, in much the same manner indicated under the discussion of frequency distribution curves above.

Both the frequency distribution curve and the control chart are intended to give, each in its particular manner, what might be called a composite picture of the condition of the general batch of work being analyzed.

The chart might be considered, from a purely mechanical viewpoint, as a series of small frequency distributions of batches of parts the several successive samples represent, as shown in Fig. 15-9. In this case the control limits apply collectively to the whole group of frequency distribution curves, in other words.

If a frequency distribution and its curve were made of the entire lot of pieces, then the larger curve, as Fig. 15-9 also shows, would equal the summation of the smaller curves.

In order to secure the value of 3 sigma and determine where the 3 sigma limits should appear on the chart, however, the mathematician makes use of a value known as *range*. He is able to convert range into 3 sigma values, where small samples are used, by formulae that need not be gone into here.

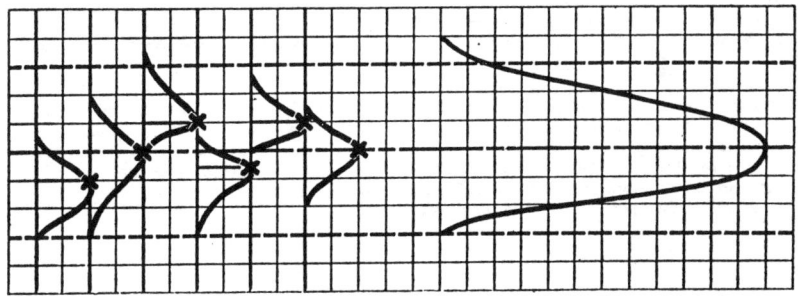

Fig. 15-9.

In dimensional quality control, range is the difference between the greatest and smallest dimension observed in each sample taken.

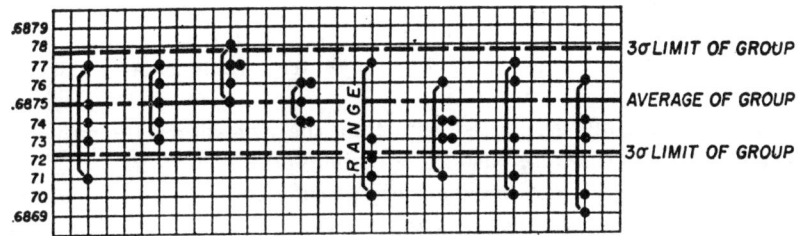

Fig. 15-10.

Fig. 15-10 shows a chart where each of the 5 readings from the samples is plotted in proper position. It illustrates for each sample taken the highest and lowest reading and the spread, or range, between them as well as the variation in range from sample to sample.

Fig. 15-11 may help the layman, in an unmathematical way, to appreciate the relationship of "Range" and "Averages" and their effect during the run of an operation. Consider the solid stone walls lining the road as tolerance limits, the edges of the paved road along the shoulder

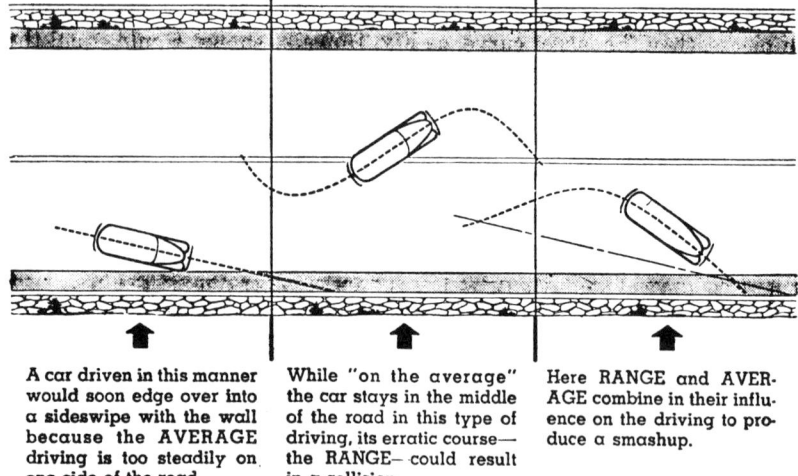

| A car driven in this manner would soon edge over into a sideswipe with the wall because the AVERAGE driving is too steadily on one side of the road. | While "on the average" the car stays in the middle of the road in this type of driving, its erratic course— the RANGE— could result in a collision. | Here RANGE and AVERAGE combine in their influence on the driving to produce a smashup. |

Fig. 15-11.

as control (3 σ) limits and the white line in the middle of the road as average.

As can be seen, the chart system embraces the average and the use of 3 sigma limits in much the same manner as was observed on the frequency distribution curve of Fig. 15-5. In the case of the control chart system, the limits are calculated from a group of samples and the limits and average found apply only to and with the particular group of samples.

It is respectfully suggested here that the elementary dimensional quality control procedure outlined from here on be absorbed and practiced. Not only will the foregoing explanation seem even still clearer but, after such a brief experience, anyone who has not previously delved very deeply into statistical quality control can more readily turn to the theory and philosophy involved, as adequately covered in treatises.

FIRST STEPS IN APPLYING STATISTICAL QUALITY CONTROL TO DIMENSIONS

As stated before, a complete knowledge of statistical control of quality involves a careful study of mathematical theory, but for practical purposes the system outlined hereafter offers a simple, direct, and entirely workable method of controlling dimensional quality more accurately and efficiently than by older methods.

The general idea of statistical quality control of dimensions is the taking of sample lots of workpieces while they are being produced. Then, by immediately recording the actual dimensions of these samples, obtain the data which, when used with predetermined formulas, will give you an accurate graphic report on the dimensional quality or accuracy of the work coming out of the machines. It warns you when the quality is departing from the desired standards.

The facts obtained will also frequently furnish information which will suggest improvements in the machine setup or, on the other hand, changes in the original dimensional specifications.

Not only should the information be collected and analyzed in a systematic manner, but it is recommended that it be presented graphically as on a chart. In fact the chart is an imperative part of the system of analyzing the dimensional quality of an operation. In addition, control limits for the chart are obtained from a few simple calculations made from the data obtained. These are directly related to the 3 sigma limits previously spoken of. Control limits are used as a basis for judging the significance of variations in the operation; they are measures, standards in a sense, for comparison purposes.

Consequently, as will be seen on succeeding pages, a *work sheet* is suggested on which data is compiled and transferred to a *quality control shop analysis chart*.

Before proceeding to the actual dimensional quality control application, however, a plan of action should be determined. Suggestions contained in the following sections are intended as guide posts in the preparatory work.

Choice of Operation and Dimension. The locale for the initial dimensional quality control application should be considered first. A decision must be made as to which operation is to be observed and which dimension studied and plotted. If quality control is new, preference should be given an operation that is to run steadily over a period of time. Availability or accessibility of the machine influences the decision. In other words, start quality control if possible where the inspector can secure the workpieces and make his measurements with a minimum of interference to the operator or other personnel and equipment in the department. A turning, boring, grinding, or draw press operation is probably preferable.

As for the dimension, or characteristic, select a single dimension only, at the very start. Choose perhaps a "critical" dimension (especially in a multi-dimensional operation), or one normally hardest to keep under control or that has the greatest effect on some future assembly. Since the statistical method depends upon knowing accurately the degrees of variation between the several workpieces, it is important that an indicating gage be used.

SAMPLING PROCEDURE LEADING TO THE FORMATION AND USE OF QUALITY CONTROL CHARTS

The size of sample, the method and frequency of sampling are important because the statistical system for determining the quality level of the output is based on the intelligent interpretation of the information the sampling supplies.

Size of Sample. Five (5) pieces selected at random for checking measurements or observations will be found a handy, and common, sample size at any one reading. A smaller sample size is usually not recommended.

Sample sizes of 10 or 15 units are occasionally preferred for specific purposes.

Small samples at short intervals are much more to be preferred over large, less frequent samples.

After the statistical method of dimensional quality control has been learned, applied, and practiced for a period, a sample size of 5, for instance, can be paralleled with a special study using larger sample sizes, the general results compared and a decision made as to which sample size better protects the quality with minimum cost of inspection and sampling.

Frequency of Sampling. A convenient time unit is the usual basis for a sampling schedule. Generally samples are taken not more often than every quarter hour or perhaps not less often than once an hour. There are, however, many situations where the sample needs to be secured only every other hour or even once per shift. Or the time unit may be discarded in favor of the production unit as, for instance, 5 pieces sampled out of each pound produced.

The amount of sampling required on any particular job is probably better determined from a study. After several sets of observations have been made and the charts plotted from them, more frequent sampling may appear necessary in order to disclose operating variations sooner.

Examining a frequency distribution analysis of, say, a hundred-piece random selection from the run is a common method of checking the adequacy of the sampling procedure.

INSPECTION AND QUALITY CONTROL

Sampling can well be done at intervals sufficiently irregular to preclude any influence due to periodic factors in the process or prediction of the selection time.

How Samples are Taken. The method of selecting samples should be consistent. One practice is that of taking units directly from the machine at each inspection, for example the last 5 pieces run as the inspector approaches. This method gives the best information for the tool setter; and it indicates the correct condition of the machine, operator, and process.

Another system of sampling takes a random sample from the batch manufactured since the last visit. It is a check on the work made between visits and gives a picture of the overall quality of output. Either way, be consistent; sample in the same order. Be careful never to allow any portion of a sample to contain pieces from work that has been previously sampled. During any one observation period the sample size should not be varied. Interpretation can be undermined and judgment unintentionally warped where the sample fails to present an honest cross-section of the type of information desired.

HOW TO USE THE WORK SHEET

The next step is to tabulate, average, and plot the results of measurements, readings, or observations. Examples of a work sheet and a chart form obtainable from Federal Products Corporation are shown in Figs. 15-12 and 15-13.

This discussion presupposes that a decision has been reached as to what dimensional quality variation of the product is to be analyzed by these means.

The selected dimension and its tolerances are to be written in the space provided.

This explanation, and the illustrations, are based on 15 samplings or sub-groups of 5 pieces each taken every half hour. The work sheet is similarly laid out. The same general methods can be adapted, of course, to other sizes and frequencies of sampling.

Record the Sample Dimensions. Make the required measurements or observations on 5 sample pieces. Record them as illustrated in the first column of the facsimile WORK SHEET; Fig. 15-12. The 5 readings are to be averaged as shown. Repeat this operation at predetermined time intervals, recording each lot of 5 sample readings in the sub-group numbered columns provided on the work sheet.

A minimum of 10 sequences of readings, or sub-groups, should be

DIMENSIONAL QUALITY CONTROL
WORK SHEET
AVERAGES AND RANGES

PART NUMBER PL 3421 A	DESCRIPTION Plunger - finished O.D.				
LOT NUMBER 17	ORDER NO. 22-185	MACHINE NO. G 4		DEPT. E-GM	
OPERATOR Jones	SHIFT 2	DATE 9-18-45		INSPECTOR Smith	

SAMPLE OR SUB-GROUP NUMBER		1	2	3	4	5	AVERAGES
DIMENSION AND TOLERANCES: .6875 ± .0005		.6875	.6877	.6875	.6875	.6877	.6874
	SAMPLE READINGS OR MEASUREMENTS X	.6877	.6879	.6877	.6877	.6874	.6877
		.6873	.6877	.6877	.6873	.6874	.6876
							.6875
		.6871	.6875	.6876	.6875	.6878	.6874
							.6875
		.6874	.6878	.6874	.6873	.6877	.6876
							.6877
SUM OF SAMPLE READINGS		3.4370	3.4386	3.4379	3.4373	3.4380	.6878
							.6879
AVERAGE of each Sample X (Plot on Chart)		.6874	.6877	.6876	.6875	.6876	.6876
							.6875
LARGEST VALUE		.6877	.6879	.6877	.6877	.6878	.6876
							10.3139 TOTAL
SMALLEST VALUE		.6871	.6875	.6874	.6873	.6874	.6876
RANGE of each Sample R (Plot on Chart)		.0006	.0004	.0003	.0004	.0004	GRAND AVERAGE X

6.	7	8	9	10	11	12	13	14	15	RANGES
.6871	.6877	.6876	.6875	.6878	.6880	.6883	.6874	.6876	.6875	.0006
										.0004
.6874	.6878	.6877	.6876	.6880	.6873	.6879	.6876	.6877	.6877	.0003
										.0004
.6871	.6875	.6875	.6876	.6876	.6879	.6880	.6877	.6875	.6876	.0004
										.0007
.6872	.6873	.6875	.6875	.6877	.6882	.6879	.6875	.6875	.6876	.0006
										.0003
.6878	.6872	.6878	.6877	.6876	.6874	.6876	.6876	.6873	.6876	.0002
										.0004
3.4366	3.4375	3.4381	3.4379	3.4387	3.4388	3.4397	3.4378	3.4376	3.4380	.0009
										.0007
.6873	.6875	.6876	.6876	.6877	.6878	.6879	.6876	.6875	.6876	.0003
										.0004
.6878	.6878	.6878	.6877	.6880	.6882	.6883	.6877	.6877	.6877	.0002
										.0068
.6871	.6872	.6875	.6875	.6876	.6873	.6876	.6874	.6873	.6875	TOTAL
										.0004
.0007	.0006	.0003	.0002	.0004	.0009	.0007	.0003	.0004	.0002	GRAND AVERAGE R

LIMITS FOR RANGES CHART = $D_4\bar{R}$ and $D_3\bar{R}$	LIMITS FOR AVERAGES CHART = $\bar{X} \pm A_2\bar{R}$
2.114 x .0004 = .00085	.6876 ± .577 x .0004
UPPER: .00085	UPPER: .6878
LOWER: 0	LOWER: .6874

Fig. 15-12.

recorded—more than 25 is usually unnecessary—producing thus a series of Averages, ($\bar{X}$), one for each sub-group.

Simultaneously record the highest and lowest reading or measurement from each sample of 5, and subtract one from the other. This difference is known as the Range, (R), within each sample group.

Next transfer these Averages ($\bar{X}$) to the upper right hand column headed AVERAGES, and average these, as illustrated, and secure the Grand Average figure, ($\bar{\bar{X}}$).

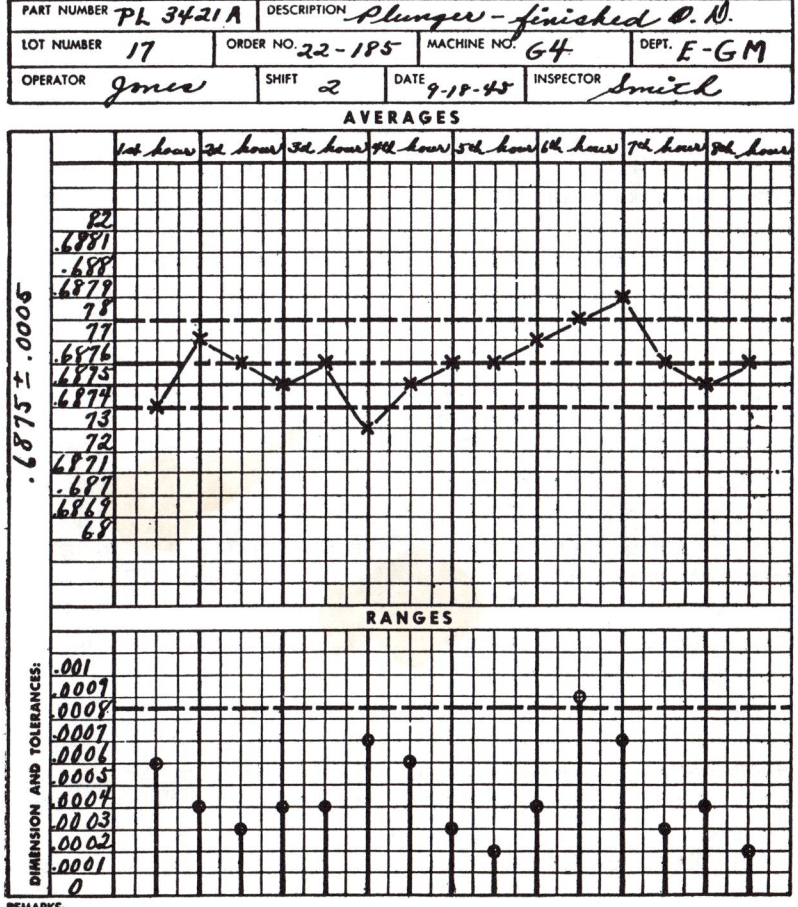

Fig. 15-13.

The Grand Average ($\bar{X}$), as its name implies, marks the practical mean of the group under consideration, or of the day's work.

Next transfer the Ranges (R's) to the column headed RANGES, and add and average them to secure an average of Range, ($\bar{R}$).

There are available now the data necessary for putting boundary lines on the map, or, to use more exact terminology, for establishing control limits for the quality control chart.

HOW TO OBTAIN THE CONTROL LIMITS FOR THE CHART

Control limits are of course an essential part of the quality control chart. (These have also been referred to earlier as 3 sigma limits.) They represent a limiting control, a standard of comparison or a verification, of the collective quality of a group and they are computed from observations made on the individual units of the group. As dimensions they are limits outside of which the collective quality of any sub-division of the main groups does not digress or drift unless the elements of the sub-division are too much at variance with the majority.

There are two sections of the quality control shop analysis chart, one for recording and analyzing averages and the other for range. Hence, two sets of calculations are required to obtain the control limits, as illustrated at the bottom of the sample work sheet, Fig. 15-12, and as amplified below. The figures used are taken from the work sheet.

To Obtain the Grand Average. On the chart for averages, one control line is the Grand Average ($\bar{X}$) which, as its name implies, marks the middle of the road or the practical mean of the group or day's work under consideration. In this example, see Fig. 15-2, it has been calculated, as above, and recorded as $\bar{X}$ in the upper right-hand column summarizing the averages. Its value, in this example, is .6876.

As the sample chart, Fig. 15-13, shows, this value, when drawn on the chart, seldom necessarily coincides with the mean dimension arbitrarily established on the blueprint.

An upper control limit for averages and a lower control limit are also used. These are determined by adding to and subtracting from the Grand Average ($\bar{X}$), a figure obtained by multiplying the Average Range ($\bar{R}$) by a suitable *limiting constant* designated as A_2. List of A_2 values is given below.

To Obtain the Value of the Upper Control Limit. In this example, multiply the Average Range ($\bar{R}$), .0004, by the A_2 "Constant" —.577, and get .0002308. Discard the last two digits as irrelevant and add the .00023 to .6876, the Grand Average ($\bar{\bar{X}}$) which gives .6878. This is the *Upper Limit*. The actual formula is expressed as follows:

$$\text{Upper Control Limit} = \bar{\bar{X}} + A_2 R.$$

To Obtain the Lower Control Limit. This is arrived at similarly by subtracting .00023 from .6876 and get .6874.

The formula used for this limit is:

$$\text{Lower Control Limit} = \bar{\bar{X}} - A_2 \bar{R}.$$

The upper and lower control limits can be seen on the illustrative chart for averages, see **Fig. 15-13**.

The *constant* used varies with the sample size. A condensed list of them is shown directly below. These are known as *limiting constants* and their use saves rather complex formulas and tedious calculations.

Sample Size or Number of Observations	5	8	10	12	15
Value of Limiting Constant (A_2)	0.577	0.373	0.308	0.266	0.223

Since the sample size in this particular problem is 5 the constant used is, of course, .577.

Looking at the above in briefer manner, this is what we did:

The actual mathematical formula used is:

$$\text{Upper and Lower Control Limits} = \bar{\bar{X}} \pm A_2 \bar{R}.$$

Here it is worked out as described above.

Upper Control Limit =
$\bar{\bar{X}}$, or .6876, $+ A_2$ or .577, $\times \bar{R}$, or .0004, $= .6878$
Lower Control Limit =
$\bar{\bar{X}}$, or .6876 $- A_2$ or .577, $\times \bar{R}$, or .0004 $= .6874$

To Obtain Control Limits for Range. The chart for range must also be bounded by upper and lower control lines, as in the case of the chart for averages. As a general practice the average range is not drawn on the chart. Here again two simple formulas are used.

The upper limit for range is calculated by multiplying the Average Range ($\bar{R}$), in this example .0004, by another limiting constant, designated as D_4. In this particular case D_4 is 2.114. The result is an upper limit value (for range) of .00085.

The lower limit is obtained in the same way except a constant D_3 is used.

In this case the constant, referred to as D_4, varies with the sample size in the same manner as the A_2 constant described in the paragraphs above. The values of D_4 are shown in the condensed list here.

Size of Sample or Number of Observations	5	8	10	12	15
Value of D_4	2.114	1.864	1.777	1.717	1.652
Value of D_3	0	0.136	0.223	0.284	0.348

The mathematical formulas used are as follows:

Upper Control Limit for Range = $D_4 \times R$ or $2.114 \times .0004 = .00085$
Lower Control Limit for Range = $D_3 \times R$ or $0 \times .0004 = 0$

In this example, with a sample size of 5, the value of the constant D_3 is 0 as the table of constants shows. Hence the lower control limit is 0.

HOW TO CONSTRUCT AND PLOT THE QUALITY CONTROL SHOP ANALYSIS CHART

The chart might be considered the most important element in the mechanics of a quality control method. While all the information actually needed for judgment, comparison, and action appears on the work sheet it is nowhere near as usable as when it is put in graphic form. The chart is the newsreel of the system.

It is usually divided into two sections, the chart for averages and the chart for range. As the sample control chart, Fig. 15-13, illustrates, all relevant data should be filled in to complete the margin and headings. The chart and work sheet should correspond exactly in this respect.

Establish the Mean or "O" Line. In our example and on the sample control chart, Fig. 15-13, both have been used. The chart 0 line represents the blueprint mean of .6875.

The Grand Average ($\bar{X}$) line has been drawn in at .6876 as a dash and dot line.

On the chart for Range ($\bar{R}$) no mean or nominal line is used; simply control limit lines. (See next section.)

Note: Since the location of the Grand Average line ($\bar{X}$) on the chart does not necessarily coincide with the mean or nominal blueprint or specification figure, it is probably better practice for the sake of more complete interpretation of the chart to establish the nominal blueprint value as the mean or 0 line of the chart itself, and let the $\bar{X}$ line fall where the calculation of the Grand Average ($\bar{X}$) places it in relation to the blueprint mean.

INSPECTION AND QUALITY CONTROL 475

Establish the Upper and Lower Control Limits. We have already calculated the upper control limit as .6878. Therefore, this is drawn in on the control chart for averages at that point as a line in red ink or red pencil. Color is suggested because control lines mark the danger point at which the dimension goes out of control.

Next draw in the lower control limit line in red on the chart for averages at .6874, as previously calculated.

Now, on the range chart draw in the upper limit. This has already been calculated at .00085 and is also drawn in red as shown.

With a sample size of 5 the lower Range ($\bar{R}$) limit is, as has been shown, at 0.

PLOTTING THE CHART

The charts are now ready for plotting the individual averages obtained on the work sheets. Each vertical line on the control chart illustrated represents time intervals of every quarter-hour (15 min.). Other time units can be chosen to suit your particular requirements. Or, units of quantity can be substituted in place of time units.

In the illustration small x's have been used for plotting the various calculated values of Averages ($\bar{X}$). And, in a similar manner, a small o plots the proper ordinate for each Range figure.

The trend of averages can be better observed if the small x's are connected. On the range chart the observations can be emphasized by drawing lines (sometimes with colored pencil) from each o to the base.

Note: It is usually better to do the actual charting as the observations and calculations of Averages ($\bar{X}$) and Range (R) are made, i.e., as in the example, every half hour.

The Grand Average ($\bar{\bar{X}}$) line and the several control lines cannot be marked on the chart, of course, until the results of 15 sets of readings are known, i.e., not until, in this example, the work sheet is completed.

Applying the basic theories of quality control at this point, it may be re-emphasized that controls may be established after 10 sets of readings are taken. To set up controls on a lesser number of sub-groups is likely to introduce errors. On the other hand, more than 25 sets of readings are usually unnecessary for the accurate determination of controls.

All of which means that an undesirable trend in the work, an out of control condition or a definite basis for action cannot be so accurately determined until after the entire batch of work, represented by the fifteen odd samples, has been completed. To get around this condition, since it is admittedly desirable to take corrective action or at least segregate suspected work immediately the sample discloses it, the controls calculated from one group of readings are sometimes used for the next chart, for the succeeding shift's or day's work, for example.

If the operation is continuous, another practice is to average the controls from a series of charts to designate a set of controls for the charts to be used thereafter, for a period at least, to analyze the particular operation.

Whenever the process, method or engineering, material, or machine is changed, a new determination of control limits must be obtained.*

QUALITY CONTROL AND STATISTICAL SAMPLING

A system of quality control which is rapid, economical, and capable of maintaining reasonable standards of perfection is highly desirable in mass production industry. Considerable research and investigation has been done by large volume production manufacturers to determine causes of variation in finished products and in the raw materials obtained from existing production methods. If such causes and conditions of variations can be discovered, they may be controlled or eliminated, and products of better quality may be expected at no increase in cost.

Applying a statistical sampling procedure to quality control is one of the simplest and most effective methods employed by modern industry. Inspected products and the performance they give are systematically observed for the purpose of revising either the specifications or the methods of inspection. Carefully kept inspection records may influence the manufacturer to change present methods of manufacture or convince him of the advisability of using new materials for his product. Finally, by adopting statistical techniques, the manufacturer is able to analyze conditions producing a wide variation in the quality of the product, and thus incorporate an efficient yet economical system of control.

The problems involved in quality control depend upon various factors. Certain products, due to their functional requirements, need 100 per cent inspection of all component parts to be used in the assembly of the final product. For other products, only a fraction of all the parts can be inspected and a determination made of the relative number of defective parts. If the percentage of defective parts seems small enough to make it economically feasible, the whole lot may be passed and the defectives rejected on the assembly line. A good example is the manufacturing and inspection of nails, rivets, screws, and bolts; 100 per

---

*Conclusion of extracted material, see footnote page 460.

cent inspection of these parts is not necessary and is unwise. However, in the manufacture of ball bearings, a 100 per cent inspection, including accurate gaging and visual examination, is necessary for successful performance of the assembled product.

Statistical methods of sampling have been applied in various ways, and a number of sampling procedures have been devised by large volume-production manufacturers for the elimination of defectives. In some cases the object is the establishment of an average outgoing quality with a given minimum of inspection, which consists of drawing sample pieces at random. If in this random sampling more than the allowable number of defectives is found, the entire lot is thoroughly inspected, and all defective parts are replaced or corrected. On the other hand, the lot is accepted with no further inspection if the sample has less than the maximum defectives permitted. This method allows the development of a statistical relationship between the sample size and the minimum number of defectives which may be allowed and yet insure standards at a minimum cost.

AUTOMATIC PROCESS INSPECTION AND QUALITY CONTROL

Automatic control is not limited to actual manufacturing operations, but can also be successfully applied in the inspection of workpieces throughout the various phases of production without interfering with the process. For an example, a checking gage can be introduced in a grinding operation to check the accuracy of a dimension without interference with the grinding wheel. In other machining operations proper gages can be so located with reference to the work so that the dimensional accuracy can be checked and controlled throughout the operation without interfering with the motion of the tools. This is truly quality control of the work from the rough to the finished state.

Other methods of automatic inspection and quality control consist of gaging each workpiece at the completion of the processing or machining operation and automatic reading of the results will then cause tool adjustment, so that the workpiece heads towards the safe center of tolerance rather than the upper or lower limits where a slight deviation may create trouble. As tool wear occurs, it can be gradually advanced, keeping workpieces within a narrow tolerance band. Furthermore, when the tool has worn beyond desirable limits for work quality or regrind of the tool, a new tool or cutting edge can be brought into play, the tool slide retracted, and again self-adjusted to find the center of tolerance.

The ultimate result of an approach of this type to processing or machining is a finished product of higher quality. Machines and tools will be developed along these lines, whereby workpieces will be manufactured with fairly constant dimensional accuracy, thus making the assembly of the component parts an easy task. The cost of workpieces because of such operation is naturally reduced, the quality of the work is increased, less scrap will result, and a machine operator of less exacting skills may be chosen for the work.

Chapter XVI

TOOL ENGINEERING MATHEMATICS

ANALYSIS OF MANUFACTURING COSTS

One basic problem always confronts the tool engineer—that of manufacturing any kind of product at the lowest possible cost. This statement implies (1) the tool engineer has the knowledge and experience necessary to manufacture any kind of product, and (2) he has the ability to estimate costs with a view towards economical production.

Indeed, the good tool engineer does have these attributes; moreover, because of the wide range of modern production equipment, tools, and methods at his disposal, he is in a splendid position to select the most suitable equipment and arrange it to cut costs.

The tool engineer is only indirectly concerned with the financial phases of the business which include *assets, liabilities* and *income*. Also, the expenses incurred in *administration, promotion,* and *selling* are the duty of management to handle. He is, however, directly concerned with the phase of *expense*—the cost of operating a business, but only with that part of this phase which is directly incurred in *production*. The *cost of production*, the tool engineer's prime concern financially, is known as *manufacturing cost*. This is broken into three main groups:

Fixed costs,
Direct costs,
Indirect costs.

Fixed costs. These costs include the *initial investment* for production equipment and such charges as *depreciation, interest,* and *taxes and insurance* on the investment. When a capital investment is made on production equipment a satisfactory return is expected in a reasonable length of time. An obligation is incurred against a portion of the equipment's earnings to pay off and protect the original outlay of capital.

Fixed cost can be expressed in equation form. Let I_a represent the cost for production equipment, transportation, handling, and accessories. Let I_b represent the cost of setting up the equipment on the premises and testing. Then, the initial investment, $I_c = I_a + I_b$, assuming that the equipment being replaced has no monetary value. If the replaced equipment is salvageable, then the net initial investment $I_d = I_b = R$, R represents the realizable value of the replaced equipment. (Subscripts a, b, c and d, have been used for simplicity. In the following formulas, other subscripts will be used to denote specific conditions).

Depreciation. Production equipment diminishes in value because of prolonged use, physical depreciation, and obsolescence. Several methods are practiced in computing depreciation. For simplicity, the *straight-line* method is frequently used. Here the assumption is made that the decrease in value of production equipment is the same from year to year over the life of the equipment. This can be readily formulated as follows:

I_N = initial investment for equipment in dollars.
N = number of years of useful life of the equipment.
I_n = value of equipment at the end of n years.
D = annual percentage allowance for depreciation.
D_c = annual cost of depreciation in dollars.
R_N = salvage value in dollars at the end of N years of useful life.
$D_n \times$ accrued depreciation in dollars to n years. (n denotes any number of years during the useful life).

Since $D = \dfrac{I}{N}$ and $D_c = \dfrac{(I_N - R_N)}{N}$, we can combine these equations to eliminate N.

Then:
$$D_c = (I_N - R_N) D.$$

If the salvage value is negligible, then:
$$D_c = I_N D, \text{ and } D_n = {}_nD_c = {}_nD (I_N - R_N),$$
and finally,
$$I_n = I_N - {}_nD (I_N - R_N)$$

Interest, Taxes, and Insurance. Interest is money paid for the use of money, usually expressed as a percentage of the principal per unit of time. The principal is the capital worth of the initial investment. The annual interest charge on depreciated production equipment is assessed on the undepreciated balance.

Taxes and insurance are also assessed on the undepreciated balance since they are proportional to the current value of the initial investment.

Average annual interest rate is computed as follows: (Assuming a linear depreciation and neglecting the equipment's salvage value.)

I_N = the initial investment in dollars.
a = the annual interest rate in per cent (estimated).
N = the years of useful equipment life (estimated).

Then, I_{Na} = the interest payment on the initial investment at the end of the first year. (Note that the payment is charged at year's end, although calculated at the beginning of the year).

$\dfrac{I_{Na}}{N}$ = the interest payment at the end of the last year estimated to be the useful life of the equipment.

$$\dfrac{\left(I_{Na} + \dfrac{I_{Na}}{N}\right)}{2} = \text{the average interest payment.}$$

Rearranged: $\quad \dfrac{I_{Na}}{2}\left(\dfrac{N+1}{N}\right)$

Since average rate is $\dfrac{average\ payment}{initial\ investment}$, dividing above expression by I_N will result in $\dfrac{a}{2}\left(\dfrac{N+1}{N}\right) = A$. A is the average annual interest rate.

Taxes and insurance can be formulated in the same fashion as interest. Then, if b represents the annual tax and insurance rate in per cent, the following expressions can be formulated:

$$\dfrac{I_N b + \dfrac{I_N b}{N}}{2} = \text{the average tax and insurance payment.}$$

Rearranged: $\quad \dfrac{I_N b}{2}\left(\dfrac{N+1}{N}\right)$

Dividing by I_N: $\quad \dfrac{b}{2}\left(\dfrac{N+1}{N}\right) = B.$

B is the average annual tax and insurance rate.

Direct Costs. These costs are incurred in the operation of production equipment. They include costs of direct labor, direct material, power for the equipment, lubrication, and coolants. Since direct costs are in proportion to the volume of production, it is a convenient way of grouping them. They are also called operating expenses.

Direct labor costs are the wages paid workers for actual labor performed in manufacturing the products. Direct labor can be estimated fairly accurately from past experience with similar equipment and layout. Job evaluation rates are established for each class of work on the basis of time cards and job tickets.

Direct material costs are the costs for those materials which are actually consumed in making the product. They are determined from the invoices of purchased material and usually include transportation and handling charges.

Indirect Costs. These are costs which do not have a direct relationship to volume of production. Certain labor and material costs are billed as indirect costs and are often called *overhead*. Indirect labor costs are the wages paid workers who are not directly associated with the manufacturing process. Supervisors, inspectors, stockroom clerks, and truckers are charged to indirect labor. Indirect material costs are the costs for materials and supplies which are not considered a part of the product or production equipment. Costs for maintenance of buildings and grounds, building rental payments, lighting, heat, water, etc., are indirect costs.

The total cost of operation per year in dollars, O, and the unit cost, U, may be summed up in the following manner:
From previous statements,

I_d = net investment in equipment in dollars.
D = annual percentage allowance for depreciation.
A = average annual interest rate.
B = average annual tax and insurance rates.
Y = the average annual total fixed charges in dollars.

Then: $Y = I_d (D + A + B)$.

If,

M = annual total cost of materials in dollars.
L = annual total cost of direct labor in dollars.
C = annual total cost of maintenance in dollars.
E = annual total cost of power and supplies in dollars.
F = annual total cost of space allotted for machines.
T = annual total cost of indirect labor for machines.

Then the total yearly cost, $O = Y + M + L + C + E + F + T$.

The unit cost, or the cost per piece, $U = \dfrac{O}{P}$, where P = the total number of parts produced per year.

ECONOMIC SELECTION OF MACHINE EQUIPMENT

The most obvious reason for selecting machine equipment occurs in the case of a new installation. In some respects this kind of selection is easier than when equipment must be replaced or added, because limiting factors can usually be altered, if necessary, to suit the proposed equipment's design.

Selecting machine equipment to replace that which is currently operating can be more difficult. The present installation may be worn, appear obsolete, or production requirements may have been above the installation's capacity. Replacement with different equipment will likely be limited by such factors as floor space, equipment weight, power, associated equipment, and others which are influenced by the existing conditions.

Another reason for selecting machine equipment is to add to the present installation. Normally, the equipment currently operating is adequate in all respects except capacity. To increase capacity is primarily a matter of adding manufacturing units of the same type of equipment until the desired total output is reached. Floor space and power requirements can be a problem in equipment supplementing. Also, there is a danger of unbalancing the process.

Although each situation has its own problems, in all cases the technique of selection becomes one of locating eligible types of equipment, comparing each type against others on the basis of cost and function, and systematically arriving at the most economical equipment. Each type of cost and function is analyzed separately by comparing a feature of one unit of manufacturing equipment with a similar feature of other units.

The selection of a satisfactory piece or pieces of equipment from the points of view discussed above, must meet management's approval, which is based on how long the equipment's earnings will take to pay off the initial outlay of money. If the best equipment obtainable cannot repay the initial investment in what it considers a reasonable length of time, management will look on the transaction with disfavor. To buy the equipment despite its predicted shortcomings is a risk that is usually not taken.

FACTORS INFLUENCING THE INVESTMENT

Determining how long an equipment unit will take to pay for itself depends on a number of factors. The machine's mechanical performance is one factor; its durability another. The selection on the basis of *equipment life*, which is one of the most important factors, can be determined by simple calculations. This will be shown in the following text material. Another important factor is the analysis of the market with a view toward accurately estimating the demand for the manufactured product. If an accurate estimate indicates the demand will be great then a large volume of the product can safely be manufactured, which will reduce the unit cost, increase the unit profit, and assure a total profit so the initial investment can be protected.

Equipment life is the length of time in which a machine unit can repay its original costs. There is a correlation between equipment life and depreciation. Physical depreciation is due to deterioration and impairment from prolonged use. Corresponding to physical depreciation is *physical life*, the period during which the equipment will function satisfactorily, before excessive maintenance costs make it unprofitable. Functional depreciation is also due to obsolescence. Corresponding to functional depreciation is *serviceable life*, the length of time before obsolescence makes machine equipment unprofitable to operate. Machines of new designs, capable of higher productivity, are often the cause of functional obsolescence of machines currently used, which should be replaced, even though they are not worn out or physically obsolete.

Critical life, or minimum life is the minimum length of time in which a machine unit can repay its original costs. There are no standard or definite periods of critical life. However, with the constantly increasing complexity of manufacturing machinery, the shorter the critical life, the better. Manufacturers today allow a period of from one to three years for automated machinery and from three to five years for partly automated machinery. For extremely special manufacturing equipment capable of high productivity even less than one year is allowed. By the same token, for more or less standard machinery, a period of over five years is allowed.

Computing Critical Life. In computing the critical life of a new manufacturing unit there is always a comparison with another unit—either one previously used which it will replace, or another new unit of a different design but in the same general category.
Let:

X = critical life of new machine equipment, in years.
P = total number of units of the products made per year.
V_1 = all costs except the initial investment, per unit of product manufactured by the equipment to be replaced (old equipment), in dollars.
V_2 = all costs, except the initial investment, per unit of product the new equipment will manufacture, in dollars.
I_N = initial investment in new equipment, in dollars.
A = current annual interest rate.
B = current annual insurance and tax rate.
R_N = estimated salvage value of new equipment at the end of its equipment life, in dollars.

The basic equation is:
$$X\left[P(V_1 - V_2) - I_N(A + B)\right] = I_N - R_N$$

Note the following number of points:
1. The total number of units of the product manufactured yearly will be the same for both old and new manufacturing equipment.
2. Current interest, tax and insurance rates will be accurate enough for computing their average values.
3. The expression $P(V_1 - V_2)$, since it represents a difference in costs, need not include any costs that are the same, for both old equipment and new.
4. The types of cost included in V_1 and V_2 are equivalent.

Solving for critical life,
$$X = \frac{I_N - R_N}{P(V_1 - V_2) - I_N(A + B)}$$

Computing Tooling Costs. Selecting tools is approached in much the same way as equipment. In many cases, a cost by cost comparison between tools is conducted to determine the most economical tool, although in considering other than machine tools, the critical nature of costs is greatly reduced. Since the analysis of costs presented previously covers all types of manufacturing equipment, we will not adopt that general information to apply here. Instead, we will present a number of simple formulas which are used by tool engineers in solving problems concerning tools other than machine tools. These formulas are intended to supply answers to one or more of four basic questions the tool engineer is likely to ask about tooling requirements. These questions are:

1. What minimum quantity of parts must be produced to pay off a tool based on the savings in direct labor the tool will give over the present one? The cost of the tool and the estimated savings per part are known. For example, how many parts must be produced to justify buying a $600 tool to save $0.02 on the direct labor charge of each part?

2. What maximum sum can be spent on a tool based on the savings in direct labor the tool will give over the present one for a stipulated quantity of parts? The estimated savings per part and quantity of parts are known. For example, how much can be spent on a tool for a specified 50,000 parts, if the tool can save $0.04 on the direct labor charge of each part?

486   MATHEMATICS FOR INDUSTRY

3. What minimum length of time will a tool take to pay for itself based on the savings in direct labor the tool will give over the present one, for a stipulated production rate and carrying charges? The tool cost, estimated savings per part, rate of production, and fixed charges are known. For example, how long will a tool costing $400 take to pay for itself if there are savings of $0.03 per part on direct labor, 200 parts per hour, and an annual fixed charge of $5,000?

4. What maximum profit can be earned by a tool based on the savings the tool will give over the present one, for a stipulated quantity of parts? The tool's cost, estimated savings per part, and quantity of parts are known. For example, what profit will a $300 tool earn if it will save in direct labor cost $0.04 a part on 25,000 parts?

The first step is to establish formulas which will answer these questions. Symbols which are not necessarily related to the previous formulas, will be specified for the various factors to be formulated. They are:

N = number of parts processed annually, or per single run of less than a year's duration.

A = annual percentage allowance for interest on the initial investment.

B = annual percentage allowance for insurance and taxes.

C = annual percentage allowance for maintenance.

H = number of years required to pay off the investment with the tool's savings.

I = estimated cost of the tool ready to run, including costs incurred in its design, material and toolroom costs, in dollars.

$\dfrac{I}{H}$ = annual percentage allowance for depreciation, figured on a straight-line basis.

Y = annual cost of setups, in dollars. The expense of putting the machine into normal running condition is included here.

Z = the fixed charges plus set-up costs = $I\left(A+B+C+\dfrac{1}{H}\right) + Y$.

The annual savings a proposed tool will give over the present one, are represented by the following symbols:

S = annual total savings in direct labor costs, in dollars. These savings are equal to the number of parts processed annually, N, multiplied by the estimated savings per part in direct labor the proposed tool will give over the present one(s).

T = annual total savings in labor overhead, in dollars. These savings are equal to the annual total savings in direct labor costs, S, multiplied by the rate of overhead on the labor saved, t.

P = annual gross operating profit in excess of fixed charges and set-up costs, in dollars.

The formulas can be developed thus: To break even, the annual total savings in direct labor costs, plus labor overhead, must equal the annual total fixed charges and set-up costs for a comparable period. Thus:

$$S + T = Z$$

or,

$$NS + NST = I\left(A + B + C + \frac{I}{H}\right) + Y,$$

which becomes:

$$NS\,(I + T) = I\left(A + B + C + \frac{I}{H}\right) + Y.$$

Solving for N,

$$N = \frac{I\left(A + B + C + \frac{I}{H}\right) + Y}{S\,(I + T)}$$

Solving for I,

$$I = \frac{NS\,(I + T) - Y}{\left(A + B + C + \frac{I}{H}\right)}$$

Solving for H,

$$H = \frac{I}{NS\,(I + T) - I\,(A + B + C) - Y}$$

Solving for P,

P, the gross annual profit, is equal to the difference between $(S + T)$ and Z. Solving finally for P,

$$P = NS\,(I + T) - I\left(A + B + C + \frac{I}{H}\right) - Y$$

Factors, such as interest, insurance, taxes, maintenance, and depreciation, do not vary appreciably for normal periods of time in any one industrial organization, assuming a fairly stable management policy exists, and tool engineers often combine these near constants into a single factor of convenience. In many industrial concerns costs for tools are estimated as a percentage of machine equipment costs based on past knowledge and experience. Thirty-five per cent is a common percentage assigned in high production plants. For relatively low production, however, a tool cost analysis as in machine equipment selections, should be made. Dimensional and finish tolerances also affect how tool costs should be determined.

Notice that solving for symbols N, I, H, and P supplies answers to basic questions, 1, 2, 3 and 4, respectively.

PRACTICE PROBLEMS APPLICABLE IN THE SHOP AND TOOLROOM

1. Cost Comparing Problem

A manufacturer of bevel side gears has a situation confronting him of economically broaching splines in his product. Upon a recent increase in gear production potential, due to a sizeable increase in demand, the broaching facilities of the plant were found to be inadequate. Until the increase in production potential, three horizontal pull-type broaching machines had satisfactorily met the demand, but these machines have seen extensive service and are not useable in the new production schedule. The manufacturer has decided to replace the present machines with newer and more productive machines.

Broaching Machine Choices

The tool engineering department of the plant has conducted a thorough examination of the market, computed and compared costs, and has finally narrowed the search to two choices, A and B:

Choice A Broaching Machine: A double-spindle, vertical, hydraulic machine of the pull-type, a general class of machine in which the broaches are drawn or pulled through the work.

This machine is equipped with fixtures for broaching automobile connecting rods and bearing cups, but with proper re-tooling, it can be adapted for broaching gear splines.

The equipment can be tooled-up for a broaching production rate of 550 parts per hour. The cost of one machine with tooling is $23,000. In order to meet the anticipated demand of 4,000,000 parts annually, it will be necessary to purchase five machines. Enough flexibility is built into this type machine, so that with a minimum of re-tooling, different size gears can be broached and different broaching operations performed.

Choice B Broaching Machine: A four-station, vertical, pull-up type machine.

The machine is equipped with a gravity chute feed and loading slide so all the operator has to do is keep the four stations in the chute filled with parts. The other operations are fully automatic. This type broaching equipment also provides adequate flexibility with a minimum of re-tooling.

The broaching production rate for the machine is 1000 parts per hour. Its cost with tooling is $35,000. In order to meet the annual demand of 4,000,000 parts, three machines will be needed, although together, the three have a capacity well in excess of the four million quota.

In computing the costs, three costs will not enter into our calculations: the salvage value of the old machines, which is considered negligible, and the costs of direct material and indirect labor, which would be the same for both installations. The number of years of useful life for each type of equipment is estimated to be three years ($N = 3$). Other costs besides the initial investment, are tabulated as follows:

Interest = 10% per annum for both type machines.
Taxes and Insurance = 3% per annum for both type machines.
Direct Labor = $2.35 per hr. per man for both type machines.
Only one man is required to operate either type machine.
Maintenance = $275 per annum for each Choice A machine.
Maintenance = $480 per annum for each Choice B machine.
Power & Supplies = $310 per annum for each Choice A machine.
Power & Supplies = $415 per annum for each Choice B machine.
Floor space = $60 per annum for each Choice A machine.
Floor space = $90 per annum for each Choice B machine.

a. Which type machines should be chosen and how much in savings per year will be realized by using the chosen machines?

b. Calculate the unit costs for broaching splines on gears on each choice machine, and compare. Show the savings per part realized on chosen machines.

2. Critical Life Problem

Let us use the same machines from the previous problem, adding data where necessary, to create a new problem.

Let us also assume that the five double-spindle broaching machines have been in operation for several years, and that due to their age and increasing inefficiency, management feels that a new installation is necessary. Let us assume further, that the three, four-station broaching machines have already been shown to be the best in the field, except for the matter of critical equipment life.

Data from previous problem:

The total investment of new manufacturing
  equipment.................................. = $105,000
The number of units of the product yearly......... = 4,000,000
Current annual interest rate..................... = 10%
Current tax and insurance rate.................. = 3%

Additional data:

Costs per part produced yearly by old
  equipment, less the initial investment........... V_1 = $0.024
Costs per part produced yearly by new
  equipment, less the initial investment........... V_2 = $0.012
Estimated value of new equipment at the
  end of its equipment life (per machine).......... R_N = $5,000

Solve for the critical equipment life of the three four-station broaching machines.

3. Tool Economy Problems

A. *Solving for Number of Parts I.* A manufacturer is planning to purchase a fixture for tapping an aluminum die casting.

The fixture is a horizontal indexing device designed for tapping twelve equally spaced ¼"-20 through holes in the die casting. A microswitch arrangement integral with the fixture automatically controls the complete cycle of tapping the twelve holes. The hourly production rate is estimated to be 100 parts or 1200 holes, an appreciable improvement over the present arrangement, in which a simple manually-operated fixture without automatic control is used.

The final selection of the fixture under consideration is based on the answer to this question: How many parts must be processed using this fixture in one run or lot each year, to return its cost in two years out of the savings gained over the present fixture?

The following data is known:

The estimated cost of the fixture, together with its
  microswitch control ready for operation............. $375.00
The estimated savings in direct labor cost per part...... $0.005
The percentage overhead on direct labor savings........ 30%
The estimated cost of each set-up..................... $6.00

The annual percentage allowance for interest on the invested capital.................................	6%
The annual percentage allowance for insurance and taxes...	3%
The annual estimated allowance for maintenance.......	8%
The estimated equipment life........................	2 years
The annual percentage allowance for depreciation on a straight-line basis...........................	50%

B. *Solving for Number of Parts II.* If the parts in Problem 1 are put through in 8 runs or lots per year, how many parts must be processed annually to return the cost of the fixture out of savings in two years?

Notice that increasing the number of set-ups per year may increase the number of parts that must be processed annually. However, it is sometimes more economical to have multiple lots, despite a higher manufacturing cost per part, because of storage and other limiting factors.

C. *Solving for Lot Size.* Let us assume that the fixture in the foregoing problem is to pay for itself in a single lot, in a year. How large should that lot be?

Notice that demanding a shorter equipment life means more parts must be processed annually.

D. *Solving for Investment I.* The fixture is of the vertical clamping type, used in a drilling machine that drills a 23/64" diameter hole ln a stainless steel casting. The estimated drilling time is 12 seconds, giving a production rate of 240 parts per hour.

A spring-loaded locator in the fixture positions the parts; the same fixture can also be used for a 3/8" reaming operation.

How large a capital investment is justified in a fixture like this for a single run of 35,000 castings, at an estimated savings in direct labor cost of $0.01 per part? Let us assume that all values are the same as in Problem C, except that the savings per part in direct labor = $0.01 and the set-up cost = $10.

E. *Solving for Investment II.* How much of an initial investment is justified in a fixture similar to that in Problem D, only one of more rugged design, so that it will process 50,000 castings per year in eight runs? A savings in direct labor of $0.007 per casting is estimated. The fixture's cost is to be paid off in two years. Assume all other values are the same as those in Problem D.

F. **Solving for Production Time.** How long will a fixture similar to the one in Problem D take to pay for itself, estimating its cost at $430, with a single set-up cost of $80? Assume all other values are the same as those in Problem E.

G. **Solving for Profit.** Assume that the fixture in Problem F costs $300 instead of $430. What will be the gross annual profit, if the fixture meets all other conditions of Problem F, including the depreciation allowance percentage of $\frac{1}{1.42}$ or 70.5%?

ANSWERS TO PROBLEMS IN SELF-EVALUATING TEST

Problem
1. 8-3/4″
2. 7-9/64″
3. 1-5/64″
4. A. 7-37/64″
   B. 35/64″
5. B. 27/32″
6. A. 7-5/16″
7. 5-5/8″
8. 3-1/16″
9. 6-13/64″
10. 6-31/64″
11. 6-3/8″
12. 5-55/64″
13. 6-1/8″
14. 1-37/53 per ft
15. 1-43/53 per ft
16. 2-2/53 per ft
17. 1-43/53 per ft
18. 2-14/53 per ft
19. 1-2/7 per ft
20. 40-53/70 lb of alum.
    3-2/105 lb of silicon
    1-107/210 lb of iron
21. 52-5/16″
22. 4-256/327
23. 5-23/90
24. 503-299/560
25. 889-144481/162656
26. 43-3/5 lb of alum.
    2-68/75 lb of copper
    1-211/225 lb of other elements
27. 31-32/425 lb of copper
    1-137/425 lb of beryllium
    281/425 lb of other elements
28. A. 2.401″
    B. 2.027″
29. A. 2.425″
    B. 2.051″
30. A. 2.365″
    B. 1.991″
31. Rad. .260″
    A. 2.391″
32. A. .9375
    B. 1.3258″
    C. .4166
    D. .5469″
33. *For* $d = 1/2″$
    A. .7500″
    B. .7071″
    C. .3333″
    D. .4375″
    For $d = 9/16″$
    A. .8437
    B. .7955
    C. .3750
    D. .4922
    For $d = 3/4″$
    A. 1.125″
    B. 1.0606″
    C. .4999″
    D. .6562″
34. 3.486″

35. 3.485″
36. 57/64
  29/32″
37. 21704.5998117
38. .01103
39. .890625
  .609375
40. 27/32
41. 12/25
42. 722.
43. $12\sqrt{3}$
44. 14.812
45. 2.9
46. $-576\,ax^5y^2-.744a^2x^6y^4+1116a^3x^4y^3+868a^4x^5y^5+816a^4\times2y^2+536a^5x^3y^4-629a^6\times y^3$.
47. $12mxy^3+16nx^2y-18mnx^3y^4$.
48. $(a-24)(a-8)$
49. .0000007206
50. 6,530,000
51. 327.122

52. $h = \dfrac{\pi d^2 - 4A}{4b}$

  $d = \dfrac{2\sqrt{bh+A}}{\pi}$

53. 1.9677 sq in.

54. $h = \dfrac{4V}{\pi(D^2-d^2)}$

  $D = \sqrt{\dfrac{4V}{\pi h}+d^2}$

  $d = \sqrt{D^2-\dfrac{4V}{\pi h}}$

55. 40.566 cu in.
56. $\angle ACB$ 37°
  $\angle ACD$ 143°
  $\angle DCE$ 37°
  $\angle CAF$ 127°
57. Indicate graphically on figure.
58. X 14°
  Y 104°
  Z 76°
  U 45°
  V 76°
59. R 2.952″
  r 1.640″

APPENDIX OF USEFUL TABLES

TABLE 1

MATHEMATICAL SHORT-CUTS

Circle

Circumference	= diam. ×3.1416, or diam. ÷0.3183.
Diameter	= circum. ×0.3183, or circum. ÷3.1416.
Radius	= circum. ×0.15915, or circum. ÷6.28318.
Side of inscribed square	= diam. ×0.7071.
	= circum. ×0.2251, or circum. ÷4.4428.
Side of equal square	= diam. ×0.8862, or diam. ÷1.1284.
	= circum. ×0.2821, or circum. ÷3.545.
Area	= circum. × ¼ diam.
	= sq. of diam. ×0.7854.
	= sq. of circum. ×.07958.
	= sq. ½ diam. ×3.1416.

Square

A side ×1.4142	= diam. of circumscribing circle.
A side ×4.4428	= circum. of circumscribing circle.
A side ×1.1284	= diam. of equal circle.
A side ×3.545	= circum. of equal circle.
Area in sq. in. ×1.273	= circular ins. of equal circle

Sphere or Globe

Surface area	= diam. ×circum.
	= sq. of diam. ×3.1416.
	= 4×sq. of radius ×3.1416.

TABLE 2

Squares, Cubes, Square Roots, Cube Roots, and Reciprocals of Numbers from 1 to 50

No.	Square	Cube	Sq. Root	Cube Root	Reciprocal
1	1	1	1.00000	1.00000	1.0000000
2	4	8	1.41421	1.25992	0.5000000
3	9	27	1.73205	1.44225	0.3333333
4	16	64	2.00000	1.58740	0.2500000
5	25	125	2.23607	1.70998	0.2000000
6	36	216	2.44949	1.81712	0.1666667
7	49	343	2.64575	1.91293	0.1428571
8	64	512	2.82843	2.00000	0.1250000
9	81	729	3.00000	2.08008	0.1111111
10	100	1,000	3.16228	2.15443	0.1000000
11	121	1,331	3.31662	2.22398	0.0909091
12	144	1,728	3.46410	2.28943	0.0833333
13	169	2,197	3.60555	2.35133	0.0769231
14	196	2,744	3.74166	2.41014	0.0714286
15	225	3,375	3.87298	2.46621	0.0666667
16	256	4,096	4.00000	2.51984	0.0625000
17	289	4,913	4.12311	2.57128	0.0588235
18	324	5,832	4.24264	2.62074	0.0555556
19	361	6,859	4.35890	2.66840	0.0526316
20	400	8,000	4.47214	2.71442	0.0500000
21	441	9,261	4.58258	2.75892	0.0476190
22	484	10,648	4.69042	2.80204	0.0454545
23	529	12,167	4.79583	2.84387	0.0434783
24	576	13,824	4.89898	2.88450	0.0416667
25	625	15,625	5.00000	2.92402	0.0400000
26	676	17,576	5.09902	2.96250	0.0384615
27	729	19,683	5.19615	3.00000	0.0370370
28	784	21,952	5.29150	3.03659	0.0357143
29	841	24,389	5.38516	3.07232	0.0344828
30	900	27,000	5.47723	3.10723	0.0333333
31	961	29,791	5.56776	3.14138	0.0322581
32	1,024	32,768	5.65685	3.17480	0.0312500
33	1,089	35,937	5.74456	3.20753	0.0303030
34	1,156	39,304	5.83095	3.23961	0.0294118
35	1,225	42,875	5.91608	3.27107	0.0285714
36	1,296	46,656	6.00000	3.30193	0.0277778
37	1,369	50,653	6.08276	3.33222	0.0270270
38	1,444	54,872	6.16441	3.36198	0.0263158
39	1,521	59,319	6.24500	3.39121	0.0256410
40	1,600	64,000	6.32456	3.41995	0.0250000
41	1,681	68,921	6.40312	3.44822	0.0243902
42	1,764	74,088	6.48074	3.47603	0.0238095
43	1,849	79,507	6.55744	3.50340	0.0232558
44	1,936	85,184	6.63325	3.53035	0.0227273
45	2,025	91,125	6.70820	3.55689	0.0222222
46	2,116	97,336	6.78233	3.58305	0.0217391
47	2,209	103,823	6.85565	3.60883	0.0212766
48	2,304	110,592	6.92820	3.63424	0.0208333
49	2,401	117,649	7.00000	3.65931	0.0204082
50	2,500	125,000	7.07107	3.68403	0.0200000
No.	Square	Cube	Sq. Root	Cube Root	Reciprocal

TABLE 2—Continued

Squares, Cubes, Square Roots, Cube Roots, and Reciprocals of Numbers from 51 to 100

No.	Square	Cube	Sq. Root	Cube Root	Reciprocal
51	2,601	132,651	7.14143	3.70843	0.0196078
52	2,704	140,608	7.21110	3.73251	0.0192308
53	2,809	148,877	7.28011	3.75629	0.0188679
54	2,916	157,464	7.34847	3.77976	0.0185185
55	3,025	166,375	7.41620	3.80295	0.0181818
56	3,136	175,616	7.48331	3.82586	0.0178571
57	3,249	185,193	7.54983	3.84850	0.0175439
58	3,364	195,112	7.61577	3.87088	0.0172414
59	3,481	205,379	7.68115	3.89300	0.0169492
60	3,600	216,000	7.74597	3.91487	0.0166667
61	3,721	226,981	7.81025	3.93650	0.0163934
62	3,844	238,328	7.87401	3.95789	0.0161290
63	3,969	250,047	7.93725	3.97906	0.0158730
64	4,096	262,144	8.00000	4.00000	0.0156250
65	4,225	274,625	8.06226	4.02073	0.0153846
66	4,356	287,496	8.12404	4.04124	0.0151515
67	4,489	300,763	8.18535	4.06155	0.0149254
68	4,624	314,432	8.24621	4.08166	0.0147059
69	4,761	328,509	8.30662	4.10157	0.0144928
70	4,900	343,000	8.36660	4.12129	0.0142857
71	5,041	357,911	8.42615	4.14082	0.0140845
72	5,184	373,248	8.48528	4.16017	0.0138889
73	5,329	389,017	8.54400	4.17934	0.0136986
74	5,476	405,224	8.60233	4.19834	0.0135135
75	5,625	421,875	8.66025	4.21716	0.0133333
76	5,776	438,976	8.71780	4.23582	0.0131579
77	5,929	456,533	8.77496	4.25432	0.0129870
78	6,084	474,552	8.83176	4.27266	0.0128205
79	6,241	493,039	8.88819	4.29084	0.0126582
80	6,400	512,000	8.94427	4.30887	0.0125000
81	6,561	531,441	9.00000	4.32675	0.0123457
82	6,724	551,368	9.05539	4.34448	0.0121951
83	6,889	571,787	9.11043	4.36207	0.0120482
84	7,056	592,704	9.16515	4.37952	0.0119048
85	7,225	614,125	9.21954	4.39683	0.0117647
86	7,396	636,056	9.27362	4.41400	0.0116279
87	7,569	658,503	9.32738	4.43105	0.0114943
88	7,744	681,472	9.38083	4.44797	0.0113636
89	7,921	704,969	9.43398	4.46475	0.0112360
90	8,100	729,000	9.48683	4.48140	0.0111111
91	8,281	753,571	9.53939	4.49794	0.0109890
92	8,464	778,688	9.59166	4.51436	0.0108696
93	8,649	804,357	9.64365	4.53065	0.0107527
94	8,836	830,584	9.69536	4.54684	0.0106383
95	9,025	857,375	9.74679	4.56290	0.0105263
96	9,216	884,736	9.79796	4.57886	0.0104167
97	9,409	912,673	9.84886	4.59470	0.0103093
98	9,604	941,192	9.89949	4.61044	0.0102041
99	9,801	970,299	9.94987	4.62607	0.0101010
100	10,000	1,000,000	10.00000	4.64159	0.0100000
No.	Square	Cube	Sq. Root	Cube Root	Reciprocal

TABLE 3
Decimal Equivalents of Common Fractions of an Inch

Common Fractions	Decimal Equivalents	Common Fractions	Decimal Equivalents	Common Fractions	Decimal Equivalents	Common Fractions	Decimal Equivalents
1/64	.015625	17/64	.265625	33/64	.515625	49/64	.765625
1/32	.03125	9/32	.28125	17/32	.53125	25/32	.78125
3/64	.046875	19/64	.296875	35/64	.546875	51/64	.796875
1/16	.0625	5/16	.3125	9/16	.5625	13/16	.8125
5/64	.078125	21/64	.328125	37/64	.578125	53/64	.828125
3/32	.09375	11/32	.34375	19/32	.59375	27/32	.84375
7/64	.109375	23/64	.359375	39/64	.609375	55/64	.859375
1/8	.125	3/8	.375	5/8	.625	7/8	.875
9/64	.140625	25/64	.390625	41/64	.640625	57/64	.890625
5/32	.15625	13/32	.40625	21/32	.65625	29/32	.90625
11/64	.171875	27/64	.421875	43/64	.671875	59/64	.921875
3/16	.1875	7/16	.4375	11/16	.6875	15/16	.9375
13/64	.203125	29/64	.453125	45/64	.703125	61/64	.953125
7/32	.21875	15/32	.46875	23/32	.71875	31/32	.96875
15/64	.234375	31/64	.484375	47/64	.734375	63/64	.984375
1/4	.25	1/2	.50	3/4	.75		

TABLE 4

Square and Cube Roots of Decimal Numbers from .01 to 1

Decimal	Square Root	Cube Root	Decimal	Square Root	Cube Root
0.01	0.1000	0.2154	0.51	0.7141	0.7990
0.02	0.1414	0.2714	0.52	0.7211	0.8041
0.03	0.1732	0.3107	0.53	0.7280	0.8093
0.04	0.2000	0.3420	0.54	0.7348	0.8143
0.05	0.2236	0.3684	0.55	0.7416	0.8193
0.06	0.2449	0.3915	0.56	0.7483	0.8243
0.07	0.2646	0.4121	0.57	0.7550	0.8291
0.08	0.2828	0.4309	0.58	0.7616	0.8340
0.09	0.3000	0.4481	0.59	0.7681	0.8387
0.10	0.3162	0.4642	0.60	0.7746	0.8434
0.11	0.3317	0.4791	0.61	0.7810	0.8481
0.12	0.3464	0.4932	0.62	0.7874	0.8527
0.13	0.3606	0.5066	0.63	0.7937	0.8573
0.14	0.3742	0.5192	0.64	0.8000	0.8618
0.15	0.3873	0.5313	0.65	0.8062	0.8662
0.16	0.4000	0.5429	0.66	0.8124	0.8707
0.17	0.4123	0.5540	0.67	0.8185	0.8750
0.18	0.4243	0.5646	0.68	0.8246	0.8794
0.19	0.4359	0.5749	0.69	0.8307	0.8837
0.20	0.4472	0.5848	0.70	0.8367	0.8879
0.21	0.4583	0.5944	0.71	0.8426	0.8921
0.22	0.4690	0.6037	0.72	0.8485	0.8963
0.23	0.4796	0.6127	0.73	0.8544	0.9004
0.24	0.4899	0.6214	0.74	0.8602	0.9045
0.25	0.5000	0.6300	0.75	0.8660	0.9086
0.26	0.5099	0.6383	0.76	0.8718	0.9126
0.27	0.5196	0.6463	0.77	0.8775	0.9166
0.28	0.5292	0.6542	0.78	0.8832	0.9205
0.29	0.5385	0.6619	0.79	0.8888	0.9244
0.30	0.5477	0.6694	0.80	0.8944	0.9283
0.31	0.5568	0.6768	0.81	0.9000	0.9322
0.32	0.5657	0.6840	0.82	0.9055	0.9360
0.33	0.5745	0.6910	0.83	0.9110	0.9398
0.34	0.5831	0.6980	0.84	0.9165	0.9435
0.35	0.5916	0.7047	0.85	0.9220	0.9473
0.36	0.6000	0.7114	0.86	0.9274	0.9510
0.37	0.6083	0.7179	0.87	0.9327	0.9546
0.38	0.6164	0.7243	0.88	0.9381	0.9583
0.39	0.6245	0.7306	0.89	0.9434	0.9619
0.40	0.6325	0.7368	0.90	0.9487	0.9655
0.41	0.6403	0.7429	0.91	0.9539	0.9691
0.42	0.6481	0.7489	0.92	0.9592	0.9726
0.43	0.6557	0.7548	0.93	0.9644	0.9761
0.44	0.6633	0.7606	0.94	0.9695	0.9796
0.45	0.6708	0.7663	0.95	0.9747	0.9830
0.46	0.6782	0.7719	0.96	0.9798	0.9865
0.47	0.6856	0.7775	0.97	0.9849	0.9899
0.48	0.6928	0.7830	0.98	0.9899	0.9933
0.49	0.7000	0.7884	0.99	0.9950	0.9967
0.50	0.7071	0.7937	1.00	1.0000	1.0000

TABLE 5

Squares of Mixed Numbers from 1/64 to 6 by 64ths

Fraction	0	1	2	3	4	5
1/64	0.00024	1.03149	4.06274	9.09399	16.12524	25.15649
1/32	0.00098	1.06348	4.12598	9.18848	16.25098	25.31348
3/64	0.00220	1.09595	4.18970	9.28345	16.37720	25.47095
1/16	0.00391	1.12891	4.25391	9.37891	16.50391	25.62891
5/64	0.00610	1.16235	4.31860	9.47485	16.63110	25.78735
3/32	0.00879	1.19629	4.38379	9.57129	16.75879	25.94629
7/64	0.01196	1.23071	4.44946	9.66821	16.88696	26.10571
1/8	0.01562	1.26562	4.51562	9.76562	17.01562	26.26562
9/64	0.01978	1.30103	4,58228	9.86353	17.14478	26.42603
5/32	0.02441	1.33691	4.64941	9.96191	17.27441	26.58691
11/64	0.02954	1.37329	4.71704	10.06079	17.40454	26.74829
3/16	0.03516	1.41016	4.78516	10.16016	17.53516	26.91016
13/64	0.04126	1.44751	4.85376	10.26001	17.66626	27.07251
7/32	0.04785	1.48535	4.92285	10.36035	17.79785	27.23585
15/64	0.05493	1.52368	4.99243	10.46118	17.92993	27.39868
1/4	0.06250	1.56250	5.06250	10.56250	18.06250	27.56250
17/64	0.07056	1.60181	5.13306	10.66431	18.19556	27.72681
9/32	0.07910	1.64160	5.20410	10.76660	18.32910	27.89160
19/64	0.08813	1.68188	5.27563	10.86938	18.46313	28.05688
5/16	0.09766	1.72266	5.34766	10.97266	18.59766	28.22266
21/64	0.10767	1.76392	5.42017	11.07642	18.73267	28.38892
11/32	0.11816	1.80566	5.49316	11.18066	18.86816	28.55566
23/64	0.12915	1.84790	5.56663	11.28540	19.00415	28.72290
3/8	0.14062	1.89062	5.64062	11.39062	19.14062	28.89062
25/64	0.15259	1.93384	5.71509	11.49634	19.27759	29.05884
13/32	0.16504	1.97754	5.79004	11.60254	19.41504	29.22754
27/64	0.17798	2.02173	5.86548	11.70923	19.55298	29.39673
7/16	0.19141	2.06641	5.94141	11.81641	19.69141	29.56641
29/64	0.20532	2.11157	6.01782	11.92407	19.83032	29.73657
15/32	0.21973	2.15723	6.09473	12.03223	19.96973	29.90723
31/64	0.23462	2.20337	6.17212	12.14087	20.10962	30.07837
1/2	0.25000	2.25000	6.25000	12.25000	20.25000	30.25000
33/64	0.26587	2.29712	6.32837	12.35962	20.39087	30.42212
17/32	0.28223	2.34473	6.40723	12.46973	20.53223	30.59473
35/64	0.29907	2.39282	6.48657	12.58032	20.67407	30.76782
9/16	0.31641	2.44141	6.56641	12.69141	20 81641	30.94141
37/64	0.33423	2.49048	6.64673	12.80298	20.95923	31.11548
19/32	0.35254	2.54004	6.72754	12.91504	21.10254	31.29004
39/64	0.37134	2.59009	6.80884	13.02759	21.24634	31.46509
5/8	0.39062	2.64062	6.89062	13.14062	21.39062	31.64062
41/64	0.41040	2.69165	6.97290	13.25415	21.53540	31.81665
21/32	0.43066	2.74316	7.05566	13.36816	21.68066	31.99316
43/64	0.45142	2.79517	7.13892	13.48267	21.82642	32.17017
11/16	0.47266	2.84766	7.22266	13.59766	21.97266	32.34766
45/64	0.49438	2.90063	7.30688	13.71313	22.11938	32.52563
23/32	0.51660	2.95410	7.39160	13.82910	22.26660	32.70410
47/64	0.53931	3.00806	7.47681	13.94556	22.41431	32.88306
3/4	0.56250	3.06250	7.56250	14.06250	22.56250	33.06250
49/64	0.58618	3.11743	7.64868	14.17993	22.71118	33.24243
25/32	0.61035	3.17285	7.73535	14.29785	22.86035	33.42285
51/64	0.63501	3.22876	7.82251	14.41626	23.01001	33.60376
13/16	0.66016	3.28516	7.91016	14.53516	23.16016	33.78516
53/64	0.68579	3.34204	7.99829	14.65454	23.31079	33.96704
27/32	0.71191	3.39941	8.08691	14.77441	23.46191	34.14941
55/64	0.73853	3.45728	8.17603	14.89478	23.61363	34.33228
7/8	0.76562	3.51562	8.26562	15.01562	23.76562	34.51562
57/64	0.79321	3.57446	8.35571	15.13696	23.91821	34.69946
29/32	0.82129	3.63379	8.44629	15.25879	24.07129	34.88379
59/64	0.84985	3.69360	8.53735	15.38110	24.22485	35.06860
15/16	0.87891	3.75391	8.62891	15.50391	24.37891	35.25391
61/64	0.90845	3.81470	8.72095	15.62720	24.53345	35.43970
31/32	0.93848	3.87598	8.81348	15.75098	24.68848	35.62598
63/64	0.96899	3.93774	8.90649	15.87524	24.84399	35.81274

TABLE 5—*Continued*

Squares of Mixed Numbers from 6 to 12 by 64ths

Fraction		6	7	8	9	10	11
1/64		36.18774	49.21899	64.25024	81.28149	100.31274	121.34399
1/32		36.37598	49.43848	64.50098	81.56348	100.62598	121.68848
	3/64	36.56470	49.65845	64.75220	81.84595	100.93970	122.03345
1/16		36.75391	49.87891	65.00391	82.12891	101.25391	122.37891
	5/64	36.94360	50.09985	65.25610	82.41235	101.56860	122.72485
3/32		37.13379	50.32129	65.50879	82.69629	101.88379	123.07129
	7/64	37.32446	50.54321	65.76196	82.98071	102.19946	123.41821
1/8		37.51562	50.76562	66.01562	83.26562	102.51562	123.76562
9/64		37.70728	50.98853	66.26978	83.55103	102.83228	124.11353
	5/32	37.89941	51.21191	66.52441	83.83691	103.14941	124.46191
11/64		38.09204	51.43579	66.77954	84.12329	103.46704	124.81079
	3/16	38.28516	51.66016	67.03516	84.41016	103.78516	125.16016
13/64		38.47876	51.88501	67.29126	84.69751	104.10376	125.51001
	7/32	38.67285	52.11035	67.54785	84.98535	104.42285	125.86035
15/64		38.86743	52.33618	67.80493	85.27368	104.74243	126.21110
1/4		39.06250	52.56250	68.06250	85.56250	105.06250	126.56250
17/64		39.25806	52.78931	68.32056	85.85181	105.38306	126.91431
	9/32	39.45410	53.01660	68.57910	86.14160	105.70410	127.26660
19/64		39.65063	53.24438	68.83813	86.43188	106.02563	127.61938
	5/16	39.84766	53.47266	69.09766	86.72266	106.34766	127.97266
21/64		40.04517	53.70142	69.35767	87.01392	106.67017	128.32642
	11/32	40.24316	53.93066	69.61816	87.30566	106.99316	128.68066
23/64		40.44165	54.16040	69.87915	87.59790	107.31665	129.03540
3/8		40.64062	54.39062	70.14062	87.89062	107.64062	129.39062
25/64		40.84009	54.62134	70.40259	88.18384	107.96509	129.74634
	13/32	41.04004	54.85254	70.66504	88.47754	108.29004	130.10254
27/64		41.24048	55.08423	70.92798	88.77173	108.61548	130.45923
	7/16	41.44141	55.31641	71.19141	89.06641	108.94141	130.81641
29/64		41.64282	55.54907	71.45532	89.36157	109.26782	131.17407
	15/32	41.84473	55.78223	71.71973	89.65723	109.59473	131.53223
31/64		42.04712	56.01587	71.98462	89.95337	109.92212	131.89087
1/2		42.25000	56.25000	72.25000	90.25000	110.25000	132.25000
33/64		42.45337	56.48462	72.51587	90.54712	110.57837	132.60962
	17/32	42.65723	56.71973	72.78223	90.84473	110.90723	132.96973
35/64		42.86157	56.95532	73.04907	91.14282	111.23657	133.33032
	9/16	43.06641	57.19141	73.31641	91.44141	111.56641	133.69141
37/64		43.27173	57.42798	73.58423	91.74048	111.89673	134.05298
	19/32	43.47754	57.66504	73.85254	92.04004	112.22754	134.41504
39/64		43.68384	57.90259	74.12134	92.34009	112.55884	134.77759
5/8		43.89062	58.14062	74.39062	92.64062	112.89062	135.14062
41/64		44.09790	58.37915	74.66040	92.94165	113.22290	135.50415
	21/32	44.30566	58.61816	74.93066	93.24316	113.55566	135.86816
43/64		44.51392	58.85767	75.20142	93.54517	113.88892	136.23267
	11/16	44.72266	59.09766	75.47266	93.84766	114.22266	136.59766
45/64		44.93188	59.33813	75.74438	94.15063	114.55688	136.96313
	23/32	45.14160	59.57910	76.01660	94.45410	114.89160	137.32910
47/64		45.35181	59.82056	76.28931	94.75806	115.22681	137.69556
3/4		45.56250	60.06250	76.56250	95.06250	115.56250	138.06250
49/64		45.77368	60.30493	76.83618	95.36743	115.89868	138.42993
	25/32	45.98535	60.54785	77.11035	95.67285	116.23535	138.79785
51/64		46.19751	60.79126	77.38501	95.97876	116.57251	139.16626
	13/16	46.41016	61.03516	77.66016	96.28516	116.91016	139.53516
53/64		46.62329	61.27954	77.93579	96.59204	117.24829	139.90454
	27/32	46.83691	61.52441	78.21191	96.89941	117.58691	140.27441
55/64		47.05103	61.76978	78.48853	97.20728	117.92603	140.64478
7/8		47.26562	62.01562	78.76562	97.51562	118.26562	141.01562
57/64		47.48071	62.26196	79.04321	97.82446	118.60571	141.38696
	29/32	47.69629	62.50879	79.32129	98.13379	118.94629	141.75879
59/64		47.91235	62.75610	79.59985	98.44360	119.28735	142.13110
	15/16	48.12891	63.00391	79.87891	98.75391	119.62891	142.50391
61/64		48.34595	63.25220	80.15845	99.06470	119.97095	142.87720
	31/32	48.56348	63.50098	80.43848	99.37598	120.31348	143.25098
63/64		48.78149	63.75024	80.71899	99.68774	120.65649	143.62524

TABLE 6

Decimal Equivalents, Squares, Cubes, Square and Cube Roots, Circumferences and Areas of Circles, from 1/64 to 1 Inch in Diameter

Dia.	Dec. Equiv.	Square	Square Root	Cube	Cube Root	Circle Circum.	Circle Area
1/64	.015625	.0002441	.125	.000003815	.25	.04909	.000192
1/32	.03125	.0009766	.176777	.000030518	.31498	.09817	.000767
3/64	.046875	.0021973	.216506	.000102997	.36056	.14726	.001726
1/16	.0625	.0039063	.25	.00024414	.39685	.19635	.003068
5/64	.078125	.0061035	.279508	.00047684	.42749	.24544	.004794
3/32	.09375	.0087891	.306186	.00082397	.45428	.29452	.006903
7/64	.109375	.0119629	.330719	.0013084	.47823	.34361	.009396
1/8	.125	.015625	.353553	.0019531	.5	.39270	.012272
9/64	.140625	.0197754	.375	.0027809	.52002	.44179	.015532
5/32	.15625	.0244141	.395285	.0038147	.53861	.49087	.019175
11/64	.171875	.0295410	.414578	.0050774	.55600	.53996	.023201
3/16	.1875	.0351563	.433013	.0065918	.57236	.58905	.027611
13/64	.203125	.0412598	.450694	.0083809	.58783	.63814	.032405
7/32	.21875	.0478516	.467707	.010468	.60254	.68722	.037583
15/64	.234375	.0549316	.484123	.012875	.61655	.73631	.043143
1/4	.25	.0625	.5	.015625	.62996	.78540	.049087
17/64	.265625	.0705566	.515388	.018742	.64282	.83449	.055415
9/32	.28125	.0791016	.530330	.022247	.65519	.88357	.062126
19/64	.296875	.0881348	.544862	.026165	.66710	.93266	.069221
5/16	.3125	.0976562	.559017	.030518	.67860	.98175	.076699
21/64	.328125	.107666	.572822	.035328	.68973	1.03084	.084561
11/32	.34375	.118164	.586302	.040619	.70051	1.07992	.092806
23/64	.359375	.129150	.599479	.046413	.71097	1.12901	.101434
3/8	.375	.140625	.612372	.052734	.72112	1.17810	.110445
25/64	.390625	.1525879	.625	.059605	.73100	1.22718	.119842
13/32	.40625	.1650351	.637377	.067047	.74062	1.27627	.129621
27/64	.421875	.1779785	.649519	.075085	.75	1.32536	.139784
7/16	.4375	.1914063	.661438	.083740	.75915	1.37445	.150330
29/64	.453125	.2053223	.673146	.093037	.76808	1.42353	.161260
15/32	.46875	.2197266	.684653	.102997	.77681	1.47262	.172573
31/64	.484375	.2346191	.695971	.113644	.78535	1.52171	.184269
1/2	.5	.25	.707107	.125	.79370	1.57080	.196350
33/64	.515625	.265869	.718070	.137089	.80188	1.61988	.208813
17/32	.53125	.282227	.728869	.149933	.80990	1.66897	.221660
35/64	.546875	.299072	.739510	.163555	.81777	1.71806	.234891
9/16	.5625	.316406	.75	.177979	.82548	1.76715	.248505
37/64	.578125	.334229	.760345	.193226	.83306	1.81623	.262502
19/32	.59375	.352539	.770552	.209320	.84049	1.86532	.276884
39/64	.609375	.371338	.780625	.226284	.84780	1.91441	.291648
5/8	.625	.390625	.790569	.244141	.85499	1.96350	.306796
41/64	.640625	.410400	.800391	.262913	.86205	2.01258	.322328
21/32	.65625	.430664	.810093	.282623	.86901	2.06167	.338243
43/64	.671875	.451416	.819680	.303295	.87585	2.11076	.354541
11/16	.6875	.472656	.829156	.324951	.88259	2.15984	.371223
45/64	.703125	.494385	.838525	.347614	.88922	2.20893	.388289
23/32	.71875	.516602	.847791	.371307	.89576	2.25802	.405737
47/64	.734375	.539307	.856957	.396053	.90221	2.30711	.423570
3/4	.75	.5625	.866025	.421875	.90856	2.35619	.441786
49/64	.765625	.586182	.875	.448795	.91483	2.40528	.460386
25/32	.78125	.610352	.883883	.476837	.92101	2.45437	.479369
51/64	.796875	.635010	.892679	.506023	.92711	2.50346	.498736
13/16	.8125	.660156	.901388	.536377	.93313	2.55254	.518486
53/64	.828125	.685791	.910014	.567921	.93907	2.60163	.538619
27/32	.84375	.711914	.918559	.600677	.94494	2.65072	.559136
55/64	.859375	.738525	.927024	.634670	.95074	2.69981	.580036
7/8	.875	.765625	.935414	.669922	.95647	2.74889	.601320
57/64	.890625	.793213	.943729	.706455	.96213	2.79798	.622988
29/32	.90625	.821289	.951972	.744293	.96772	2.84707	.645039
59/64	.921875	.849854	.960143	.783459	.97325	2.89616	.667473
15/16	.9375	.878906	.968246	.823975	.97872	2.94524	.690291
61/64	.953125	.908447	.976281	.865864	.98412	2.99433	.713493
31/32	.96875	.938477	.984251	.909149	.98947	3.04342	.737078
63/64	.984375	.968994	.992157	.953854	.99476	3.09251	.761046
1	1	1	1	1	1	3.14159	.785398

TABLE 7

Inches and Equivalents in Millimeters

In.	Mm.	In.	Mm.	In.	Mm.	In.	Mm.
1	25.4	26	660.4	51	1295.4	76	1930.4
2	50.8	27	685.8	52	1320.8	77	1955.8
3	76.2	28	711.2	53	1346.2	78	1981.2
4	101.6	29	736.6	54	1371.6	79	2006.6
5	127.0	30	762.0	55	1397.0	80	2032.0
6	152.4	31	787.4	56	1422.4	81	2057.4
7	177.8	32	812.8	57	1447.8	82	2082.8
8	203.2	33	838.2	58	1473.2	83	2108.2
9	228.6	34	863.6	59	1498.6	84	2133.6
10	254.0	35	889.0	60	1524.0	85	2159.0
11	279.4	36	914.4	61	1549.4	86	2184.4
12	304.8	37	939.8	62	1574.8	87	2209.8
13	330.2	38	965.2	63	1600.2	88	2235.2
14	355.6	39	990.6	64	1625.6	89	2260.6
15	381.0	40	1016.0	65	1651.0	90	2286.0
16	406.4	41	1041.4	66	1676.4	91	2311.4
17	431.8	42	1066.8	67	1701.8	92	2336.8
18	457.2	43	1092.2	68	1727.2	93	2362.2
19	482.6	44	1117.6	69	1752.6	94	2387.6
20	508.0	45	1143.0	70	1778.0	95	2413.0
21	533.4	46	1168.4	71	1803.4	96	2438.4
22	558.8	47	1193.8	72	1828.8	97	2463.8
23	584.2	48	1219.2	73	1854.2	98	2489.2
24	609.6	49	1244.6	74	1879.6	99	2514.6
25	635.0	50	1270.0	75	1905.0	100	2540.0

Millimeters and Equivalents in Inches

Mm.	In.	Mm.	In.	Mm.	In.	Mm.	In.
1	0.039370	26	1.023622	51	2.007874	76	2.992126
2	0.078740	27	1.062992	52	2.047244	77	3.031496
3	0.118110	28	1.102362	53	2.086614	78	3.070866
4	0.157480	29	1.141732	54	2.125984	79	3.110236
5	0.196850	30	1.181102	55	2.165354	80	3.149606
6	0.236220	31	1.220472	56	2.204724	81	3.188976
7	0.275591	32	1.259843	57	2.244094	82	3.228346
8	0.314961	33	1.299213	58	2.283465	83	3.267717
9	0.354331	34	1.338583	59	2.322835	84	3.307087
10	0.393701	35	1.377953	60	2.362205	85	3.346457
11	0.433071	36	1.417323	61	2.401575	86	3.385827
12	0.472441	37	1.456693	62	2.440945	87	3.425197
13	0.511811	38	1.496063	63	2.480315	88	3.464567
14	0.551181	39	1.535433	64	2.519685	89	3.503937
15	0.590551	40	1.574803	65	2.559055	90	3.543307
16	0.629921	41	1.614173	66	2.598425	91	3.582677
17	0.669291	42	1.653543	67	2.637795	92	3.622047
18	0.708661	43	1.692913	68	2.677165	93	3.661417
19	0.748031	44	1.732283	69	2.716535	94	3.700787
20	0.787402	45	1.771654	70	2.755906	95	3.740157
21	0.826772	46	1.811024	71	2.795276	96	3.779528
22	0.866142	47	1.850394	72	2.834646	97	3.818898
23	0.905512	48	1.889764	73	2.874016	98	3.858268
24	0.944882	49	1.929134	74	2.913386	99	3.897638
25	0.984252	50	1.968504	75	2.952756	100	3.937008

TABLE 8
Specific Gravities and Weights of Commonly Used Materials

The Basis for Specific Gravities Is Pure Water at 62° F. Weight of 1 Cu. Ft. = 62.355 Lb.	Average Specific Gravity. Water = 1	Average Weight of 1 Cu. In. in Lb.	Average Weight of 1 Cu. Ft. in Lb.
Air, atmospheric at 60°F., under pressure of 14.7 p.s.i.	.00123		.0765
Aluminum	2.56	.0924	159.7
Antimony	6.71	.2422	418.7
Bismuth	9.80	.3538	611.5
Brass:			
80 copper + 20 zinc	8.60	.3105	536.6
70 copper + 30 zinc	8.40	.3032	524.1
60 copper + 40 zinc	8.36	.3018	521.7
50 copper + 50 zinc	8.20	.2960	511.6
Bronze	8.85	.3195	552.2
Chromium	6.50	.2347	405.6
Copper	8.82	.3184	550.4
Gold	19.32	.6975	1,205.6
Iron:			
cast	7.20	.2600	449.2
wrought	7.85	.2834	489.8
Lead	11.37	.4105	709.5
Magnesium	1.74	.0628	108.6
Maple	.70	.0250	43.7
Molybdenum	8.56	.3090	534.2
Nickel	8.80	.3177	549.1
Pine:			
white	.50	.0180	31.2
yellow	.65	.0230	40.6
Platinum, rolled	22.67	.8184	1,414.6
Silver	10.53	.3802	657.1
Steel	7.80	.2816	486.7
Tin	7.29	.2632	454.8
Tungsten	18.77	.6776	1,171.2
Vanadium	5.50	.1986	343.2
Zinc:			
cast	6.86	.2476	428.1
rolled	7.15	.2581	446.1

TABLE 9
Moduli of Elasticity of Common Materials

Materials	Moduli of Elasticity in psi.
Aluminum	11,000,000
Brass:	
cast	9,000,000
annealed wire	14,000,000
Brick	2,000,000
Bronze	10,000,000
Bronze, Tobin	4,500,000
Granite	7,000,000
Cast iron	12,000,000
Wrought iron	27,000,000
Lead	1,000,000
Steel:	
castings	30,000,000
structural	29,000,000
Stone	6,000,000
Timber	1,500,000
Tin	4,000,000
Zinc	13,000,000

TABLE 10
Factors of Safety

Materials	Steady Loads	Loads Varying from 0 to Max. in One Direction	Loads Varying from 0 to Max. in Both Directions	Suddenly Varying Loads and Shocks
Cast iron	6	10	15	20
Wrought iron	4	6	8	12
Steel	5	6	8	12
Wood	8	10	15	20
Brick	15	20	25	30
Stone	15	20	25	30
Leather	7	10	15	20
Soft metals and alloys	6	8	10	15

TABLE 11
Ultimate Strengths of Common Materials

MATERIALS	Ultimate Tensile Strengths in $p.s.i.$	Ultimate Compressive Strengths in $p.s.i.$	Ultimate Shearing Strengths in $p.s.i.$
Aluminum, cast	15,000 to 36,000	12,000 and up	12,000
Aluminum bronze, cast	65,000 to 80,000	120,000	
Wrought aluminum	50,000 to 75,000	50,000 and up	
Brass			
cast	25,000 to 27,000	30,000	
sheets and rods	40,000 to 87,000	42,000 to 117,000	Up to 36,000
Brick	400	3,000	1,000
Bronze	30,000 to 100,000	42,000 to 147,000	
Cement	400 to 500	2,000 to 3,000	
Copper	36,000 to 60,000	32,000 to 40,000	
Cast iron	16,000 to 35,000	80,000 to 150,000	17,000 to 38,000
Cast iron, high test and malleable	45,000 to 54,000	200,000	48,000 to 50,000
Lead	2,200 to 3,000		
Magnesium	28,000 to 40,000	7,000 to 28,000	
Steel:			
cast	70,000	70,000	60,000
carbon	56,000 to 120,000	56,000 to 120,000	42,000 to 90,000
structural	60,000 to 70,000	60,000 to 70,000	45,000
alloy	120,000 to 300,000	120,000 to 225,000	95,000 to 165,000
Wrought iron	48,000	46,000	40,000
Stone		6,000	1,500
Timber	10,000	8,000	{500 along grain, 3,000 across grain}
Tin	3,500 to 11,000	6,000	
Zinc	4,000 to 16,000	18,000	

0° Natural Trigonometric Functions 179°

M	Sine	Cosine	Tan.	Cotan.	Secant	Cosec.	Vrs. Sin.	Vrs. Cos.	M
0	0.00000	1.0000	0.00000	Infinite	1.0000	Infinite	0.00000	1.00000	60
1	.00029	.0000	.00029	3437.7	.0000	3437.7	.00000	0.99971	59
2	.00058	.0000	.00058	1718.9	.0000	1718.9	.00000	.99942	58
3	.00087	.0000	.00087	1145.9	.0000	1145.9	.00000	.99913	57
4	.00116	.0000	.00116	859.44	.0000	859.44	.00000	.99884	56
5	0.00145	1.0000	0.00145	687.55	1.0000	687.55	0.00000	0.99854	55
6	.00174	.0000	.00174	572.96	.0000	572.96	.00000	.99825	54
7	.00204	.0000	.00204	491.11	.0000	491.11	.00000	.99796	53
8	.00233	.0000	.00233	429.72	.0000	429.72	.00000	.99767	52
9	.00262	.0000	.00262	381.97	.0000	381.97	.00000	.99738	51
10	0.00291	0.99999	0.00291	343.77	1.0000	343.77	0.00000	0.99709	50
11	.00320	.99999	.00320	312.52	.0000	312.52	.00000	.99680	49
12	.00349	.99999	.00349	286.48	.0000	286.48	.00001	.99651	48
13	.00378	.99999	.00378	264.44	.0000	264.44	.00001	.99622	47
14	.00407	.99999	.00407	245.55	.0000	245.55	.00001	.99593	46
15	0.00436	0.99999	0.00436	229.18	1.0000	229.18	0.00001	0.99564	45
16	.00465	.99999	.00465	214.86	.0000	214.86	.00001	.99534	44
17	.00494	.99999	.00494	202.22	.0000	202.22	.00001	.99505	43
18	.00524	.99999	.00524	190.98	.0000	190.99	.00001	.99476	42
19	.00553	.99998	.00553	180.93	.0000	180.93	.00001	.99447	41
20	0.00582	0.99998	0.00582	171.88	1.0000	171.89	0.00002	0.99418	40
21	.00611	.99998	.00611	163.70	.0000	163.70	.00002	.99389	39
22	.00640	.99998	.00640	156.26	.0000	156.26	.00002	.99360	38
23	.00669	.99998	.00669	149.46	.0000	149.47	.00002	.99331	37
24	.00698	.99997	.00698	143.24	.0000	143.24	.00002	.99302	36
25	0.00727	0.99997	0.00727	137.51	1.0000	137.51	0.00003	0.99273	35
26	.00756	.99997	.00756	132.22	.0000	132.22	.00003	.99244	34
27	.00785	.99997	.00785	127.32	.0000	127.32	.00003	.99215	33
28	.00814	.99997	.00814	122.77	.0000	122.78	.00003	.99185	32
29	.00843	.99996	.00844	118.54	.0000	118.54	.00003	.99156	31
30	0.00873	0.99996	0.00873	114.59	1.0000	114.59	0.00004	0.99127	30
31	.00902	.99996	.00902	110.89	.0000	110.90	.00004	.99098	29
32	.00931	.99996	.00931	107.43	.0000	107.43	.00004	.99069	28
33	.00960	.99995	.00960	104.17	.0000	104.17	.00005	.99040	27
34	.00989	.99995	.00989	101.11	.0000	101.11	.00005	.99011	26
35	0.01018	0.99995	0.01018	98.218	1.0000	98.223	0.00005	0.98982	25
36	.01047	.99994	.01047	95.489	.0000	95.495	.00005	.98953	24
37	.01076	.99994	.01076	92.908	.0000	92.914	.00006	.98924	23
38	.01105	.99994	.01105	90.463	.0001	90.469	.00006	.98895	22
39	.01134	.99993	.01134	88.143	.0001	88.149	.00006	.98865	21
40	0.01163	0.99993	0.01164	85.940	1.0001	85.946	0.00007	0.98836	20
41	.01193	.99993	.01193	83.843	.0001	83.849	.00007	.98807	19
42	.01222	.99992	.01222	81.847	.0001	81.853	.00007	.98778	18
43	.01251	.99992	.01251	79.943	.0001	79.950	.00008	.98749	17
44	.01280	.99992	.01280	78.126	.0001	78.133	.00008	.98720	16
45	0.01309	0.99991	0.01309	76.390	1.0001	76.396	0.00008	0.98691	15
46	.01338	.99991	.01338	74.729	.0001	74.736	.00009	.98662	14
47	.01367	.99991	.01367	73.139	.0001	73.146	.00009	.98633	13
48	.01396	.99990	.01396	71.615	.0001	71.622	.00010	.98604	12
49	.01425	.99990	.01425	70.153	.0001	70.160	.00010	.98575	11
50	0.01454	0.99989	0.01454	68.750	1.0001	68.757	0.00010	0.98546	10
51	.01483	.99989	.01484	67.402	.0001	67.409	.00011	.98516	9
52	.01512	.99988	.01513	66.105	.0001	66.113	.00011	.98487	8
53	.01542	.99988	.01542	64.858	.0001	64.866	.00012	.98458	7
54	.01571	.99988	.01571	63.657	.0001	63.664	.00012	.98429	6
55	0.01600	0.99987	0.01600	62.499	1.0001	62.507	0.00013	0.98400	5
56	.01629	.99987	.01629	61.383	.0001	61.391	.00013	.98371	4
57	.01658	.99987	.01658	60.306	.0001	60.314	.00014	.98342	3
58	.01687	.99986	.01687	59.266	.0001	59.274	.00014	.98313	2
59	.01716	.99985	.01716	58.261	.0001	58.270	.00015	.98284	1
60	0.01745	0.99985	0.01745	57.290	1.0001	57.299	0.00015	0.98255	0
M	Cosine	Sine	Cotan.	Tan.	Cosec.	Secant	Vrs. Cos.	Vrs. Sin.	M

90° 89°

1°   Natural Trigonometric Functions   178°

M	Sine	Cosine	Tan.	Cotan.	Secant	Cosec.	Vrs. Sin.	Vrs. Cos.	M
0	0.01745	0.99985	0.01745	57.290	1.0001	57.299	0.00015	0.98255	60
1	.01774	.99984	.01775	56.350	.0001	56.359	.00016	.98226	59
2	.01803	.99984	.01804	55.441	.0001	55.450	.00016	.98196	58
3	.01832	.99983	.01833	54.561	.0002	54.570	.00017	.98167	57
4	.01861	.99983	.01862	53.708	.0002	53.718	.00017	.98138	56
5	0.01891	0.99982	0.01891	52.882	1.0002	52.891	0.00018	0.98109	55
6	.01920	.99981	.01920	52.081	.0002	52.090	.00018	.98080	54
7	.01949	.99981	.01949	51.303	.0002	51.313	.00019	.98051	53
8	.01978	.99980	.01978	50.548	.0002	50.558	.00019	.98022	52
9	.02007	.99980	.02007	49.816	.0002	49.826	.00020	.97993	51
10	0.02036	0.99979	0.02036	49.104	1.0002	49.114	0.00021	0.97964	50
11	.02065	.99979	.02066	48.412	.0002	48.422	.00021	.97935	49
12	.02094	.99978	.02095	47.739	.0002	47.750	.00022	.97906	48
13	.02123	.99977	.02124	47.085	.0002	47.096	.00022	.97877	47
14	.02152	.99977	.02153	46.449	.0002	46.460	.00023	.97847	46
15	0.02181	0.99976	0.02182	45.829	1.0002	45.840	0.00024	0.97818	45
16	.02210	.99975	.02211	45.226	.0002	45.237	.00024	.97789	44
17	.02240	.99975	.02240	44.638	.0002	44.650	.00025	.97760	43
18	.02269	.99974	.02269	44.066	.0002	44.077	.00026	.97731	42
19	.02298	.99974	.02298	43.508	.0003	43.520	.00026	.97702	41
20	0.02326	0.99973	0.02327	42.964	1.0003	42.976	0.00027	0.97673	40
21	.02356	.99972	.02357	42.433	.0003	42.445	.00028	.97644	39
22	.02385	.99971	.02386	41.916	.0003	41.928	.00028	.97615	38
23	.02414	.99971	.02415	41.410	.0003	41.423	.00029	.97586	37
24	.02443	.99970	.02444	40.917	.0003	40.930	.00030	.97557	36
25	0.02472	0.99969	0.02473	40.436	1.0003	40.448	0.00030	0.97528	35
26	.02501	.99969	.02502	39.965	.0003	39.978	.00031	.97499	34
27	.02530	.99968	.02531	39.506	.0003	39.518	.00032	.97469	33
28	.02559	.99967	.02560	39.057	.0003	39.069	.00033	.97440	32
29	.02589	.99966	.02589	38.618	.0003	38.631	.00033	.97411	31
30	0.02618	0.99966	0.02618	38.188	1.0003	38.201	0.00034	0.97382	30
31	.02647	.99965	.02648	37.769	.0003	37.782	.00035	.97353	29
32	.02676	.99964	.02677	37.358	.0003	37.371	.00036	.97324	28
33	.02705	.99963	.02706	36.956	.0004	36.969	.00036	.97295	27
34	.02734	.99963	.02735	36.563	.0004	36.576	.00037	.97266	26
35	0.02763	0.99962	0.02764	36.177	1.0004	36.191	0.00038	0.97237	25
36	.02792	.99961	.02793	35.800	.0004	35.814	.00039	.97208	24
37	.02821	.99960	.02822	35.431	.0004	35.445	.00040	.97179	23
38	.02850	.99959	.02851	35.069	.0004	35.084	.00041	.97150	22
39	.02879	.99958	.02880	34.715	.0004	34.729	.00041	.97121	21
40	0.02908	0.99958	0.02910	34.368	1.0004	34.382	0.00042	0.97091	20
41	.02937	.99957	.02939	34.027	.0004	34.042	.00043	.97062	19
42	.02967	.99956	.02968	33.693	.0004	33.708	.00044	.97033	18
43	.02996	.99955	.02997	33.366	.0004	33.381	.00045	.97004	17
44	.03025	.99954	.03026	33.045	.0004	33.060	.00046	.96975	16
45	0.03054	0.99953	0.03055	32.730	1.0005	32.745	0.00046	0.96946	15
46	.03083	.99952	.03084	32.421	.0005	32.437	.00047	.96917	14
47	.03112	.99951	.03113	32.118	.0005	32.134	.00048	.96888	13
48	.03141	.99951	.03143	31.820	.0005	31.836	.00049	.96859	12
49	.03170	.99950	.03172	31.528	.0005	31.544	.00050	.96830	11
50	0.03199	0.99949	0.03201	31.241	1.0005	31.257	0.00051	0.96801	10
51	.03228	.99948	.03230	30.960	.0005	30.976	.00052	.96772	9
52	.03257	.99947	.03259	30.683	.0005	30.699	.00053	.96743	8
53	.03286	.99946	.03288	30.411	.0005	30.428	.00054	.96713	7
54	.03315	.99945	.03317	30.145	.0005	30.161	.00055	.96684	6
55	0.03344	0.99944	0.03346	29.882	1.0005	29.899	0.00056	0.96655	5
56	.03374	.99943	.03375	29.624	.0006	29.641	.00057	.96626	4
57	.03403	.99942	.03405	29.371	.0006	29.388	.00058	.96597	3
58	.03432	.99941	.03434	29.122	.0006	29.139	.00059	.96568	2
59	.03461	.99940	.03463	28.877	.0006	28.894	.00060	.96539	1
60	0.03490	0.99939	0.03492	28.636	1.0006	28.654	0.00061	0.96510	0
M	Cosine	Sine	Cotan.	Tan.	Cosec.	Secant	Vrs. Cos.	Vrs. Sin.	M

91°   **88°**

2° Natural Trigonometric Functions 177°

M	Sine	Cosine	Tan.	Cotan.	Secant	Cosec.	Vrs. Sin.	Vrs. Cos.	M
0	0.03490	0.99939	0.03492	28.636	1.0006	28.654	0.00061	0.96510	60
1	.03519	.99938	.03521	28.399	.0006	28.417	.00062	.96481	59
2	.03548	.99937	.03550	28.166	.0006	28.184	.00063	.96452	58
3	.03577	.99936	.03579	27.937	.0006	27.955	.00064	.96423	57
4	.03606	.99935	.03608	27.712	.0006	27.730	.00065	.96394	56
5	0.03635	0.99934	0.03638	27.490	1.0007	27.508	0.00066	0.96365	55
6	.03664	.99933	.03667	27.271	.0007	27.290	.00067	.96336	54
7	.03693	.99932	.03696	27.056	.0007	27.075	.00068	.96306	53
8	.03722	.99931	.03725	26.845	.0007	26.864	.00069	.96277	52
9	.03751	.99930	.03754	26.637	.0007	26.655	.00070	.96248	51
10	0.03781	0.99928	0.03783	26.432	1.0007	26.450	0.00071	0.96219	50
11	.03810	.99927	.03812	26.230	.0007	26.249	.00073	.96190	49
12	.03839	.99926	.03842	26.031	.0007	26.050	.00074	.96161	48
13	.03868	.99925	.03871	25.835	.0007	25.854	.00075	.96132	47
14	.03897	.99924	.03900	25.642	.0008	25.661	.00076	.96103	46
15	0.03926	0.99923	0.03929	25.452	1.0008	25.471	0.00077	0.96074	45
16	.03955	.99922	.03958	25.264	.0008	25.284	.00078	.96045	44
17	.03984	.99921	.03987	25.080	.0008	25.100	.00079	.96016	43
18	.04013	.99919	.04016	24.898	.0008	24.918	.00080	.95987	42
19	.04042	.99918	.04045	24.718	.0008	24.739	.00082	.95958	41
20	0.04071	0.99917	0.04075	24.542	1.0008	24.562	0.00083	0.95929	40
21	.04100	.99916	.04104	24.367	.0008	24.388	.00084	.95900	39
22	.04129	.99915	.04133	24.196	.0008	24.216	.00085	.95870	38
23	.04158	.99913	.04162	24.026	.0009	24.047	.00086	.95841	37
24	.04187	.99912	.04191	23.859	.0009	23.880	.00088	.95812	36
25	0.04217	0.99911	0.04220	23.694	1.0009	23.716	0.00089	0.95783	35
26	.04246	.99910	.04249	23.532	.0009	23.553	.00090	.95754	34
27	.04275	.99908	.04279	23.372	.0009	23.393	.00091	.95725	33
28	.04304	.99907	.04308	23.214	.0009	23.235	.00093	.95696	32
29	.04333	.99906	.04337	23.058	.0009	23.079	.00094	.95667	31
30	0.04362	0.99905	0.04366	22.904	1.0009	22.925	0.00095	0.95638	30
31	.04391	.99903	.04395	22.752	.0010	22.774	.00096	.95609	29
32	.04420	.99902	.04424	22.602	.0010	22.624	.00098	.95580	28
33	.04449	.99901	.04453	22.454	.0010	22.476	.00099	.95551	27
34	.04478	.99900	.04483	22.308	.0010	22.330	.00100	.95522	26
35	0.04507	0.99898	0.04512	22.164	1.0010	22.186	0.00102	0.95493	25
36	.04536	.99897	.04541	22.022	.0010	22.044	.00103	.95464	24
37	.04565	.99896	.04570	21.881	.0010	21.904	.00104	.95435	23
38	.04594	.99894	.04599	21.742	.0010	21.765	.00106	.95405	22
39	.04623	.99893	.04628	21.606	.0011	21.629	.00107	.95376	21
40	0.04652	0.99892	0.04657	21.470	1.0011	21.494	0.00108	0.95347	20
41	.04681	.99890	.04687	21.337	.0011	21.360	.00110	.95318	19
42	.04711	.99889	.04716	21.205	.0011	21.228	.00111	.95289	18
43	.04740	.99888	.04745	21.075	.0011	21.098	.00112	.95260	17
44	.04769	.99886	.04774	20.946	.0011	20.970	.00114	.95231	16
45	0.04798	0.99885	0.04803	20.819	1.0011	20.843	0.00115	0.95202	15
46	.04827	.99883	.04832	20.693	.0012	20.717	.00116	.95173	14
47	.04856	.99882	.04862	20.569	.0012	20.593	.00118	.95144	13
48	.04885	.99881	.04891	20.446	.0012	20.471	.00119	.95115	12
49	.04914	.99879	.04920	20.325	.0012	20.350	.00121	.95086	11
50	0.04943	0.99878	0.04949	20.205	1.0012	20.230	0.00122	0.95057	10
51	.04972	.99876	.04978	20.087	.0012	20.112	.00124	.95028	9
52	.05001	.99875	.05007	19.970	.0012	19.995	.00125	.94999	8
53	.05030	.99873	.05037	19.854	.0013	19.880	.00127	.94970	7
54	.05059	.99872	.05066	19.740	.0013	19.766	.00128	.94941	6
55	0.05088	0.99870	0.05095	19.627	1.0013	19.653	0.00129	0.94912	5
56	.05117	.99869	.05124	19.515	.0013	19.541	.00131	.94883	4
57	.05146	.99867	.05153	19.405	.0013	19.431	.00132	.94853	3
58	.05175	.99866	.05182	19.296	.0013	19.322	.00134	.94824	2
59	.05204	.99864	.05212	19.188	.0013	19.214	.00135	.94795	1
60	0.05234	0.99863	0.05241	19.081	1.0014	19.107	0.00137	0.94766	0
M	Cosine	Sine	Cotan.	Tan.	Cosec.	Secant	Vrs. Cos.	Vrs. Sin.	M

92°   87°

3° Natural Trigonometric Functions 176°

M	Sine	Cosine	Tan.	Cotan.	Secant	Cosec.	Vrs. Sin.	Vrs. Cos.	M
0	0.05234	0.99863	0.05241	19.081	1.0014	19.107	0.00137	0.94766	60
1	.05263	.99861	.05270	18.975	.0014	19.002	.00138	.94737	59
2	.05292	.99860	.05299	18.871	.0014	18.897	.00140	.94708	58
3	.05321	.99858	.05328	18.768	.0014	18.794	.00142	.94679	57
4	.05350	.99857	.05357	18.665	.0014	18.692	.00143	.94650	56
5	0.05379	0.99855	0.05387	18.564	1.0014	18.591	0.00145	0.94621	55
6	.05408	.99854	.05416	18.464	.0015	18.491	.00146	.94592	54
7	.05437	.99852	.05445	18.365	.0015	18.393	.00148	.94563	53
8	.05466	.99850	.05474	18.268	.0015	18.295	.00149	.94534	52
9	.05495	.99849	.05503	18.171	.0015	18.198	.00151	.94505	51
10	0.05524	0.99847	0.05532	18.075	1.0015	18.103	0.00153	0.94476	50
11	.05553	.99846	.05562	17.980	.0015	18.008	.00154	.94447	49
12	.05582	.99844	.05591	17.886	.0016	17.914	.00156	.94418	48
13	.05611	.99842	.05620	17.793	.0016	17.821	.00157	.94389	47
14	.05640	.99841	.05649	17.701	.0016	17.730	.00159	.94360	46
15	0.05669	0.99839	0.05678	17.610	1.0016	17.639	0.00161	0.94331	45
16	.05698	.99837	.05707	17.520	.0016	17.549	.00162	.94302	44
17	.05727	.99836	.05737	17.431	.0016	17.460	.00164	.94273	43
18	.05756	.99834	.05766	17.343	.0017	17.372	.00166	.94244	42
19	.05785	.99832	.05795	17.256	.0017	17.285	.00167	.94214	41
20	0.05814	0.99831	0.05824	17.169	1.0017	17.198	0.00169	0.94185	40
21	.05843	.99829	.05853	17.084	.0017	17.113	.00171	.94156	39
22	.05872	.99827	.05883	16.999	.0017	.17.028	.00172	.94127	38
23	.05902	.99826	.05912	16.915	.0017	16.944	.00174	.94098	37
24	.05931	.99824	.05941	16.832	.0018	16.861	.00176	.94069	36
25	0.05960	0.99822	0.05970	16.750	1.0018	16.779	0.00178	0.94040	35
26	.05989	.99820	.05999	16.668	.0018	16.698	.00179	.94011	34
27	.06018	.99819	.06029	16.587	.0018	16.617	.00181	.93982	33
28	.06047	.99817	.06058	16.507	.0018	16.538	.00183	.93953	32
29	.06076	.99815	.06087	16.428	.0018	16.459	.00185	.93924	31
30	0.06105	0.99813	0.06116	16.350	1.0019	16.380	0.00186	0.93895	30
31	.06134	.99812	.06145	16.272	.0019	16.303	.00188	.93866	29
32	.06163	.99810	.06175	16.195	.0019	16.226	.00190	.93837	28
33	.06192	.99808	.06204	16.119	.0019	16.150	.00192	.93808	27
34	.06221	.99806	.06233	16.043	.0019	16.075	.00194	.93777	26
35	0.06250	0.99804	0.06262	15.969	1.0019	16.000	0.00195	0.93750	25
36	.06279	.99803	.06291	15.894	.0020	15.926	.00197	.93721	24
37	.06308	.99801	.06321	15.821	.0020	15.853	.00199	.93692	23
38	.06337	.99799	.06350	15.748	.0020	15.780	.00201	.93663	22
39	.06366	.99797	.06379	15.676	.0020	15.708	.00203	.93634	21
40	0.06395	0.99795	0.06408	15.605	1.0020	15.637	0.00205	0.93605	20
41	.06424	.99793	.06437	15.534	.0021	15.566	.00206	.93576	19
42	.06453	.99791	.06467	15.464	.0021	15.496	.00208	.93547	18
43	.06482	.99790	.06496	15.394	.0021	15.427	.00210	.93518	17
44	.06511	.99788	.06525	15.325	.0021	15.358	.00212	.93489	16
45	0.06540	0.99786	0.06554	15.257	1.0021	15.290	0.00214	0.93460	15
46	.06569	.99784	.06583	15.189	.0022	15.222	.00216	.93431	14
47	.06598	.99782	.06613	15.122	.0022	15.155	.00218	.93402	13
48	.06627	.99780	.06642	15.056	.0022	15.089	.00220	.93373	12
49	.06656	.99778	.06671	14.990	.0022	15.023	.00222	.93343	11
50	0.06685	0.99776	0.06700	14.924	1.0022	14.958	0.00224	0.93314	10
51	.06714	.99774	.06730	14.860	.0023	14.893	.00226	.93285	9
52	.06743	.99772	.06759	14.795	.0023	14.829	.00228	.93256	8
53	.06772	.99770	.06788	14.732	.0023	14.765	.00230	.93227	7
54	.06801	.99768	.06817	14.668	.0023	14.702	.00231	.93198	6
55	0.06830	0.99766	0.06846	14.606	1.0023	14.640	0.00233	0.93169	5
56	.06859	.99764	.06876	14.544	.0024	14.578	.00235	.93140	4
57	.06888	.99762	.06905	14.482	.0024	14.517	.00237	.93111	3
58	.06918	.99760	.06934	14.421	.0024	14.456	.00239	.93082	2
59	.06947	.99758	.06963	14.361	.0024	14.395	.00241	.93053	1
60	0.06976	0.99756	0.06993	14.301	1.0024	14.335	0.00243	0.93024	0
M	Cosine	Sine	Cotan.	Tan.	Cosec.	Secant	Vrs. Cos.	Vrs. Sin.	M

93° 86°

4°   Natural Trigonometric Functions   175°

M	Sine	Cosine	Tan.	Cotan.	Secant	Cosec.	Vrs. Sin.	Vrs. Cos.	M
0	0.06976	0.99756	0.06993	14.301	1.0024	14.335	0.00243	0.93024	60
1	.07005	.99754	.07022	14.241	.0025	14.276	.00246	.92995	59
2	.07034	.99752	.07051	14.182	.0025	14.217	.00248	.92966	58
3	.07063	.99750	.07080	14.123	.0025	14.159	.00250	.92937	57
4	.07092	.99748	.07110	14.065	.0025	14.101	.00252	.92908	56
5	0.07121	0.99746	0.07139	14.008	1.0025	14.043	0.00254	0.92879	55
6	.07150	.99744	.07168	13.951	.0026	13.986	.00256	.92850	54
7	.07179	.99742	.07197	13.894	.0026	13.930	.00258	.92821	53
8	.07208	.99740	.07226	13.838	.0026	13.874	.00260	.92792	52
9	.07237	.99738	.07256	13.782	.0026	13.818	.00262	.92763	51
10	0.07266	0.99736	0.07285	13.727	1.0026	13.763	0.00264	0.92734	50
11	.07295	.99733	.07314	13.672	.0027	13.708	.00266	.92705	49
12	.07324	.99731	.07343	13.617	.0027	13.654	.00268	.92676	48
13	.07353	.99729	.07373	13.563	.0027	13.600	.00271	.92647	47
14	.07382	.99727	.07402	13.510	.0027	13.547	.00273	.92618	46
15	0.07411	0.99725	0.07431	13.457	1.0027	13.494	0.00275	0.92589	45
16	.07440	.99723	.07460	13.404	.0028	13.441	.00277	.92560	44
17	.07469	.99721	.07490	13.351	.0028	13.389	.00279	.92531	43
18	.07498	.99718	.07519	13.299	.0028	13.337	.00281	.92502	42
19	.07527	.99716	.07548	13.248	.0028	13.286	.00284	.92473	41
20	0.07556	0.99714	0.07577	13.197	1.0029	13.235	0.00286	0.92444	40
21	.07585	.99712	.07607	13.146	.0029	13.184	.00288	.92415	39
22	.07614	.99710	.07636	13.096	.0029	13.134	.00290	.92386	38
23	.07643	.99707	.07665	13.046	.0029	13.084	.00292	.92357	37
24	.07672	.99705	.07694	12.996	.0029	13.034	.00295	.92328	36
25	0.07701	0.99703	0.07724	12.947	1.0030	12.985	0.00297	0.92299	35
26	.07730	.99701	.07753	12.898	.0030	12.937	.00299	.92270	34
27	.07759	.99693	.07782	12.849	.0030	12.888	.00301	.92241	33
28	.07788	.99696	.07812	12.801	.0030	12.840	.00304	.92212	32
29	.07817	.99694	.07841	12.754	.0031	12.793	.00306	.92183	31
30	0.07846	0.99692	0.07870	12.706	1.0031	12.745	0.00308	0.92154	30
31	.07875	.99689	.07899	12.659	.0031	12.698	.00310	.92125	29
32	.07904	.99687	.07929	12.612	.0031	12.652	.00313	.92096	28
33	.07933	.99685	.07958	12.566	.0032	12.606	.00315	.92067	27
34	.07962	.99682	.07987	12.520	.0032	12.560	.00317	.92038	26
35	0.07991	0.99680	0.08016	12.474	1.0032	12.514	0.00320	0.92009	25
36	.08020	.99678	.08046	12.429	.0032	12.469	.00322	.91980	24
37	.08049	.99675	.08075	12.384	.0032	12.424	.00324	.91951	23
38	.08078	.99673	.08104	12.339	.0033	12.379	.00327	.91922	22
39	.08107	.99671	.08134	12.295	.0033	12.335	.00329	.91893	21
40	0.08136	0.99668	0.08163	12.250	1.0033	12.291	0.00331	0.91864	20
41	.08165	.99666	.08192	12.207	.0033	12.248	.00334	.91835	19
42	.08194	.99664	.08221	12.163	.0034	12.204	.00336	.91806	18
43	.08223	.99661	.08251	12.120	.0034	12.161	.00339	.91777	17
44	.08252	.99659	.08280	12.077	.0034	12.118	.00341	.91748	16
45	0.08281	0.99656	0.08309	12.035	1.0034	12.076	0.00343	0.91719	15
46	.08310	.99654	.08339	11.992	.0035	12.034	.00346	.91690	14
47	.08339	.99652	.08368	11.950	.0035	11.992	.00348	.91661	13
48	.08368	.99649	.08397	11.909	.0035	11.950	.00351	.91632	12
49	.08397	.99647	.08426	11.867	.0035	11.909	.00353	.91603	11
50	0.08426	0.99644	0.08456	11.826	1.0036	11.868	0.00356	0.91574	10
51	.08455	.99642	.08485	11.785	.0036	11.828	.00358	.91545	9
52	.08484	.99639	.08514	11.745	.0036	11.787	.00360	.91516	8
53	.08513	.99637	.08544	11.704	.0036	11.747	.00363	.91487	7
54	.08542	.99634	.08573	11.664	.0037	11.707	.00365	.91458	6
55	0.08571	0.99632	0.08602	11.625	1.0037	11.668	0.00368	0.91429	5
56	.08600	.99629	.08632	11.585	.0037	11.628	.00370	.91400	4
57	.08629	.99627	.08661	11.546	.0037	11.589	.00373	.91371	3
58	.08658	.99624	.08690	11.507	.0038	11.550	.00375	.91342	2
59	.08687	.99622	.08719	11.468	.0038	11.512	.00378	.91313	1
60	0.08715	0.99619	0.08749	11.430	1.0038	11.474	0.00380	0.91284	0
M	Cosine	Sine	Cotan.	Tan.	Cosec.	Secant	Vrs. Cos.	Vrs. Sin.	M

94°   85°

5° Natural Trigonometric Functions 174°

M	Sine	Cosine	Tan.	Cotan.	Secant	Cosec.	Vrs. Sin.	Vrs. Cos.	M
0	0.08715	0.99619	0.08749	11.430	1.0038	11.474	0.00380	0.91284	60
1	.08744	.99617	.08778	11.392	.0038	11.436	.00383	.91255	59
2	.08773	.99614	.08807	11.354	.0039	11.398	.00386	.91226	58
3	.08802	.99612	.08837	11.316	.0039	11.360	.00388	.91197	57
4	.08831	.99609	.08866	11.279	.0039	11.323	.00391	.91168	56
5	0.08860	0.99607	0.08895	11.242	1.0039	11.286	0.00393	0.91139	55
6	.08889	.99604	.08925	11.205	.0040	11.249	.00396	.91110	54
7	.08918	.99601	.08954	11.168	.0040	11.213	.00398	.91082	53
8	.08947	.99599	.08983	11.132	.0040	11.176	.00401	.91053	52
9	.08976	.99596	.09013	11.095	.0040	11.140	.00404	.91024	51
10	0.09005	0.99594	0.09042	11.059	1.0041	11.104	0.00406	0.90995	50
11	.09034	.99591	.09071	11.024	.0041	11.069	.00409	.90966	49
12	.09063	.99588	.09101	10.988	.0041	11.033	.00411	.90937	48
13	.09092	.99586	.09130	10.953	.0041	10.998	.00414	.90908	47
14	.09121	.99583	.09159	10.918	.0042	10.963	.00417	.90879	46
15	0.09150	0.99580	0.09189	10.883	1.0042	10.929	0.00419	0.90850	45
16	.09179	.99578	.09218	10.848	.0042	10.894	.00422	.90821	44
17	.09208	.99575	.09247	10.814	.0043	10.860	.00425	.90792	43
18	.09237	.99572	.09277	10.780	.0043	10.826	.00427	.90763	42
19	.09266	.99570	.09306	10.746	.0043	10.792	.00430	.90734	41
20	0.09295	0.99567	0.09335	10.712	1.0043	10.758	0.00433	0.90705	40
21	.09324	.99564	.09365	10.678	.0044	10.725	.00436	.90676	39
22	.09353	.99562	.09394	10.645	.0044	10.692	.00438	.90647	38
23	.09382	.99559	.09423	10.612	.0044	10.659	.00441	.90618	37
24	.09411	.99556	.09453	10.579	.0044	10.626	.00444	.90589	36
25	0.09440	0.99553	0.09482	10.546	1.0045	10.593	0.00446	0.90560	35
26	.09469	.99551	.09511	10.514	.0045	10.561	.00449	.90531	34
27	.09498	.99548	.09541	10.481	.0045	10.529	.00452	.90502	33
28	.09527	.99545	.09570	10.449	.0046	10.497	.00455	.90473	32
29	.09556	.99542	.09599	10.417	.0046	10.465	.00458	.90444	31
30	0.09584	0.99540	0.09629	10.385	1.0046	10.433	0.00460	0.90415	30
31	.09613	.99537	.09658	10.354	.0046	10.402	.00463	.90386	29
32	.09642	.99534	.09688	10.322	.0047	10.371	.00466	.90357	28
33	.09671	.99531	.09717	10.291	.0047	10.340	.00469	.90328	27
34	.09700	.99528	.09746	10.260	.0047	10.309	.00472	.90300	26
35	0.09729	0.99525	0.09776	10.229	1.0048	10.278	0.00474	0.90271	25
36	.09758	.99523	.09805	10.199	.0048	10.248	.00477	.90242	24
37	.09787	.99520	.09834	10.168	.0048	10.217	.00480	.90213	23
38	.09816	.99517	.09864	10.138	.0048	10.187	.00483	.90184	22
39	.09845	.99514	.09893	10.108	.0049	10.157	.00486	.90155	21
40	0.09874	0.99511	0.09922	10.078	1.0049	10.127	0.00489	0.90126	20
41	.09903	.99508	.09952	10.048	.0049	10.098	.00491	.90097	19
42	.09932	.99505	.09981	10.019	.0050	10.068	.00494	.90068	18
43	.09961	.99503	.10011	9.9893	.0050	10.039	.00497	.90039	17
44	.09990	.99500	.10040	9.9601	.0050	10.010	.00500	.90010	16
45	0.10019	0.99497	0.10069	9.9310	1.0050	9.9812	0.00503	0.89981	15
46	.10048	.99494	.10099	9.9021	.0051	9.9525	.00506	.89952	14
47	.10077	.99491	.10128	9.8734	.0051	9.9239	.00509	.89923	13
48	.10106	.99488	.10158	9.8448	.0051	9.8955	.00512	.89894	12
49	.10134	.99485	.10187	9.8164	.0052	9.8672	.00515	.89865	11
50	0.10163	0.99482	0.10216	9.7882	1.0052	9.8391	0.00518	0.89836	10
51	.10192	.99479	.10246	9.7601	.0052	9.8112	.00521	.89807	9
52	.10221	.99476	.10275	9.7322	.0053	9.7834	.00524	.89779	8
53	.10250	.99473	.10305	9.7044	.0053	9.7558	.00527	.89750	7
54	.10279	.99470	.10334	9.6768	.0053	9.7283	.00530	.89721	6
55	0.10308	0.99467	0.10363	9.6493	1.0053	9.7010	0.00533	0.89692	5
56	.10337	.99464	.10393	9.6220	.0054	9.6739	.00536	.89663	4
57	.10366	.99461	.10422	9.5949	.0054	9.6469	.00539	.89634	3
58	.10395	.99458	.10452	9.5679	.0054	9.6200	.00542	.89605	2
59	.10424	.99455	.10481	9.5411	.0055	9.5933	.00545	.89576	1
60	0.10453	0.99452	0.10510	9.5144	1.0055	9.5668	0.00548	0.89547	0
M	Cosine	Sine	Cotan.	Tan.	Cosec.	Secant	Vrs. Cos.	Vrs. Sin.	M

95° 84°

6° Natural Trigonometric Functions 173°

M	Sine	Cosine	Tan.	Cotan.	Secant	Cosec.	Vrs. Sin.	Vrs. Cos.	M
0	0.10453	0.99452	0.10510	9.5144	1.0055	9.5668	0.00548	0.89547	60
1	.10482	.99449	.10540	.4878	.0055	.5404	.00551	.89518	59
2	.10511	.99446	.10569	.4614	.0056	.5141	.00554	.89489	58
3	.10540	.99443	.10599	.4351	.0056	.4880	.00557	.89460	57
4	.10568	.99440	.10628	.4090	.0056	.4620	.00560	.89431	56
5	0.10597	0.99437	0.10657	9.3831	1.0057	9.4362	0.00563	0.89402	55
6	.10626	.99434	.10687	.3572	.0057	.4105	.00566	.89373	54
7	.10655	.99431	.10716	.3315	.0057	.3850	.00569	.89345	53
8	.10684	.99428	.10746	.3060	.0057	.3596	.00572	.89316	52
9	.10713	.99424	.10775	.2806	.0058	.3343	.00575	.89287	51
10	0.10742	0.99421	0.10805	9.2553	1.0058	9.3092	0.00579	0.89258	50
11	.10771	.99418	.10834	.2302	.0058	.2842	.00582	.89229	49
12	.10800	.99415	.10863	.2051	.0059	.2593	.00585	.89200	48
13	.10829	.99412	.10893	.1803	.0059	.2346	.00588	.89171	47
14	.10858	.99409	.10922	.1555	.0059	.2100	.00591	.89142	46
15	0.10887	0.99406	0.10952	9.1309	1.0060	9.1855	0.00594	0.89113	45
16	.10916	.99402	.10981	.1064	.0060	.1612	.00597	.89084	44
17	.10944	.99399	.11011	.0821	.0060	.1370	.00601	.89055	43
18	.10973	.99396	.11040	.0579	.0061	.1129	.00604	.89026	42
19	.11002	.99393	.11069	.0338	.0061	.0890	.00607	.88998	41
20	0.11031	0.99390	0.11099	9.0098	1.0061	9.0651	0.00610	0.88969	40
21	.11060	.99386	.11128	8.9860	.0062	.0414	.00613	.88940	39
22	.11089	.99383	.11158	.9623	.0062	.0179	.00617	.88911	38
23	.11118	.99380	.11187	.9387	.0062	8.9944	.00620	.88882	37
24	.11147	.99377	.11217	.9152	.0063	.9711	.00623	.88853	36
25	0.11176	0.99373	0.11246	8.8918	1.0063	8.9479	0.00626	0.88824	35
26	.11205	.99370	.11276	.8686	.0063	.9248	.00630	.88795	34
27	.11234	.99367	.11305	.8455	.0064	.9018	.00633	.88766	33
28	.11262	.99364	.11335	.8225	.0064	.8790	.00636	.88737	32
29	.11291	.99360	.11364	.7996	.0064	.8563	.00639	.88708	31
30	0.11320	0.99357	0.11393	8.7769	1.0065	8.8337	0.00643	0.88680	30
31	.11349	.99354	.11423	.7542	.0065	.8112	.00646	.88651	29
32	.11378	.99350	.11452	.7317	.0065	.7888	.00649	.88622	28
33	.11407	.99347	.11482	.7093	.0066	.7665	.00653	.88593	27
34	.11436	.99344	.11511	.6870	.0066	.7444	.00656	.88564	26
35	0.11465	0.99341	0.11541	8.6648	1.0066	8.7223	0.00659	0.88535	25
36	.11494	.99337	.11570	.6427	.0067	.7004	.00663	.88506	24
37	.11523	.99334	.11600	.6208	.0067	.6786	.00666	.88477	23
38	.11551	.99330	.11629	.5989	.0067	.6569	.00669	.88448	22
39	.11580	.99327	.11659	.5772	.0068	.6353	.00673	.88420	21
40	0.11609	0.99324	0.11688	8.5555	1.0068	8.6138	0.00676	0.88391	20
41	.11638	.99320	.11718	.5340	.0068	.5924	.00679	.88362	19
42	.11667	.99317	.11747	.5126	.0069	.5711	.00683	.88333	18
43	.11696	.99314	.11777	.4913	.0069	.5499	.00686	.88304	17
44	.11725	.99310	.11806	.4701	.0069	.5289	.00690	.88272	16
45	0.11754	0.99307	0.11836	8.4489	1.0070	8.5079	0.00693	0.88246	15
46	.11783	.99303	.11865	.4279	.0070	.4871	.00696	.88217	14
47	.11811	.99300	.11895	.4070	.0070	.4663	.00700	.88188	13
48	.11840	.99296	.11924	.3862	.0071	.4457	.00703	.88160	12
49	.11869	.99293	.11954	.3655	.0071	.4251	.00707	.88131	11
50	0.11898	0.99290	0.11983	8.3449	1.0071	8.4046	0.00710	0.88102	10
51	.11927	.99286	.12013	.3244	.0072	.3843	.00714	.88073	9
52	.11956	.99283	.12042	.3040	.0072	.3640	.00717	.88044	8
53	.11985	.99279	.12072	.2837	.0073	.3439	.00721	.88015	7
54	.12014	.99276	.12101	.2635	.0073	.3238	.00724	.87986	6
55	0.12042	0.99272	0.12131	8.2434	1.0073	8.3039	0.00728	0.87957	5
56	.12071	.99269	.12160	.2234	.0074	.2840	.00731	.87928	4
57	.12100	.99265	.12190	.2035	.0074	.2642	.00735	.87900	3
58	.12129	.99262	.12219	.1837	.0074	.2446	.00738	.87871	2
59	.12158	.99258	.12249	.1640	.0075	.2250	.00742	.87842	1
60	0.12187	0.99255	0.12278	8.1443	1.0075	8.2055	0.00745	0.87813	0
M	Cosine	Sine	Cotan.	Tan.	Cosec.	Secant	Vrs. Cos.	Vrs. Sin.	M

96°   83°

7°   Natural Trigonometric Functions   172°

M	Sine	Cosine	Tan.	Cotan.	Secant	Cosec.	Vrs. Sin.	Vrs. Cos.	M
0	0.12187	0.99255	0.12278	8.1443	1.0075	8.2055	0.00745	0.87813	60
1	.12216	.99251	.12308	.1248	.0075	.1861	.00749	.87787	59
2	.12245	.99247	.12337	.1053	.0076	.1668	.00752	.87755	58
3	.12273	.99244	.12367	.0860	.0076	.1476	.00756	.87726	57
4	.12302	.99240	.12396	.0667	.0076	.1285	.00760	.87697	56
5	0.12331	0.99237	0.12426	8.0476	1.0077	8.1094	0.00763	0.87669	55
6	.12360	.99233	.12456	.0285	.0077	.0905	.00767	.87640	54
7	.12389	.99229	.12485	.0095	.0078	.0717	.00770	.87611	53
8	.12418	.99226	.12515	7.9906	.0078	.0529	.00774	.87582	52
9	.12447	.99222	.12544	.9717	.0078	.0342	.00778	.87553	51
10	0.12476	0.99219	0.12574	7.9530	1.0079	8.0156	0.00781	0.87524	50
11	.12504	.99215	.12603	.9344	.0079	7.9971	.00785	.87495	49
12	.12533	.99211	.12633	.9158	.0079	.9787	.00788	.87467	48
13	.12562	.99208	.12662	.8973	.0080	.9604	.00792	.87438	47
14	.12591	.99204	.12692	.8789	.0080	.9421	.00796	.87409	46
15	0.12620	0.99200	0.12722	7.8606	1.0080	7.9240	0.00799	0.87380	45
16	.12649	.99197	.12751	.8424	.0081	.9059	.00803	.87351	44
17	.12678	.99193	.12781	.8243	.0081	.8879	.00807	.87322	43
18	.12706	.99189	.12810	.8062	.0082	.8700	.00810	.87293	42
19	.12735	.99186	.12840	.7882	.0082	.8522	.00814	.87265	41
20	0.12764	0.99182	0.12869	7.7703	1.0082	7.8344	0.00818	0.87236	40
21	.12793	.99178	.12899	.7525	.0083	.8168	.00822	.87207	39
22	.12822	.99174	.12928	.7348	.0083	.7992	.00825	.87178	38
23	.12851	.99171	.12958	.7171	.0084	.7817	.00829	.87149	37
24	.12879	.99167	.12988	.6996	.0084	.7642	.00833	.87120	36
25	0.12908	0.99163	0.13017	7.6821	1.0084	7.7469	0.00837	0.87091	35
26	.12937	.99160	.13047	.6646	.0085	.7296	.00840	.87063	34
27	.12966	.99156	.13076	.6473	.0085	.7124	.00844	.87034	33
28	.12995	.99152	.13106	.6300	.0085	.6953	.00848	.87005	32
29	.13024	.99148	.13136	.6129	.0086	.6783	.00852	.86976	31
30	0.13053	0.99144	0.13165	7.5957	1.0086	7.6613	0.00855	0.86947	30
31	.13081	.99141	.13195	.5787	.0087	.6444	.00859	.86918	29
32	.13110	.99137	.13224	.5617	.0087	.6276	.00863	.86890	28
33	.13139	.99133	.13254	.5449	.0087	.6108	.00867	.86861	27
34	.13168	.99129	.13284	.5280	.0088	.5942	.00871	.86832	26
35	0.13197	0.99125	0.13313	7.5113	1.0088	7.5776	0.00875	0.86803	25
36	.13226	.99121	.13343	.4946	.0089	.5611	.00878	.86774	24
37	.13254	.99118	.13372	.4780	.0089	.5446	.00882	.86745	23
38	.13283	.99114	.13402	.4615	.0089	.5282	.00886	.86717	22
39	.13312	.99110	.13432	.4451	.0090	.5119	.00890	.86688	21
40	0.13341	0.99106	0.13461	7.4287	1.0090	7.4957	0.00894	0.86659	20
41	.13370	.99102	.13491	.4124	.0090	.4795	.00898	.86630	19
42	.13399	.99098	.13520	.3961	.0091	.4634	.00902	.86601	18
43	.13427	.99094	.13550	.3800	.0091	.4474	.00905	.86572	17
44	.13456	.99090	.13580	.3639	.0092	.4315	.00909	.86544	16
45	0.13485	0.99086	0.13609	7.3479	1.0092	7.4156	0.00913	0.86515	15
46	.13514	.99083	.13639	.3319	.0092	.3998	.00917	.86486	14
47	.13543	.99079	.13669	.3160	.0093	.3840	.00921	.86457	13
48	.13571	.99075	.13698	.3002	.0093	.3683	.00925	.86428	12
49	.13600	.99071	.13728	.2844	.0094	.3527	.00929	.86400	11
50	0.13629	0.99067	0.13757	7.2687	1.0094	7.3372	0.00933	0.86371	10
51	.13658	.99063	.13787	.2531	.0094	.3217	.00937	.86342	9
52	.13687	.99059	.13817	.2375	.0095	.3063	.00941	.86313	8
53	.13716	.99055	.13846	.2220	.0095	.2909	.00945	.86284	7
54	.13744	.99051	.13876	.2066	.0096	.2757	.00949	.86255	6
55	0.13773	0.99047	0.13906	7.1912	1.0096	7.2604	0.00953	0.86227	5
56	.13802	.99043	.13935	.1759	.0097	.2453	.00957	.86198	4
57	.13831	.99039	.13965	.1607	.0097	.2302	.00961	.86169	3
58	.13860	.99035	.13995	.1455	.0097	.2152	.00965	.86140	2
59	.13888	.99031	.14024	.1304	.0098	.2002	.00969	.86111	1
60	0.13917	0.99027	0.14054	7.1154	1.0098	7.1853	0.00973	0.86083	0
M	Cosine	Sine	Cotan.	Tan.	Cosec.	Secant	Vrs. Cos.	Vrs. Sin.	M

97°   82°

8° Natural Trigonometric Functions 171°

M	Sine	Cosine	Tan.	Cotan.	Secant	Cosec.	Vrs. Sin.	Vrs. Cos.	M
0	0.13917	0.99027	0.14054	7.1154	1.0098	7.1853	0.00973	0.86083	60
1	.13946	.99023	.14084	.1004	.0099	.1704	.00977	.86054	59
2	.13975	.99019	.14113	.0854	.0099	.1557	.00981	.86025	58
3	.14004	.99015	.14143	.0706	.0099	.1409	.00985	.85996	57
4	.14032	.99010	.14173	.0558	.0100	.1263	.00989	.85967	56
5	0.14061	0.99006	0.14202	7.0410	1.0100	7.1117	0.00993	0.85939	55
6	.14090	.99002	.14232	.0264	.0101	.0972	.00998	.85910	54
7	.14119	.98998	.14262	.0117	.0101	.0827	.01002	.85881	53
8	.14148	.98994	.14291	6.9972	.0102	.0683	.01006	.85852	52
9	.14176	.98990	.14321	.9827	.0102	.0539	.01010	.85823	51
10	0.14205	0.98986	0.14351	6.9682	1.0102	7.0396	0.01014	0.85795	50
11	.14234	.98982	.14380	.9538	.0103	.0254	.01018	.85766	49
12	.14263	.98978	.14410	.9395	.0103	.0112	.01022	.85737	48
13	.14292	.98973	.14440	.9252	.0104	6.9971	.01026	.85708	47
14	.14320	.98969	.14470	.9110	.0104	.9830	.01031	.85679	46
15	0.14349	0.98965	0.14499	6.8969	1.0104	6.9690	0.01035	0.85651	45
16	.14378	.98961	.14529	.8828	.0105	.9550	.01039	.85622	44
17	.14407	.98957	.14559	.8687	.0105	.9411	.01043	.85593	43
18	.14436	.98952	.14588	.8547	.0106	.9273	.01047	.85564	42
19	.14464	.98948	.14618	.8408	.0106	.9135	.01052	.85536	41
20	0.14493	0.98944	0.14648	6.8269	1.0107	6.8998	0.01056	0.85507	40
21	.14522	.98940	.14677	.8131	.0107	.8861	.01060	.85478	39
22	.14551	.98936	.14707	.7993	.0107	.8725	.01064	.85449	38
23	.14579	.98931	.14737	.7856	.0108	.8589	.01068	.85420	37
24	.14608	.98927	.14767	.7720	.0108	.8454	.01073	.85392	36
25	0.14637	0.98923	0.14796	6.7584	1.0109	6.8320	0.01077	0.85363	35
26	.14666	.98919	.14826	.7448	.0109	.8185	.01081	.85334	34
27	.14695	.98914	.14856	.7313	.0110	.8052	.01085	.85305	33
28	.14723	.98910	.14886	.7179	.0110	.7919	.01090	.85277	32
29	.14752	.98906	.14915	.7045	.0111	.7787	.01094	.85248	31
30	0.14781	0.98901	0.14945	6.6911	1.0111	6.7655	0.01098	0.85219	30
31	.14810	.98897	.14975	.6779	.0111	.7523	.01103	.85190	29
32	.14838	.98893	.15004	.6646	.0112	.7392	.01107	.85161	28
33	.14867	.98889	.15034	.6514	.0112	.7262	.01111	.85133	27
34	.14896	.98884	.15064	.6383	.0113	.7132	.01116	.85104	26
35	0.14925	0.98880	0.15094	6.6252	1.0113	6.7003	0.01120	0.85075	25
36	.14953	.98876	.15123	.6122	.0114	.6874	.01124	.85046	24
37	.14982	.98871	.15153	.5992	.0114	.6745	.01129	.85018	23
38	.15011	.98867	.15183	.5863	.0115	.6617	.01133	.84989	22
39	.15040	.98862	.15213	.5734	.0115	.6490	.01137	.84960	21
40	0.15068	0.98858	0.15243	6.5605	1.0115	6.6363	0.01142	0.84931	20
41	.15097	.98854	.15272	.5478	.0116	.6237	.01146	.84903	19
42	.15126	.98849	.15302	.5350	.0116	.6111	.01151	.84874	18
43	.15155	.98845	.15332	.5223	.0117	.5985	.01155	.84845	17
44	.15183	.98840	.15362	.5097	.0117	.5860	.01159	.84816	16
45	0.15212	0.98836	0.15391	6.4971	1.0118	6.5736	0.01164	0.84788	15
46	.15241	.98832	.15421	.4845	.0118	.5612	.01168	.84759	14
47	.15270	.98827	.15451	.4720	.0119	.5488	.01173	.84730	13
48	.15298	.98823	.15481	.4596	.0119	.5365	.01177	.84701	12
49	.15328	.98818	.15511	.4472	.0119	.5243	.01182	.84672	11
50	0.15356	0.98814	0.15540	6.4348	1.0120	6.5121	0.01186	0.84644	10
51	.15385	.98809	.15570	.4225	.0120	.4999	.01190	.84615	9
52	.15413	.98805	.15600	.4103	.0121	.4878	.01195	.84586	8
53	.15442	.98800	.15630	.3980	.0121	.4757	.01199	.84558	7
54	.15471	.98796	.15659	.3859	.0122	.4637	.01204	.84529	6
55	0.15500	0.98791	0.15689	6.3737	1.0122	6.4517	0.01208	0.84500	5
56	.15528	.98787	.15719	.3616	.0123	.4398	.01213	.84471	4
57	.15557	.98782	.15749	.3496	.0123	.4279	.01217	.84443	3
58	.15586	.98778	.15779	.3376	.0124	.4160	.01222	.84414	2
59	.15615	.98773	.15809	.3257	.0124	.4042	.01227	.84385	1
60	0.15643	0.98769	0.15838	6.3137	1.0125	6.3924	0.01231	0.84356	0
M	Cosine	Sine	Cotan.	Tan.	Cosec.	Secant	Vrs. Cos.	Vrs. Sin.	M

98° 81°

9° Natural Trigonometric Functions 170°

M	Sine	Cosine	Tan.	Cotan.	Secant	Cosec.	Vrs. Sin.	Vrs. Cos.	M
0	0.15643	0.98769	0.15838	6.3137	1.0125	6.3924	0.01231	0.84356	60
1	.15672	.98764	.15868	.3019	.0125	.3807	.01236	.84328	59
2	.15701	.98760	.15898	.2901	.0125	.3690	.01240	.84299	58
3	.15730	.98755	.15928	.2783	.0126	.3574	.01245	.84270	57
4	.15758	.98750	.15958	.2665	.0126	.3458	.01249	.84242	56
5	0.15787	0.98746	0.15987	6.2548	1.0127	6.3343	0.01254	0.84213	55
6	.15816	.98741	.16017	.2432	.0127	.3228	.01259	.84184	54
7	.15844	.98737	.16047	.2316	.0128	.3113	.01263	.84155	53
8	.15873	.98732	.16077	.2200	.0128	.2999	.01268	.84127	52
9	.15902	.98727	.16107	.2085	.0129	.2885	.01272	.84098	51
10	0.15931	0.98723	0.16137	6.1970	1.0129	6.2772	0.01277	0.84069	50
11	.15959	.98718	.16167	.1856	.0130	.2659	.01282	.84041	49
12	.15988	.98714	.16196	.1742	.0130	.2546	.01286	.84012	48
13	.16017	.98709	.16226	.1628	.0131	.2434	.01291	.83983	47
14	.16045	.98704	.16256	.1515	.0131	.2322	.01296	.83954	46
15	0.16074	0.98700	0.16286	6.1402	1.0132	6.2211	0.01300	0.83926	45
16	.16103	.98695	.16316	.1290	.0132	.2100	.01305	.83897	44
17	.16132	.98690	.16346	.1178	.0133	.1990	.01310	.83868	43
18	.16160	.98685	.16376	.1066	.0133	.1880	.01314	.83840	42
19	.16189	.98681	.16405	.0955	.0134	.1770	.01319	.83811	41
20	0.16218	0.98676	0.16435	6.0844	1.0134	6.1661	0.01324	0.83782	40
21	.16246	.98671	.16465	.0734	.0135	.1552	.01328	.83753	39
22	.16275	.98667	.16495	.0624	.0135	.1443	.01333	.83725	38
23	.16304	.98662	.16525	.0514	.0136	.1335	.01338	.83696	37
24	.16333	.98657	.16555	.0405	.0136	.1227	.01343	.83667	36
25	0.16361	0.98652	0.16585	6.0296	1.0136	6.1120	0.01347	0.83639	35
26	.16390	.98648	.16615	.0188	.0137	.1013	.01352	.83610	34
27	.16419	.98643	.16644	.0080	.0137	.0906	.01357	.83581	33
28	.16447	.98638	.16674	5.9972	.0138	.0800	.01362	.83553	32
29	.16476	.98633	.16704	.9865	.0138	.0694	.01367	.83524	31
30	0.16505	0.98628	0.16734	5.9758	1.0139	6.0588	0.01371	0.83495	30
31	.16533	.98624	.16764	.9651	.0139	.0483	.01376	.83466	29
32	.16562	.98619	.16794	.9545	.0140	.0379	.01381	.83438	28
33	.16591	.98614	.16824	.9439	.0140	.0274	.01386	.83409	27
34	.16619	.98609	.16854	.9333	.0141	.0170	.01391	.83380	26
35	0.16648	0.98604	0.16884	5.9228	1.0141	6.0066	0.01395	0.83352	25
36	.16677	.98600	.16914	.9123	.0142	5.9963	.01400	.83323	24
37	.16705	.98595	.16944	.9019	.0142	.9860	.01405	.83294	23
38	.16734	.98590	.16973	.8915	.0143	.9758	.01410	.83266	22
39	.16763	.98585	.17003	.8811	.0143	.9655	.01415	.83237	21
40	0.16791	0.98580	0.17033	5.8708	1.0144	5.9554	0.01420	0.83208	20
41	.16820	.98575	.17063	.8605	.0144	.9452	.01425	.83180	19
42	.16849	.98570	.17093	.8502	.0145	.9351	.01430	.83151	18
43	.16878	.98565	.17123	.8400	.0145	.9250	.01434	.83122	17
44	.16906	.98560	.17153	.8298	.0145	.9150	.01439	.83094	16
45	0.16935	0.98556	0.17183	5.8196	1.0146	5.9049	0.01444	0.83065	15
46	.16964	.98551	.17213	.8095	.0147	.8950	.01449	.83036	14
47	.16992	.98546	.17243	.7994	.0147	.8850	.01454	.83008	13
48	.17021	.98541	.17273	.7894	.0148	.8751	.01459	.82979	12
49	.17050	.98536	.17303	.7794	.0148	.8652	.01464	.82950	11
50	0.17078	0.98531	0.17333	5.7694	1.0149	5.8554	0.01469	0.82922	10
51	.17107	.98526	.17363	.7594	.0150	.8456	.01474	.82893	9
52	.17136	.98521	.17393	.7495	.0150	.8358	.01479	.82864	8
53	.17164	.98516	.17423	.7396	.0151	.8261	.01484	.82836	7
54	.17193	.98511	.17453	.7297	.0151	.8163	.01489	.82807	6
55	0.17221	0.98506	0.17483	5.7199	1.0152	5.8067	0.01494	0.82778	5
56	.17250	.98501	.17513	.7101	.0152	.7970	.01499	.82750	4
57	.17279	.98496	.17543	.7004	.0153	.7874	.01504	.82721	3
58	.17307	.98491	.17573	.6906	.0153	.7778	.01509	.82692	2
59	.17336	.98486	.17603	.6809	.0154	.7683	.01514	.82664	1
60	0.17365	0.98481	0.17633	5.6713	1.0154	5.7588	0.01519	0.82635	0
M	Cosine	Sine	Cotan.	Tan.	Cosec.	Secant	Vrs. Cos.	Vrs. Sin.	M

99° 80°

10°   Natural Trigonometric Functions   169°

M	Sine	Cosine	Tan.	Cotan.	Secant	Cosec.	Vrs. Sin.	Vrs. Cos.	M
0	0.17365	0.98481	0.17633	5.6713	1.0154	5.7588	0.01519	0.82635	60
1	.17393	.98476	.17663	.6616	.0155	.7493	.01524	.82606	59
2	.17422	.98471	.17693	.6520	.0155	.7398	.01529	.82578	58
3	.17451	.98465	.17723	.6425	.0156	.7304	.01534	.82549	57
4	.17479	.98460	.17753	.6329	.0156	.7210	.01539	.82521	56
5	0.17508	0.98455	0.17783	5.6234	1.0157	5.7117	0.01544	0.82492	55
6	.17537	.98450	.17813	.6140	.0157	.7023	.01550	.82463	54
7	.17565	.98445	.17843	.6045	.0158	.6930	.01555	.82435	53
8	.17594	.98440	.17873	.5951	.0158	.6838	.01560	.82406	52
9	.17622	.98435	.17903	.5857	.0159	.6745	.01565	.82377	51
10	0.17651	0.98430	0.17933	5.5764	1.0159	5.6653	0.01570	0.82349	50
11	.17680	.98425	.17963	.5670	.0160	.6561	.01575	.82320	49
12	.17708	.98419	.17993	.5578	.0160	.6470	.01580	.82291	48
13	.17737	.98414	.18023	.5485	.0161	.6379	.01585	.82263	47
14	.17766	.98409	.18053	.5393	.0162	.6288	.01591	.82234	46
15	0.17794	0.98404	0.18083	5.5301	1.0162	5.6197	0.01596	0.82206	45
16	.17823	.98399	.18113	.5209	.0163	.6107	.01601	.82177	44
17	.17852	.98394	.18143	.5117	.0163	.6017	.01606	.82148	43
18	.17880	.98388	.18173	.5026	.0164	.5928	.01611	.82120	42
19	.17909	.98383	.18203	.4936	.0164	.5838	.01617	.82091	41
20	0.17937	0.98378	0.18233	5.4845	1.0165	5.5749	0.01622	0.82062	40
21	.17966	.98373	.18263	.4755	.0165	.5660	.01627	.82034	39
22	.17995	.98368	.18293	.4665	.0166	.5572	.01632	.82005	38
23	.18023	.98362	.18323	.4575	.0166	.5484	.01638	.81977	37
24	.18052	.98357	.18353	.4486	.0167	.5396	.01643	.81948	36
25	0.18080	0.98352	0.18383	5.4396	1.0167	5.5308	0.01648	0.81919	35
26	.18109	.98347	.18413	.4308	.0168	.5221	.01653	.81891	34
27	.18138	.98341	.18444	.4219	.0169	.5134	.01659	.81862	33
28	.18166	.98336	.18474	.4131	.0169	.5047	.01664	.81834	32
29	.18195	.98331	.18504	.4043	.0170	.4960	.01669	.81805	31
30	0.18223	0.98325	0.18534	5.3955	1.0170	5.4874	0.01674	0.81776	30
31	.18252	.98320	.18564	.3868	.0171	.4788	.01680	.81748	29
32	.18281	.98315	.18594	.3780	.0171	.4702	.01685	.81719	28
33	.18309	.98309	.18624	.3694	.0172	.4617	.01690	.81691	27
34	.18338	.98304	.18654	.3607	.0172	.4532	.01696	.81662	26
35	0.18366	0.98299	0.18684	5.3521	1.0173	5.4447	0.01701	0.81633	25
36	.18395	.98293	.18714	.3434	.0174	.4362	.01706	.81605	24
37	.18424	.98288	.18745	.3349	.0174	.4278	.01712	.81576	23
38	.18452	.98283	.18775	.3263	.0175	.4194	.01717	.81548	22
39	.18481	.98277	.18805	.3178	.0175	.4110	.01722	.81519	21
40	0.18509	0.98272	0.18835	5.3093	1.0176	5.4026	0.01728	0.81490	20
41	.18538	.98267	.18865	.3008	.0176	.3943	.01733	.81462	19
42	.18567	.98261	.18895	.2923	.0177	.3860	.01739	.81433	18
43	.18595	.98256	.18925	.2839	.0177	.3777	.01744	.81405	17
44	.18624	.98250	.18955	.2755	.0178	.3695	.01749	.81376	16
45	0.18652	0.98245	0.18985	5.2671	1.0179	5.3612	0.01755	0.81348	15
46	.18681	.98240	.19016	.2588	.0179	.3530	.01760	.81319	14
47	.18709	.98234	.19046	.2505	.0180	.3449	.01766	.81290	13
48	.18738	.98229	.19076	.2422	.0180	.3367	.01771	.81262	12
49	.18767	.98223	.19106	.2339	.0181	.3286	.01777	.81233	11
50	0.18795	0.98218	0.19136	5.2257	1.0181	5.3205	0.01782	0.81205	10
51	.18824	.98212	.19166	.2174	.0182	.3124	.01788	.81176	9
52	.18852	.98207	.19197	.2092	.0182	.3044	.01793	.81147	8
53	.18881	.98201	.19227	.2011	.0183	.2963	.01799	.81119	7
54	.18909	.98196	.19257	.1929	.0184	.2883	.01804	.81090	6
55	0.18938	0.98190	0.19287	5.1848	1.0184	5.2803	0.01810	0.81062	5
56	.18967	.98185	.19317	.1767	.0185	.2724	.01815	.81033	4
57	.18995	.98179	.19347	.1686	.0185	.2645	.01821	.81005	3
58	.19024	.98174	.19378	.1606	.0186	.2566	.01826	.80976	2
59	.19052	.98168	.19408	.1525	.0186	.2487	.01832	.80948	1
60	0.19081	0.98163	0.19438	5.1445	1.0187	5.2408	0.01837	0.80919	0
M	Cosine	Sine	Cotan.	Tan.	Cosec.	Secant	Vrs. Cos.	Vrs. Sin.	M

100°   79°

11° Natural Trigonometric Functions 168°

M	Sine	Cosine	Tan.	Cotan.	Secant	Cosec.	Vrs. Sin.	Vrs. Cos.	M
0	0.19081	0.98163	0.19438	5.1445	1.0187	5.2408	0.01837	0.80919	60
1	.19109	.98157	.19468	.1366	.0188	.2330	.01843	.80890	59
2	.19138	.98152	.19498	.1286	.0188	.2252	.01848	.80862	58
3	.19166	.98146	.19529	.1207	.0189	.2174	.01854	.80833	57
4	.19195	.98140	.19559	.1128	.0189	.2097	.01859	.80805	56
5	0.19224	0.98135	0.19589	5.1049	1.0190	5.2019	0.01865	0.80776	55
6	.19252	.98129	.19619	.0970	.0191	.1942	.01871	.80748	54
7	.19281	.98124	.19649	.0892	.0191	.1865	.01876	.80719	53
8	.19309	.98118	.19680	.0814	.0192	.1788	.01882	.80691	52
9	.19338	.98112	.19710	.0736	.0192	.1712	.01887	.80662	51
10	0.19366	0.98107	0.19740	5.0658	1.0193	5.1636	0.01893	0.80634	50
11	.19395	.98101	.19770	.0581	.0193	.1560	.01899	.80605	49
12	.19423	.98095	.19800	.0504	.0194	.1484	.01904	.80576	48
13	.19452	.98090	.19831	.0427	.0195	.1409	.01910	.80548	47
14	.19480	.98084	.19861	.0350	.0195	.1333	.01916	.80519	46
15	0.19509	0.98078	0.19891	5.0273	1.0196	5.1258	0.01921	0.80491	45
16	.19537	.98073	.19921	.0197	.0196	.1183	.01927	.80462	44
17	.19566	.98067	.19952	.0121	.0197	.1109	.01933	.80434	43
18	.19595	.98061	.19982	.0045	.0198	.1034	.01938	.80405	42
19	.19623	.98056	.20012	4.9969	.0198	.0960	.01944	.80377	41
20	0.19652	0.98050	0.20042	4.9894	1.0199	5.0886	0.01950	0.80348	40
21	.19680	.98044	.20073	.9819	.0199	.0812	.01956	.80320	39
22	.19709	.98039	.20103	.9744	.0200	.0739	.01961	.80291	38
23	.19737	.98033	.20133	.9669	.0201	.0666	.01967	.80263	37
24	.19766	.98027	.20163	.9594	.0201	.0593	.01973	.80234	36
25	0.19794	0.98021	0.20194	4.9520	1.0202	5.0520	0.01979	0.80206	35
26	.19823	.98016	.20224	.9446	.0202	.0447	.01984	.80177	34
27	.19851	.98010	.20254	.9372	.0203	.0375	.01990	.80149	33
28	.19880	.98004	.20285	.9298	.0204	.0302	.01996	.80120	32
29	.19908	.97998	.20315	.9225	.0204	.0230	.02002	.80092	31
30	0.19937	0.97992	0.20345	4.9151	1.0205	5.0158	0.02007	0.80063	30
31	.19965	.97987	.20375	.9078	.0205	.0087	.02013	.80035	29
32	.19994	.97981	.20406	.9006	.0206	.0015	.02019	.80006	28
33	.20022	.97975	.20436	.8933	.0207	4.9944	.02025	.79978	27
34	.20051	.97969	.20466	.8860	.0207	.9873	.02031	.79949	26
35	0.20079	0.97963	0.20497	4.8788	1.0208	4.9802	0.02037	0.79921	25
36	.20108	.97957	.20527	.8716	.0208	.9732	.02042	.79892	24
37	.20136	.97952	.20557	.8644	.0209	.9661	.02048	.79863	23
38	.20165	.97946	.20588	.8573	.0210	.9591	.02054	.79835	22
39	.20193	.97940	.20618	.8501	.0210	.9521	.02060	.79807	21
40	0.20222	0.97934	0.20648	4.8430	1.0211	4.9452	0.02066	0.79778	20
41	.20250	.97928	.20679	.8359	.0211	.9382	.02072	.79750	19
42	.20279	.97922	.20709	.8288	.0212	.9313	.02078	.79721	18
43	.20307	.97916	.20739	.8217	.0213	.9243	.02084	.79693	17
44	.20336	.97910	.20770	.8147	.0213	.9175	.02089	.79664	16
45	0.20364	0.97904	0.20800	4.8077	1.0214	4.9106	0.02095	0.79636	15
46	.20393	.97899	.20830	.8007	.0215	.9037	.02101	.79607	14
47	.20421	.97893	.20861	.7937	.0215	.8969	.02107	.79579	13
48	.20450	.97887	.20891	.7867	.0216	.8901	.02113	.79550	12
49	.20478	.97881	.20921	.7798	.0216	.8833	.02119	.79522	11
50	0.20506	0.97875	0.20952	4.7728	1.0217	4.8765	0.02125	0.79493	10
51	.20535	.97869	.20982	.7659	.0218	.8697	.02131	.79465	9
52	.20563	.97863	.21012	.7591	.0218	.8630	.02137	.79436	8
53	.20592	.97857	.21043	.7522	.0219	.8563	.02143	.79408	7
54	.20620	.97851	.21073	.7453	.0220	.8496	.02149	.79379	6
55	0.20649	0.97845	0.21104	4.7385	1.0220	4.8429	0.02155	0.79351	5
56	.20677	.97839	.21134	.7317	.0221	.8362	.02161	.79323	4
57	.20706	.97833	.21164	.7249	.0221	.8296	.02167	.79294	3
58	.20734	.97827	.21195	.7181	.0222	.8229	.02173	.79266	2
59	.20763	.97821	.21225	.7114	.0223	.8163	.02179	.79237	1
60	0.20791	0.97815	0.21256	4.7046	1.0223	4.8097	0.02185	0.79209	0
M	Cosine	Sine	Cotan.	Tan.	Cosec.	Secant	Vrs. Cos.	Vrs. Sin.	M

101° 78°

12° Natural Trigonometric Functions 167°

M	Sine	Cosine	Tan.	Cotan.	Secant	Cosec.	Vrs. Sin.	Vrs. Cos.	M
0	0.20791	0.97815	0.21256	4.7046	1.0223	4.8097	0.02185	0.79209	60
1	.20820	.97809	.21286	.6979	.0224	.8032	.02191	.79180	59
2	.20848	.97803	.21316	.6912	.0225	.7966	.02197	.79152	58
3	.20876	.97797	.21347	.6845	.0225	.7901	.02203	.79123	57
4	.20905	.97790	.21377	.6778	.0226	.7835	.02209	.79105	56
5	0.20933	0.97784	0.21408	4.6712	1.0226	4.7770	0.02215	0.79066	55
6	.20962	.97778	.21438	.6646	.0227	.7706	.02222	.79038	54
7	.20990	.97772	.21468	.6580	.0228	.7641	.02228	.79010	53
8	.21019	.97766	.21499	.6514	.0228	.7576	.02234	.78981	52
9	.21047	.97760	.21529	.6448	.0229	.7512	.02240	.78953	51
10	0.21076	0.97754	0.21560	4.6382	1.0230	4.7448	0.02246	0.78924	50
11	.21104	.97748	.21590	.6317	.0230	.7384	.02252	.78896	49
12	.21132	.97741	.21621	.6252	.0231	.7320	.02258	.78867	48
13	.21161	.97735	.21651	.6187	.0232	.7257	.02264	.78839	47
14	.21189	.97729	.21682	.6122	.0232	.7193	.02271	.78811	46
15	0.21218	0.97723	0.21712	4.6057	1.0233	4.7130	0.02277	0.78782	45
16	.21246	.97717	.21742	.5993	.0234	.7067	.02283	.78754	44
17	.21275	.97711	.21773	.5928	.0234	.7004	.02289	.78725	43
18	.21303	.97704	.21803	.5864	.0235	.6942	.02295	.78697	42
19	.21331	.97698	.21834	.5800	.0235	.6879	.02302	.78668	41
20	0.21360	0.97692	0.21864	4.5736	1.0236	4.6817	0.02308	0.78640	40
21	.21388	.97686	.21895	.5673	.0237	.6754	.02314	.78612	39
22	.21417	.97680	.21925	.5609	.0237	.6692	.02320	.78583	38
23	.21445	.97673	.21956	.5546	.0238	.6631	.02326	.78555	37
24	.21473	.97667	.21986	.5483	.0239	.6569	.02333	.78526	36
25	0.21502	0.97661	0.22017	4.5420	1.0239	4.6507	0.02339	0.78508	35
26	.21530	.97655	.22047	.5357	.0240	.6446	.02345	.78470	34
27	.21559	.97648	.22078	.5294	.0241	.6385	.02351	.78441	33
28	.21587	.97642	.22108	.5232	.0241	.6324	.02358	.78413	32
29	.21615	.97636	.22139	.5169	.0242	.6263	.02364	.78384	31
30	0.21644	0.97630	0.22169	4.5107	1.0243	4.6201	0.02370	0.78356	30
31	.21672	.97623	.22200	.5045	.0243	.6142	.02377	.78328	29
32	.21701	.97617	.22230	.4983	.0244	.6081	.02383	.78299	28
33	.21729	.97611	.22261	.4921	.0245	.6021	.02389	.78271	27
34	.21757	.97604	.22291	.4860	.0245	.5961	.02396	.78242	26
35	0.21786	0.97598	0.22322	4.4799	1.0246	4.5901	0.02402	0.78214	25
36	.21814	.97592	.22353	.4737	.0247	.5841	.02408	.78186	24
37	.21843	.97585	.22383	.4676	.0247	.5782	.02415	.78154	23
38	.21871	.97579	.22414	.4615	.0248	.5722	.02421	.78129	22
39	.21899	.97573	.22444	.4555	.0249	.5663	.02427	.78100	21
40	0.21928	0.97566	0.22475	4.4494	1.0249	4.5604	0.02434	0.78072	20
41	.21956	.97560	.22505	.4434	.0250	.5545	.02440	.78043	19
42	.21985	.97553	.22536	.4373	.0251	.5486	.02446	.78015	18
43	.22013	.97547	.22566	.4313	.0251	.5428	.02453	.77987	17
44	.22041	.97541	.22597	.4253	.0252	.5369	.02459	.77959	16
45	0.22070	0.97534	0.22628	4.4194	1.0253	4.5311	0.02466	0.77930	15
46	.22098	.97528	.22658	.4134	.0253	.5253	.02472	.77902	14
47	.22126	.97521	.22689	.4074	.0254	.5195	.02479	.77873	13
48	.22155	.97515	.22719	.4015	.0255	.5137	.02485	.77845	12
49	.22183	.97508	.22750	.3956	.0255	.5079	.02491	.77817	11
50	0.22211	0.97502	0.22781	4.3897	1.0256	4.5021	0.02498	0.77788	10
51	.22240	.97495	.22811	.3838	.0257	.4964	.02504	.77760	9
52	.22268	.97489	.22842	.3779	.0257	.4907	.02511	.77732	8
53	.22297	.97483	.22872	.3721	.0258	.4850	.02517	.77703	7
54	.22325	.97476	.22903	.3662	.0259	.4793	.02524	.77675	6
55	0.22353	0.97470	0.22934	4.3604	1.0260	4.4736	0.02530	0.77647	5
56	.22382	.97463	.22964	.3546	.0260	.4679	.02537	.77618	4
57	.22410	.97457	.22995	.3488	.0261	.4623	.02543	.77590	3
58	.22438	.97450	.23025	.3430	.0262	.4566	.02550	.77561	2
59	.22467	.97443	.23056	.3372	.0262	.4510	.02556	.77533	1
60	0.22495	0.97437	0.23087	4.3315	1.0263	4.4454	0.02563	0.77505	0
M	Cosine	Sine	Cotan.	Tan.	Cosec.	Secant	Vrs. Cos.	Vrs. Sin.	M

102° 77°

13°   Natural Trigonometric Functions   **166°**

M	Sine	Cosine	Tan.	Cotan.	Secant	Cosec.	Vrs. Sin.	Vrs. Cos.	M
0	0.22495	0.97437	0.23087	4.3315	1.0263	4.4454	0.02563	0.77505	60
1	.22523	.97430	.23117	.3257	.0264	.4398	.02569	.77476	59
2	.22552	.97424	.23148	.3200	.0264	.4342	.02576	.77448	58
3	.22580	.97417	.23179	.3143	.0265	.4287	.02583	.77420	57
4	.22608	.97411	.23209	.3086	.0266	.4231	.02589	.77391	56
5	0.22637	0.97404	0.23240	4.3029	1.0266	4.4176	0.02596	0.77363	55
6	.22665	.97398	.23270	.2972	.0267	.4121	.02602	.77335	54
7	.22693	.97391	.23301	.2916	.0268	.4065	.02609	.77306	53
8	.22722	.97384	.23332	.2859	.0268	.4011	.02616	.77278	52
9	.22750	.97378	.23363	.2803	.0269	.3956	.02622	.77250	51
10	0.22778	0.97371	0.23393	4.2747	1.0270	4.3901	0.02629	0.77221	50
11	.22807	.97364	.23424	.2691	.0271	.3847	.02635	.77193	49
12	.22835	.97358	.23455	.2635	.0271	.3792	.02642	.77165	48
13	.22863	.97351	.23485	.2579	.0272	.3738	.02649	.77136	47
14	.22892	.97344	.23516	.2524	.0273	.3684	.02655	.77108	46
15	0.22920	0.97338	0.23547	4.2468	1.0273	4.3630	0.02662	0.77080	45
16	.22948	.97331	.23577	.2413	.0274	.3576	.02669	.77052	44
17	.22977	.97324	.23608	.2358	.0275	.3522	.02675	.77023	43
18	.23005	.97318	.23639	.2303	.0276	.3469	.02682	.76995	42
19	.23033	.97311	.23670	.2248	.0276	.3415	.02689	.76967	41
20	0.23061	0.97304	0.23700	4.2193	1.0277	4.3362	0.02695	0.76938	40
21	.23090	.97298	.23731	.2139	.0278	.3309	.02702	.76910	39
22	.23118	.97291	.23762	.2084	.0278	.3256	.02709	.76882	38
23	.23146	.97284	.23793	.2030	.0279	.3203	.02716	.76853	37
24	.23175	.97277	.23823	.1976	.0280	.3150	.02722	.76825	36
25	0.23203	0.97271	0.23854	4.1921	1.0280	4.3098	0.02729	0.76797	35
26	.23231	.97264	.23885	.1867	.0281	.3045	.02736	.76769	34
27	.23260	.97257	.23916	.1814	.0282	.2993	.02743	.76740	33
28	.23288	.97250	.23946	.1760	.0283	.2941	.02749	.76712	32
29	.23316	.97244	.23977	.1706	.0283	.2888	.02756	.76684	31
30	0.23344	0.97237	0.24008	4.1653	1.0284	4.2836	0.02763	0.76655	30
31	.23373	.97230	.24039	.1600	.0285	.2785	.02770	.76627	29
32	.23401	.97223	.24069	.1546	.0285	.2733	.02777	.76599	28
33	.23429	.97216	.24100	.1493	.0286	.2681	.02783	.76571	27
34	.23458	.97210	.24131	.1440	.0287	.2630	.02790	.76542	26
35	0.23486	0.97203	0.24162	4.1388	1.0288	4.2579	0.02797	0.76514	25
36	.23514	.97196	.24192	.1335	.0288	.2527	.02804	.76486	24
37	.23542	.97189	.24223	.1282	.0289	.2476	.02811	.76457	23
38	.23571	.97182	.24254	.1230	.0290	.2425	.02818	.76429	22
39	.23599	.97175	.24285	.1178	.0291	.2375	.02824	.76401	21
40	0.23627	0.97169	0.24316	4.1126	1.0291	4.2324	0.02831	0.76373	20
41	.23655	.97162	.24346	.1073	.0292	.2273	.02838	.76344	19
42	.23684	.97155	.24377	.1022	.0293	.2223	.02845	.76316	18
43	.23712	.97148	.24408	.0970	.0293	.2173	.02852	.76288	17
44	.23740	.97141	.24439	.0918	.0294	.2122	.02859	.76260	16
45	0.23768	0.97134	0.24470	4.0867	1.0295	4.2072	0.02866	0.76231	15
46	.23797	.97127	.24501	.0815	.0296	.2022	.02873	.76203	14
47	.23825	.97120	.24531	.0764	.0296	.1972	.02880	.76175	13
48	.23853	.97113	.24562	.0713	.0297	.1923	.02886	.76147	12
49	.23881	.97106	.24593	.0662	.0298	.1873	.02893	.76118	11
50	0.23910	0.97099	0.24624	4.0611	1.0299	4.1824	0.02900	0.76090	10
51	.23938	.97092	.24655	.0560	.0299	.1774	.02907	.76062	9
52	.23966	.97086	.24686	.0509	.0300	.1725	.02914	.76034	8
53	.23994	.97079	.24717	.0458	.0301	.1676	.02921	.76005	7
54	.24023	.97072	.24747	.0408	.0302	.1627	.02928	.75977	6
55	0.24051	0.97065	0.24778	4.0358	1.0302	4.1578	0.02935	0.75949	5
56	.24079	.97058	.24809	.0307	.0303	.1529	.02942	.75921	4
57	.24107	.97051	.24840	.0257	.0304	.1481	.02949	.75892	3
58	.24136	.97044	.24871	.0207	.0305	.1432	.02956	.75864	2
59	.24164	.97037	.24902	.0157	.0305	.1384	.02963	.75836	1
60	0.24192	0.97029	0.24933	4.0108	1.0306	4.1336	0.02970	0.75808	0
M	Cosine	Sine	Cotan.	Tan.	Cosec.	Secant	Vrs. Cos.	Vrs. Sin.	M

103°   **76°**

14° Natural Trigonometric Functions 165°

M	Sine	Cosine	Tan.	Cotan.	Secant	Cosec.	Vrs. Sin.	Vrs. Cos.	M
0	0.24192	0.97029	0.24933	4.0108	1.0306	4.1336	0.02970	0.75808	60
1	.24220	.97022	.24964	.0058	.0307	.1287	.02977	.75779	59
2	.24249	.97015	.24995	.0009	.0308	.1239	.02984	.75751	58
3	.24277	.97008	.25025	3.9959	.0308	.1191	.02991	.75723	57
4	.24305	.97001	.25056	.9910	.0309	.1144	.02999	.75695	56
5	0.24333	0.96994	0.25087	3.9861	1.0310	4.1096	0.03006	0.75667	55
6	.24361	.96987	.25118	.9812	.0311	.1048	.03013	.75638	54
7	.24390	.96980	.25149	.9763	.0311	.1001	.03020	.75610	53
8	.24418	.96973	.25180	.9714	.0312	.0953	.03027	.75582	52
9	.24446	.96966	.25211	.9665	.0313	.0906	.03034	.75554	51
10	0.24474	0.96959	0.25242	3.9616	1.0314	4.0859	0.03041	0.75526	50
11	.24502	.96952	.25273	9568	.0314	.0812	.03048	.75497	49
12	.24531	.96944	.25304	.9520	.0315	.0765	.03055	.75469	48
13	.24559	.96937	.25335	.9471	.0316	.0718	.03063	.75441	47
14	.24587	.96930	.25366	.9423	.0317	.0672	.03070	.75413	46
15	0.24615	0.96923	0.25397	3.9375	1.0317	4.0625	0.03077	0.75385	45
16	.24643	.96916	.25428	.9327	.0318	.0579	.03084	.75356	44
17	.24672	.96909	.25459	.9279	.0319	.0532	.03091	.75328	43
18	.24700	.96901	.25490	.9231	.0320	.0486	.03098	.75300	42
19	.24728	.96894	.25521	.9184	.0320	.0440	.03106	.75272	41
20	0.24756	0.96887	0.25552	3.9136	1.0321	4.0394	0.03113	0.75244	40
21	.24784	.96880	.25583	.9089	.0322	.0348	.03120	.75215	39
22	.24813	.96873	.25614	.9042	.0323	.0302	.03127	.75187	38
23	.24841	.96865	.25645	.8994	.0323	.0256	.03134	.75159	37
24	.24869	.96858	.25676	.8947	.0324	.0211	.03142	.75131	36
25	0.24897	0.96851	0.25707	3.8900	1.0325	4.0165	0.03149	0.75103	35
26	.24925	.96844	.25738	.8853	.0326	.0120	.03156	.75075	34
27	.24953	.96836	.25769	.8807	.0327	.0074	.03163	.75046	33
28	.24982	.96829	.25800	.8760	.0327	.0029	.03171	.75018	32
29	.25010	.96822	.25831	.8713	.0328	3.9984	.03178	.74990	31
30	0.25038	0.96815	0.25862	3.8667	1.0329	3.9939	0.03185	0.74962	30
31	.25066	.96807	.25893	.8621	.0330	.9894	.03192	.74934	29
32	.25094	.96800	.25924	.8574	.0330	.9850	.03200	.74906	28
33	.25122	.96793	.25955	.8528	.0331	.9805	.03207	.74877	27
34	.25151	.96785	.25986	.8482	.0332	.9760	.03214	.74849	26
35	0.25179	0.96778	0.26017	3.8436	1.0333	3.9716	0.03222	0.74821	25
36	.25207	.96771	.26048	.8390	.0334	.9672	.03229	.74793	24
37	.25235	.96763	.26079	.8345	.0334	.9627	.03236	.74765	23
38	.25263	.96756	.26110	.8299	.0335	.9583	.03244	.74737	22
39	.25291	.96749	.26141	.8254	.0336	.9539	.03251	.74709	21
40	0.25319	0.96741	0.26172	3.8208	1.0337	3.9495	0.03258	0.74680	20
41	.25348	.96734	.26203	.8163	.0338	.9451	.03266	.74652	19
42	.25376	.96727	.26234	.8118	.0338	.9408	.03273	.74624	18
43	.25404	.96719	.26266	.8073	.0339	.9364	.03281	.74596	17
44	.25432	.96712	.26297	.8027	.0340	.9320	.03288	.74568	16
45	0.25460	0.96704	0.26328	3.7983	1.0341	3.9277	0.03295	0.74540	15
46	.25488	.96697	.26359	.7938	.0341	.9234	.03303	.74512	14
47	.25516	.96690	.26390	.7893	.0342	.9190	.03310	.74483	13
48	.25544	.96682	.26421	.7848	.0343	.9147	.03318	.74455	12
49	.25573	.96675	.26452	.7804	.0344	.9104	.03325	.74427	11
50	0.25601	0.96667	0.26483	3.7759	1.0345	3.9061	0.03332	0.74399	10
51	.25629	.96660	.26514	.7715	.0345	.9018	.03340	.74371	9
52	.25657	.96652	.26546	.7671	.0346	.8976	.03347	.74344	8
53	.25685	.96645	.26577	.7627	.0347	.8933	.03355	.74315	7
54	.25713	.96638	.26608	.7583	.0348	.8890	.03362	.74287	6
55	0.25741	0.96630	0.26639	3.7539	1.0349	3.8848	0.03370	0.74259	5
56	.25769	.96623	.26670	.7495	.0349	.8805	.03377	.74230	4
57	.25798	.96615	.26701	.7451	.0350	.8763	.03385	.74202	3
58	.25826	.96608	.26732	.7407	.0351	.8721	.03392	.74174	2
59	.25854	.96600	.26764	.7364	.0352	.8679	.03400	.74146	1
60	0.25882	0.96592	0.26795	3.7320	1.0353	3.8637	0.03407	0.74118	0
M	Cosine	Sine	Cotan.	Tan.	Cosec.	Secant	Vrs. Cos.	Vrs. Sin.	M

104° 75°

15°   Natural Trigonometric Functions   164°

M	Sine	Cosine	Tan.	Cotan.	Secant	Cosec.	Vrs. Sin.	Vrs. Cos.	M
0	0.25882	0.96592	0.26795	3.7320	1.0353	3.8637	0.03407	0.74118	60
1	.25910	.96585	.26826	.7277	.0353	.8595	.03415	.74090	59
2	.25938	.96577	.26857	.7234	.0354	.8553	.03422	.74062	58
3	.25966	.96570	.26888	.7191	.0355	.8512	.03430	.74034	57
4	.25994	.96562	.26920	.7147	.0356	.8470	.03438	.74006	56
5	0.26022	0.96555	0.26951	3.7104	1.0357	3.8428	0.03445	0.73978	55
6	.26050	.96547	.26982	.7062	.0358	.8387	.03453	.73949	54
7	.26078	.96540	.27013	.7019	.0358	.8346	.03460	.73921	53
8	.26107	.96532	.27044	.6976	.0359	.8304	.03468	.73893	52
9	.26135	.96524	.27076	.6933	.0360	.8263	.03475	.73865	51
10	0.26163	0.96517	0.27107	3.6891	1.0361	3.8222	0.03483	0.73837	50
11	.26191	.96509	.27138	.6848	.0362	.8181	.03491	.73809	49
12	.26219	.96502	.27169	.6806	.0362	.8140	.03498	.73781	48
13	.26247	.96494	.27201	.6764	.0363	.8100	.03506	.73753	47
14	.26275	.96486	.27232	.6722	.0364	.8059	.03514	.73725	46
15	0.26303	0.96479	0.27263	3.6679	1.0365	3.8018	0.03521	0.73697	45
16	.26331	.96471	.27294	.6637	.0366	.7978	.03529	.73669	44
17	.26359	.96463	.27326	.6596	.0367	.7937	.03536	.73641	43
18	.26387	.96456	.27357	.6554	.0367	.7897	.03544	.73613	42
19	.26415	.96448	.27388	.6512	.0368	.7857	.03552	.73585	41
20	0.26443	0.96440	0.27419	3.6470	1.0369	3.7816	0.03560	0.73556	40
21	.26471	.96433	.27451	.6429	.0370	.7776	.03567	.73528	39
22	.26499	.96425	.27482	.6387	.0371	.7736	.03575	.73500	38
23	.26527	.96417	.27513	.6346	.0371	.7697	.03583	.73472	37
24	.26556	.96409	.27544	.6305	.0372	.7657	.03590	.73444	36
25	0.26584	0.96402	0.27576	3.6263	1.0373	3.7617	0.03598	0.73416	35
26	.26612	.96394	.27607	.6222	.0374	.7577	.03606	.73388	34
27	.26640	.96386	.27638	.6181	.0375	.7538	.03614	.73360	33
28	.26668	.96378	.27670	.6140	.0376	.7498	.03621	.73332	32
29	.26696	.96371	.27701	.6100	.0376	.7459	.03629	.73304	31
30	0.26724	0.96363	0.27732	3.6059	1.0377	3.7420	0.03637	0.73276	30
31	.26752	.96355	.27764	.6018	.0378	.7380	.03645	.73248	29
32	.26780	.96347	.27795	.5977	.0379	.7341	.03652	.73220	28
33	.26808	.96340	.27826	.5937	.0380	.7302	.03660	.73192	27
34	.26836	.96332	.27858	.5896	.0381	.7263	.03668	.73164	26
35	0.26864	0.96324	0.27889	3.5856	1.0382	3.7224	0.03676	0.73136	25
36	.26892	.96316	.27920	.5816	.0382	.7186	.03684	.73108	24
37	.26920	.96308	.27952	.5776	.0383	.7147	.03691	.73080	23
38	.26948	.96301	.27983	.5736	.0384	.7108	.03699	.73052	22
39	.26976	.96293	.28014	.5696	.0385	.7070	.03707	.73024	21
40	0.27004	0.96285	0.28046	3.5656	1.0386	3.7031	0.03715	0.72996	20
41	.27032	.96277	.28077	.5616	.0387	.6993	.03723	.72968	19
42	.27060	.96269	.28109	.5576	.0387	.6955	.03731	.72940	18
43	.27088	.96261	.28140	.5536	.0388	.6917	.03739	.72912	17
44	.27116	.96253	.28171	.5497	.0389	.6878	.03746	.72884	16
45	0.27144	0.96245	0.28203	3.5457	1.0390	3.6840	0.03754	0.72856	15
46	.27172	.96238	.28234	.5418	.0391	.6802	.03762	.72828	14
47	.27200	.96230	.28266	.5378	.0392	.6765	.03770	.72800	13
48	.27228	.96222	.28297	.5339	.0393	.6727	.03778	.72772	12
49	.27256	.96214	.28328	.5300	.0393	.6689	.03786	.72744	11
50	0.27284	0.96206	0.28360	3.5261	1.0394	3.6651	0.03794	0.72716	10
51	.27312	.96198	.28391	.5222	.0395	.6614	.03802	.72688	9
52	.27340	.96190	.28423	.5183	.0396	.6576	.03810	.72660	8
53	.27368	.96182	.28454	.5144	.0397	.6539	.03818	.72632	7
54	.27396	.96174	.28486	.5105	.0398	.6502	.03826	.72604	6
55	0.27424	0.96166	0.28517	3.5066	1.0399	3.6464	0.03834	0.72576	5
56	.27452	.96158	.28549	.5028	.0399	.6427	.03842	.72548	4
57	.27480	.96150	.28580	.4989	.0400	.6390	.03850	.72520	3
58	.27508	.96142	.28611	.4951	.0401	.6353	.03858	.72492	2
59	.27536	.96134	.28643	.4912	.0402	.6316	.03866	.72464	1
60	0.27564	0.96126	0.28674	3.4874	1.0403	3.6279	0.03874	0.72436	0
M	Cosine	Sine	Cotan.	Tan.	Cosec.	Secant	Vrs. Cos.	Vrs. Sin.	M

105°   74°

16°   Natural Trigonometric Functions   **163°**

M	Sine	Cosine	Tan.	Cotan.	Secant	Cosec.	Vrs. Sin.	Vrs. Cos.	M
0	0.27564	0.96126	0.28674	3.4874	1.0403	3.6279	0.03874	0.72436	60
1	.27592	.96118	.28706	.4836	.0404	.6243	.03882	.72408	59
2	.27620	.96110	.28737	.4798	.0405	.6206	.03890	.72380	58
3	.27648	.96102	.28769	.4760	.0406	.6169	.03898	.72352	57
4	.27675	.96094	.28800	.4722	.0406	.6133	.03906	.72324	56
5	0.27703	0.96086	0.28832	3.4684	1.0407	3.6096	0.03914	0.72296	55
6	.27731	.96078	.28863	.4646	.0408	.6060	.03922	.72268	54
7	.27759	.96070	.28895	.4608	.0409	.6024	.03930	.72240	53
8	.27787	.96062	.28926	.4570	.0410	.5987	.03938	.72213	52
9	.27815	.96054	.28958	.4533	.0411	.5951	.03946	.72185	51
10	0.27843	0.96045	0.28990	3.4495	1.0412	3.5915	0.03954	0.72157	50
11	.27871	.96037	.29021	.4458	.0413	.5879	.03962	.72129	49
12	.27899	.96029	.29053	.4420	.0413	.5843	.03971	.72101	48
13	.27927	.96021	.29084	.4383	.0414	.5807	.03979	.72073	47
14	.27955	.96013	.29116	.4346	.0415	.5772	.03987	.72045	46
15	0.27983	0.96005	0.29147	3.4308	1.0416	3.5736	0.03995	0.72017	45
16	.28011	.95997	.29179	.4271	.0417	.5700	.04003	.71989	44
17	.28039	.95989	.29210	.4234	.0418	.5665	.04011	.71961	43
18	.28067	.95980	.29242	.4197	.0419	.5629	.04019	.71933	42
19	.28094	.95972	.29274	.4160	.0420	.5594	.04028	.71905	41
20	0.28122	0.95964	0.29305	3.4124	1.0420	3.5559	0.04036	0.71877	40
21	.28150	.95956	.29337	.4087	.0421	.5523	.04044	.71849	39
22	.28178	.95948	.29368	.4050	.0422	.5488	.04052	.71822	38
23	.28206	.95940	.29400	.4014	.0423	.5453	.04060	.71794	37
24	.28234	.95931	.29432	.3977	.0424	.5418	.04069	.71766	36
25	0.28262	0.95923	0.29463	3.3941	1.0425	3.5383	0.04077	0.71738	35
26	.28290	.95915	.29495	.3904	.0426	.5348	.04085	.71710	34
27	.28318	.95907	.29526	.3868	.0427	.5313	.04093	.71682	33
28	.28346	.95898	.29558	.3832	.0428	.5279	.04101	.71654	32
29	.28374	.95890	.29590	.3795	.0428	.5244	.04110	.71626	31
30	0.28401	0.95882	0.29621	3.3759	1.0429	3.5209	0.04118	0.71608	30
31	.28429	.95874	.29653	.3723	.0430	.5175	.04126	.71570	29
32	.28457	.95865	.29685	.3687	.0431	.5140	.04134	.71543	28
33	.28485	.95857	.29716	.3651	.0432	.5106	.04143	.71515	27
34	.28513	.95849	.29748	.3616	.0433	.5072	.04151	.71487	26
35	0.28541	0.95840	0.29780	3.3580	1.0434	3.5037	0.04159	0.71459	25
36	.28569	.95832	.29811	.3544	.0435	.5003	.04168	.71431	24
37	.28597	.95824	.29843	.3509	.0436	.4969	.04176	.71403	23
38	.28624	.95816	.29875	.3473	.0437	.4935	.04184	.71375	22
39	.28652	.95807	.29906	.3438	.0438	.4901	.04193	.71347	21
40	0.28680	0.95799	0.29938	3.3402	1.0438	3.4867	0.04201	0.71320	20
41	.28708	.95791	.29970	.3367	.0439	.4833	.04209	.71292	19
42	.28736	.95782	.30001	.3332	.0440	.4799	.04218	.71264	18
43	.28764	.95774	.30033	.3296	.0441	.4766	.04226	.71236	17
44	.28792	.95765	.30065	.3261	.0442	.4732	.04234	.71208	16
45	0.28820	0.95757	0.30096	3.3226	1.0443	3.4698	0.04243	0.71180	15
46	.28847	.95749	.30128	.3191	.0444	.4665	.04251	.71152	14
47	.28875	.95740	.30160	.3156	.0445	.4632	.04260	.71125	13
48	.28903	.95732	.30192	.3121	.0446	.4598	.04268	.71097	12
49	.28931	.95723	.30223	.3087	.0447	.4565	.04276	.71069	11
50	0.28959	0.95715	0.30255	3.3052	1.0448	3.4532	0.04285	0.71041	10
51	.28987	.95707	.30287	.3017	.0448	.4498	.04293	.71013	9
52	.29014	.95698	.30319	.2983	.0449	.4465	.04302	.70985	8
53	.29042	.95690	.30350	.2948	.0450	.4432	.04310	.70958	7
54	.29070	.95681	.30382	.2914	.0451	.4399	.04319	.70930	6
55	0.29098	0.95673	0.30414	3.2879	1.0452	3.4366	0.04327	0.70902	5
56	.29126	.95664	.30446	.2845	.0453	.4334	.04335	.70874	4
57	.29154	.95656	.30478	.2811	.0454	.4301	.04344	.70846	3
58	.29181	.95647	.30509	.2777	.0455	.4268	.04352	.70818	2
59	.29209	.95639	.30541	.2742	.0456	.4236	.04361	.70791	1
60	0.29237	0.95630	0.30573	3.2708	1.0457	3.4203	0.04369	0.70763	0
M	Cosine	Sine	Cotan.	Tan.	Cosec.	Secant	Vrs. Cos.	Vrs. Sin.	M

106°   **73°**

17°   Natural Trigonometric Functions   **162°**

M	Sine	Cosine	Tan.	Cotan.	Secant	Cosec.	Vrs. Sin.	Vrs. Cos.	M
0	0.29237	0.95630	0.30573	3.2708	1.0457	3.4203	0.04369	0.70763	60
1	.29265	.95622	.30605	.2674	.0458	.4170	.04378	.70735	59
2	.29293	.95613	.30637	.2640	.0459	.4138	.04386	.70707	58
3	.29321	.95605	.30668	.2607	.0460	.4106	.04395	.70679	57
4	.29348	.95596	.30700	.2573	.0461	.4073	.04404	.70651	56
5	0.29376	0.95588	0.30732	3.2539	1.0461	3.4041	0.04412	0.70624	55
6	.29404	.95579	.30764	.2505	.0462	.4009	.04421	.70596	54
7	.29432	.95571	.30796	.2472	.0463	.3977	.04429	.70568	53
8	.29460	.95562	.30828	.2438	.0464	.3945	.04438	.70540	52
9	.29487	.95554	.30859	.2405	.0465	.3913	.04446	.70512	51
10	0.29515	0.95545	0.30891	3.2371	1.0466	3.3881	0.04455	0.70485	50
11	.29543	.95536	.30923	.2338	.0467	.3849	.04463	.70457	49
12	.29571	.95528	.30955	.2305	.0468	.3817	.04472	.70429	48
13	.29598	.95519	.30987	.2271	.0469	.3785	.04481	.70401	47
14	.29626	.95511	.31019	.2238	.0470	.3754	.04489	.70374	46
15	0.29654	0.95502	0.31051	3.2205	1.0471	3.3722	0.04498	0.70346	45
16	.29682	.95493	.31083	.2172	.0472	.3690	.04507	.70318	44
17	.29710	.95485	.31115	.2139	.0473	.3659	.04515	.70290	43
18	.29737	.95476	.31146	.2106	.0474	.3627	.04524	.70262	42
19	.29765	.95467	.31178	.2073	.0475	.3596	.04532	.70235	41
20	0.29793	0.95459	0.31210	3.2041	1.0476	3.3565	0.04541	0.70207	40
21	.29821	.95450	.31242	.2008	.0477	.3534	.04550	.70179	39
22	.29848	.95441	.31274	.1975	.0478	.3502	.04558	.70151	38
23	.29876	.95433	.31306	.1942	.0478	.3471	.04567	.70124	37
24	.29904	.95424	.31338	.1910	.0479	.3440	.04576	.70096	36
25	0.29932	0.95415	0.31370	3.1877	1.0480	3.3409	0.04585	0.70068	35
26	.29959	.95407	.31402	.1845	.0481	.3378	.04593	.70040	34
27	.29987	.95398	.31434	.1813	.0482	.3347	.04602	.70013	33
28	.30015	.95389	.31466	.1780	.0483	.3316	.04611	.69982	32
29	.30043	.95380	.31498	.1748	.0484	.3286	.04619	.69957	31
30	0.30070	0.95372	0.31530	3.1716	1.0485	3.3255	0.04628	0.69929	30
31	.30098	.95363	.31562	.1684	.0486	.3224	.04637	.69902	29
32	.30126	.95354	.31594	.1652	.0487	.3194	.04646	.69874	28
33	.30154	.95345	.31626	.1620	.0488	.3163	.04654	.69846	27
34	.30181	.95337	.31658	.1588	.0489	.3133	.04663	.69818	26
35	0.30209	0.95328	0.31690	3.1556	1.0490	3.3102	0.04672	0.69791	25
36	.30237	.95319	.31722	.1524	.0491	.3072	.04681	.69763	24
37	.30265	.95310	.31754	.1492	.0492	.3042	.04690	.69735	23
38	.30292	.95301	.31786	.1460	.0493	.3011	.04698	.69707	22
39	.30320	.95293	.31818	.1429	.0494	.2981	.04707	.69680	21
40	0.30348	0.95284	0.31850	3.1397	1.0495	3.2951	0.04716	0.69652	20
41	.30375	.95275	.31882	.1366	.0496	.2921	.04725	.69624	19
42	.30403	.95266	.31914	.1334	.0497	.2891	.04734	.69597	18
43	.30431	.95257	.31946	.1303	.0498	.2861	.04743	.69569	17
44	.30459	.95248	.31978	.1271	.0499	.2831	.04751	.69541	16
45	0.30486	0.95239	0.32010	3.1240	1.0500	3.2801	0.04760	0.69513	15
46	.30514	.95231	.32042	.1209	.0501	.2772	.04769	.69486	14
47	.30542	.95222	.32074	.1177	.0502	.2742	.04778	.69458	13
48	.30569	.95213	.32106	.1146	.0503	.2712	.04787	.69430	12
49	.30597	.95204	.32138	.1115	.0504	.2683	.04796	.69403	11
50	0.30625	0.95195	0.32171	3.1084	1.0505	3.2653	0.04805	0.69375	10
51	.30653	.95186	.32203	.1053	.0506	.2624	.04814	.69347	9
52	.30680	.95177	.32235	.1022	.0507	.2594	.04823	.69320	8
53	.30708	.95168	.32267	.0991	.0508	.2565	.04832	.69292	7
54	.30736	.95159	.32299	.0960	.0509	.2535	.04840	.69264	6
55	0.30763	0.95150	0.32331	3.0930	1.0510	3.2506	0.04849	0.69237	5
56	.30791	.95141	.32363	.0899	.0511	.2477	.04858	.69209	4
57	.30819	.95132	.32395	.0868	.0512	.2448	.04867	.69181	3
58	.30846	.95124	.32428	.0838	.0513	.2419	.04876	.69154	2
59	.30874	.95115	.32460	.0807	.0514	.2390	.04885	.69126	1
60	0.30902	0.95106	0.32492	3.0777	1.0515	3.2361	0.04894	0.69098	0
M	Cosine	Sine	Cotan.	Tan.	Cosec.	Secant	Vrs. Cos.	Vrs. Sin.	M

107°   **72°**

18° Natural Trigonometric Functions 161°

M	Sine	Cosine	Tan.	Cotan.	Secant	Cosec.	Vrs. Sin.	Vrs. Cos.	M
0	0.30902	0.95106	0.32492	3.0777	1.0515	3.2361	0.04894	0.69098	60
1	.30929	.95097	.32524	.0746	.0516	.2332	.04903	.69071	59
2	.30957	.95088	.32556	.0716	.0517	.2303	.04912	.69043	58
3	.30985	.95079	.32588	.0686	.0518	.2274	.04921	.69015	57
4	.31012	.95070	.32621	.0655	.0519	.2245	.04930	.68988	56
5	0.31040	0.95061	0.32653	3.0625	1.0520	3.2216	0.04939	0.68960	55
6	.31068	.95051	.32685	.0595	.0521	.2188	.04948	.68932	54
7	.31095	.95042	.32717	.0565	.0522	.2159	.04957	.68905	53
8	.31123	.95033	.32749	.0535	.0523	.2131	.04966	.68877	52
9	.31150	.95024	.32782	.0505	.0524	.2102	.04975	.68849	51
10	0.31178	0.95015	0.32814	3.0475	1.0525	3.2074	0.04985	0.68822	50
11	.31206	.95006	.32846	.0445	.0526	.2045	.04994	.68794	49
12	.31233	.94997	.32878	.0415	.0527	.2017	.05003	.68766	48
13	.31261	.94988	.32910	.0385	.0528	.1989	.05012	.68739	47
14	.31289	.94979	.32943	.0356	.0529	.1960	.05021	.68711	46
15	0.31316	0.94970	0.32975	3.0326	1.0530	3.1932	0.05030	0.68684	45
16	.31344	.94961	.33007	.0296	.0531	.1904	.05039	.68656	44
17	.31372	.94952	.33039	.0267	.0532	.1876	.05048	.68628	43
18	.31399	.94942	.33072	.0237	.0533	.1848	.05057	.68601	42
19	.31427	.94933	.33104	.0208	.0534	.1820	.05066	.68573	41
20	0.31454	0.94924	0.33136	3.0178	1.0535	3.1792	0.05076	0.68545	40
21	.31482	.94915	.33169	.0149	.0536	.1764	.05085	.68518	39
22	.31510	.94906	.33201	.0120	.0537	.1736	.05094	.68490	38
23	.31537	.94897	.33233	.0090	.0538	.1708	.05103	.68463	37
24	.31565	.94888	.33265	.0061	.0539	.1681	.05112	.68435	36
25	0.31592	0.94878	0.33298	3.0032	1.0540	3.1653	0.05121	0.68407	35
26	.31620	.94869	.33330	.0003	.0541	.1625	.05131	.68380	34
27	.31648	.94860	.33362	2.9974	.0542	.1598	.05140	.68352	33
28	.31675	.94851	.33395	.9945	.0543	.1570	.05149	.68325	32
29	.31703	.94841	.33427	.9916	.0544	.1543	.05158	.68297	31
30	0.31730	0.94832	0.33459	2.9887	1.0545	3.1515	0.05168	0.68269	30
31	.31758	.94823	.33492	.9858	.0546	.1488	.05177	.68242	29
32	.31786	.94814	.33524	.9829	.0547	.1461	.05186	.68214	28
33	.31813	.94805	.33557	.9800	.0548	.1433	.05195	.68187	27
34	.31841	.94795	.33589	.9772	.0549	.1406	.05205	.68159	26
35	0.31868	0.94786	0.33621	2.9743	1.0550	3.1379	0.05214	0.68132	25
36	.31896	.94777	.33654	.9714	.0551	.1352	.05223	.68104	24
37	.31923	.94767	.33686	.9686	.0552	.1325	.05232	.68076	23
38	.31951	.94758	.33718	.9657	.0553	.1298	.05242	.68049	22
39	.31978	.94749	.33751	.9629	.0554	.1271	.05251	.68021	21
40	0.32006	0.94740	0.33783	2.9600	1.0555	3.1244	0.05260	0.67994	20
41	.32034	.94730	.33816	.9572	.0556	.1217	.05270	.67966	19
42	.32061	.94721	.33848	.9544	.0557	.1190	.05279	.67939	18
43	.32089	.94712	.33880	.9515	.0558	.1163	.05288	.67911	17
44	.32116	.94702	.33913	.9487	.0559	.1137	.05297	.67884	16
45	0.32144	0.94693	0.33945	2.9459	1.0560	3.1110	0.05307	0.67856	15
46	.32171	.94684	.33978	.9431	.0561	.1083	.05316	.67828	14
47	.32199	.94674	.34010	.9403	.0562	.1057	.05326	.67801	13
48	.32226	.94665	.34043	.9375	.0563	.1030	.05335	.67773	12
49	.32254	.94655	.34075	.9347	.0564	.1004	.05344	.67746	11
50	0.32282	0.94646	0.34108	2.9319	1.0566	3.0977	0.05354	0.67718	10
51	.32309	.94637	.34140	.9291	.0567	.0951	.05363	.67691	9
52	.32337	.94627	.34173	.9263	.0568	.0925	.05373	.67663	8
53	.32364	.94618	.34205	.9235	.0569	.0898	.05382	.67636	7
54	.32392	.94608	.34238	.9208	.0570	.0872	.05391	.67608	6
55	0.32419	0.94599	0.34270	2.9180	1.0571	3.0846	0.05401	0.67581	5
56	.32447	.94590	.34303	.9152	.0572	.0820	.05410	.67553	4
57	.32474	.94580	.34335	.9125	.0573	.0793	.05420	.67526	3
58	.32502	.94571	.34368	.9097	.0574	.0767	.05429	.67498	2
59	.32529	.94561	.34400	.9069	.0575	.0741	.05439	.67471	1
60	0.32557	0.94552	0.34433	2.9042	1.0576	3.0715	0.05448	0.67443	0
M	Cosine	Sine	Cotan.	Tan.	Cosec.	Secant	Vrs. Cos.	Vrs. Sin.	M

108° 71°

19° Natural Trigonometric Functions 160°

M	Sine	Cosine	Tan.	Cotan.	Secant	Cosec.	Vrs. Sin.	Vrs. Cos.	M
0	0.32557	0.94552	0.34433	2.9042	1.0576	3.0715	0.05448	0.67443	60
1	.32584	.94542	.34465	.9015	.0577	.0690	.05458	.67416	59
2	.32612	.94533	.34498	.8987	.0578	.0664	.05467	.67388	58
3	.32639	.94523	.34530	.8960	.0579	.0638	.05476	.67361	57
4	.32667	.94514	.34563	.8933	.0580	.0612	.05486	.67333	56
5	0.32694	0.94504	0.34595	2.8905	1.0581	3.0586	0.05495	0.67306	55
6	.32722	.94495	.34628	.8878	.0582	.0561	.05505	.67278	54
7	.32749	.94485	.34661	.8851	.0584	.0535	.05515	.67251	53
8	.32777	.94476	.34693	.8824	.0585	.0509	.05524	.67223	52
9	.32804	.94466	.34726	.8797	.0586	.0484	.05534	.67196	51
10	0.32832	0.94457	0.34758	2.8770	1.0587	3.0458	0.05543	0.67168	50
11	.32859	.94447	.34791	.8743	.0588	.0433	.05553	.67141	49
12	.32887	.94438	.34824	.8716	.0589	.0407	.05562	.67113	48
13	.32914	.94428	.34856	.8689	.0590	.0382	.05572	.67086	47
14	.32942	.94418	.34889	.8662	.0591	.0357	.05581	.67058	46
15	0.32969	0.94409	0.34921	2.8636	1.0592	3.0331	0.05591	0.67031	45
16	.32996	.94399	.34954	.8609	.0593	.0306	.05601	.67003	44
17	.33024	.94390	.34987	.8582	.0594	.0281	.05610	.66976	43
18	.33051	.94380	.35019	.8555	.0595	.0256	.05620	.66948	42
19	.33079	.94370	.35052	.8529	.0596	.0231	.05629	.66921	41
20	0.33106	0.94361	0.35085	2.8502	1.0598	3.0206	0.05639	0.66894	40
21	.33134	.94351	.35117	.8476	.0599	.0181	.05649	.66866	39
22	.33161	.94341	.35150	.8449	.0600	.0156	.05658	.66839	38
23	.33189	.94332	.35183	.8423	.0601	.0131	.05668	.66811	37
24	.33216	.94322	.35215	.8396	.0602	.0106	.05678	.66784	36
25	0.33243	0.94313	0.35248	2.8370	1.0603	3.0081	0.05687	0.66756	35
26	.33271	.94303	.35281	.8344	.0604	.0056	.05697	.66729	34
27	.33298	.94293	.35314	.8318	.0605	.0031	.05707	.66701	33
28	.33326	.94283	.35346	.8291	.0606	.0007	.05716	.66674	32
29	.33353	.94274	.35379	.8265	.0607	2.9982	.05726	.66647	31
30	0.33381	0.94264	0.35412	2.8239	1.0608	2.9957	0.05736	0.66619	30
31	.33408	.94254	.35445	.8213	.0609	.9933	.05745	.66592	29
32	.33435	.94245	.35477	.8187	.0611	.9908	.05755	.66564	28
33	.33463	.94235	.35510	.8161	.0612	.9884	.05765	.66537	27
34	.33490	.94225	.35543	.8135	.0613	.9859	.05775	.66510	26
35	0.33518	0.94215	0.35576	2.8109	1.0614	2.9835	0.05784	0.66482	25
36	.33545	.94206	.35608	.8083	.0615	.9810	.05794	.66455	24
37	.33572	.94196	.35641	.8057	.0616	.9786	.05804	.66427	23
38	.33600	.94186	.35674	.8032	.0617	.9762	.05814	.66400	22
39	.33627	.94176	.35707	.8006	.0618	.9738	.05823	.66373	21
40	0.33655	0.94167	0.35739	2.7980	1.0619	2.9713	0.05833	0.66345	20
41	.33682	.94157	.35772	.7954	.0620	.9689	.05843	.66318	19
42	.33709	.94147	.35805	.7929	.0622	.9665	.05853	.66290	18
43	.33737	.94137	.35838	.7903	.0623	.9641	.05863	.66263	17
44	.33764	.94127	.35871	.7878	.0624	.9617	.05872	.66236	16
45	0.33792	0.94118	0.35904	2.7852	1.0625	2.9593	0.05882	0.66208	15
46	.33819	.94108	.35936	.7827	.0626	.9569	.05892	.66181	14
47	.33846	.94098	.35969	.7801	.0627	.9545	.05902	.66153	13
48	.33874	.94088	.36002	.7776	.0628	.9521	.05912	.66126	12
49	.33901	.94078	.36035	.7751	.0629	.9497	.05922	.66099	11
50	0.33928	0.94068	0.36068	2.7725	1.0630	2.9474	0.05932	0.66071	10
51	.33956	.94058	.36101	.7700	.0632	.9450	.05941	.66044	9
52	.33983	.94049	.36134	.7675	.0633	.9426	.05951	.66017	8
53	.34011	.94039	.36167	.7650	.0634	.9402	.05961	.65989	7
54	.34038	.94029	.36199	.7625	.0635	.9379	.05971	.65962	6
55	0.34065	0.94019	0.36232	2.7600	1.0636	2.9355	0.05981	0.65935	5
56	.34093	.94009	.36265	.7575	.0637	.9332	.05991	.65907	4
57	.34120	.93999	.36298	.7550	.0638	.9308	.06001	.65880	3
58	.34147	.93989	.36331	.7525	.0639	.9285	.06011	.65853	2
59	.34175	.93979	.36364	.7500	.0641	.9261	.06021	.65825	1
60	0.34202	0.93969	0.36397	2.7475	1.0642	2.9238	0.06031	0.65798	0
M	Cosine	Sine	Cotan.	Tan.	Cosec.	Secant	Vrs. Cos.	Vrs. Sin.	M

109° 70°

20° Natural Trigonometric Functions 159°

M	Sine	Cosine	Tan.	Cotan.	Secant	Cosec.	Vrs. Sin.	Vrs. Cos.	M
0	0.34202	0.93969	0.36397	2.7475	1.0642	2.9238	0.06031	0.65798	60
1	.34229	.93959	.36430	.7450	.0643	.9215	.06041	.65771	59
2	.34257	.93949	.36463	.7425	.0644	.9191	.06051	.65743	58
3	.34284	.93939	.36496	.7400	.0645	.9168	.06061	.65716	57
4	.34311	.93929	.36529	.7376	.0646	.9145	.06071	.65689	56
5	0.34339	0.93919	0.36562	2.7351	1.0647	2.9122	0.06080	0.65661	55
6	.34366	.93909	.36595	.7326	.0648	.9098	.06090	.65634	54
7	.34393	.93899	.36628	.7302	.0650	.9075	.06100	.65607	53
8	.34421	.93889	.36661	.7277	.0651	.9052	.06110	.65579	52
9	.34448	.93879	.36694	.7252	.0652	.9029	.06121	.65552	51
10	0.34475	0.93869	0.36727	2.7228	1.0653	2.9006	0.06131	0.65525	50
11	.34502	.93859	.36760	.7204	.0654	.8983	.06141	.65497	49
12	.34530	.93849	.36793	.7179	.0655	.8960	.06151	.65470	48
13	.34557	.93839	.36826	.7155	.0656	.8937	.06161	.65443	47
14	.34584	.93829	.36859	.7130	.0658	.8915	.06171	.65415	46
15	0.34612	0.93819	0.36892	2.7106	1.0659	2.8892	0.06181	0.65388	45
16	.34639	.93809	.36925	.7082	.0660	.8869	.06191	.65361	44
17	.34666	.93799	.36958	.7058	.0661	.8846	.06201	.65334	43
18	.34693	.93789	.36991	.7033	.0662	.8824	.06211	.65306	42
19	.34721	.93779	.37024	.7009	.0663	.8801	.06221	.65279	41
20	0.34748	0.93769	0.37057	2.6985	1.0664	2.8778	0.06231	0.65252	40
21	.34775	.93758	.37090	.6961	.0666	.8756	.06241	.65225	39
22	.34803	.93748	.37123	.6937	.0667	.8733	.06251	.65197	38
23	.34830	.93738	.37156	.6913	.0668	.8711	.06262	.65170	37
24	.34857	.93728	.37190	.6889	.0669	.8688	.06272	.65143	36
25	0.34884	0.93718	0.37223	2.6865	1.0670	2.8666	0.06282	0.65115	35
26	.34912	.93708	.37256	.6841	.0671	.8644	.06292	.65088	34
27	.34939	.93698	.37289	.6817	.0673	.8621	.06302	.65061	33
28	.34966	.93687	.37322	.6794	.0674	.8599	.06312	.65034	32
29	.34993	.93677	.37355	.6770	.0675	.8577	.06323	.65006	31
30	0.35021	0.93667	0.37388	2.6746	1.0676	2.8554	0.06333	0.64979	30
31	.35048	.93657	.37422	.6722	.0677	.8532	.06343	.64952	29
32	.35075	.93647	.37455	.6699	.0678	.8510	.06353	.64925	28
33	.35102	.93637	.37488	.6675	.0679	.8488	.06363	.64897	27
34	.35130	.93626	.37521	.6652	.0681	.8466	.06373	.64870	26
35	0.35157	0.93616	0.37554	2.6628	1.0682	2.8444	0.06384	0.64843	25
36	.35184	.93606	.37587	.6604	.0683	.8422	.06394	.64816	24
37	.35211	.93596	.37621	.6581	.0684	.8400	.06404	.64789	23
38	.35239	.93585	.37654	.6558	.0685	.8378	.06414	.64761	22
39	.35266	.93575	.37687	.6534	.0686	.8356	.06425	.64734	21
40	0.35293	0.93565	0.37720	2.6511	1.0688	2.8334	0.06435	0.64707	20
41	.35320	.93555	.37754	.6487	.0689	.8312	.06445	.64680	19
42	.35347	.93544	.37787	.6464	.0690	.8290	.06456	.64652	18
43	.35375	.93534	.37820	.6441	.0691	.8269	.06466	.64625	17
44	.35402	.93524	.37853	.6418	.0692	.8247	.06476	.64598	16
45	0.35429	0.93513	0.37887	2.6394	1.0694	2.8225	0.06486	0.64571	15
46	.35456	.93503	.37920	.6371	.0695	.8204	.06497	.64544	14
47	.35483	.93493	.37953	.6348	.0696	.8182	.06507	.64516	13
48	.35511	.93482	.37986	.6325	.0697	.8160	.06517	.64489	12
49	.35538	.93472	.38020	.6302	.0698	.8139	.06528	.64462	11
50	0.35565	0.93462	0.38053	2.6279	1.0699	2.8117	0.06538	0.64435	10
51	.35592	.93451	.38086	.6256	.0701	.8096	.06548	.64408	9
52	.35619	.93441	.38120	.6233	.0702	.8074	.06559	.64380	8
53	.35647	.93431	.38153	.6210	.0703	.8053	.06569	.64353	7
54	.35674	.93420	.38186	.6187	.0704	.8032	.06579	.64326	6
55	0.35701	0.93410	0.38220	2.6164	1.0705	2.8010	0.06590	0.64299	5
56	.35728	.93400	.38253	.6142	.0707	.7989	.06600	.64272	4
57	.35755	.93389	.38286	.6119	.0708	.7968	.06611	.64245	3
58	.35782	.93379	.38320	.6096	.0709	.7947	.06621	.64217	2
59	.35810	.93368	.38353	.6073	.0710	.7925	.06631	.64190	1
60	0.35837	0.93358	0.38386	2.6051	1.0711	2.7904	0.06642	0.64163	0
M	Cosine	Sine	Cotan.	Tan.	Cosec.	Secant	Vrs. Cos.	Vrs. Sin.	M

110°   69°

21° Natural Trigonometric Functions 158°

M	Sine	Cosine	Tan.	Cotan.	Secant	Cosec.	Vrs. Sin.	Vrs. Cos.	M
0	0.35837	0.93358	0.38386	2.6051	1.0711	2.7904	0.06642	0.64163	60
1	.35864	.93348	.38420	6028	.0713	.7883	.06652	.64136	59
2	.35891	.93337	.38453	6006	.0714	.7862	.06663	.64109	58
3	.35918	.93327	.38486	5983	.0715	.7841	.06673	.64082	57
4	.35945	.93316	.38520	5960	.0716	.7820	.06684	.64055	56
5	0.35972	0.93306	0.38553	2.5938	1.0717	2.7799	0.06694	0.64027	55
6	.36000	.93295	.38587	5916	.0719	.7778	.06705	.64000	54
7	.36027	.93285	.38620	5893	.0720	.7757	.06715	.63973	53
8	.36054	.93274	.38654	5871	.0721	.7736	.06726	.63946	52
9	.36081	.93264	.38687	5848	.0722	.7715	.06736	.63919	51
10	0.36108	0.93253	0.38720	2.5826	1.0723	2.7694	0.06747	0.63892	50
11	.36135	.93243	.38754	5804	.0725	.7674	.06757	.63865	49
12	.36162	.93232	.38787	5781	.0726	.7653	.06768	.63837	48
13	.36189	.93222	.38821	5759	.0727	.7632	.06778	.63810	47
14	.36217	.93211	.38854	5737	.0728	.7611	.06789	.63783	46
15	0.36244	0.93201	0.38888	2.5715	1.0729	2.7591	0.06799	0.63756	45
16	.36271	.93190	.38921	5693	.0731	.7570	.06810	.63729	44
17	.36298	.93180	.38955	5671	.0732	.7550	.06820	.63702	43
18	.36325	.93169	.38988	5649	.0733	.7529	.06831	.63675	42
19	.36352	.93158	.39022	5627	.0734	.7509	.06841	.63648	41
20	0.36379	0.93148	0.39055	2.5605	1.0736	2.7488	0.06852	0.63621	40
21	.36406	.93137	.39089	5583	.0737	.7468	.06863	.63593	39
22	.36433	.93127	.39122	5561	.0738	.7447	.06873	.63566	38
23	.36460	.93116	.39156	5539	.0739	.7427	.06884	.63539	37
24	.36488	.93105	.39189	5517	.0740	.7406	.06894	.63512	36
25	0.36515	0.93095	0.39223	2.5495	1.0742	2.7386	0.06905	0.63485	35
26	.36542	.93084	.39257	5473	.0743	.7366	.06916	.63458	34
27	.36569	.93074	.39290	5451	.0744	.7346	.06926	.63431	33
28	.36596	.93063	.39324	5430	.0745	.7325	.06937	.63404	32
29	.36623	.93052	.39357	5408	.0747	.7305	.06947	.63377	31
30	0.36650	0.93042	0.39391	2.5386	1.0748	2.7285	0.06958	0.63350	30
31	.36677	.93031	.39425	5365	.0749	.7265	.06969	.63323	29
32	.36704	.93020	.39458	5343	.0750	.7245	.06979	.63296	28
33	.36731	.93010	.39492	5322	.0751	.7225	.06990	.63269	27
34	.36758	.92999	.39525	5300	.0753	.7205	.07001	.63242	26
35	0.36785	0.92988	0.39559	2.5278	1.0754	2.7185	0.07012	0.63214	25
36	.36812	.92978	.39593	5257	.0755	.7165	.07022	.63187	24
37	.36839	.92967	.39626	5236	.0756	.7145	.07033	.63160	23
38	.36866	.92956	.39660	5214	.0758	.7125	.07044	.63133	22
39	.36893	.92945	.39694	5193	.0759	.7105	.07054	.63106	21
40	0.36921	0.92935	0.39727	2.5171	1.0760	2.7085	0.07065	0.63079	20
41	.36948	.92924	.39761	5150	.0761	.7065	.07076	.63052	19
42	.36975	.92913	.39795	5129	.0763	.7045	.07087	.63025	18
43	.37002	.92902	.39828	5108	.0764	.7026	.07097	.62998	17
44	.37029	.92892	.39862	5086	.0765	.7006	.07108	.62971	16
45	0.37056	0.92881	0.39896	2.5065	1.0766	2.6986	0.07119	0.62944	15
46	.37083	.92870	.39930	5044	.0768	.6967	.07130	.62917	14
47	.37110	.92859	.39963	5023	.0769	.6947	.07141	.62890	13
48	.37137	.92848	.39997	5002	.0770	.6927	.07151	.62863	12
49	.37164	.92838	.40031	.4981	.0771	.6908	.07162	.62836	11
50	0.37191	0.92827	0.40065	2.4960	1.0773	2.6888	0.07173	0.62809	10
51	.37218	.92816	.40098	.4939	.0774	.6869	.07184	.62782	9
52	.37245	.92805	.40132	.4918	.0775	.6849	.07195	.62755	8
53	.37272	.92794	.40166	.4897	.0776	.6830	.07205	.62728	7
54	.37299	.92784	.40200	.4876	.0778	.6810	.07216	.62701	6
55	0.37326	0.92773	0.40233	2.4855	1.0779	2.6791	0.07227	0.62674	5
56	.37353	.92762	.40267	.4834	.0780	.6772	.07238	.62647	4
57	.37380	.92751	.40301	.4813	.0781	.6752	.07249	.62620	3
58	.37407	.92740	.40335	.4792	.0783	.6733	.07260	.62593	2
59	.37434	.92729	.40369	.4772	.0784	.6714	.07271	.62566	1
60	0.37461	0.92718	0.40403	2.4751	1.0785	2.6695	0.07282	0.62539	0
M	Cosine	Sine	Cotan.	Tan.	Cosec.	Secant	Vrs. Cos.	Vrs. Sin.	M

111° 68°

22°   Natural Trigonometric Functions   157°

M	Sine	Cosine	Tan.	Cotan.	Secant	Cosec.	Vrs. Sin.	Vrs. Cos.	M
0	0.37461	0.92718	0.40403	2.4751	1.0785	2.6695	0.07282	0.62539	60
1	.37488	.92707	.40436	.4730	.0787	.6675	.07292	.62512	59
2	.37514	.92696	.40470	.4709	.0788	.6656	.07303	.62485	58
3	.37541	.92686	.40504	.4689	.0789	.6637	.07314	.62458	57
4	.37568	.92675	.40538	.4668	.0790	.6618	.07325	.62431	56
5	0.37595	0.92664	0.40572	2.4647	1.0792	2.6599	0.07336	0.62404	55
6	.37622	.92653	.40606	.4627	.0793	.6580	.07347	.62377	54
7	.37649	.92642	.40640	.4606	.0794	.6561	.07358	.62351	53
8	.37676	.92631	.40673	.4586	.0795	.6542	.07369	.62324	52
9	.37703	.92620	.40707	.4565	.0797	.6523	.07380	.62297	51
10	0.37730	0.92609	0.40741	2.4545	1.0798	2.6504	0.07391	0.62270	50
11	.37757	.92598	.40775	.4525	.0799	.6485	.07402	.62243	49
12	.37784	.92587	.40809	.4504	.0801	.6466	.07413	.62216	48
13	.37811	.92576	.40843	.4484	.0802	.6447	.07424	.62189	47
14	.37838	.92565	.40877	.4463	.0803	.6428	.07435	.62162	46
15	0.37865	0.92554	0.40911	2.4443	1.0804	2.6410	0.07446	0.62135	45
16	.37892	.92543	.40945	.4423	.0806	.6391	.07457	.62108	44
17	.37919	.92532	.40979	.4403	.0807	.6372	.07468	.62081	43
18	.37946	.92521	.41013	.4382	.0808	.6353	.07479	.62054	42
19	.37972	.92510	.41047	.4362	.0810	.6335	.07490	.62027	41
20	0.37999	0.92499	0.41081	2.4342	1.0811	2.6316	0.07501	0.62000	40
21	.38026	.92488	.41115	.4322	.0812	.6297	.07512	.61974	39
22	.38053	.92477	.41149	.4302	.0813	.6279	.07523	.61947	38
23	.38080	.92466	.41183	.4282	.0815	.6260	.07534	.61920	37
24	.38107	.92455	.41217	.4262	.0816	.6242	.07545	.61893	36
25	0.38134	0.92443	0.41251	2.4242	1.0817	2.6223	0.07556	0.61866	35
26	.38161	.92432	.41285	.4222	.0819	.6205	.07567	.61839	34
27	.38188	.92421	.41319	.4202	.0820	.6186	.07579	.61812	33
28	.38214	.92410	.41353	.4182	.0821	.6168	.07590	.61785	32
29	.38241	.92399	.41387	.4162	.0823	.6150	.07601	.61758	31
30	0.38268	0.92388	0.41421	2.4142	1.0824	2.6131	0.07612	0.61732	30
31	.38295	.92377	.41455	.4122	.0825	.6113	.07623	.61705	29
32	.38322	.92366	.41489	.4102	.0826	.6095	.07634	.61678	28
33	.38349	.92354	.41524	.4083	.0828	.6076	.07645	.61651	27
34	.38376	.92343	.41558	.4063	.0829	.6058	.07657	.61624	26
35	0.38403	0.92332	0.41592	2.4043	1.0830	2.6040	0.07668	0.61597	25
36	.38429	.92321	.41626	.4023	.0832	.6022	.07679	.61570	24
37	.38456	.92310	.41660	.4004	.0833	.6003	.07690	.61544	23
38	.38483	.92299	.41694	.3984	.0834	.5985	.07701	.61517	22
39	.38510	.92287	.41728	.3964	.0836	.5967	.07712	.61490	21
40	0.38537	0.92276	0.41762	2.3945	1.0837	2.5949	0.07724	0.61463	20
41	.38564	.92265	.41797	.3925	.0838	.5931	.07735	.61436	19
42	.38591	.92254	.41831	.3906	.0840	.5913	.07746	.61409	18
43	.38617	.92242	.41865	.3886	.0841	.5895	.07757	.61382	17
44	.38644	.92231	.41899	.3867	.0842	.5877	.07769	.61356	16
45	0.38671	0.92220	0.41933	2.3847	1.0844	2.5859	0.07780	0.61329	15
46	.38698	.92209	.41968	.3828	.0845	.5841	.07791	.61302	14
47	.38725	.92197	.42002	.3808	.0846	.5823	.07802	.61275	13
48	.38751	.92186	.42036	.3789	.0847	.5805	.07814	.61248	12
49	.38778	.92175	.42070	.3770	.0849	.5787	.07825	.61222	11
50	0.38805	0.92164	0.42105	2.3750	1.0850	2.5770	0.07836	0.61195	10
51	.38832	.92152	.42139	.3731	.0851	.5752	.07847	.61168	9
52	.38859	.92141	.42173	.3712	.0853	.5734	.07859	.61141	8
53	.38886	.92130	.42207	.3692	.0854	.5716	.07870	.61114	7
54	.38912	.92118	.42242	.3673	.0855	.5699	.07881	.61088	6
55	0.38939	0.92107	0.42276	2.3654	1.0857	2.5681	0.07893	0.61061	5
56	.38966	.92096	.42310	.3635	.0858	.5663	.07904	.61034	4
57	.38993	.92084	.42344	.3616	.0859	.5646	.07915	.61007	3
58	.39019	.92073	.42379	.3597	.0861	.5628	.07927	.60980	2
59	.39046	.92062	.42413	.3577	.0862	.5610	.07938	.60954	1
60	0.39073	0.92050	0.42447	2.3558	1.0864	2.5593	0.07949	0.60927	0
M	Cosine	Sine	Cotan.	Tan.	Cosec.	Secant	Vrs. Cos.	Vrs. Sin.	M

112°   67°

23° Natural Trigonometric Functions 156°

M	Sine	Cosine	Tan.	Cotan.	Secant	Cosec.	Vrs. Sin.	Vrs. Cos.	M
0	0.39073	0.92050	0.42447	2.3558	1.0864	2.5593	0.07949	0.60927	60
1	.39100	.92039	.42482	.3539	.0865	.5575	.07961	.60900	59
2	.39126	.92028	.42516	.3520	.0866	.5558	.07972	.60873	58
3	.39153	.92016	.42550	.3501	.0868	.5540	.07984	.60846	57
4	.39180	.92005	.42585	.3482	.0869	.5523	.07995	.60820	56
5	0.39207	0.91993	0.42619	2.3463	1.0870	2.5506	0.08006	0.60793	55
6	.39234	.91982	.42654	.3445	.0872	.5488	.08018	.60766	54
7	.39260	.91971	.42688	.3426	.0873	.5471	.08029	.60739	53
8	.39287	.91959	.42722	.3407	.0874	.5453	.08041	.60713	52
9	.39314	.91948	.42757	.3388	.0876	.5436	.08052	.60686	51
10	0.39341	0.91936	0.42791	2.3369	1.0877	2.5419	0.08063	0.60659	50
11	.39367	.91925	.42826	.3350	.0878	.5402	.08075	.60632	49
12	.39394	.91913	.42860	.3332	.0880	.5384	.08086	.60606	48
13	.39421	.91902	.42894	.3313	.0881	.5367	.08098	.60579	47
14	.39448	.91891	.42929	.3294	.0882	.5350	.08109	.60552	46
15	0.39474	0.91879	0.42963	2.3276	1.0884	2.5333	0.08121	0.60526	45
16	.39501	.91868	.42998	.3257	.0885	.5316	.08132	.60499	44
17	.39528	.91856	.43032	.3238	.0886	.5299	.08144	.60472	43
18	.39554	.91845	.43067	.3220	.0888	.5281	.08155	.60445	42
19	.39581	.91833	.43101	.3201	.0889	.5264	.08167	.60419	41
20	0.39608	0.91822	0.43136	2.3183	1.0891	2.5247	0.08178	0.60392	40
21	.39635	.91810	.43170	.3164	.0892	.5230	.08190	.60365	39
22	.39661	.91798	.43205	.3145	.0893	.5213	.08201	.60339	38
23	.39688	.91787	.43239	.3127	.0895	.5196	.08213	.60312	37
24	.39715	.91775	.43274	.3109	.0896	.5179	.08224	.60285	36
25	0.39741	0.91764	0.43308	2.3090	1.0897	2.5163	0.08236	0.60258	35
26	.39768	.91752	.43343	.3072	.0899	.5146	.08248	.60232	34
27	.39795	.91741	.43377	.3053	.0900	.5129	.08259	.60205	33
28	.39821	.91729	.43412	.3035	.0902	.5112	.08271	.60178	32
29	.39848	.91718	.43447	.3017	.0903	.5095	.08282	.60152	31
30	0.39875	0.91706	0.43481	2.2998	1.0904	2.5078	0.08294	0.60125	30
31	.39901	.91694	.43516	.2980	.0906	.5062	.08306	.60098	29
32	.39928	.91683	.43550	.2962	.0907	.5045	.08317	.60072	28
33	.39955	.91671	.43585	.2944	.0908	.5028	.08329	.60045	27
34	.39981	.91659	.43620	.2925	.0910	.5011	.08340	.60018	26
35	0.40008	0.91648	0.43654	2.2907	1.0911	2.4995	0.08352	0.59992	25
36	.40035	.91636	.43689	.2889	.0913	.4978	.08364	.59965	24
37	.40061	.91625	.43723	.2871	.0914	.4961	.08375	.59938	23
38	.40088	.91613	.43758	.2853	.0915	.4945	.08387	.59912	22
39	.40115	.91601	.43793	.2835	.0917	.4928	.08399	.59885	21
40	0.40141	0.91590	0.43827	2.2817	1.0918	2.4912	0.08410	0.59858	20
41	.40168	.91578	.43862	.2799	.0920	.4895	.08422	.59832	19
42	.40195	.91566	.43897	.2781	.0921	.4879	.08434	.59805	18
43	.40221	.91554	.43932	.2763	.0922	.4862	.08445	.59778	17
44	.40248	.91543	.43966	.2745	.0924	.4846	.08457	.59752	16
45	0.40275	0.91531	0.44001	2.2727	1.0925	2.4829	0.08469	0.59725	15
46	.40301	.91519	.44036	.2709	.0927	.4813	.08480	.59699	14
47	.40328	.91508	.44070	.2691	.0928	.4797	.08492	.59672	13
48	.40354	.91496	.44105	.2673	.0929	.4780	.08504	.59645	12
49	.40381	.91484	.44140	.2655	.0931	.4764	.08516	.59619	11
50	0.40408	0.91472	0.44175	2.2637	1.0932	2.4748	0.08527	0.59592	10
51	.40434	.91461	.44209	.2619	.0934	.4731	.08539	.59566	9
52	.40461	.91449	.44244	.2602	.0935	.4715	.08551	.59539	8
53	.40487	.91437	.44279	.2584	.0936	.4699	.08563	.59512	7
54	.40514	.91425	.44314	.2566	.0938	.4683	.08575	.59486	6
55	0.40541	0.91414	0.44349	2.2548	1.0939	2.4666	0.08586	0.59459	5
56	.40567	.91402	.44383	.2531	.0941	.4650	.08598	.59433	4
57	.40594	.91390	.44418	.2513	.0942	.4634	.08610	.59406	3
58	.40620	.91378	.44453	.2495	.0943	.4618	.08622	.59379	2
59	.40647	.91366	.44488	.2478	.0945	.4602	.08634	.59353	1
60	0.40674	0.91354	0.44523	2.2460	1.0946	2.4586	0.08645	0.59326	0
M	Cosine	Sine	Cotan.	Tan.	Cosec.	Secant	Vrs. Cos.	Vrs. Sin.	M

113°   66°

24° Natural Trigonometric Functions 155°

M	Sine	Cosine	Tan.	Cotan.	Secant	Cosec.	Vrs. Sin.	Vrs. Cos.	M
0	0.40674	0.91354	0.44523	2.2460	1.0946	2.4586	0.08645	0.59326	60
1	.40700	.91343	.44558	.2443	.0948	.4570	.08657	.59300	59
2	.40727	.91331	.44593	.2425	.0949	.4554	.08669	.59273	58
3	.40753	.91319	.44627	.2408	.0951	.4538	.08681	.59247	57
4	.40780	.91307	.44662	.2390	.0952	.4522	.08693	.59220	56
5	0.40806	0.91295	0.44697	2.2373	1.0953	2.4506	0.08705	0.59193	55
6	.40833	.91283	.44732	.2355	.0955	.4490	.08716	.59167	54
7	.40860	.91271	.44767	.2338	.0956	.4474	.08728	.59140	53
8	.40886	.91260	.44802	.2320	.0958	.4458	.08740	.59114	52
9	.40913	.91248	.44837	.2303	.0959	.4442	.08752	.59087	51
10	0.40939	0.91236	0.44872	2.2286	1.0961	2.4426	0.08764	0.59061	50
11	.40966	.91224	.44907	.2268	.0962	.4411	.08776	.59034	49
12	.40992	.91212	.44942	.2251	.0963	.4395	.08788	.59008	48
13	.41019	.91200	.44977	2234	.0965	.4379	.08800	.58981	47
14	.41045	.91188	.45012	.2216	0966	.4363	.08812	.58955	46
15	0.41072	0.91176	0.45047	2.2199	1.0968	2.4347	0.08824	0.58928	45
16	.41098	.91164	.45082	.2182	.0969	.4332	.08836	.58901	44
17	.41125	.91152	.45117	.2165	.0971	.4316	.08848	.58875	43
18	.41151	.91140	.45152	2147	.0972	.4300	.08860	.58848	42
19	.41178	.91128	45187	2130	.0973	.4285	.08872	.58822	41
20	0.41204	0.91116	0.45222	2.2113	1.0975	2.4269	0.08884	0.58795	40
21	.41231	.91104	45257	.2096	.0976	.4254	.08896	.58769	39
22	.41257	.91092	.45292	2079	0978	.4238	.08908	.58742	38
23	.41284	.91080	.45327	.2062	.0979	.4222	.08920	58716	37
24	.41310	.91068	.45362	2045	.0981	.4207	.08932	.58689	36
25	0.41337	0.91056	0.45397	2.2028	1.0982	2.4191	0.08944	0.58663	35
26	.41363	.91044	45432	.2011	.0984	.4176	.08956	.58636	34
27	.41390	.91032	45467	1994	.0985	.4160	.08968	.58610	33
28	.41416	.91020	.45502	.1977	.0986	.4145	.08980	.58584	32
29	.41443	.91008	45537	.1960	.0988	.4130	.08992	.58557	31
30	0.41469	0.90996	0.45573	2.1943	1.0989	2.4114	0.09004	0.58531	30
31	.41496	.90984	.45608	.1926	.0991	.4099	.09016	.58504	29
32	.41522	.90972	45643	.1909	.0992	.4083	.09028	.58478	28
33	.41549	.90960	.45678	1892	.0994	.4068	.09040	.58451	27
34	.41575	90948	45713	.1875	.0995	.4053	.09052	.58425	26
35	0.41602	0.90936	0.45748	2.1859	1.0997	2.4037	0.09064	0.58398	25
36	.41628	.90924	.45783	.1842	.0998	.4022	.09076	.58372	24
37	.41654	.90911	.45819	.1825	.1000	.4007	.09088	.58345	23
38	.41681	.90899	.45854	.1808	.1001	.3992	.09101	.58319	22
39	.41707	.90887	.45889	.1792	.1003	.3976	.09113	.58292	21
40	0.41734	0.90875	0.45924	2.1775	1.1004	2.3961	0.09125	0.58266	20
41	.41760	.90863	.45960	1758	.1005	.3946	.09137	.58240	19
42	.41787	.90851	.45995	.1741	.1007	.3931	.09149	.58213	18
43	.41813	90839	46030	.1725	.1008	.3916	.09161	.58187	17
44	.41839	.90826	.46065	.1708	.1010	.3901	.09173	.58160	16
45	0.41866	0.90814	0.46101	2.1692	1.1011	2.3886	0.09186	0.58134	15
46	.41892	.90802	.46136	.1675	.1013	.3871	.09198	.58108	14
47	.41919	.90790	.46171	.1658	.1014	.3856	.09210	.58081	13
48	.41945	.90778	.46206	.1642	.1016	.3841	.09222	.58055	12
49	.41972	.90765	.46242	.1625	.1017	.3826	.09234	.58028	11
50	0.41998	0.90753	0.46277	2.1609	1.1019	2.3811	0.09247	0.58002	10
51	.42024	.90741	.46312	.1592	.1020	.3796	.09259	57975	9
52	.42051	.90729	.46348	.1576	.1022	.3781	.09271	.57949	8
53	.42077	.90717	.46383	.1559	.1023	.3766	.09283	.57923	7
54	.42103	.90704	46418	.1543	.1025	.3751	.09296	.57896	6
55	0.42130	0.90692	0.46454	2.1527	1.1026	2.3736	0.09308	0.57870	5
56	.42156	.90680	.46489	1510	.1028	.3721	.09320	.57844	4
57	.42183	.90668	.46524	1494	.1029	.3706	.09332	.57817	3
58	.42209	.90655	.46560	1478	.1031	.3691	.09345	.57791	2
59	.42235	.90643	.46595	.1461	.1032	.3677	.09357	.57764	1
60	0.42262	0.90631	0.46631	2.1445	1.1034	2.3662	0.09369	0.57738	0
M	Cosine	Sine	Cotan.	Tan.	Cosec.	Secant	Vrs. Cos.	Vrs. Sin.	M

114°   65°

25° Natural Trigonometric Functions 154°

M	Sine	Cosine	Tan.	Cotan.	Secant	Cosec.	Vrs. Sin.	Vrs. Cos.	M
0	0.42262	0.90631	0.46631	2.1445	1.1034	2.3662	0.09369	0.57738	60
1	.42288	.90618	.46666	.1429	.1035	.3647	.09381	.57712	59
2	.42314	.90606	.46702	.1412	.1037	.3632	.09394	.57685	58
3	.42341	.90594	.46737	.1396	.1038	.3618	.09406	.57659	57
4	.42367	.90581	.46772	.1380	.1040	.3603	.09418	.57633	56
5	0.42394	0.90569	0.46808	2.1364	1.1041	2.3588	0.09431	0.57606	55
6	.42420	.90557	.46843	.1348	.1043	.3574	.09443	.57580	54
7	.42446	.90544	.46879	.1331	.1044	.3559	.09455	.57554	53
8	.42473	.90532	.46914	.1315	.1046	.3544	.09468	.57527	52
9	.42499	.90520	.46950	.1299	.1047	.3530	.09480	.57501	51
10	0.42525	0.90507	0.46985	2.1283	1.1049	2.3515	0.09492	0.57475	50
11	.42552	.90495	.47021	.1267	.1050	.3501	.09505	.57448	49
12	.42578	.90483	.47056	.1251	.1052	.3486	.09517	.57422	48
13	.42604	.90470	.47092	.1235	.1053	.3472	.09530	.57396	47
14	.42630	.90458	.47127	.1219	.1055	.3457	.09542	.57369	46
15	0.42657	0.90445	0.47163	2.1203	1.1056	2.3443	0.09554	0.57343	45
16	.42683	.90433	.47199	.1187	.1058	.3428	.09567	.57317	44
17	.42709	.90421	.47234	.1171	.1059	.3414	.09579	.57290	43
18	.42736	.90408	.47270	.1155	.1061	.3399	.09592	.57264	42
19	.42762	.90396	.47305	.1139	.1062	.3385	.09604	.57238	41
20	0.42788	0.90383	0.47341	2.1123	1.1064	2.3371	0.09617	0.57212	40
21	.42815	.90371	.47376	.1107	.1065	.3356	.09629	.57185	39
22	.42841	.90358	.47412	.1092	.1067	.3342	.09641	.57159	38
23	.42867	.90346	.47448	.1076	.1068	.3328	.09654	.57133	37
24	.42893	.90333	.47483	.1060	.1070	.3313	.09666	.57106	36
25	0.42920	0.90321	0.47519	2.1044	1.1072	2.3299	0.09679	0.57080	35
26	.42946	.90308	.47555	.1028	.1073	.3285	.09691	.57054	34
27	.42972	.90296	.47590	.1013	.1075	.3271	.09704	.57028	33
28	.42998	.90283	.47626	.0997	.1076	.3256	.09716	.57001	32
29	.43025	.90271	.47662	.0981	.1078	.3242	.09729	.56975	31
30	0.43051	0.90258	0.47697	2.0965	1.1079	2.3228	0.09741	0.56949	30
31	.43077	.90246	.47733	.0950	.1081	.3214	.09754	.56923	29
32	.43104	.90233	.47769	.0934	.1082	.3200	.09766	.56896	28
33	.43130	.90221	.47805	.0918	.1084	.3186	.09779	.56870	27
34	.43156	.90208	.47840	.0903	.1085	.3172	.09792	.56844	26
35	0.43182	0.90196	0.47876	2.0887	1.1087	2.3158	0.09804	0.56818	25
36	.43208	.90183	.47912	.0872	.1088	.3143	.09817	.56791	24
37	.43235	.90171	.47948	.0856	.1090	.3129	.09829	.56765	23
38	.43261	.90158	.47983	.0840	.1092	.3115	.09842	.56739	22
39	.43287	.90145	.48019	.0825	.1093	.3101	.09854	.56713	21
40	0.43313	0.90133	0.48055	2.0809	1.1095	2.3087	0.09867	0.56686	20
41	.43340	.90120	.48091	.0794	.1096	.3073	.09880	.56660	19
42	.43366	.90108	.48127	.0778	.1098	.3059	.09892	.56634	18
43	.43392	.90095	.48162	.0763	.1099	.3046	.09905	.56608	17
44	.43418	.90082	.48198	.0747	.1101	.3032	.09917	.56582	16
45	0.43444	0.90070	0.48234	2.0732	1.1102	2.3018	0.09930	0.56555	15
46	.43471	.90057	.48270	.0717	.1104	.3004	.09943	.56529	14
47	.43497	.90044	.48306	.0701	.1106	.2990	.09955	.56503	13
48	.43523	.90032	.48342	.0686	.1107	.2976	.09968	.56477	12
49	.43549	.90019	.48378	.0671	.1109	.2962	.09981	.56451	11
50	0.43575	0.90006	0.48414	2.0655	1.1110	2.2949	0.09993	0.56424	10
51	.43602	.89994	.48449	.0640	.1112	.2935	.10006	.56398	9
52	.43628	.89981	.48485	.0625	.1113	.2921	.10019	.56372	8
53	.43654	.89968	.48521	.0609	.1115	.2907	.10031	.56346	7
54	.43680	.89956	.48557	.0594	.1116	.2894	.10044	.56320	6
55	0.43706	0.89943	0.48593	2.0579	1.1118	2.2880	0.10057	0.56294	5
56	.43732	.89930	.48629	.0564	.1120	.2866	.10070	.56267	4
57	.43759	.89918	.48665	.0548	.1121	.2853	.10082	.56241	3
58	.43785	.89905	.48701	.0533	.1123	.2839	.10095	.56215	2
59	.43811	.89892	.48737	.0518	.1124	.2825	.10108	.56189	1
60	0.43837	0.89879	0.48773	2.0503	1.1126	2.2812	0.10121	0.56163	0
M	Cosine	Sine	Cotan.	Tan.	Cosec.	Secant	Vrs. Cos.	Vrs. Sin.	M

115° 64°

26° Natural Trigonometric Functions 153°

M	Sine	Cosine	Tan.	Cotan.	Secant	Cosec.	Vrs. Sin.	Vrs. Cos.	M
0	0.43837	0.89879	0.48773	2.0503	1.1126	2.2812	0.10121	0.56163	60
1	.43863	.89867	.48809	.0488	.1127	.2798	.10133	.56137	59
2	.43889	.89854	.48845	.0473	.1129	.2784	.10146	.56111	58
3	.43915	.89841	.48881	.0458	.1131	.2771	.10159	.56084	57
4	.43942	.89828	.48917	.0443	.1132	.2757	.10172	.56058	56
5	0.43968	0.89815	0.48953	2.0427	1.1134	2.2744	0.10184	0.56032	55
6	.43994	.89803	.48989	.0412	.1135	.2730	.10197	.56006	54
7	.44020	.89790	.49025	.0397	.1137	.2717	.10210	.55980	53
8	.44046	.89777	.49062	.0382	.1139	.2703	.10223	.55954	52
9	.44072	.89764	.49098	.0367	.1140	.2690	.10236	.55928	51
10	0.44098	0.89751	0.49134	2.0352	1.1142	2.2676	0.10248	0.55902	50
11	.44124	.89739	.49170	.0338	.1143	.2663	.10261	.55875	49
12	.44150	.89726	.49206	.0323	.1145	.2650	.10274	.55849	48
13	.44177	.89713	.49242	.0308	.1147	.2636	.10287	.55823	47
14	.44203	.89700	.49278	.0293	.1148	.2623	.10300	.55797	46
15	0.44229	0.89687	0.49314	2.0278	1.1150	2.2610	0.10313	0.55771	45
16	.44255	.89674	.49351	.0263	.1151	.2596	.10326	.55745	44
17	.44281	.89661	.49387	.0248	.1153	.2583	.10338	.55719	43
18	.44307	.89649	.49423	.0233	.1155	.2570	.10351	.55693	42
19	.44333	.89636	.49459	.0219	.1156	.2556	.10364	.55667	41
20	0.44359	0.89623	0.49495	2.0204	1.1158	2.2543	0.10377	0.55641	40
21	.44385	.89610	.49532	.0189	.1159	.2530	.10390	.55615	39
22	.44411	.89597	.49568	.0174	.1161	.2517	.10403	.55589	38
23	.44437	.89584	.49604	.0159	.1163	.2503	.10416	.55562	37
24	.44463	.89571	.49640	.0145	.1164	.2490	.10429	.55536	36
25	0.44489	0.89558	0.49677	2.0130	1.1166	2.2477	0.10442	0.55510	35
26	.44516	.89545	.49713	.0115	.1167	.2464	.10455	.55484	34
27	.44542	.89532	.49749	.0101	.1169	.2451	.10468	.55458	33
28	.44568	.89519	.49785	.0086	.1171	.2438	.10481	.55432	32
29	.44594	.89506	.49822	.0071	.1172	.2425	.10493	.55406	31
30	0.44620	0.89493	0.49858	2.0057	1.1174	2.2411	0.10506	0.55380	30
31	.44646	.89480	.49894	.0042	.1176	.2398	.10519	.55354	29
32	.44672	.89467	.49931	.0028	.1177	.2385	.10532	.55328	28
33	.44698	.89454	.49967	.0013	.1179	.2372	.10545	.55302	27
34	.44724	.89441	.50003	1.9998	.1180	.2359	.10558	.55276	26
35	0.44750	0.89428	0.50040	1.9984	1.1182	2.2346	0.10571	0.55250	25
36	.44776	.89415	.50076	.9969	.1184	.2333	.10584	.55224	24
37	.44802	.89402	.50113	.9955	.1185	.2320	.10598	.55198	23
38	.44828	.89389	.50149	.9940	.1187	.2307	.10611	.55172	22
39	.44854	.89376	.50185	.9926	.1189	.2294	.10624	.55146	21
40	0.44880	0.89363	0.50222	1.9912	1.1190	2.2282	0.10637	0.55120	20
41	.44906	.89350	.50258	.9897	.1192	.2269	.10650	.55094	19
42	.44932	.89337	.50295	.9883	.1193	.2256	.10663	.55068	18
43	.44958	.89324	.50331	.9868	.1195	.2243	.10676	.55042	17
44	.44984	.89311	.50368	.9854	.1197	.2230	.10689	.55016	16
45	0.45010	0.89298	0.50404	1.9840	1.1198	2.2217	0.10702	0.54990	15
46	.45036	.89285	.50441	.9825	.1200	.2204	.10715	.54964	14
47	.45062	.89272	.50477	.9811	.1202	.2192	.10728	.54938	13
48	.45088	.89258	.50514	.9797	.1203	.2179	.10741	.54912	12
49	.45114	.89245	.50550	.9782	.1205	.2166	.10754	.54886	11
50	0.45140	0.89232	0.50587	1.9768	1.1207	2.2153	0.10768	0.54860	10
51	.45166	.89219	.50623	.9754	.1208	.2141	.10781	.54834	9
52	.45191	.89206	.50660	.9739	.1210	.2128	.10794	.54808	8
53	.45217	.89193	.50696	.9725	.1212	.2115	.10807	.54782	7
54	.45243	.89180	.50733	.9711	.1213	.2103	.10820	.54756	6
55	0.45269	0.89166	0.50769	1.9697	1.1215	2.2090	0.10833	0.54730	5
56	.45295	.89153	.50806	.9683	.1217	.2077	.10846	.54705	4
57	.45321	.89140	.50843	.9668	.1218	.2065	.10860	.54679	3
58	.45347	.89127	.50879	.9654	.1220	.2052	.10873	.54653	2
59	.45373	.89114	.50916	.9640	.1222	.2039	.10886	.54627	1
60	0.45399	0.89101	0.50952	1.9626	1.1223	2.2027	0.10899	0.54601	0
M	Cosine	Sine	Cotan.	Tan.	Cosec.	Secant	Vrs. Cos.	Vrs. Sin.	M

116°   63°

27° Natural Trigonometric Functions 152°

M	Sine	Cosine	Tan.	Cotan.	Secant	Cosec.	Vrs. Sin.	Vrs. Cos.	M
0	0.45399	0.89101	0.50952	1.9626	1.1223	2.2027	0.10899	0.54601	60
1	.45425	.89087	.50989	.9612	.1225	.2014	.10912	.54575	59
2	.45451	.89074	.51026	.9598	.1226	.2002	.10926	.54549	58
3	.45477	.89061	.51062	.9584	.1228	.1989	.10939	.54523	57
4	.45503	.89048	.51099	.9570	.1230	.1977	.10952	.54497	56
5	0.45528	0.89034	0.51136	1.9556	1.1231	2.1964	0.10965	0.54471	55
6	.45554	.89021	.51172	.9542	.1233	.1952	.10979	.54445	54
7	.45580	.89008	.51209	.9528	.1235	.1939	.10992	.54420	53
8	.45606	.88995	.51246	.9514	.1237	.1927	.11005	.54394	52
9	.45632	.88981	.51283	.9500	.1238	.1914	.11018	.54368	51
10	0.45658	0.88968	0.51319	1.9486	1.1240	2.1902	0.11032	0.54342	50
11	.45684	.88955	.51356	.9472	.1242	.1889	.11045	.54316	49
12	.45710	.88942	.51393	.9458	.1243	.1877	.11058	.54290	48
13	.45736	.88928	.51430	.9444	.1245	.1865	.11072	.54264	47
14	.45761	.88915	.51466	.9430	.1247	.1852	.11085	.54238	46
15	0.45787	0.88902	0.51503	1.9416	1.1248	2.1840	0.11098	0.54213	45
16	.45813	.88888	.51540	.9402	.1250	.1828	.11112	.54187	44
17	.45839	.88875	.51577	.9388	.1252	.1815	.11125	.54161	43
18	.45865	.88862	.51614	.9375	.1253	.1803	.11138	.54135	42
19	.45891	.88848	.51651	.9361	.1255	.1791	.11152	.54109	41
20	0.45917	0.88835	0.51687	1.9347	1.1257	2.1778	0.11165	0.54083	40
21	.45942	.88822	.51724	.9333	.1258	.1766	.11178	.54057	39
22	.45968	.88808	.51761	.9319	.1260	.1754	.11192	.54032	38
23	.45994	.88795	.51798	.9306	.1262	.1742	.11205	.54006	37
24	.46020	.88781	.51835	.9292	.1264	.1730	.11218	.53980	36
25	0.46046	0.88768	0.51872	1.9278	1.1265	2.1717	0.11232	0.53954	35
26	.46072	.88755	.51909	.9264	.1267	.1705	.11245	.53928	34
27	.46097	.88741	.51946	.9251	.1269	.1693	.11259	.53902	33
28	.46123	.88728	.51983	.9237	.1270	.1681	.11272	.53877	32
29	.46149	.88714	.52020	.9223	.1272	.1669	.11285	.53851	31
30	0.46175	0.88701	0.52057	1.9210	1.1274	2.1657	0.11299	0.53825	30
31	.46201	.88688	.52094	.9196	.1275	.1645	.11312	.53799	29
32	.46226	.88674	.52131	.9182	.1277	.1633	.11326	.53773	28
33	.46252	.88661	.52168	.9169	.1279	.1620	.11339	.53748	27
34	.46278	.88647	.52205	.9155	.1281	.1608	.11353	.53722	26
35	0.46304	0.88634	0.52242	1.9142	1.1282	2.1596	0.11366	0.53696	25
36	.46330	.88620	.52279	.9128	.1284	.1584	.11380	.53670	24
37	.46355	.88607	.52316	.9115	.1286	.1572	.11393	.53645	23
38	.46381	.88593	.52353	.9101	.1287	.1560	.11407	.53619	22
39	.46407	.88580	.52390	.9088	.1289	.1548	.11420	.53593	21
40	0.46433	0.88566	0.52427	1.9074	1.1291	2.1536	0.11434	0.53567	20
41	.46458	.88553	.52464	.9061	.1293	.1525	.11447	.53541	19
42	.46484	.88539	.52501	.9047	.1294	.1513	.11461	.53516	18
43	.46510	.88526	.52538	.9034	.1296	.1501	.11474	.53490	17
44	.46536	.88512	.52575	.9020	.1298	.1489	.11488	.53464	16
45	0.46561	0.88499	0.52612	1.9007	1.1299	2.1477	0.11501	0.53438	15
46	.46587	.88485	.52650	.8993	.1301	.1465	.11515	.53413	14
47	.46613	.88472	.52687	.8980	.1303	.1453	.11528	.53387	13
48	.46639	.88458	.52724	.8967	.1305	.1441	.11542	.53361	12
49	.46664	.88444	.52761	.8953	.1306	.1430	.11555	.53336	11
50	0.46690	0.88431	0.52798	1.8940	1.1308	2.1418	0.11569	0.53310	10
51	.46716	.88417	.52836	.8927	.1310	.1406	.11583	.53284	9
52	.46741	.88404	.52873	.8913	.1312	.1394	.11596	.53258	8
53	.46767	.88390	.52910	.8900	.1313	.1382	.11610	.53233	7
54	.46793	.88376	.52947	.8887	.1315	.1371	.11623	.53207	6
55	0.46819	0.88363	0.52984	1.8873	1.1317	2.1359	0.11637	0.53181	5
56	.46844	.88349	.53022	.8860	.1319	.1347	.11651	.53156	4
57	.46870	.88336	.53059	.8847	.1320	.1335	.11664	.53130	3
58	.46896	.88322	.53096	.8834	.1322	.1324	.11678	.53104	2
59	.46921	.88308	.53134	.8820	.1324	.1312	.11691	.53078	1
60	0.46947	0.88295	0.53171	1.8807	1.1326	2.1300	0.11705	0.53053	0
M	Cosine	Sine	Cotan.	Tan.	Cosec.	Secant	Vrs. Cos.	Vrs. Sin.	M

117° 62°

28° Natural Trigonometric Functions 151°

M	Sine	Cosine	Tan.	Cotan.	Secant	Cosec.	Vrs. Sin.	Vrs. Cos.	M
0	0.46947	0.88295	0.53171	1.8807	1.1326	2.1300	0.11705	0.53053	60
1	.46973	.88281	.53208	.8794	.1327	.1289	.11719	.53027	59
2	.46998	.88267	.53245	.8781	.1329	.1277	.11732	.53001	58
3	.47024	.88254	.53283	.8768	.1331	.1266	.11746	.52976	57
4	.47050	.88240	.53320	.8754	.1333	.1254	.11760	.52950	56
5	0.47075	0.88226	0.53358	1.8741	1.1334	2.1242	0.11774	0.52924	55
6	.47101	.88213	.53395	.8728	.1336	.1231	.11787	.52899	54
7	.47127	.88199	.53432	.8715	.1338	.1219	.11801	.52873	53
8	.47152	.88185	.53470	.8702	.1340	.1208	.11815	.52847	52
9	.47178	.88171	.53507	.8689	.1341	.1196	.11828	.52822	51
10	0.47204	0.88158	0.53545	1.8676	1.1343	2.1185	0.11842	0.52796	50
11	.47229	.88144	.53582	.8663	.1345	.1173	.11856	.52770	49
12	.47255	.88130	.53619	.8650	.1347	.1162	.11870	.52745	48
13	.47281	.88117	.53657	.8637	.1349	.1150	.11883	.52719	47
14	.47306	.88103	.53694	.8624	.1350	.1139	.11897	.52694	46
15	0.47332	0.88089	0.53732	1.8611	1.1352	2.1127	0.11911	0.52668	45
16	.47357	.88075	.53769	.8598	.1354	.1116	.11925	.52642	44
17	.47383	.88061	.53807	.8585	.1356	.1104	.11938	.52617	43
18	.47409	.88048	.53844	.8572	.1357	.1093	.11952	.52591	42
19	.47434	.88034	.53882	.8559	.1359	.1082	.11966	.52565	41
20	0.47460	0.88020	0.53919	1.8546	1.1361	2.1070	0.11980	0.52540	40
21	.47486	.88006	.53957	.8533	.1363	.1059	.11994	.52514	39
22	.47511	.87992	.53995	.8520	.1365	.1048	.12007	.52489	38
23	.47537	.87979	.54032	.8507	.1366	.1036	.12021	.52463	37
24	.47562	.87965	.54070	.8495	.1368	.1025	.12035	.52437	36
25	0.47588	0.87951	0.54107	1.8482	1.1370	2.1014	0.12049	0.52412	35
26	.47613	.87937	.54145	.8469	.1372	.1002	.12063	.52386	34
27	.47639	.87923	.54183	.8456	.1373	.0991	.12077	.52361	33
28	.47665	.87909	.54220	.8443	.1375	.0980	.12090	.52335	32
29	.47690	.87895	.54258	.8430	.1377	.0969	.12104	.52310	31
30	0.47716	0.87882	0.54295	1.8418	1.1379	2.0957	0.12118	0.52284	30
31	.47741	.87868	.54333	.8405	.1381	.0946	.12132	.52258	29
32	.47767	.87854	.54371	.8392	.1382	.0935	.12146	.52233	28
33	.47792	.87840	.54409	.8379	.1384	.0924	.12160	.52207	27
34	.47818	.87826	.54446	.8367	.1386	.0912	.12174	.52182	26
35	0.47844	0.87812	0.54484	1.8354	1.1388	2.0901	0.12188	0.52156	25
36	.47869	.87798	.54522	.8341	.1390	.0890	.12202	.52131	24
37	.47895	.87784	.54559	.8329	.1391	.0879	.12216	.52105	23
38	.47920	.87770	.54597	.8316	.1393	.0868	.12229	.52080	22
39	.47946	.87756	.54635	.8303	.1395	.0857	.12243	.52054	21
40	0.47971	0.87742	0.54673	1.8291	1.1397	2.0846	0.12257	0.52029	20
41	.47997	.87728	.54711	.8278	.1399	.0835	.12271	.52003	19
42	.48022	.87715	.54748	.8265	.1401	.0824	.12285	.51978	18
43	.48048	.87701	.54786	.8253	.1402	.0812	.12299	.51952	17
44	.48073	.87687	.54824	.8240	.1404	.0801	.12313	.51927	16
45	0.48099	0.87673	0.54862	1.8227	1.1406	2.0790	0.12327	0.51901	15
46	.48124	.87659	.54900	.8215	.1408	.0779	.12341	.51876	14
47	.48150	.87645	.54937	.8202	.1410	.0768	.12355	.51850	13
48	.48175	.87631	.54975	.8190	.1411	.0757	.12369	.51825	12
49	.48201	.87617	.55013	.8177	.1413	.0746	.12383	.51799	11
50	0.48226	0.87603	0.55051	1.8165	1.1415	2.0735	0.12397	0.51774	10
51	.48252	.87588	.55089	.8152	.1417	.0725	.12411	.51748	9
52	.48277	.87574	.55127	.8140	.1419	.0714	.12425	.51723	8
53	.48303	.87560	.55165	.8127	.1421	.0703	.12439	.51697	7
54	.48328	.87546	.55203	.8115	.1422	.0692	.12453	.51672	6
55	0.48354	0.87532	0.55241	1.8102	1.1424	2.0681	0.12468	0.51646	5
56	.48379	.87518	.55279	.8090	.1426	.0670	.12482	.51621	4
57	.48405	.87504	.55317	.8078	.1428	.0659	.12496	.51595	3
58	.48430	.87490	.55355	.8065	.1430	.0648	.12510	.51570	2
59	.48455	.87476	.55393	.8053	.1432	.0637	.12524	.51544	1
60	0.48481	0.87462	0.55431	1.8040	1.1433	2.0627	0.12538	0.51519	0
M	Cosine	Sine	Cotan.	Tan.	Cosec.	Secant	Vrs. Cos.	Vrs. Sin.	M

118°   61°

29°   Natural Trigonometric Functions   **150°**

M	Sine	Cosine	Tan.	Cotan.	Secant	Cosec.	Vrs. Sin.	Vrs. Cos.	M
0	0.48481	0.87462	0.55431	1.8040	1.1433	2.0627	0.12538	0.51519	60
1	.48506	.87448	.55469	.8028	.1435	.0616	.12552	.51493	59
2	.48532	.87434	.55507	.8016	.1437	.0605	.12566	.51468	58
3	.48557	.87420	.55545	.8003	.1439	.0594	.12580	.51443	57
4	.48583	.87405	.55583	.7991	.1441	.0583	.12594	.51417	56
5	0.48608	0.87391	0.55621	1.7979	1.1443	2.0573	0.12609	0.51392	55
6	.48633	.87377	.55659	.7966	.1445	.0562	.12623	.51366	54
7	.48659	.87363	.55697	.7954	.1446	.0551	.12637	.51341	53
8	.48684	.87349	.55735	.7942	.1448	.0540	.12651	.51316	52
9	.48710	.87335	.55774	.7930	.1450	.0530	.12665	.51290	51
10	0.48735	0.87320	0.55812	1.7917	1.1452	2.0519	0.12679	0.51265	50
11	.48760	.87306	.55850	.7905	.1454	.0508	.12694	.51239	49
12	.48786	87292	55888	.7893	.1456	.0498	.12708	.51214	48
13	.48811	.87278	55926	.7881	.1458	.0487	.12722	.51189	47
14	.48837	.87264	.55964	.7868	.1459	.0476	.12736	.51163	46
15	0.48862	0.87250	0.56003	1.7856	1.1461	2.0466	0.12750	0.51138	45
16	.48887	.87235	.56041	.7844	.1463	.0455	.12765	.51112	44
17	48913	.87221	.56079	.7832	.1465	.0444	.12779	.51087	43
18	.48938	.87207	.56117	.7820	.1467	.0434	.12793	.51062	42
19	.48964	.87193	.56156	.7808	.1469	.0423	.12807	.51036	41
20	0.48989	0.87178	0.56194	1.7795	1.1471	2.0413	0.12821	0.51011	40
21	.49014	.87164	.56232	.7783	.1473	.0402	.12836	.50986	39
22	.49040	.87150	.56270	.7771	.1474	.0392	.12850	.50960	38
23	.49065	.87136	.56309	.7759	.1476	.0381	.12864	.50935	37
24	.49090	.87121	.56347	.7747	.1478	.0370	.12879	.50910	36
25	0.49116	0.87107	0.56385	1.7735	1.1480	2.0360	0.12893	0.50884	35
26	.49141	.87093	.56424	.7723	.1482	.0349	.12907	.50859	34
27	.49166	.87078	.56462	.7711	.1484	.0339	.12921	.50834	33
28	.49192	.87064	.56500	.7699	.1486	.0329	.12936	.50808	32
29	.49217	.87050	.56539	.7687	.1488	.0318	.12950	.50783	31
30	0.49242	0.87035	0.56577	1.7675	1.1489	2.0308	0.12964	0.50758	30
31	.49268	.87021	.56616	.7663	.1491	.0297	.12979	.50732	29
32	.49293	.87007	.56654	.7651	.1493	.0287	.12993	.50707	28
33	.49318	.86992	.56692	.7639	.1495	.0276	.13007	.50682	27
34	.49343	.86978	.56731	.7627	.1497	.0266	.13022	.50656	26
35	0.49369	0.86964	0.56769	1.7615	1.1499	2.0256	0.13036	0.50631	25
36	.49394	.86949	.56808	.7603	.1501	.0245	.13050	.50606	24
37	.49419	.86935	.56846	.7591	.1503	.0235	.13065	.50580	23
38	.49445	.86921	.56885	.7579	.1505	.0224	.13079	.50555	22
39	.49470	.86906	.56923	.7567	.1507	:0214	.13094	.50530	21
40	0.49495	0.86892	0.56962	1.7555	1.1508	2.0204	0.13108	0.50505	20
41	.49521	.86877	.57000	.7544	.1510	.0194	.13122	.50479	19
42	.49546	.86863	.57039	.7532	.1512	.0183	.13137	.50454	18
43	.49571	.86849	.57077	.7520	.1514	.0173	.13151	.50429	17
44	.49596	.86834	.57116	.7508	.1516	.0163	.13166	.50404	16
45	0.49622	0.86820	0.57155	1.7496	1.1518	2.0152	0.13180	0.50378	15
46	.49647	.86805	.57193	.7484	.1520	.0142	.13194	.50353	14
47	.49672	.86791	.57232	.7473	.1522	.0132	.13209	.50328	13
48	.49697	.86776	.57270	.7461	.1524	.0122	.13223	.50303	12
49	.49723	.86762	.57309	.7449	.1526	.0111	.13238	.50277	11
50	0.49748	0.86748	0.57348	1.7437	1.1528	2.0101	0.13252	0.50252	10
51	.49773	.86733	.57386	.7426	.1530	.0091	.13267	.50227	9
52	.49798	.86719	.57425	.7414	.1531	.0081	.13281	.50202	8
53	.49823	.86704	.57464	.7402	.1533	.0071	.13296	.50176	7
54	.49849	.86690	.57502	.7390	.1535	.0061	.13310	.50151	6
55	0.49874	0.86675	0.57541	1.7379	1.1537	2.0050	0.13325	0.50126	5
56	.49899	.86661	.57580	.7367	.1539	.0040	.13339	.50101	4
57	.49924	.86646	.57619	.7355	.1541	.0030	.13354	.50076	3
58	.49950	.86632	.57657	.7344	.1543	.0020	.13368	.50050	2
59	.49975	.86617	.57696	.7332	.1545	.0010	.13383	.50025	1
60	0.50000	0.86603	0.57735	1.7320	1.1547	2.0000	0.13397	0.50000	0
M	Cosine	Sine	Cotan.	Tan.	Cosec.	Secant	Vrs. Cos.	Vrs. Sin.	M

119°   **60°**

Natural Trigonometric Functions — 30° / 149°

M	Sine	Cosine	Tan.	Cotan.	Secant	Cosec.	Vrs. Sin.	Vrs. Cos.	M
0	0.50000	0.86603	0.57735	1.7320	1.1547	2.0000	0.13397	0.50000	60
1	.50025	.86588	.57774	.7309	.1549	1.9990	.13412	.49975	59
2	.50050	.86573	.57813	.7297	.1551	.9980	.13426	.49950	58
3	.50075	.86559	.57851	.7286	.1553	.9970	.13441	.49924	57
4	.50101	.86544	.57890	.7274	.1555	.9960	.13456	.49899	56
5	0.50126	0.86530	0.57929	1.7262	1.1557	1.9950	0.13470	0.49874	55
6	.50151	.86515	.57968	.7251	.1559	.9940	.13485	.49849	54
7	.50176	.86500	.58007	.7239	.1561	.9930	.13499	.49824	53
8	.50201	.86486	.58046	.7228	.1562	.9920	.13514	.49799	52
9	.50226	.86471	.58085	.7216	.1564	.9910	.13529	.49773	51
10	0.50252	0.86457	0.58123	1.7205	1.1566	1.9900	0.13543	0.49748	50
11	.50277	.86442	.58162	.7193	.1568	.9890	.13558	.49723	49
12	.50302	.86427	.58201	.7182	.1570	.9880	.13572	.49698	48
13	.50327	.86413	.58240	.7170	.1572	.9870	.13587	.49673	47
14	.50352	.86398	.58279	.7159	.1574	.9860	.13602	.49648	46
15	0.50377	0.86383	0.58318	1.7147	1.1576	1.9850	0.13616	0.49623	45
16	.50402	.86369	.58357	.7136	.1578	.9840	.13631	.49597	44
17	.50428	.86354	.58396	.7124	.1580	.9830	.13646	.49572	43
18	.50453	.86339	.58435	.7113	.1582	.9820	.13660	.49547	42
19	.50478	.86325	.58474	.7101	.1584	.9811	.13675	.49522	41
20	0.50503	0.86310	0.58513	1.7090	1.1586	1.9801	0.13690	0.49497	40
21	.50528	.86295	.58552	.7079	.1588	.9791	.13704	.49472	39
22	.50553	.86281	.58591	.7067	.1590	.9781	.13719	.49447	38
23	.50578	.86266	.58630	.7056	.1592	.9771	.13734	.49422	37
24	.50603	.86251	.58670	.7044	.1594	.9761	.13749	.49397	36
25	0.50628	0.86237	0.58709	1.7033	1.1596	1.9752	0.13763	0.49371	35
26	.50653	.86222	.58748	.7022	.1598	.9742	.13778	.49346	34
27	.50679	.86207	.58787	.7010	.1600	.9732	.13793	.49321	33
28	.50704	.86192	.58826	.6999	.1602	.9722	.13807	.49296	32
29	.50729	.86178	.58865	.6988	.1604	.9713	.13822	.49271	31
30	0.50754	0.86163	0.58904	1.6977	1.1606	1.9703	0.13837	0.49246	30
31	.50779	.86148	.58944	.6965	.1608	.9693	.13852	.49221	29
32	.50804	.86133	.58983	.6954	.1610	.9683	.13867	.49196	28
33	.50829	.86118	.59022	.6943	.1612	.9674	.13881	.49171	27
34	.50854	.86104	.59061	.6931	.1614	.9664	.13896	.49146	26
35	0.50879	0.86089	0.59100	1.6920	1.1616	1.9654	0.13911	0.49121	25
36	.50904	.86074	.59140	.6909	.1618	.9645	.13926	.49096	24
37	.50929	.86059	.59179	.6898	.1620	.9635	.13941	.49071	23
38	.50954	.86044	.59218	.6887	.1622	.9625	.13955	.49046	22
39	.50979	.86030	.59258	.6875	.1624	.9616	.13970	.49021	21
40	0.51004	0.86015	0.59297	1.6864	1.1626	1.9606	0.13985	0.48996	20
41	.51029	.86000	.59336	.6853	.1628	.9596	.14000	.48971	19
42	.51054	.85985	.59376	.6842	.1630	.9587	.14015	.48946	18
43	.51079	.85970	.59415	.6831	.1632	.9577	.14030	.48921	17
44	.51104	.85955	.59454	.6820	.1634	.9568	.14044	.48896	16
45	0.51129	0.85941	0.59494	1.6808	1.1636	1.9558	0.14059	0.48871	15
46	.51154	.85926	.59533	.6797	.1638	.9549	.14074	.48846	14
47	.51179	.85911	.59572	.6786	.1640	.9539	.14089	.48821	13
48	.51204	.85896	.59612	.6775	.1642	.9530	.14104	.48796	12
49	.51229	.85881	.59651	.6764	.1644	.9520	.14119	.48771	11
50	0.51254	0.85866	0.59691	1.6753	1.1646	1.9510	0.14134	0.48746	10
51	.51279	.85851	.59730	.6742	.1648	.9501	.14149	.48721	9
52	.51304	.85836	.59770	.6731	.1650	.9491	.14164	.48696	8
53	.51329	.85821	.59809	.6720	.1652	.9482	.14178	.48671	7
54	.51354	.85806	.59849	.6709	.1654	.9473	.14193	.48646	6
55	0.51379	0.85791	0.59888	1.6698	1.1656	1.9463	0.14208	0.48621	5
56	.51404	.85777	.59928	.6687	.1658	.9454	.14223	.48596	4
57	.51429	.85762	.59967	.6676	.1660	.9444	.14238	.48571	3
58	.51454	.85747	.60007	.6665	.1662	.9435	.14253	.48546	2
59	.51479	.85732	.60046	.6654	.1664	.9425	.14268	.48521	1
60	0.51504	0.85717	0.60086	1.6643	1.1666	1.9416	0.14283	0.48496	0
M	Cosine	Sine	Cotan.	Tan.	Cosec.	Secant	Vrs. Cos.	Vrs. Sin.	M

120°   59°

31° Natural Trigonometric Functions 148°

M	Sine	Cosine	Tan.	Cotan.	Secant	Cosec.	Vrs. Sin.	Vrs. Cos.	M
0	0.51504	0.85717	0.60086	1.6643	1.1666	1.9416	0.14283	0.48496	60
1	.51529	.85702	.60126	.6632	.1668	.9407	.14298	.48471	59
2	.51554	.85687	.60165	.6621	.1670	.9397	.14313	.48446	58
3	.51578	.85672	.60205	.6610	.1672	.9388	.14328	.48421	57
4	.51603	.85657	.60244	.6599	.1674	.9378	.14343	.48396	56
5	0.51628	0.85642	0.60284	1.6588	1.1676	1.9369	0.14358	0.48371	55
6	.51653	.85627	.60324	.6577	.1678	.9360	.14373	.48347	54
7	.51678	.85612	.60363	.6566	.1681	.9350	.14388	.48322	53
8	.51703	.85597	.60403	.6555	.1683	.9341	.14403	.48297	52
9	.51728	.85582	.60443	.6544	.1685	.9332	.14418	.48272	51
10	0.51753	0.85566	0.60483	1.6534	1.1687	1.9322	0.14433	0.48247	50
11	.51778	.85551	.60522	.6523	.1689	.9313	.14448	.48222	49
12	.51803	.85536	.60562	.6512	.1691	.9304	.14463	.48197	48
13	.51827	.85521	.60602	.6501	.1693	.9295	.14479	.48172	47
14	.51852	.85506	.60642	.6490	.1695	.9285	.14494	.48147	46
15	0.51877	0.85491	0.60681	1.6479	1.1697	1.9276	0.14509	0.48123	45
16	.51902	.85476	.60721	.6469	.1699	.9267	.14524	.48098	44
17	.51927	.85461	.60761	.6458	.1701	.9258	.14539	.48073	43
18	.51952	.85446	.60801	.6447	.1703	.9248	.14554	.48048	42
19	.51977	.85431	.60841	.6436	.1705	.9239	.14569	.48023	41
20	0.52002	0.85416	0.60881	1.6425	1.1707	1.9230	0.14584	0.47998	40
21	.52026	.85400	.60920	.6415	.1709	.9221	.14599	.47973	39
22	.52051	.85385	.60960	.6404	.1712	.9212	.14615	.47949	38
23	.52076	.85370	.61000	.6393	.1714	.9203	.14630	.47924	37
24	.52101	.85355	.61040	.6383	.1716	.9193	.14645	.47899	36
25	0.52126	0.85340	0.61080	1.6372	1.1718	1.9184	0.14660	0.47874	35
26	.52151	.85325	.61120	.6361	.1720	.9175	.14675	.47849	34
27	.52175	.85309	.61160	.6350	.1722	.9166	.14690	.47824	33
28	.52200	.85294	.61200	.6340	.1724	.9157	.14706	.47800	32
29	.52225	.85279	.61240	.6329	.1726	.9148	.14721	.47775	31
30	0.52250	0.85264	0.61280	1.6318	1.1728	1.9139	0.14736	0.47750	30
31	.52275	.85249	.61320	.6308	.1730	.9130	.14751	.47725	29
32	.52299	.85234	.61360	.6297	.1732	.9121	.14766	.47700	28
33	.52324	.85218	.61400	.6286	.1734	.9112	.14782	.47676	27
34	.52349	.85203	.61440	.6276	.1737	.9102	.14797	.47651	26
35	0.52374	0.85188	0.61480	1.6265	1.1739	1.9093	0.14812	0.47626	25
36	.52398	.85173	.61520	.6255	.1741	.9084	.14827	.47601	24
37	.52423	.85157	.61560	.6244	.1743	.9075	.14842	.47577	23
38	.52448	.85142	.61601	.6233	.1745	.9066	.14858	.47552	22
39	.52473	.85127	.61641	.6223	.1747	.9057	.14873	.47527	21
40	0.52498	0.85112	0.61681	1.6212	1.1749	1.9048	0.14888	0.47502	20
41	.52522	.85096	.61721	.6202	.1751	.9039	.14904	.47477	19
42	.52547	.85081	.61761	.6191	.1753	.9030	.14919	.47453	18
43	.52572	.85066	.61801	.6181	.1756	.9021	.14934	.47428	17
44	.52597	.85050	.61842	.6170	.1758	.9013	.14949	.47403	16
45	0.52621	0.85035	0.61882	1.6160	1.1760	1.9004	0.14965	0.47379	15
46	.52646	.85020	.61922	.6149	.1762	.8995	.14980	.47354	14
47	.52671	.85004	.61962	.6139	.1764	.8986	.14995	.47329	13
48	.52695	.84989	.62003	.6128	.1766	.8977	.15011	.47304	12
49	.52720	.84974	.62043	.6118	.1768	.8968	.15026	.47280	11
50	0.52745	0.84959	0.62083	1.6107	1.1770	1.8959	0.15041	0.47255	10
51	.52770	.84943	.62123	.6097	.1772	.8950	.15057	.47230	9
52	.52794	.84928	.62164	.6086	.1775	.8941	.15072	.47205	8
53	.52819	.84912	.62204	.6076	.1777	.8932	.15087	.47181	7
54	.52844	.84897	.62244	.6066	.1779	.8924	.15103	.47156	6
55	0.52868	0.84882	0.62285	1.6055	1.1781	1.8915	0.15118	0.47131	5
56	.52893	.84866	.62325	.6045	.1783	.8906	.15133	.47107	4
57	.52918	.84851	.62366	.6034	.1785	.8897	.15149	.47082	3
58	.52942	.84836	.62406	.6024	.1787	.8888	.15164	.47057	2
59	.52967	.84820	.62446	.6014	.1790	.8879	.15180	.47033	1
60	0.52992	0.84805	0.62487	1.6003	1.1792	1.8871	0.15195	0.47008	0
M	Cosine	Sine	Cotan.	Tan.	Cosec.	Secant	Vrs. Cos.	Vrs. Sin.	M

121° 58°

32°   Natural Trigonometric Functions   147°

M	Sine	Cosine	Tan.	Cotan.	Secant	Cosec.	Vrs. Sin.	Vrs. Cos	M
0	0.52992	0.84805	0.62487	1.6003	1.1792	1.8871	0.15195	0.47008	60
1	.53016	.84789	.62527	.5993	.1794	.8862	.15211	.46983	59
2	.53041	.84774	.62568	.5983	.1796	.8853	.15226	.46959	58
3	.53066	.84758	.62608	.5972	.1798	.8844	.15241	.46934	57
4	.53090	.84743	.62649	.5962	.1800	.8836	.15257	.46909	56
5	0.53115	0.84728	0.62689	1.5952	1.1802	1.8827	0.15272	0.46885	55
6	.53140	.84712	.62730	.5941	.1805	.8818	.15288	.46860	54
7	.53164	.84697	62770	.5931	.1807	.8809	.15303	46835	53
8	.53189	.84681	.62811	.5921	.1809	.8801	.15319	46811	52
9	.53214	.84666	.62851	.5910	.1811	.8792	.15334	.46786	51
10	0.53238	0.84650	0.62892	1.5900	1.1813	1.8783	0.15350	0.46762	50
11	.53263	.84635	.62933	.5890	.1815	.8775	.15365	.46737	49
12	.53288	.84619	.62973	.5880	.1818	.8766	.15381	.46712	48
13	.53312	.84604	.63014	.5869	.1820	.8757	.15396	.46688	47
14	.53337	.84588	.63055	.5859	.1822	.8749	.15412	.46663	46
15	0.53361	0.84573	0.63095	1.5849	1.1824	1.8740	0.15427	0.46638	45
16	.53386	.84557	.63136	.5839	.1826	.8731	.15443	.46614	44
17	.53411	.84542	.63177	.5829	.1828	.8723	.15458	.46589	43
18	.53435	.84526	.63217	.5818	.1831	.8714	.15474	.46565	42
19	.53460	.84511	.63258	.5808	.1833	.8706	.15489	.46540	41
20	0.53484	0.84495	0.63299	1.5798	1.1835	1.8697	0.15505	0.46516	40
21	.53509	.84479	.63339	.5788	.1837	.8688	.15520	.46491	39
22	.53533	.84464	.63380	.5778	.1839	.8680	.15536	.46466	38
23	.53558	.84448	.63421	.5768	.1841	.8671	.15552	.46442	37
24	.53583	.84433	.63462	.5757	.1844	.8663	.15567	.46417	36
25	0.53607	0.84417	0.63503	1.5747	1.1846	1.8654	0.15583	0.46393	35
26	.53632	.84402	.63543	.5737	.1848	.8646	.15598	.46368	34
27	.53656	.84386	.63584	.5727	.1850	.8637	.15614	.46344	33
28	.53681	.84370	.63625	.5717	.1852	.8629	.15630	.46319	32
29	.53705	.84355	.63666	.5707	.1855	.8620	.15645	.46294	31
30	0.53730	0.84339	0.63707	1.5697	1.1857	1.8611	0.15661	0.46270	30
31	.53754	.84323	.63748	.5687	.1859	.8603	.15676	.46245	29
32	.53779	.84308	.63789	.5677	.1861	.8595	.15692	.46221	28
33	.53803	.84292	.63830	.5667	.1863	.8586	.15708	.46196	27
34	.53828	.84276	.63871	.5657	.1866	.8578	.15723	.46172	26
35	0.53852	0.84261	0.63912	1.5646	1.1868	1.8569	0.15739	0.46147	25
36	.53877	.84245	.63953	.5636	.1870	.8561	.15755	.46123	24
37	.53901	.84229	.63994	.5626	.1872	.8552	.15770	.46098	23
38	.53926	.84214	.64035	.5616	.1874	.8544	.15786	.46074	22
39	.53950	.84198	.64076	.5606	.1877	.8535	.15802	.46049	21
40	0.53975	0.84182	0.64117	1.5596	1.1879	1.8527	0.15817	0.46025	20
41	.53999	.84167	.64158	.5586	.1881	.8519	.15833	.46000	19
42	.54024	.84151	.64199	.5577	.1883	.8510	.15849	.45976	18
43	.54048	.84135	.64240	.5567	.1886	.8502	.15865	.45951	17
44	.54073	.84120	.64281	.5557	.1888	.8493	.15880	.45927	16
45	0.54097	0.84104	0.64322	1.5547	1.1890	1.8485	0.15896	0.45902	15
46	.54122	.84088	.64363	.5537	.1892	.8477	.15912	.45878	14
47	.54146	.84072	.64404	.5527	.1894	.8468	.15927	.45854	13
48	.54171	.84057	.64446	.5517	.1897	.8460	.15943	.45829	12
49	.54195	.84041	.64487	.5507	.1899	.8452	.15959	.45805	11
50	0.54220	0.84025	0.64528	1.5497	1.1901	1.8443	0.15975	0.45780	10
51	.54244	.84009	.64569	.5487	.1903	.8435	.15991	.45756	9
52	.54268	.83993	.64610	.5477	.1906	.8427	.16006	.45731	8
53	.54293	.83978	.64652	.5467	.1908	.8418	.16022	.45707	7
54	.54317	.83962	.64693	.5458	.1910	.8410	.16038	.45682	6
55	0.54342	0.83946	0.64734	1.5448	1.1912	1.8402	0.16054	0.45658	5
56	.54366	.83930	.64775	.5438	.1915	.8394	.16070	.45634	4
57	.54391	.83914	.64817	.5428	.1917	.8385	.16085	.45609	3
58	.54415	.83899	.64858	.5418	.1919	.8377	.16101	.45585	2
59	.54439	.83883	.64899	.5408	.1921	.8369	.16117	.45560	1
60	0.54464	0.83867	0.64941	1.5399	1.1924	1.8361	0.16133	0.45536	0
M	Cosine	Sine	Cotan.	Tan.	Cosec.	Secant	Vrs. Cos.	Vrs. Sin.	M

122°   57°

33°   Natural Trigonometric Functions   **146°**

M	Sine	Cosine	Tan.	Cotan.	Secant	Cosec.	Vrs. Sin.	Vrs. Cos.	M
0	0.54464	0.83867	0.64941	1.5399	1.1924	1.8361	0.16133	0.45536	60
1	.54488	.83851	.64982	.5389	.1926	.8352	.16149	.45512	59
2	.54513	.83835	.65023	.5379	.1928	.8344	.16165	.45487	58
3	.54537	.83819	.65065	.5369	.1930	.8336	.16180	.45463	57
4	.54561	.83804	.65106	.5359	.1933	.8328	.16196	.45438	56
5	0.54586	0.83788	0.65148	1.5350	1.1935	1.8320	0.16212	0.45414	55
6	.54610	.83772	.65189	.5340	.1937	.8311	.16228	.45390	54
7	.54634	.83756	.65231	.5330	.1939	.8303	.16244	.45365	53
8	.54659	.83740	.65272	.5320	.1942	.8295	.16260	.45341	52
9	.54683	.83724	.65314	.5311	.1944	.8287	.16276	.45317	51
10	0.54708	0.83708	0.65355	1.5301	1.1946	1.8279	0.16292	0.45292	50
11	.54732	.83692	.65397	.5291	.1948	.8271	.16308	.45268	49
12	.54756	.83676	.65438	.5282	.1951	.8263	.16323	.45244	48
13	.54781	.83660	.65480	.5272	.1953	.8255	.16339	.45219	47
14	.54805	.83644	.65521	.5262	.1955	.8246	.16355	.45195	46
15	0.54829	0.83629	0.65563	1.5252	1.1958	1.8238	0.16371	0.45171	45
16	.54854	.83613	.65604	.5243	.1960	.8230	.16387	.45146	44
17	.54878	.83597	.65646	.5233	.1962	.8222	.16403	.45122	43
18	.54902	.83581	.65688	.5223	.1964	.8214	.16419	.45098	42
19	.54926	.83565	.65729	.5214	.1967	.8206	.16435	.45073	41
20	0.54951	0.83549	0.65771	1.5204	1.1969	1.8198	0.16451	0.45049	40
21	.54975	.83533	.65813	.5195	.1971	.8190	.16467	.45025	39
22	.54999	.83517	.65854	.5185	.1974	.8182	.16483	.45000	38
23	.55024	.83501	.65896	.5175	.1976	.8174	.16499	.44976	37
24	.55048	.83485	.65938	.5166	.1978	.8166	.16515	.44952	36
25	0.55072	0.83469	0.65980	1.5156	1.1980	1.8158	0.16531	0.44928	35
26	.55097	.83453	.66021	.5147	.1983	.8150	.16547	.44903	34
27	.55121	.83437	.66063	.5137	.1985	.8142	.16563	.44879	33
28	.55145	.83421	.66105	.5127	.1987	.8134	.16579	.44855	32
29	.55169	.83405	.66147	.5118	.1990	.8126	.16595	.44830	31
30	0.55194	0.83388	0.66188	1.5108	1.1992	1.8118	0.16611	0.44806	30
31	.55218	.83372	.66230	.5099	.1994	.8110	.16627	.44782	29
32	.55242	.83356	.66272	.5089	.1997	.8102	.16643	.44758	28
33	.55266	.83340	.66314	.5080	.1999	.8094	.16660	.44733	27
34	.55291	.83324	.66356	.5070	.2001	.8086	.16676	.44709	26
35	0.55315	0.83308	0.66398	1.5061	1.2004	1.8078	0.16692	0.44685	25
36	.55339	.83292	.66440	.5051	.2006	.8070	.16708	.44661	24
37	.55363	.83276	.66482	.5042	.2008	.8062	.16724	.44637	23
38	.55388	.83260	.66524	.5032	.2010	.8054	.16740	.44612	22
39	.55412	.83244	.66566	.5023	.2013	.8047	.16756	.44588	21
40	0.55436	0.83228	0.66608	1.5013	1.2015	1.8039	0.16772	0.44564	20
41	.55460	.83211	.66650	.5004	.2017	.8031	.16788	.44540	19
42	.55484	.83195	.66692	.4994	.2020	.8023	.16804	.44515	18
43	.55509	.83179	.66734	.4985	.2022	.8015	.16821	.44491	17
44	.55533	.83163	.66776	.4975	.2024	.8007	.16837	.44467	16
45	0.55557	0.83147	0.66818	1.4966	1.2027	1.7999	0.16853	0.44443	15
46	.55581	.83131	.66860	.4957	.2029	.7992	.16869	.44419	14
47	.55605	.83115	.66902	.4947	.2031	.7984	.16885	.44395	13
48	.55629	.83098	.66944	.4938	.2034	.7976	.16901	.44370	12
49	.55654	.83082	.66986	.4928	.2036	.7968	.16918	.44346	11
50	0.55678	0.83066	0.67028	1.4919	1.2039	1.7960	0.16934	0.44322	10
51	.55702	.83050	.67071	.4910	.2041	.7953	.16950	.44298	9
52	.55726	.83034	.67113	.4900	.2043	.7945	.16966	.44274	8
53	.55750	.83017	.67155	.4891	.2046	.7937	.16982	.44250	7
54	.55774	.83001	.67197	.4881	.2048	.7929	.16999	.44225	6
55	0.55799	0.82985	0.67239	1.4872	1.2050	1.7921	0.17015	0.44201	5
56	.55823	.82969	.67282	.4863	.2053	.7914	.17031	.44177	4
57	.55847	.82952	.67324	.4853	.2055	.7906	.17047	.44153	3
58	.55871	.82936	.67366	.4844	.2057	.7898	.17064	.44129	2
59	.55895	.82920	.67408	.4835	.2060	.7891	.17080	.44105	1
60	0.55919	0.82904	0.67451	1.4826	1.2062	1.7883	0.17096	0.44081	0
M	Cosine	Sine	Cotan.	Tan.	Cosec.	Secant	Vrs. Cos.	Vrs. Sin.	M

123°   **56°**

34°　　　Natural Trigonometric Functions　　　145°

M	Sine	Cosine	Tan.	Cotan.	Secant	Cosec.	Vrs. Sin.	Vrs. Cos.	M
0	0.55919	0.82904	0.67451	1.4826	1.2062	1.7883	0.17096	0.44081	60
1	.55943	.82887	.67493	.4816	.2064	.7875	.17112	.44057	59
2	.55967	.82871	.67535	.4807	.2067	.7867	.17129	.44032	58
3	.55992	.82855	.67578	.4798	.2069	.7860	.17145	.44008	57
4	.56016	.82839	.67620	.4788	.2072	.7852	.17161	.43984	56
5	0.56040	0.82822	0.67663	1.4779	1.2074	1.7844	0.17178	0.43960	55
6	.56064	.82806	.67705	.4770	.2076	.7837	.17194	.43936	54
7	.56088	.82790	.67747	.4761	.2079	.7829	.17210	.43912	53
8	.56112	.82773	.67790	.4751	.2081	.7821	.17227	.43888	52
9	.56136	.82757	.67832	.4742	.2083	.7814	.17243	.43864	51
10	0.56160	0.82741	0.67875	1.4733	1.2086	1.7806	0.17259	0.43840	50
11	.56184	.82724	.67917	.4724	.2088	.7798	.17276	.43816	49
12	.56208	.82708	.67960	.4714	.2091	.7791	.17292	.43792	48
13	.56232	.82692	.68002	.4705	.2093	.7783	.17308	.43768	47
14	.56256	.82675	.68045	.4696	.2095	.7776	.17325	.43743	46
15	0.56280	0.82659	0.68087	1.4687	1.2098	1.7768	0.17341	0.43719	45
16	.56304	.82643	.68130	.4678	.2100	.7760	.17357	.43695	44
17	.56328	.82626	.68173	.4669	.2103	.7753	.17374	.43671	43
18	.56353	.82610	.68215	.4659	.2105	.7745	.17390	.43647	42
19	56377	.82593	.68258	.4650	.2107	.7738	.17406	.43623	41
20	0.56401	0.82577	0.68301	1.4641	1.2110	1.7730	0.17423	0.43599	40
21	.56425	.82561	.68343	.4632	.2112	.7723	.17439	.43575	39
22	.56449	.82544	.68386	.4623	.2115	.7715	.17456	.43551	38
23	.56473	.82528	.68429	.4614	.2117	.7708	.17472	.43527	37
24	.56497	.82511	.68471	.4605	.2119	.7700	.17489	.43503	36
25	0.56521	0.82495	0.68514	1.4595	1.2122	1.7693	0.17505	0.43479	35
26	.56545	.82478	.68557	.4586	.2124	.7685	.17521	.43455	34
27	.56569	.82462	.68600	.4577	.2127	.7678	.17538	.43431	33
28	.56593	.82445	.68642	.4568	.2129	.7670	.17554	.43407	32
29	.56617	.82429	.68685	.4559	.2132	.7663	.17571	.43383	31
30	0.56641	0.82413	0.68728	1.4550	1.2134	1.7655	0.17587	0.43359	30
31	.56664	.82396	.68771	.4541	.2136	.7648	.17604	.43335	29
32	.56688	.82380	.68814	.4532	.2139	.7640	.17620	.43311	28
33	.56712	.82363	.68857	.4523	.2141	.7633	.17637	.43287	27
34	.56736	.82347	.68899	.4514	.2144	.7625	.17653	.43263	26
35	0.56760	0.82330	0.68942	1.4505	1.2146	1.7618	0.17670	0.43239	25
36	.56784	.82314	.68985	.4496	.2149	.7610	.17686	.43216	24
37	.56808	.82297	.69028	.4487	.2151	.7603	.17703	.43192	23
38	.56832	.82280	.69071	.4478	.2153	7596	.17719	.43168	22
39	.56856	.82264	.69114	.4469	.2156	.7588	.17736	.43144	21
40	0.56880	0.82247	0.69157	1.4460	1.2158	1.7581	0.17752	0.43120	20
41	.56904	.82231	.69200	.4451	.2161	.7573	.17769	.43096	19
42	.56928	.82214	.69243	.4442	.2163	.7566	.17786	.43072	18
43	.56952	.82198	.69286	.4433	.2166	.7559	.17802	.43048	17
44	.56976	.82181	.69329	.4424	.2168	.7551	.17819	.43024	16
45	0.57000	0.82165	0.69372	1.4415	1.2171	1.7544	0.17835	0.43000	15
46	.57023	.82148	.69415	.4406	.2173	.7537	.17852	.42976	14
47	.57047	.82131	.69459	.4397	.2175	.7529	.17868	.42952	13
48	.57071	.82115	.69502	.4388	.2178	.7522	.17885	.42929	12
49	.57095	.82098	.69545	.4379	.2180	.7514	.17902	.42905	11
50	0.57119	0.82082	0.69588	1.4370	1.2183	1.7507	0.17918	0.42881	10
51	.57143	.82065	.69631	.4361	.2185	.7500	.17935	.42857	9
52	.57167	.82048	.69674	.4352	.2188	.7493	.17951	.42833	8
53	.57191	.82032	.69718	.4343	.2190	.7485	.17968	.42809	7
54	.57214	.82015	.69761	.4335	.2193	.7478	.17985	.42785	6
55	0.57238	0.81998	0.69804	1.4326	1.2195	1.7471	0.18001	0.42761	5
56	.57262	.81982	.69847	.4317	.2198	.7463	.18018	.42738	4
57	.57286	.81965	.69891	.4308	.2200	.7456	.18035	.42714	3
58	57310	.81948	.69934	.4299	.2203	.7449	18051	.42690	2
59	.57334	81932	.69977	.4290	.2205	.7442	.18068	.42666	1
60	0.57358	0.81915	0.70021	1.4281	1.2208	1.7434	0.18085	0.42642	0
M	Cosine	Sine	Cotan.	Tan.	Cosec.	Secant	Vrs. Cos.	Vrs. Sin.	M

124°　　　　　　　　　　　　　　　　　　　　　　　55°

35° Natural Trigonometric Functions 144°

M	Sine	Cosine	Tan.	Cotan.	Secant	Cosec.	Vrs. Sin.	Vrs. Cos.	M
0	0.57358	0.81915	0.70021	1.4281	1.2208	1.7434	0.18085	0.42642	60
1	.57381	.81898	.70064	.4273	.2210	.7427	.18101	.42618	59
2	.57405	.81882	.70107	.4264	.2213	.7420	.18118	.42595	58
3	.57429	.81865	.70151	.4255	.2215	.7413	.18135	.42571	57
4	.57453	.81848	.70194	.4246	.2218	.7405	.18151	.42547	56
5	0.57477	0.81832	0.70238	1.4237	1.2220	1.7398	0.18168	0.42523	55
6	.57500	.81815	.70281	.4228	.2223	.7391	.18185	.42499	54
7	.57524	.81798	.70325	.4220	.2225	.7384	.18202	.42476	53
8	.57548	.81781	.70368	.4211	.2228	.7377	.18218	.42452	52
9	.57572	.81765	.70412	.4202	.2230	.7369	.18235	.42428	51
10	0.57596	0.81748	0.70455	1.4193	1.2233	1.7362	0.18252	0.42404	50
11	.57619	.81731	.70499	.4185	.2235	.7355	.18269	.42380	49
12	.57643	.81714	.70542	.4176	.2238	.7348	.18285	.42357	48
13	.57667	.81698	.70586	.4167	.2240	.7341	.18302	.42333	47
14	.57691	.81681	.70629	.4158	.2243	.7334	.18319	.42309	46
15	0.57714	0.81664	0.70673	1.4150	1.2245	1.7327	0.18336	0.42285	45
16	.57738	.81647	.70717	.4141	.2248	.7319	.18353	.42262	44
17	.57762	.81630	.70760	.4132	.2250	.7312	.18369	.42238	43
18	.57786	.81614	.70804	.4123	.2253	.7305	.18386	.42214	42
19	.57809	.81597	.70848	.4115	.2255	.7298	.18403	.42190	41
20	0.57833	0.81580	0.70891	1.4106	1.2258	1.7291	0.18420	0.42167	40
21	.57857	.81563	.70935	.4097	.2260	.7284	.18437	.42143	39
22	.57881	.81546	.70979	.4089	.2263	.7277	.18453	.42119	38
23	.57904	.81530	.71022	.4080	.2265	.7270	.18470	.42096	37
24	.57928	.81513	.71066	.4071	.2268	.7263	.18487	.42072	36
25	0.57952	0.81496	0.71110	1.4063	1.2270	1.7256	0.18504	0.42048	35
26	.57975	.81479	.71154	.4054	.2273	.7249	.18521	.42024	34
27	.57999	.81462	.71198	.4045	.2276	.7242	.18538	.42001	33
28	.58023	.81445	.71241	.4037	.2278	.7234	.18555	.41977	32
29	.58047	.81428	.71285	.4028	.2281	.7227	.18571	.41953	31
30	0.58070	0.81411	0.71329	1.4019	1.2283	1.7220	0.18588	0.41930	30
31	.58094	.81395	.71373	.4011	.2286	.7213	.18605	.41906	29
32	.58118	.81378	.71417	.4002	.2288	.7206	.18622	.41882	28
33	.58141	.81361	.71461	.3994	.2291	.7199	.18639	.41859	27
34	.58165	.81344	.71505	.3985	.2293	.7192	.18656	.41835	26
35	0.58189	0.81327	0.71549	1.3976	1.2296	1.7185	0.18673	0.41811	25
36	.58212	.81310	.71593	.3968	.2298	.7178	.18690	.41788	24
37	.58236	.81293	.71637	.3959	.2301	.7171	.18707	.41764	23
38	.58259	.81276	.71681	.3951	.2304	.7164	.18724	.41740	22
39	.58283	.81259	.71725	.3942	.2306	.7157	.18741	.41717	21
40	0.58307	0.81242	0.71769	1.3933	1.2309	1.7151	0.18758	0.41693	20
41	.58330	.81225	.71813	.3925	.2311	.7144	.18775	.41669	19
42	.58354	.81208	.71857	.3916	.2314	.7137	.18792	.41646	18
43	.58378	.81191	.71901	.3908	.2316	.7130	.18809	.41622	17
44	.58401	.81174	.71945	.3899	.2319	.7123	.18826	.41599	16
45	0.58425	0.81157	0.71990	1.3891	1.2322	1.7116	0.18843	0.41575	15
46	.58448	.81140	.72034	.3882	.2324	.7109	.18860	.41551	14
47	.58472	.81123	.72078	.3874	.2327	.7102	.18877	.41528	13
48	.58496	.81106	.72122	.3865	.2329	.7095	.18894	.41504	12
49	.58519	.81089	.72166	.3857	.2332	.7088	.18911	.41481	11
50	0.58543	0.81072	0.72211	1.3848	1.2335	1.7081	0.18928	0.41457	10
51	.58566	.81055	.72255	.3840	.2337	.7075	.18945	.41433	9
52	.58590	.81038	.72299	.3831	.2340	.7068	.18962	.41410	8
53	.58614	.81021	.72344	.3823	.2342	.7061	.18979	.41386	7
54	.58637	.81004	.72388	.3814	.2345	.7054	.18996	.41363	6
55	0.58661	0.80987	0.72432	1.3806	1.2348	1.7047	0.19013	0.41339	5
56	.58684	.80970	.72477	.3797	.2350	.7040	.19030	.41316	4
57	.58708	.80953	.72521	.3789	.2353	.7033	.19047	.41292	3
58	.58731	.80936	.72565	.3781	.2355	.7027	.19064	.41268	2
59	.58755	.80919	.72610	.3772	.2358	.7020	.19081	.41245	1
60	0.58778	0.80902	0.72654	1.3764	1.2361	1.7013	0.19098	0.41221	0
M	Cosine	Sine	Cotan.	Tan.	Cosec.	Secant	Vrs. Cos.	Vrs. Sin.	M

125° 54°

36°   Natural Trigonometric Functions   143°

M	Sine	Cosine	Tan.	Cotan.	Secant	Cosec.	Vrs. Sin.	Vrs. Cos	M
0	0.58778	0.80902	0.72654	1.3764	1.2361	1.7013	0.19098	0.41221	60
1	.58802	.80885	.72699	.3755	.2363	.7006	.19115	.41198	59
2	.58825	.80867	.72743	.3747	.2366	.6999	.19132	.41174	58
3	.58849	.80850	.72788	.3738	.2368	.6993	.19150	.41151	57
4	.58873	.80833	.72832	.3730	.2371	.6986	.19167	.41127	56
5	0.58896	0.80816	0.72877	1.3722	1.2374	1.6979	0.19184	0.41104	55
6	.58920	.80799	.72921	.3713	.2376	.6972	.19201	.41080	54
7	.58943	.80782	.72966	.3705	.2379	.6965	.19218	.41057	53
8	.58967	.80765	.73010	.3697	.2382	.6959	.19235	.41033	52
9	.58990	.80747	.73055	.3688	.2384	.6952	.19252	.41010	51
10	0.59014	0.80730	0.73100	1.3680	1.2387	1.6945	0.19270	0.40986	50
11	.59037	.80713	.73144	.3672	.2389	.6938	.19287	.40963	49
12	.59060	.80696	.73189	.3663	.2392	.6932	.19304	.40939	48
13	.59084	.80679	.73234	.3655	.2395	.6925	.19321	.40916	47
14	.59107	.80662	.73278	.3647	.2397	.6918	.19338	.40892	46
15	0.59131	0.80644	0.73323	1.3638	1.2400	1.6912	0.19355	0.40869	45
16	.59154	.80627	.73368	.3630	.2403	.6905	.19373	.40845	44
17	.59178	.80610	.73412	.3622	.2405	.6898	.19390	.40822	43
18	.59201	.80593	.73457	.3613	.2408	.6891	.19407	.40799	42
19	.59225	.80576	.73502	.3605	.2411	.6885	.19424	.40775	41
20	0.59248	0.80558	0.73547	1.3597	1.2413	1.6878	0.19442	0.40752	40
21	.59272	.80541	.73592	.3588	.2416	.6871	.19459	.40728	39
22	.59295	.80524	.73637	.3580	.2419	.6865	.19476	.40705	38
23	.59318	.80507	.73681	.3572	.2421	.6858	.19493	.40681	37
24	.59342	.80489	.73726	.3564	.2424	.6851	.19511	.40658	36
25	0.59365	0.80472	0.73771	1.3555	1.2427	1.6845	0.19528	0.40635	35
26	.59389	.80455	.73816	.3547	.2429	.6838	.19545	.40611	34
27	.59412	.80437	.73861	.3539	.2432	.6831	.19562	.40588	33
28	.59435	.80420	.73906	.3531	.2435	.6825	.19580	.40564	32
29	.59459	.80403	.73951	.3522	.2437	.6818	.19597	.40541	31
30	0.59482	0.80386	0.73996	1.3514	1.2440	1.6812	0.19614	0.40518	30
31	.59506	.80368	.74041	.3506	.2443	.6805	.19632	.40494	29
32	.59529	.80351	.74086	.3498	.2445	.6798	.19649	.40471	28
33	.59552	.80334	.74131	.3489	.2448	.6792	.19666	.40447	27
34	.59576	.80316	.74176	.3481	.2451	.6785	.19683	.40424	26
35	0.59599	0.80299	0.74221	1.3473	1.2453	1.6779	0.19701	0.40401	25
36	.59622	.80282	.74266	.3465	.2456	.6772	.19718	.40377	24
37	.59646	.80264	.74312	.3457	.2459	.6766	.19736	.40354	23
38	.59669	.80247	.74357	.3449	.2461	.6759	.19753	.40331	22
39	.59692	.80230	.74402	.3440	.2464	.6752	.19770	.40307	21
40	0.59716	0.80212	0.74447	1.3432	1.2467	1.6746	0.19788	0.40284	20
41	.59739	.80195	.74492	.3424	.2470	.6739	.19805	.40261	19
42	.59762	.80177	.74538	.3416	.2472	.6733	.19822	.40237	18
43	.59786	.80160	.74583	.3408	.2475	.6726	.19840	.40214	17
44	.59809	.80143	.74628	.3400	.2478	.6720	.19857	.40191	16
45	0.59832	0.80125	0.74673	1.3392	1.2480	1.6713	0.19875	0.40167	15
46	.59856	.80108	.74719	.3383	.2483	.6707	.19892	.40144	14
47	.59879	.80090	.74764	.3375	.2486	.6700	.19909	.40121	13
48	.59902	.80073	.74809	.3367	.2488	.6694	.19927	.40098	12
49	.59926	.80056	.74855	.3359	.2491	.6687	.19944	.40074	11
50	0.59949	0.80038	0.74900	1.3351	1.2494	1.6681	0.19962	0.40051	10
51	.59972	.80021	.74946	.3343	.2497	.6674	.19979	.40028	9
52	.59995	.80003	.74991	.3335	.2499	.6668	.19997	.40004	8
53	.60019	.79986	.75037	.3327	.2502	.6661	.20014	.39981	7
54	.60042	.79968	.75082	.3319	.2505	.6655	.20031	.39958	6
55	0.60065	0.79951	0.75128	1.3311	1.2508	1.6648	0.20049	0.39935	5
56	.60088	.79933	.75173	.3303	.2510	.6642	.20066	.39911	4
57	.60112	.79916	.75219	.3294	.2513	.6636	.20084	.39888	3
58	.60135	.79898	.75264	.3286	.2516	.6629	.20101	.39865	2
59	.60158	.79881	.75310	.3278	.2519	.6623	.20119	.39842	1
60	0.60181	0.79863	0.75355	1.3270	1.2521	1.6616	0.20136	0.39818	0
M	Cosine	Sine	Cotan.	Tan.	Cosec.	Secant	Vrs. Cos.	Vrs. Sin.	M

126°   53°

37°   Natural Trigonometric Functions   142°

M	Sine	Cosine	Tan.	Cotan.	Secant	Cosec.	Vrs. Sin.	Vrs. Cos.	M
0	0.60181	0.79863	0.75355	1.3270	1.2521	1.6616	0.20136	0.39818	60
1	.60205	.79846	.75401	.3262	.2524	.6610	.20154	.39795	59
2	.60228	.79828	.75447	.3254	.2527	.6603	.20171	.39772	58
3	.60251	.79811	.75492	.3246	.2530	.6597	.20189	.39749	57
4	.60274	.79793	.75538	.3238	.2532	.6591	.20206	.39726	56
5	0.60298	0.79776	0.75584	1.3230	1.2535	1.6584	0.20224	0.39702	55
6	.60320	.79758	.75629	.3222	.2538	.6578	.20242	.39679	54
7	.60344	.79741	.75675	.3214	.2541	.6572	.20259	.39656	53
8	.60367	.79723	.75721	.3206	.2543	.6565	.20277	.39633	52
9	.60390	.79706	.75767	.3198	.2546	.6559	.20294	.39610	51
10	0.60413	0.79688	0.75812	1.3190	1.2549	1.6552	0.20312	0.39586	50
11	.60437	.79670	.75858	.3182	.2552	.6546	.20329	.39563	49
12	.60460	.79653	.75904	.3174	.2554	.6540	.20347	.39540	48
13	.60483	.79635	.75950	.3166	.2557	.6533	.20365	.39517	47
14	.60506	.79618	.75996	.3159	.2560	.6527	.20382	.39494	46
15	0.60529	0.79600	0.76042	1.3151	1.2563	1.6521	0.20400	0.39471	45
16	.60552	.79582	.76088	.3143	.2565	.6514	.20417	.39447	44
17	.60576	.79565	.76134	.3135	.2568	.6508	.20435	.39424	43
18	.60599	.79547	.76179	.3127	.2571	.6502	.20453	.39401	42
19	.60622	.79530	.76225	.3119	.2574	.6496	.20470	.39378	41
20	0.60645	0.79512	0.76271	1.3111	1.2577	1.6489	0.20488	0.39355	40
21	.60668	.79494	.76317	.3103	.2579	.6483	.20505	.39332	39
22	.60691	.79477	.76364	.3095	.2582	.6477	.20523	.39309	38
23	.60714	.79459	.76410	.3087	.2585	.6470	.20541	.39285	37
24	.60737	.79441	.76456	.3079	.2588	.6464	.20558	.39262	36
25	0.60761	0.79424	0.76502	1.3071	1.2591	1.6458	0.20576	0.39239	35
26	.60784	.79406	.76548	.3064	.2593	.6452	.20594	.39216	34
27	.60807	.79388	.76594	.3056	.2596	.6445	.20611	.39193	33
28	.60830	.79371	.76640	.3048	.2599	.6439	.20629	.39170	32
29	.60853	.79353	.76686	.3040	.2602	.6433	.20647	.39147	31
30	0.60876	0.79335	0.76733	1.3032	1.2605	1.6427	0.20665	0.39124	30
31	.60899	.79318	.76779	.3024	.2607	.6420	.20682	.39101	29
32	.60922	.79300	.76825	.3016	.2610	.6414	.20700	.39078	28
33	.60945	.79282	.76871	.3009	.2613	.6408	.20718	.39055	27
34	.60968	.79264	.76918	.3001	.2616	.6402	.20735	.39031	26
35	0.60991	0.79247	0.76964	1.2993	1.2619	1.6396	0.20753	0.39008	25
36	.61014	.79229	.77010	.2985	.2622	.6389	.20771	.38985	24
37	.61037	.79211	.77057	.2977	.2624	.6383	.20789	.38962	23
38	.61061	.79193	.77103	.2970	.2627	.6377	.20806	.38939	22
39	.61084	.79176	.77149	.2962	.2630	.6371	.20824	.38916	21
40	0.61107	0.79158	0.77196	1.2954	1.2633	1.6365	0.20842	0.38893	20
41	.61130	.79140	.77242	.2946	.2636	.6359	.20860	.38870	19
42	.61153	.79122	.77289	.2938	.2639	.6352	.20878	.38847	18
43	.61176	.79104	.77335	.2931	.2641	.6346	.20895	.38824	17
44	.61199	.79087	.77382	.2923	.2644	.6340	.20913	.38801	16
45	0.61222	0.79069	0.77428	1.2915	1.2647	1.6334	0.20931	0.38778	15
46	.61245	.79051	.77475	.2907	.2650	.6328	.20949	.38755	14
47	.61268	.79033	.77521	.2900	.2653	.6322	.20967	.38732	13
48	.61290	.79015	.77568	.2892	.2656	.6316	.20984	.38709	12
49	.61314	.78998	.77614	.2884	.2659	.6309	.21002	.38686	11
50	0.61337	0.78980	0.77661	1.2876	1.2661	1.6303	0.21020	0.38663	10
51	.61360	.78962	.77708	.2869	.2664	.6297	.21038	.38640	9
52	.61383	.78944	.77754	.2861	.2667	.6291	.21056	.38617	8
53	.61405	.78926	.77801	.2853	.2670	.6285	.21074	.38594	7
54	.61428	.78908	.77848	.2845	.2673	.6279	.21091	.38571	6
55	0.61451	0.78890	0.77895	1.2838	1.2676	1.6273	0.21109	0.38548	5
56	.61474	.78873	.77941	.2830	.2679	.6267	.21127	.38525	4
57	.61497	.78855	.77988	.2822	.2681	.6261	.21145	.38503	3
58	.61520	.78837	.78035	.2815	.2684	.6255	.21163	.38480	2
59	.61543	.78819	.78082	.2807	.2687	.6249	.21181	.38457	1
60	0.61566	0.78801	0.78128	1.2799	1.2690	1.6243	0.21199	0.38434	0
M	Cosine	Sine	Cotan.	Tan.	Cosec.	Secant	Vrs. Cos.	Vrs. Sin.	M

127°   52°

38° Natural Trigonometric Functions 141°

M	Sine	Cosine	Tan.	Cotan.	Secant	Cosec.	Vrs. Sin.	Vrs. Cos.	M
0	0.61566	0.78801	0.78128	1.2799	1.2690	1.6243	0.21199	0.38434	60
1	.61589	.78783	.78175	.2792	.2693	.6237	.21217	.38411	59
2	.61612	.78765	.78222	.2784	.2696	.6231	.21235	.38388	58
3	.61635	.78747	.78269	.2776	.2699	.6224	.21253	.38365	57
4	.61658	.78729	.78316	.2769	.2702	.6218	.21271	.38342	56
5	0.61681	0.78711	0.78363	1.2761	1.2705	1.6212	0.21288	0.38319	55
6	.61703	.78693	.78410	.2753	.2707	.6206	.21306	.38296	54
7	.61726	.78675	.78457	.2746	.2710	.6200	.21324	.38273	53
8	.61749	.78657	.78504	.2738	.2713	.6194	.21342	.38251	52
9	.61772	.78640	.78551	.2730	.2716	.6188	.21360	.38228	51
10	0.61795	0.78622	0.78598	1.2723	1.2719	1.6182	0.21378	0.38205	50
11	.61818	.78604	.78645	.2715	.2722	.6176	.21396	.38182	49
12	.61841	.78586	.78692	.2708	.2725	.6170	.21414	.38159	48
13	.61864	.78568	.78739	.2700	.2728	.6164	.21432	.38136	47
14	.61886	.78550	.78786	.2692	.2731	.6159	.21450	.38113	46
15	0.61909	0.78532	0.78834	1.2685	1.2734	1.6153	0.21468	0.38091	45
16	.61932	.78514	.78881	.2677	.2737	.6147	.21486	.38068	44
17	.61955	.78496	.78928	.2670	.2739	.6141	.21504	.38045	43
18	.61978	.78478	.78975	.2662	.2742	.6135	.21522	.38022	42
19	.62001	.78460	.79022	.2655	.2745	.6129	.21540	.37999	41
20	0.62023	0.78441	0.79070	1.2647	1.2748	1.6123	0.21558	0.37976	40
21	.62046	.78423	.79117	.2639	.2751	.6117	.21576	.37954	39
22	.62069	.78405	.79164	.2632	.2754	.6111	.21594	.37931	38
23	.62092	.78387	.79212	.2624	.2757	.6105	.21612	.37908	37
24	.62115	.78369	.79259	.2617	.2760	.6099	.21631	.37885	36
25	0.62137	0.78351	0.79306	1.2609	1.2763	1.6093	0.21649	0.37862	35
26	.62160	.78333	.79354	.2602	.2766	.6087	.21667	.37840	34
27	.62183	.78315	.79401	.2594	.2769	.6081	.21685	.37817	33
28	.62206	.78297	.79449	.2587	.2772	.6077	.21703	.37794	32
29	.62229	.78279	.79496	.2579	.2775	.6070	.21721	.37771	31
30	0.62251	0.78261	0.79543	1.2572	1.2778	1.6064	0.21739	0.37748	30
31	.62274	.78243	.79591	.2564	.2781	.6058	.21757	.37726	29
32	.62297	.78224	.79639	.2557	.2784	.6052	.21775	.37703	28
33	.62320	.78206	.79686	.2549	.2787	.6046	.21793	.37680	27
34	.62342	.78188	.79734	.2542	.2790	.6040	.21812	.37657	26
35	0.62365	0.78170	0.79781	1.2534	1.2793	1.6034	0.21830	0.37635	25
36	.62388	.78152	.79829	.2527	.2795	.6029	.21848	.37612	24
37	.62411	.78134	.79876	.2519	.2798	.6023	.21866	.37589	23
38	.62433	.78116	.79924	.2512	.2801	.6017	.21884	.37566	22
39	.62456	.78097	.79972	.2504	.2804	.6011	.21902	.37544	21
40	0.62479	0.78079	0.80020	1.2497	1.2807	1.6005	0.21921	0.37521	20
41	.62501	.78061	.80067	.2489	.2810	.6000	.21939	.37498	19
42	.62524	.78043	.80115	.2482	.2813	.5994	.21957	.37476	18
43	.62547	.78025	.80163	.2475	.2816	.5988	.21975	.37453	17
44	.62570	.78007	.80211	.2467	.2819	.5982	.21993	.37430	16
45	0.62592	0.77988	0.80258	1.2460	1.2822	1.5976	0.22011	0.37408	15
46	.62615	.77970	.80306	.2452	.2825	.5971	.22030	.37385	14
47	.62638	.77952	.80354	.2445	.2828	.5965	.22048	.37362	13
48	.62660	.77934	.80402	.2437	.2831	.5959	.22066	.37340	12
49	.62683	.77915	.80450	.2430	.2834	.5953	.22084	.37317	11
50	0.62706	0.77897	0.80498	1.2423	1.2837	1.5947	0.22103	0.37294	10
51	.62728	.77879	.80546	.2415	.2840	.5942	.22121	.37272	9
52	.62751	.77861	.80594	.2408	.2843	.5936	.22139	.37249	8
53	.62774	.77842	.80642	.2400	.2846	.5930	.22157	.37226	7
54	.62796	.77824	.80690	.2393	.2849	.5924	.22176	.37204	6
55	0.62819	0.77806	0.80738	1.2386	1.2852	1.5919	0.22194	0.37181	5
56	.62841	.77788	.80786	.2378	.2855	.5913	.22212	.37158	4
57	.62864	.77769	.80834	.2371	.2858	.5907	.22230	.37136	3
58	.62887	.77751	.80882	.2364	.2861	.5901	.22249	.37113	2
59	.62909	.77733	.80930	.2356	.2864	.5896	.22267	.37090	1
60	0.62932	0.77715	0.80978	1.2349	1.2867	1.5890	0.22285	0.37068	0
M	Cosine	Sine	Cotan.	Tan.	Cosec.	Secant	Vrs. Cos.	Vrs. Sin.	M

128°   51°

39° Natural Trigonometric Functions 140°

M	Sine	Cosine	Tan.	Cotan.	Secant	Cosec.	Vrs. Sin.	Vrs. Cos.	M
0	0.62932	0.77715	0.80978	1.2349	1.2867	1.5890	0.22285	0.37068	60
1	.62955	.77696	.81026	.2342	.2871	.5884	.22304	.37045	59
2	.62977	.77678	.81075	.2334	.2874	.5879	.22322	.37023	58
3	.63000	.77660	.81123	.2327	.2877	.5873	.22340	.37000	57
4	.63022	.77641	.81171	.2320	.2880	.5867	.22359	.36977	56
5	0.63045	0.77623	0.81219	1.2312	1.2883	1.5862	0.22377	0.36955	55
6	.63067	.77605	.81268	.2305	.2886	.5856	.22395	.36932	54
7	.63090	.77586	.81316	.2297	.2889	.5850	.22414	.36910	53
8	.63113	.77568	.81364	.2290	.2892	.5845	.22432	.36887	52
9	.63135	.77549	.81413	.2283	.2895	.5839	.22450	.36865	51
10	0.63158	0.77531	0.81461	1.2276	1.2898	1.5833	0.22469	0.36842	50
11	.63180	.77513	.81509	.2268	.2901	.5828	.22487	.36820	49
12	.63203	.77494	.81558	.2261	.2904	.5822	.22505	.36797	48
13	.63225	.77476	.81606	.2254	.2907	.5816	.22524	.36774	47
14	.63248	.77458	.81655	.2247	.2910	.5811	.22542	.36752	46
15	0.63270	0.77439	0.81703	1.2239	1.2913	1.5805	0.22561	0.36729	45
16	.63293	.77421	.81752	.2232	.2916	.5799	.22579	.36707	44
17	.63315	.77402	.81800	.2225	.2919	.5794	.22597	.36684	43
18	.63338	.77384	.81849	.2218	.2922	.5788	.22616	.36662	42
19	.63360	.77365	.81898	.2210	.2926	.5783	.22634	.36639	41
20	0.63383	0.77347	0.81946	1.2203	1.2929	1.5777	0.22653	0.36617	40
21	.63405	.77329	.81995	.2196	.2932	.5771	.22671	.36594	39
22	.63428	.77310	.82043	.2189	.2935	.5766	.22690	.36572	38
23	.63450	.77292	.82092	.2181	.2938	.5760	.22708	.36549	37
24	.63473	.77273	.82141	.2174	.2941	.5755	.22727	.36527	36
25	0.63495	0.77255	0.82190	1.2167	1.2944	1.5749	0.22745	0.36504	35
26	.63518	.77236	.82238	.2160	.2947	.5743	.22763	.36482	34
27	.63540	.77218	.82287	.2152	.2950	.5738	.22782	.36459	33
28	.63563	.77199	.82336	.2145	.2953	.5732	.22800	.36437	32
29	.63585	.77181	.82385	.2138	.2956	.5727	.22819	.36415	31
30	0.63608	0.77162	0.82434	1.2131	1.2960	1.5721	0.22837	0.36392	30
31	.63630	.77144	.82482	.2124	.2963	.5716	.22856	.36370	29
32	.63653	.77125	.82531	.2117	.2966	.5710	.22874	.36347	28
33	.63675	.77107	.82580	.2109	.2969	.5705	.22893	.36325	27
34	.63697	.77088	.82629	.2102	.2972	.5699	.22912	.36302	26
35	0.63720	0.77070	0.82678	1.2095	1.2975	1.5694	0.22930	0.36280	25
36	.63742	.77051	.82727	.2088	.2978	.5688	.22949	.36258	24
37	.63765	.77033	.82776	.2081	.2981	.5683	.22967	.36235	23
38	.63787	.77014	.82825	.2074	.2985	.5677	.22986	.36213	22
39	.63810	.76996	.82874	.2066	.2988	.5672	.23004	.36190	21
40	0.63832	0.76977	0.82923	1.2059	1.2991	1.5666	0.23023	0.36168	20
41	.63854	.76953	.82972	.2052	.2994	.5661	.23041	.36146	19
42	.63877	.76940	.83022	.2045	.2997	.5655	.23060	.36123	18
43	.63899	.76921	.83071	.2038	.3000	.5650	.23079	.36101	17
44	.63921	.76903	.83120	.2031	.3003	.5644	.23097	.36078	16
45	0.63944	0.76884	0.83169	1.2024	1.3006	1.5639	0.23116	0.36056	15
46	.63966	.76865	.83218	.2016	.3010	.5633	.23134	.36034	14
47	.63989	.76847	.83267	.2009	.3013	.5628	.23153	.36011	13
48	.64011	.76828	.83317	.2002	.3016	.5622	.23172	.35989	12
49	.64033	.76810	.83366	.1995	.3019	.5617	.23190	.35967	11
50	0.64056	0.76791	0.83415	1.1988	1.3022	1.5611	0.23209	0.35944	10
51	.64078	.76772	.83465	.1981	.3025	.5606	.23227	.35922	9
52	.64100	.76754	.83514	.1974	.3029	.5600	.23246	.35900	8
53	.64123	.76735	.83563	.1967	.3032	.5595	.23265	.35877	7
54	.64145	.76716	.83613	.1961	.3035	.5590	.23283	.35855	6
55	0.64167	0.76698	0.83662	1.1953	1.3038	1.5584	0.23302	0.35833	5
56	.64189	.76679	.83712	.1946	.3041	.5579	.23321	.35810	4
57	.64212	.76660	.83761	.1939	.3044	.5573	.23339	.35788	3
58	.64234	.76642	.83811	.1932	.3048	.5568	.23358	.35766	2
59	.64256	.76623	.83860	.1924	.3051	.5563	.23377	.35743	1
60	0.64279	0.76604	0.83910	1.1917	1.3054	1.5557	0.23395	0.35721	0
M	Cosine	Sine	Cotan.	Tan.	Cosec.	Secant	Vrs. Cos.	Vrs. Sin.	M

129° 50°

40° Natural Trigonometric Functions 139°

M	Sine	Cosine	Tan.	Cotan.	Secant	Cosec.	Vrs. Sin.	Vrs. Cos.	M
0	0.64279	0.76604	0.83910	1.1917	1.3054	1.5557	0.23395	0.35721	60
1	.64301	.76586	.83959	.1910	.3057	.5552	.23414	.35699	59
2	.64323	.76567	.84009	.1903	.3060	.5546	.23433	.35677	58
3	.64345	.76548	.84059	.1896	.3064	.5541	.23452	.35654	57
4	.64368	.76530	.84108	.1889	.3067	.5536	.23470	.35632	56
5	0.64390	0.76511	0.84158	1.1882	1.3070	1.5530	0.23489	0.35610	55
6	.64412	.76492	.84208	.1875	.3073	.5525	.23508	.35588	54
7	.64435	.76473	.84257	.1868	.3076	.5520	.23527	.35565	53
8	.64457	.76455	.84307	.1861	.3080	.5514	.23545	.35543	52
9	.64479	.76436	.84357	.1854	.3083	.5509	.23564	.35521	51
10	0.64501	0.76417	0.84407	1.1847	1.3086	1.5503	0.23583	0.35499	50
11	.64523	.76398	.84457	.1840	.3089	.5498	.23602	.35476	49
12	.64546	.76380	.84506	.1833	.3092	.5493	.23620	.35454	48
13	.64568	.76361	.84556	.1826	.3096	.5487	.23639	.35432	47
14	.64590	.76342	.84606	.1819	.3099	.5482	.23658	.35410	46
15	0.64612	0.76323	0.84656	1.1812	1.3102	1.5477	0.23677	0.35388	45
16	.64635	.76304	.84706	.1805	.3105	.5471	.23695	.35365	44
17	.64657	.76286	.84756	.1798	.3109	.5466	.23714	.35343	43
18	.64679	.76267	.84806	.1791	.3112	.5461	.23733	.35321	42
19	.64701	.76248	.84856	.1785	.3115	.5456	.23752	.35299	41
20	0.64723	0.76229	0.84906	1.1778	1.3118	1.5450	0.23771	0.35277	40
21	.64745	.76210	.84956	.1771	.3121	.5445	.23790	.35254	39
22	.64768	.76191	.85006	.1764	.3125	.5440	.23808	.35232	38
23	.64790	.76173	.85056	.1757	.3128	.5434	.23827	.35210	37
24	.64812	.76154	.85107	.1750	.3131	.5429	.23846	.35188	36
25	0.64834	0.76135	0.85157	1.1743	1.3134	1.5424	0.23865	0.35166	35
26	.64856	.76116	.85207	.1736	.3138	.5419	.23884	.35144	34
27	.64878	.76097	.85257	.1729	.3141	.5413	.23903	.35121	33
28	.64900	.76078	.85307	.1722	.3144	.5408	.23922	.35099	32
29	.64923	.76059	.85358	.1715	.3148	.5403	.23940	.35077	31
30	0.64945	0.76041	0.85408	1.1708	1.3151	1.5398	0.23959	0.35055	30
31	.64967	.76022	.85458	.1702	.3154	.5392	.23978	.35033	29
32	.64989	.76003	.85509	.1695	.3157	.5387	.23997	.35011	28
33	.65011	.75984	.85559	.1688	.3161	.5382	.24016	.34989	27
34	.65033	.75965	.85609	.1681	.3164	.5377	.24035	.34967	26
35	0.65055	0.75946	0.85660	1.1674	1.3167	1.5371	0.24054	0.34945	25
36	.65077	.75927	.85710	.1667	.3170	.5366	.24073	.34922	24
37	.65100	.75908	.85761	.1660	.3174	.5361	.24092	.34900	23
38	.65121	.75889	.85811	.1653	.3177	.5356	.24111	.34878	22
39	.65144	.75870	.85862	.1647	.3180	.5351	.24130	.34856	21
40	0.65166	0.75851	0.85912	1.1640	1.3184	1.5345	0.24149	0.34834	20
41	.65188	.75832	.85963	.1633	.3187	.5340	.24168	.34812	19
42	.65210	.75813	.86013	.1626	.3190	.5335	.24186	.34790	18
43	.65232	.75794	.86064	.1619	.3193	.5330	.24205	.34768	17
44	.65254	.75775	.86115	.1612	.3197	.5325	.24224	.34746	16
45	0.65276	0.75756	0.86165	1.1605	1.3200	1.5319	0.24243	0.34724	15
46	.65298	.75737	.86216	.1599	.3203	.5314	.24262	.34702	14
47	.65320	.75718	.86267	.1592	.3207	.5309	.24281	.34680	13
48	.65342	.75700	.86318	.1585	.3210	.5304	.24300	.34658	12
49	.65364	.75680	.86368	.1578	.3213	.5299	.24319	.34636	11
50	0.65386	0.75661	0.86419	1.1571	1.3217	1.5294	0.24338	0.34614	10
51	.65408	.75642	.86470	.1565	.3220	.5289	.24357	.34592	9
52	.65430	.75623	.86521	.1558	.3223	.5283	.24376	.34570	8
53	.65452	.75604	.86572	.1551	.3227	.5278	.24396	.34548	7
54	.65474	.75585	.86623	.1544	.3230	.5273	.24415	.34526	6
55	0.65496	0.75566	0.86674	1.1537	1.3233	1.5268	0.24434	0.34504	5
56	.65518	.75547	.86725	.1531	.3237	.5263	.24453	.34482	4
57	.65540	.75528	.86775	.1524	.3240	.5258	.24472	.34460	3
58	.65562	.75509	.86826	.1517	.3243	.5253	.24491	.34438	2
59	.65584	.75490	.86878	.1510	.3247	.5248	.24510	.34416	1
60	0.65606	0.75471	0.86929	1.1504	1.3250	1.5242	0.24529	0.34394	0
M	Cosine	Sine	Cotan.	Tan.	Cosec.	Secant	Vrs. Cos.	Vrs. Sin.	M

130°   49°

41°   Natural Trigonometric Functions   138°

M	Sine	Cosine	Tan.	Cotan.	Secant	Cosec.	Vrs. Sin.	Vrs. Cos.	M
0	0.65606	0.75471	0.86929	1.1504	1.3250	1.5242	0.24529	0.34394	60
1	.65628	.75452	.86980	.1497	.3253	.5237	.24548	.34372	59
2	.65650	.75433	.87031	.1490	.3257	.5232	.24567	.34350	58
3	.65672	.75414	.87082	.1483	.3260	.5227	.24586	.34328	57
4	.65694	.75394	.87133	.1477	.3263	.5222	.24605	.34306	56
5	0.65716	0.75375	0.87184	1.1470	1.3267	1.5217	0.24624	0.34284	55
6	.65737	.75356	.87235	.1463	.3270	.5212	.24644	.34262	54
7	.65759	.75337	.87287	.1456	.3274	.5207	.24663	.34240	53
8	.65781	.75318	.87338	.1450	.3277	.5202	.24682	.34219	52
9	.65803	.75299	.87389	.1443	.3280	.5197	.24701	.34197	51
10	0.65825	0.75280	0.87441	1.1436	1.3284	1.5192	0.24720	0.34175	50
11	.65847	.75261	.87492	.1430	.3287	.5187	.24739	.34153	49
12	.65869	.75241	.87543	.1423	.3290	.5182	.24758	.34131	48
13	.65891	.75222	.87595	.1416	.3294	.5177	.24778	.34109	47
14	.65913	.75203	.87646	.1409	.3297	.5171	.24797	.34087	46
15	0.65934	0.75184	0.87698	1.1403	1.3301	1.5166	0.24816	0.34065	45
16	.65956	.75165	.87749	.1396	.3304	.5161	.24835	.34043	44
17	.65978	.75146	.87801	.1389	.3307	.5156	.24854	.34022	43
18	.66000	.75126	.87852	.1383	.3311	.5151	.24873	.34000	42
19	.66022	.75107	.87904	.1376	.3314	.5146	.24893	.33978	41
20	0.66044	0.75088	0.87955	1.1369	1.3318	1.5141	0.24912	0.33956	40
21	.66066	.75069	.88007	.1363	.3321	.5136	.24931	.33934	39
22	.66087	.75049	.88058	.1356	.3324	.5131	.24950	.33912	38
23	.66109	.75030	.88110	.1349	.3328	.5126	.24970	.33891	37
24	.66131	.75011	.88162	.1343	.3331	.5121	.24989	.33869	36
25	0.66153	0.74992	0.88213	1.1336	1.3335	1.5116	0.25008	0.33847	35
26	.66175	.74973	.88265	.1329	.3338	.5111	.25027	.33825	34
27	.66197	.74953	.88317	.1323	.3342	.5106	.25047	.33803	33
28	.66218	.74934	.88369	.1316	.3345	.5101	.25066	.33781	32
29	.66240	.74915	.88421	.1309	.3348	.5096	.25085	.33760	31
30	0.66262	0.74895	0.88472	1.1303	1.3352	1.5092	0.25104	0.33738	30
31	.66284	.74876	.88524	.1296	.3355	.5087	.25124	.33716	29
32	.66305	.74857	.88576	.1290	.3359	.5082	.25143	.33694	28
33	.66327	.74838	.88628	.1283	.3362	.5077	.25162	.33673	27
34	.66349	.74818	.88680	.1276	.3366	.5072	.25181	.33651	26
35	0.66371	0.74799	0.88732	1.1270	1.3369	1.5067	0.25201	0.33629	25
36	.66393	.74780	.88784	.1263	.3372	.5062	.25220	.33607	24
37	.66414	.74760	.88836	.1257	.3376	.5057	.25239	.33586	23
38	.66436	.74741	.88888	.1250	.3379	.5052	.25259	.33564	22
39	.66458	.74722	.88940	.1243	.3383	.5047	.25278	.33542	21
40	0.66479	0.74702	0.88992	1.1237	1.3386	1.5042	0.25297	0.33520	20
41	.66501	.74683	.89044	.1230	.3390	.5037	.25317	.33499	19
42	.66523	.74664	.89097	.1224	.3393	.5032	.25336	.33477	18
43	.66545	.74644	.89149	.1217	.3397	.5027	.25355	.33455	17
44	.66566	.74625	.89201	.1211	.3400	.5022	.25375	.33433	16
45	0.66588	0.74606	0.89253	1.1204	1.3404	1.5018	0.25394	0.33412	15
46	.66610	.74586	.89306	.1197	.3407	.5013	.25414	.33390	14
47	.66631	.74567	.89358	.1191	.3411	.5008	.25433	.33368	13
48	.66653	.74548	.89410	.1184	.3414	.5003	.25452	.33347	12
49	.66675	.74528	.89463	.1178	.3418	.4998	.25472	.33325	11
50	0.66697	0.74509	0.89515	1.1171	1.3421	1.4993	0.25491	0.33303	10
51	.66718	.74489	.89567	.1165	.3425	.4988	.25510	.33282	9
52	.66740	.74470	.89620	.1158	.3428	.4983	.25530	.33260	8
53	.66762	.74450	.89672	.1152	.3432	.4979	.25549	.33238	7
54	.66783	.74431	.89725	.1145	.3435	.4974	.25569	.33217	6
55	0.66805	0.74412	0.89777	1.1139	1.3439	1.4969	0.25588	0.33195	5
56	.66826	.74392	.89830	.1132	.3442	.4964	.25608	.33173	4
57	.66848	.74373	.89882	.1126	.3446	.4959	.25627	.33152	3
58	.66870	.74353	.89935	.1119	.3449	.4954	.25647	.33130	2
59	.66891	.74334	.89988	.1113	.3453	.4949	.25666	.33108	1
60	0.66913	0.74314	0.90040	1.1106	1.3456	1.4945	0.25685	0.33087	0
M	Cosine	Sine	Cotan.	Tan.	Cosec.	Secant	Vrs. Cos.	Vrs. Sin.	M

131°   48°

42° Natural Trigonometric Functions 137°

M	Sine	Cosine	Tan.	Cotan.	Secant	Cosec.	Vrs. Sin.	Vrs. Cos.	M
0	0.66913	0.74314	0.90040	1.1106	1.3456	1.4945	0.25685	0.33087	60
1	.66935	.74295	.90093	.1100	.3460	.4940	.25705	.33065	59
2	.66956	.74275	.90146	.1093	.3463	.4935	.25724	.33044	58
3	.66978	.74256	.90198	.1086	.3467	.4930	.25744	.33022	57
4	.66999	.74236	.90251	.1080	.3470	.4925	.25763	.33000	56
5	0.67021	0.74217	0.90304	1.1074	1.3474	1.4921	0.25783	0.32979	55
6	.67043	.74197	.90357	.1067	.3477	.4916	.25802	.32957	54
7	.67064	.74178	.90410	.1061	.3481	.4911	.25822	.32936	53
8	.67086	.74158	.90463	.1054	.3485	.4906	.25841	.32914	52
9	.67107	.74139	.90515	.1048	.3488	.4901	.25861	.32893	51
10	0.67129	0.74119	0.90568	1.1041	1.3492	1.4897	0.25880	0.32871	50
11	.67150	.74100	.90621	.1035	.3495	.4892	.25900	.32849	49
12	.67172	.74080	.90674	.1028	.3499	.4887	.25919	.32828	48
13	.67194	.74061	.90727	.1022	.3502	.4882	.25939	.32806	47
14	.67215	.74041	.90780	.1015	.3506	.4877	.25959	.32785	46
15	0.67237	0.74022	0.90834	1.1009	1.3509	1.4873	0.25978	0.32763	45
16	.67258	.74002	.90887	.1003	.3513	.4868	.25998	.32742	44
17	.67280	.73983	.90940	.0996	.3517	.4863	.26017	.32720	43
18	.67301	.73963	.90993	.0990	.3520	.4858	.26037	.32699	42
19	.67323	.73943	.91046	.0983	.3524	.4854	.26056	.32677	41
20	0.67344	0.73924	0.91099	1.0977	1.3527	1.4849	0.26076	0.32656	40
21	.67366	.73904	.91153	.0971	.3531	.4844	.26096	.32634	39
22	.67387	.73885	.91206	.0964	.3534	.4839	.26115	.32613	38
23	.67409	.73865	.91259	.0958	.3538	.4835	.26135	.32591	37
24	.67430	.73845	.91312	.0951	.3542	.4830	.26154	.32570	36
25	0.67452	0.73826	0.91366	1.0945	1.3545	1.4825	0.26174	0.32548	35
26	.67473	.73806	.91419	.0939	.3549	.4821	.26194	.32527	34
27	.67495	.73787	.91473	.0932	.3552	.4816	.26213	.32505	33
28	.67516	.73767	.91526	.0926	.3556	.4811	.26233	.32484	32
29	.67537	.73747	.91580	.0919	.3560	.4806	.26253	.32462	31
30	0.67559	0.73728	0.91633	1.0913	1.3563	1.4802	0.26272	0.32441	30
31	.67580	.73708	.91687	.0907	.3567	.4797	.26292	.32419	29
32	.67602	.73688	.91740	.0900	.3571	.4792	.26311	.32398	28
33	.67623	.73669	.91794	.0894	.3574	.4788	.26331	.32377	27
34	.67645	.73649	.91847	.0888	.3578	.4783	.26351	.32355	26
35	0.67666	0.73629	0.91901	1.0881	1.3581	1.4778	0.26371	0.32334	25
36	.67688	.73610	.91955	.0875	.3585	.4774	.26390	.32312	24
37	.67709	.73590	.92008	.0868	.3589	.4769	.26410	.32291	23
38	.67730	.73570	.92062	.0862	.3592	.4764	.26430	.32269	22
39	.67752	.73551	.92116	.0856	.3596	.4760	.26449	.32248	21
40	0.67773	0.73531	0.92170	1.0849	1.3600	1.4755	0.26469	0.32227	20
41	.67794	.73511	.92223	.0843	.3603	.4750	.26489	.32205	19
42	.67816	.73491	.92277	.0837	.3607	.4746	.26508	.32184	18
43	.67837	.73472	.92331	.0830	.3611	.4741	.26528	.32163	17
44	.67859	.73452	.92385	.0824	.3614	.4736	.26548	.32141	16
45	0.67880	0.73432	0.92439	1.0818	1.3618	1.4732	0.26568	0.32120	15
46	.67901	.73412	.92493	.0812	.3622	.4727	.26587	.32098	14
47	.67923	.73393	.92547	.0805	.3625	.4723	.26607	.32077	13
48	.67944	.73373	.92601	.0799	.3629	.4718	.26627	.32056	12
49	.67965	.73353	.92655	.0793	.3633	.4713	.26647	.32034	11
50	0.67987	0.73333	0.92709	1.0786	1.3636	1.4709	0.26666	0.32013	10
51	.68008	.73314	.92763	.0780	.3640	.4704	.26686	.31992	9
52	.68029	.73294	.92817	.0774	.3644	.4699	.26706	.31970	8
53	.68051	.73274	.92871	.0767	.3647	.4695	.26726	.31949	7
54	.68072	.73254	.92926	.0761	.3651	.4690	.26746	.31928	6
55	0.68093	0.73234	0.92980	1.0755	1.3655	1.4686	0.26765	0.31907	5
56	.68115	.73215	.93034	.0749	.3658	.4681	.26785	.31885	4
57	.68136	.73195	.93088	.0742	.3662	.4676	.26805	.31864	3
58	.68157	.73175	.93143	.0736	.3666	.4672	.26825	.31843	2
59	.68178	.73155	.93197	.0730	.3669	.4667	.26845	.31821	1
60	0.68200	0.73135	0.93251	1.0724	1.3673	1.4663	0.26865	0.31800	0
M	Cosine	Sine	Cotan.	Tan.	Cosec.	Secant	Vrs. Cos.	Vrs. Sin.	M

132° 47°

43° Natural Trigonometric Functions 136°

M	Sine	Cosine	Tan.	Cotan.	Secant	Cosec.	Vrs. Sin.	Vrs. Cos.	M
0	0.68200	0.73135	0.93251	1.0724	1.3673	1.4663	0.26865	0.31800	60
1	.68221	.73115	.93306	.0717	.3677	.4658	.26884	.31779	59
2	.68242	.73096	.93360	.0711	.3681	.4654	.26904	.31758	58
3	.68264	.73076	.93415	.0705	.3684	.4649	.26924	.31736	57
4	.68285	.73056	.93469	.0699	.3688	.4644	.26944	.31715	56
5	0.68306	0.73036	0.93524	1.0692	1.3692	1.4640	0.26964	0.31694	55
6	.68327	.73016	.93578	.0686	.3695	.4635	.26984	.31673	54
7	.68349	.72996	.93633	.0680	.3699	.4631	.27004	.31651	53
8	.68370	.72976	.93687	.0674	.3703	.4626	.27023	.31630	52
9	.68391	.72956	.93742	.0667	.3707	.4622	.27043	.31609	51
10	0.68412	0.72937	0.93797	1.0661	1.3710	1.4617	0.27063	0.31588	50
11	.68433	.72917	.93851	.0655	.3714	.4613	.27083	.31566	49
12	.68455	.72897	.93906	.0649	.3718	.4608	.27103	.31545	48
13	.68476	.72877	.93961	.0643	.3722	.4604	.27123	.31524	47
14	.68497	.72857	.94016	.0636	.3725	.4599	.27143	.31503	46
15	0.68518	0.72837	0.94071	1.0630	1.3729	1.4595	0.27163	0.31482	45
16	.68539	.72817	.94125	.0624	.3733	.4590	.27183	.31460	44
17	.68561	.72797	.94180	.0618	.3737	.4586	.27203	.31439	43
18	.68582	.72777	.94235	.0612	.3740	.4581	.27223	.31418	42
19	.68603	.72757	.94290	.0605	.3744	.4577	.27243	.31397	41
20	0.68624	0.72737	0.94345	1.0599	1.3748	1.4572	0.27263	0.31376	40
21	.68645	.72717	.94400	.0593	.3752	.4568	.27283	.31355	39
22	.68666	.72697	.94455	.0587	.3756	.4563	.27302	.31333	38
23	.68688	.72677	.94510	.0581	.3759	.4559	.27322	.31312	37
24	.68709	.72657	.94565	.0575	.3763	.4554	.27342	.31291	36
25	0.68730	0.72637	0.94620	1.0568	1.3767	1.4550	0.27362	0.31270	35
26	.68751	.72617	.94675	.0562	.3771	.4545	.27382	.31249	34
27	.68772	.72597	.94731	.0556	.3774	.4541	.27402	.31228	33
28	.68793	.72577	.94786	.0550	.3778	.4536	.27422	.31207	32
29	.68814	.72557	.94841	.0544	.3782	.4532	.27442	.31186	31
30	0.68835	0.72537	0.94896	1.0538	1.3786	1.4527	0.27462	0.31164	30
31	.68856	.72517	.94952	.0532	.3790	.4523	.27482	.31143	29
32	.68878	.72497	.95007	.0525	.3794	.4518	.27503	.31122	28
33	.68899	.72477	.95062	.0519	.3797	.4514	.27523	.31101	27
34	.68920	.72457	.95118	.0513	.3801	.4510	.27543	.31080	26
35	0.68941	0.72437	0.95173	1.0507	1.3805	1.4505	0.27563	0.31059	25
36	.68962	.72417	.95229	.0501	.3809	.4501	.27583	.31038	24
37	.68983	.72397	.95284	.0495	.3813	.4496	.27603	.31017	23
38	.69004	.72377	.95340	.0489	.3816	.4492	.27623	.30996	22
39	.69025	.72357	.95395	.0483	.3820	.4487	.27643	.30975	21
40	0.69046	0.72337	0.95451	1.0476	1.3824	1.4483	0.27663	0.30954	20
41	.69067	.72317	.95506	.0470	.3828	.4479	.27683	.30933	19
42	.69088	.72297	.95562	.0464	.3832	.4474	.27703	.30912	18
43	.69109	.72277	.95618	.0458	.3836	.4470	.27723	.30891	17
44	.69130	.72256	.95673	.0452	.3839	.4465	.27743	.30870	16
45	0.69151	0.72236	0.95729	1.0446	1.3843	1.4461	0.27764	0.30849	15
46	.69172	.72216	.95785	.0440	.3847	.4457	.27784	.30828	14
47	.69193	.72196	.95841	.0434	.3851	.4452	.27804	.30807	13
48	.69214	.72176	.95896	.0428	.3855	.4448	.27824	.30786	12
49	.69235	.72156	.95952	.0422	.3859	.4443	.27844	.30765	11
50	0.69256	0.72136	0.96008	1.0416	1.3863	1.4439	0.27864	0.30744	10
51	.69277	.72115	.96064	.0410	.3867	.4435	.27884	.30723	9
52	.69298	.72095	.96120	.0404	.3870	.4430	.27904	.30702	8
53	.69319	.72075	.96176	.0397	.3874	.4426	.27925	.30681	7
54	.69340	.72055	.96232	.0391	.3878	.4422	.27945	.30660	6
55	0.69361	0.72035	0.96288	1.0385	1.3882	1.4417	0.27965	0.30639	5
56	.69382	.72015	.96344	.0379	.3886	.4413	.27985	.30618	4
57	.69403	.71994	.96400	.0373	.3890	.4408	.28005	.30597	3
58	.69424	.71974	.96456	.0367	.3894	.4404	.28026	.30576	2
59	.69445	.71954	.96513	.0361	.3898	.4400	.28046	.30555	1
60	0.69466	0.71934	0.96569	1.0355	1.3902	1.4395	0.28066	0.30534	0
M	Cosine	Sine	Cotan.	Tan.	Cosec.	Secant	Vrs. Cos.	Vrs. Sin.	M

44° Natural Trigonometric Functions 135°

M	Sine	Cosine	Tan.	Cotan.	Secant	Cosec.	Vrs. Sin.	Vrs. Cos.	M
0	0.69466	0.71934	0.96569	1.0355	1.3902	1.4395	0.28066	0.30534	60
1	.69487	.71914	.96625	.0349	.3905	.4391	.28086	.30513	59
2	.69508	.71893	.96681	.0343	.3909	.4387	.28106	.30492	58
3	.69528	.71873	.96738	.0337	.3913	.4382	.28127	.30471	57
4	.69549	.71853	.96794	.0331	.3917	.4378	.28147	.30450	56
5	0.69570	0.71833	0.96850	1.0325	1.3921	1.4374	0.28167	0.30430	55
6	.69591	.71813	.96907	.0319	.3925	.4370	.28187	.30409	54
7	.69612	.71792	.96963	.0313	.3929	.4365	.28208	.30388	53
8	.69633	.71772	.97020	.0307	.3933	.4361	.28228	.30367	52
9	.69654	.71752	.97076	.0301	.3937	.4357	.28248	.30346	51
10	0.69675	0.71732	0.97133	1.0295	1.3941	1.4352	0.28268	0.30325	50
11	.69696	.71711	.97189	.0289	.3945	.4348	.28289	.30304	49
12	.69716	.71691	.97246	.0283	.3949	.4344	.28309	.30283	48
13	.69737	.71671	.97302	.0277	.3953	.4339	.28329	.30263	47
14	.69758	.71650	.97359	.0271	.3957	.4335	.28349	.30242	46
15	0.69779	0.71630	0.97416	1.0265	1.3960	1.4331	0.28370	0.30221	45
16	.69800	.71610	.97472	.0259	.3964	.4327	.28390	.30200	44
17	.69821	.71589	.97529	.0253	.3968	.4322	.28410	.30179	43
18	.69841	.71569	.97586	.0247	.3972	.4318	.28431	.30158	42
19	.69862	.71549	.97643	.0241	.3976	.4314	.28451	.30138	41
20	0.69883	0.71529	0.97700	1.0235	1.3980	1.4310	0.28471	0.30117	40
21	.69904	.71508	.97756	.0229	.3984	.4305	.28492	.30096	39
22	.69925	.71488	.97813	.0223	.3988	.4301	.28512	.30075	38
23	.69945	.71468	.97870	.0218	.3992	.4297	.28532	.30054	37
24	.69966	.71447	.97927	.0212	.3996	.4292	.28553	.30034	36
25	0.69987	0.71427	0.97984	1.0206	1.4000	1.4288	0.28573	0.30013	35
26	.70008	.71406	.98041	.0200	.4004	.4284	.28593	.29992	34
27	.70029	.71386	.98098	.0194	.4008	.4280	.28614	.29971	33
28	.70049	.71366	.98155	.0188	.4012	.4276	.28634	.29950	32
29	.70070	.71345	.98212	.0182	.4016	.4271	.28654	.29930	31
30	0.70091	0.71325	0.98270	1.0176	1.4020	1.4267	0.28675	0.29909	30
31	.70112	.71305	.98327	.0170	.4024	.4263	.28695	.29888	29
32	.70132	.71284	.98384	.0164	.4028	.4259	.28716	.29867	28
33	.70153	.71264	.98441	.0158	.4032	.4254	.28736	.29847	27
34	.70174	.71243	.98499	.0152	.4036	.4250	.28756	.29826	26
35	0.70194	0.71223	0.98556	1.0146	1.4040	1.4246	0.28777	0.29805	25
36	.70215	.71203	.98613	.0141	.4044	.4242	.28797	.29785	24
37	.70236	.71182	.98671	.0135	.4048	.4238	.28818	.29764	23
38	.70257	.71162	.98728	.0129	.4052	.4233	.28838	.29743	22
39	.70277	.71141	.98786	.0123	.4056	.4229	.28859	.29722	21
40	0.70298	0.71121	0.98843	1.0117	1.4060	1.4225	0.28879	0.29702	20
41	.70319	.71100	.98901	.0111	.4065	.4221	.28899	.29681	19
42	.70339	.71080	.98958	.0105	.4069	.4217	.28920	.29660	18
43	.70360	.71059	.99016	.0099	.4073	.4212	.28940	.29640	17
44	.70381	.71039	.99073	.0093	.4077	.4208	.28961	.29619	16
45	0.70401	0.71018	0.99131	1.0088	1.4081	1.4204	0.28981	0.29598	15
46	.70422	.70998	.99189	.0082	.4085	.4200	.29002	.29578	14
47	.70443	.70977	.99246	.0076	.4089	.4196	.29022	.29557	13
48	.70463	.70957	.99304	.0070	.4093	.4192	.29043	.29536	12
49	.70484	.70936	.99362	.0064	.4097	.4188	.29063	.29516	11
50	0.70505	0.70916	0.99420	1.0058	1.4101	1.4183	0.29084	0.29495	10
51	.70525	.70895	.99478	.0052	.4105	.4179	.29104	.29475	9
52	.70546	.70875	.99536	.0047	.4109	.4175	.29125	.29454	8
53	.70566	.70854	.99593	.0041	.4113	.4171	.29145	.29433	7
54	.70587	.70834	.99651	.0035	.4117	.4167	.29166	.29413	6
55	0.70608	0.70813	0.99709	1.0029	1.4122	1.4163	0.29186	0.29392	5
56	.70628	.70793	.99767	.0023	.4126	.4159	.29207	.29372	4
57	.70649	.70772	.99825	.0017	.4130	.4154	.29228	.29351	3
58	.70669	.70752	.99884	.0012	.4134	.4150	.29248	.29330	2
59	.70690	.70731	.99942	.0006	.4138	.4146	.29269	.29310	1
60	0.70711	0.70711	1.00000	1.0000	1.4142	1.4142	0.29289	0.29289	0
M	Cosine	Sine	Cotan.	Tan.	Cosec.	Secant	Vrs. Cos.	Vrs. Sin.	M

134°   45°

Logarithms of Numbers   **100 to 150**

N.	0	1	2	3	4	5	6	7	8	9
100	00 000	00 043	00 087	00 130	00 173	00 217	00 260	00 303	00 346	00 389
101	432	475	518	561	604	647	689	732	775	817
102	860	903	945	988	01 030	01 072	01 115	01 157	01 199	01 242
103	01 284	01 326	01 368	01 410	452	494	536	578	620	662
104	703	745	787	828	870	912	953	995	02 036	02 078
105	02 119	02 160	02 202	02 243	02 284	02 325	02 366	02 408	02 449	02 490
106	531	572	613	653	694	735	776	816	857	898
107	938	979	03 020	03 060	03 100	03 141	03 181	03 222	03 262	03 302
108	03 342	03 383	423	463	503	543	583	623	663	703
109	743	783	822	862	902	941	981	04 021	04 060	04 100
110	04 139	04 179	04 218	04 258	04 297	04 336	04 376	04 415	04 454	04 493
111	532	571	611	650	689	728	766	805	844	883
112	922	961	999	05 038	05 077	05 115	05 154	05 192	05 231	05 269
113	05 308	05 346	05 385	423	461	500	538	576	614	652
114	691	729	767	805	843	881	919	956	994	06 032
115	06 070	06 108	06 145	06 183	06 221	06 258	06 296	06 333	06 371	06 408
116	446	483	521	558	595	633	670	707	744	782
117	819	856	893	930	967	07 004	07 041	07 078	07 115	07 151
118	07 188	07 225	07 262	07 299	07 335	372	409	445	482	518
119	555	591	628	664	700	737	773	809	846	882
120	07 918	07 954	07 990	08 027	08 063	08 099	08 135	08 171	08 207	08 243
121	08 279	08 314	08 350	386	422	458	493	529	565	600
122	636	672	707	743	778	814	849	885	920	955
123	991	09 026	09 061	09 096	09 132	09 167	09 202	09 237	09 272	09 307
124	09 342	377	412	447	482	517	552	587	622	656
125	09 691	09 726	09 760	09 795	09 830	09 864	09 899	09 934	09 968	10 003
126	10 037	10 072	10 106	10 140	10 175	10 209	10 243	10 278	10 312	846
127	380	415	448	483	517	551	585	619	653	687
128	721	755	789	823	857	890	924	958	992	11 025
129	11 059	11 093	11 126	11 160	11 193	11 227	11 261	11 294	11 328	361
130	11 394	11 428	11 461	11 494	11 528	11 561	11 594	11 628	11 661	11 694
131	727	760	793	827	860	893	926	959	992	12 025
132	12 057	12 090	12 123	12 156	12 189	12 222	12 254	12 287	12 320	353
133	385	418	450	483	516	548	581	613	646	678
134	711	743	775	808	840	872	905	937	969	13 001
135	13 033	13 066	13 098	13 130	13 162	13 194	13 226	13 258	13 290	13 322
136	354	386	418	450	481	513	545	577	609	640
137	672	704	735	767	799	830	862	893	925	956
138	988	14 019	14 051	14 082	14 114	14 145	14 176	14 208	14 239	14 270
139	14 302	333	364	395	426	457	489	520	551	582
140	14 613	14 644	14 675	14 706	14 737	14 768	14 799	14 829	14 860	14 891
141	922	953	984	15 014	15 045	15 076	15 106	15 137	15 168	15 198
142	15 229	15 259	15 290	321	351	382	412	442	473	503
143	534	564	594	625	655	685	715	746	776	806
144	836	866	897	927	957	987	16 017	16 047	16 077	16 107
145	16 137	16 167	16 197	16 227	16 256	16 286	16 316	16 346	16 376	16 406
146	435	465	495	524	554	584	613	643	673	702
147	732	761	791	820	850	879	909	938	967	997
148	17 026	17 056	17 085	17 114	17 143	17 173	17 202	17 231	17 260	17 290
149	319	348	377	406	435	464	493	522	551	580
150	17 609	17 638	17 667	17 696	17 725	17 754	17 783	17 811	17 840	17 869
N.	0	1	2	3	4	5	6	7	8	9

150 to 200   Logarithms of Numbers

N.	0	1	2	3	4	5	6	7	8	9
150	17 609	17 638	17 667	17 696	17 725	17 754	17 783	17 811	17 840	17 869
151	898	926	955	984	18 013	18 041	18 070	18 099	18 127	18 156
152	18 184	18 213	18 242	18 270	299	327	356	384	412	441
153	469	498	526	554	583	611	639	667	696	724
154	752	780	808	837	865	893	921	949	977	19 005
155	19 033	19 061	19 089	19 117	19 145	19 173	19 201	19 229	19 257	19 285
156	313	340	368	396	424	451	479	507	535	562
157	590	618	645	673	701	728	756	783	811	838
158	866	893	921	948	976	20 003	20 030	20 058	20 085	20 112
159	20 140	20 167	20 194	20 222	20 249	276	303	331	358	385
160	20 412	20 439	20 466	20 493	20 520	20 548	20 575	20 602	20 629	20 656
161	683	710	737	763	790	817	844	871	898	925
162	952	978	21 005	21 032	21 059	21 085	21 112	21 139	21 165	21 192
163	21 219	21 245	272	299	325	352	378	405	431	458
164	484	511	537	564	590	617	643	669	696	722
165	21 748	21 775	21 801	21 827	21 854	21 880	21 906	21 932	21 959	21 985
166	22 011	22 037	22 063	22 089	22 115	22 141	22 168	22 194	22 220	22 246
167	272	298	324	350	376	402	427	453	479	505
168	531	557	583	608	634	660	686	712	737	763
169	789	814	840	866	891	917	943	968	994	23 019
170	23 045	23 070	23 096	23 122	23 147	23 172	23 198	23 223	23 249	23 274
171	300	325	350	376	401	426	452	477	502	528
172	553	578	603	629	654	679	704	729	754	780
173	805	830	855	880	905	930	955	980	24 005	24 030
174	24 055	24 080	24 105	24 130	24 155	24 180	24 204	24 229	254	279
175	24 304	24 329	24 353	24 378	24 403	24 428	24 453	24 477	24 502	24 527
176	551	576	601	625	650	675	699	724	748	773
177	797	822	846	871	895	920	944	969	993	25 018
178	25 042	25 066	25 091	25 115	25 140	25 164	25 188	25 213	25 237	261
179	285	310	334	358	382	406	431	455	479	503
180	25 527	25 551	25 576	25 600	25 624	25 648	25 672	25 696	25 720	25 744
181	768	792	816	840	864	888	912	936	959	983
182	26 007	26 031	26 055	26 079	26 103	26 126	26 150	26 174	26 198	26 221
183	245	269	293	316	340	364	387	411	435	458
184	482	505	529	553	576	600	623	647	670	694
185	26 717	26 741	26 764	26 788	26 811	26 834	26 858	26 881	26 905	26 928
186	951	975	998	27 021	27 045	27 068	27 091	27 114	27 138	27 161
187	27 184	27 207	27 231	254	277	300	323	346	370	393
188	416	439	462	485	508	531	554	577	600	623
189	646	669	692	715	738	761	784	807	830	853
190	27 875	27 898	27 921	27 944	27 967	27 990	28 012	28 035	28 058	28 081
191	28 103	28 126	28 149	28 172	28 194	28 217	240	262	285	308
192	330	353	375	398	421	443	466	488	511	533
193	556	578	601	623	646	668	691	713	735	758
194	780	803	825	847	870	892	914	937	959	981
195	29 004	29 026	29 048	29 070	29 093	29 115	29 137	29 159	29 181	29 203
196	226	248	270	292	314	336	358	380	403	425
197	447	469	491	513	535	557	579	601	623	645
198	667	688	710	732	754	776	798	820	842	864
199	885	907	929	951	973	994	30 016	30 038	30 060	30 081
200	30 103	30 125	30 146	30 168	30 190	30 211	30 233	30 255	30 276	30 298
N.	0	1	2	3	4	5	6	7	8	9

Logarithms of Numbers **200 to 250**

N.	0	1	2	3	4	5	6	7	8	9
200	30 103	30 125	30 146	30 168	30 190	30 211	30 233	30 255	30 276	30 298
201	320	341	363	384	406	428	449	471	492	514
202	535	557	578	600	621	643	664	685	707	728
203	750	771	792	814	835	856	878	899	920	942
204	963	984	31 006	31 027	31 048	31 069	31 091	31 112	31 133	31 154
205	31 175	31 197	31 218	31 239	31 260	31 281	31 302	31 323	31 345	31 366
206	387	408	429	450	471	492	513	534	555	576
207	597	618	639	660	681	702	723	744	765	785
208	806	827	848	869	890	911	931	952	973	994
209	32 015	32 035	32 056	32 077	32 098	32 118	32 139	32 160	32 181	32 201
210	32 222	32 243	32 263	32 284	32 305	32 325	32 346	32 367	32 387	32 408
211	428	449	469	490	511	531	552	572	593	613
212	634	654	675	695	716	736	756	777	797	818
213	838	858	879	899	919	940	960	981	33 001	33 021
214	33 041	33 062	33 082	33 102	33 123	33 143	33 163	33 183	203	224
215	33 244	33 264	33 284	33 304	33 325	33 345	33 365	33 385	33 405	33 425
216	445	466	486	506	526	546	566	586	606	626
217	646	666	686	706	726	746	766	786	806	826
218	846	866	886	905	925	945	965	985	34 005	34 025
219	34 044	34 064	34 084	34 104	34 124	34 144	34 163	34 183	203	223
220	34 242	34 262	34 282	34 301	34 321	34 341	34 361	34 380	34 400	34 420
221	439	459	479	498	518	537	557	577	596	616
222	635	655	674	694	714	733	753	772	792	811
223	831	850	869	889	908	928	947	967	986	35 005
224	35 025	35 044	35 064	35 083	35 102	35 122	35 141	35 160	35 180	199
225	35 218	35 238	35 257	35 276	35 295	35 315	35 334	35 353	35 372	35 392
226	411	430	449	469	488	507	526	545	564	583
227	603	622	641	660	679	698	717	736	755	774
228	794	813	832	851	870	889	908	927	946	965
229	984	36 003	36 022	36 040	36 059	36 078	36 097	36 116	36 135	36 154
230	36 173	36 192	36 211	36 229	36 248	36 267	36 286	36 305	36 324	36 342
231	361	380	399	418	436	455	474	493	511	530
232	549	568	586	605	624	642	661	680	698	717
233	736	754	773	792	810	829	847	866	885	903
234	922	940	959	977	996	37 014	37 033	37 051	37 070	37 088
235	37 107	37 125	37 144	37 162	37 181	37 199	37 218	37 236	37 254	37 273
236	291	310	328	346	365	383	402	420	438	457
237	475	493	512	530	548	566	585	603	621	639
238	658	676	694	712	731	749	767	785	803	822
239	840	858	876	894	912	931	949	967	985	38 003
240	38 021	38 039	38 057	38 075	38 093	38 112	38 130	38 148	38 166	38 184
241	202	220	238	256	274	292	310	328	346	364
242	382	400	417	435	453	471	489	507	525	543
243	561	579	596	614	632	650	668	686	703	721
244	739	757	775	792	810	828	846	863	881	899
245	38 917	38 934	38 952	38 970	38 988	39 005	39 023	39 041	39 058	39 076
246	39 094	39 111	39 129	39 146	39 164	182	199	217	235	252
247	270	287	305	322	340	358	375	393	410	428
248	445	463	480	498	515	533	550	568	585	603
249	620	637	655	672	690	707	725	742	759	777
250	39 794	39 811	39 829	39 846	39 863	39 881	39 898	39 915	39 933	39 950
N.	0	1	2	3	4	5	6	7	8	9

250 to 300 — Logarithms of Numbers

N.	0	1	2	3	4	5	6	7	8	9
250	39 794	39 811	39 829	39 846	39 863	39 881	39 898	39 915	39 933	39 950
251	967	985	40 002	40 019	40 037	40 054	40 071	40 088	40 106	40 123
252	40 140	40 157	175	192	209	226	243	261	278	295
253	312	329	346	364	381	398	415	432	449	466
254	483	501	518	535	552	569	586	603	620	637
255	40 654	40 671	40 688	40 705	40 722	40 739	40 756	40 773	40 790	40 807
256	824	841	858	875	892	909	926	943	960	976
257	993	41 010	41 027	41 044	41 061	41 078	41 095	41 111	41 128	41 145
258	41 162	179	196	212	229	246	263	280	296	313
259	330	347	364	380	397	414	431	447	464	481
260	41 497	41 514	41 531	41 547	41 564	41 581	41 597	41 614	41 631	41 647
261	664	681	697	714	731	747	764	780	797	814
262	830	847	863	880	896	913	930	946	963	979
263	996	42 012	42 029	42 045	42 062	42 078	42 095	42 111	42 128	42 144
264	42 160	177	193	210	226	243	259	275	292	308
265	42 325	42 341	42 357	42 374	42 390	42 407	42 423	42 439	42 456	42 472
266	488	505	521	537	553	570	586	602	619	635
267	651	667	684	700	716	732	749	765	781	797
268	814	830	846	862	878	894	911	927	943	959
269	975	991	43 008	43 024	43 040	43 056	43 072	43 088	43 104	43 120
270	43 136	43 153	43 169	43 185	43 201	43 217	43 233	43 249	43 265	43 281
271	297	313	329	345	361	377	393	409	425	441
272	457	473	489	505	521	537	553	569	584	600
273	616	632	648	664	680	696	712	728	743	759
274	775	791	807	823	838	854	870	886	902	918
275	43 933	43 949	43 965	43 981	43 996	44 012	44 028	44 044	44 059	44 075
276	44 091	44 107	44 122	44 138	44 154	170	185	201	217	232
277	248	264	279	295	311	326	342	358	373	389
278	405	420	436	451	467	483	498	514	529	545
279	560	576	592	607	623	638	654	669	685	700
280	44 716	44 731	44 747	44 762	44 778	44 793	44 809	44 824	44 840	44 855
281	871	886	902	917	932	948	963	979	994	45 010
282	45 025	45 040	45 056	45 071	45 087	45 102	45 117	45 133	45 148	163
283	179	194	209	225	240	255	271	286	301	317
284	332	347	362	378	393	408	424	439	454	469
285	45 485	45 500	45 515	45 530	45 545	45 561	45 576	45 591	45 606	45 621
286	637	652	667	682	697	713	728	743	758	773
287	788	803	818	834	849	864	879	894	909	924
288	939	954	969	985	46 000	46 015	46 030	46 045	46 060	46 075
289	46 090	46 105	46 120	46 135	150	165	180	195	210	225
290	46 240	46 255	46 270	46 285	46 300	46 315	46 330	46 345	46 359	46 374
291	389	404	419	434	449	464	479	494	509	523
292	538	553	568	583	598	613	627	642	657	672
293	687	702	716	731	746	761	776	790	805	820
294	835	850	864	879	894	909	923	938	953	968
295	46 982	46 997	47 012	47 026	47 041	47 056	47 070	47 085	47 100	47 115
296	47 129	47 144	159	173	188	203	217	232	246	261
297	276	290	305	320	334	349	363	378	393	407
298	422	436	451	465	480	494	509	524	538	553
299	567	582	596	611	625	640	654	669	683	698
300	47 712	47 727	47 741	47 756	47 770	47 784	47 799	47 813	47 828	47 842
N.	0	1	2	3	4	5	6	7	8	9

Logarithms of Numbers **300 to 350**

N.	0	1	2	3	4	5	6	7	8	9
300	47 712	47 727	47 741	47 756	47 770	47 784	47 799	47 813	47 828	47 842
301	857	871	886	900	914	929	943	958	972	986
302	48 001	48 015	48 029	48 044	48 058	48 073	48 087	48 101	48 116	48 130
303	144	159	173	187	202	216	230	245	259	273
304	287	302	316	330	345	359	373	387	402	416
305	48 430	48 444	48 459	48 473	48 487	48 501	48 515	48 530	48 544	48 558
306	572	586	601	615	629	643	657	671	686	700
307	714	728	742	756	770	785	799	813	827	841
308	855	869	883	897	911	926	940	954	968	982
309	996	49 010	49 024	49 038	49 052	49 066	49 080	49 094	49 108	49 122
310	49 136	49 150	49 164	49 178	49 192	49 206	49 220	49 234	49 248	49 262
311	276	290	304	318	332	346	360	374	388	402
312	416	429	443	457	471	485	499	513	527	541
313	554	568	582	596	610	624	638	652	665	679
314	693	707	721	734	748	762	776	790	804	817
315	49 831	49 845	49 859	49 872	49 886	49 900	49 914	49 928	49 941	49 955
316	969	982	996	50 010	50 024	50 037	50 051	50 065	50 079	50 092
317	50 106	50 120	50 133	147	161	174	188	202	215	229
318	243	256	270	284	297	311	325	338	352	366
319	379	393	406	420	434	447	461	474	488	501
320	50 515	50 529	50 542	50 556	50 569	50 583	50 596	50 610	50 623	50 637
321	651	664	678	691	705	718	732	745	759	772
322	786	799	813	826	840	853	866	880	893	907
323	920	934	947	961	974	987	51 001	51 014	51 028	51 041
324	51 055	51 068	51 081	51 095	51 108	51 122	135	148	162	175
325	51 188	51 202	51 215	51 228	51 242	51 255	51 268	51 282	51 295	51 308
326	322	335	348	362	375	388	402	415	428	442
327	455	468	481	495	508	521	534	548	561	574
328	587	601	614	627	640	654	667	680	693	706
329	720	733	746	759	772	786	799	812	825	838
330	51 851	51 865	51 878	51 891	51 904	51 917	51 930	51 943	51 957	51 970
331	983	996	52 009	52 022	52 035	52 048	52 062	52 075	52 088	52 101
332	52 114	52 127	140	153	166	179	192	205	218	231
333	244	258	271	284	297	310	323	336	349	362
334	375	388	401	414	427	440	453	466	479	492
335	52 505	52 517	52 530	52 543	52 556	52 569	52 582	52 595	52 608	52 621
336	634	647	660	673	686	699	711	724	737	750
337	763	776	789	802	815	827	840	853	866	879
338	892	905	917	930	943	956	969	982	994	53 007
339	53 020	53 033	53 046	53 058	53 071	53 084	53 097	53 110	53 122	135
340	53 148	53 161	53 173	53 186	53 199	53 212	53 225	53 237	53 250	53 263
341	275	288	301	314	326	339	352	365	377	390
342	403	415	428	441	453	466	479	491	504	517
343	529	542	555	567	580	593	605	618	631	643
344	656	669	681	694	706	719	732	744	757	769
345	53 782	53 795	53 807	53 820	53 832	53 845	53 857	53 870	53 883	53 895
346	908	920	933	945	958	970	983	995	54 008	54 020
347	54 033	54 046	54 058	54 071	54 083	54 096	54 108	54 121	133	145
348	158	170	183	195	208	220	233	245	258	270
349	283	295	307	320	332	345	357	370	382	394
350	54 407	54 419	54 432	54 444	54 456	54 469	54 481	54 494	54 506	54 518
N.	0	1	2	3	4	5	6	7	8	9

350 to 400 — Logarithms of Numbers

N.	0	1	2	3	4	5	6	7	8	9
350	54 407	54 419	54 432	54 444	54 456	54 469	54 481	54 494	54 506	54 518
351	531	543	556	568	580	593	605	617	630	642
352	654	667	679	691	704	716	728	741	753	765
353	778	790	802	814	827	839	851	864	876	888
354	900	913	925	937	949	962	974	986	998	55 011
355	55 023	55 035	55 047	55 060	55 072	55 084	55 096	55 108	55 121	55 133
356	145	157	169	182	194	206	218	230	243	255
357	267	279	291	303	316	328	340	352	364	376
358	388	400	413	425	437	449	461	473	485	497
359	509	522	534	546	558	570	582	594	606	618
360	55 630	55 642	55 654	55 666	55 679	55 691	55 703	55 715	55 727	55 739
361	751	763	775	787	799	811	823	835	847	859
362	871	883	895	907	919	931	943	955	967	979
363	991	56 003	56 015	56 027	56 039	56 050	56 062	56 074	56 086	56 098
364	56 110	122	134	146	158	170	182	194	206	217
365	56 229	56 241	56 253	56 265	56 277	56 289	56 301	56 313	56 324	56 336
366	348	360	372	384	396	407	419	431	443	455
367	467	478	490	502	514	526	538	549	561	573
368	585	597	608	620	632	644	656	667	679	691
369	703	714	726	738	750	761	773	785	797	808
370	56 820	56 832	56 844	56 855	56 867	56 879	56 891	56 902	56 914	56 926
371	937	949	961	973	984	996	57 008	57 019	57 031	57 043
372	57 054	57 066	57 078	57 089	57 101	57 113	124	136	148	159
373	171	183	194	206	217	229	241	252	264	276
374	287	299	310	322	334	345	357	368	380	392
375	57 403	57 415	57 426	57 438	57 449	57 461	57 473	57 484	57 496	57 507
376	519	530	542	553	565	577	588	600	611	623
377	634	646	657	669	680	692	703	715	726	738
378	749	761	772	784	795	807	818	830	841	853
379	864	876	887	898	910	921	933	944	956	967
380	57 978	57 990	58 001	58 013	58 024	58 036	58 047	58 058	58 070	58 081
381	58 093	58 104	115	127	138	150	161	172	184	195
382	206	218	229	240	252	263	275	286	297	309
383	320	331	343	354	365	377	388	399	411	422
384	433	444	456	467	478	490	501	512	524	535
385	58 546	58 557	58 569	58 580	58 591	58 602	58 614	58 625	58 636	58 648
386	659	670	681	693	704	715	726	737	749	760
387	771	782	794	805	816	827	838	850	861	872
388	883	894	906	917	928	939	950	962	973	984
389	995	59 006	59 017	59 028	59 040	59 051	59 062	59 073	59 084	59 095
390	59 107	59 118	59 129	59 140	59 151	59 162	59 173	59 184	59 196	59 207
391	218	229	240	251	262	273	284	295	306	318
392	329	340	351	362	373	384	395	406	417	428
393	439	450	461	472	483	495	506	517	528	539
394	550	561	572	583	594	605	616	627	638	649
395	59 660	59 671	59 682	59 693	59 704	59 715	59 726	59 737	59 748	59 759
396	770	781	791	802	813	824	835	846	857	868
397	879	890	901	912	923	934	945	956	967	977
398	988	999	60 010	60 021	60 032	60 043	60 054	60 065	60 076	60 086
399	60 097	60 108	119	130	141	152	163	173	184	195
400	60 206	60 217	60 228	60 239	60 249	60 260	60 271	60 282	60 293	60 304
N.	0	1	2	3	4	5	6	7	8	9

Logarithms of Numbers **400 to 450**

N.	0	1	2	3	4	5	6	7	8	9
400	60 206	60 217	60 228	60 239	60 249	60 260	60 271	60 282	60 293	60 304
401	314	325	336	347	358	369	379	390	401	412
402	423	433	444	455	466	477	487	498	509	520
403	531	541	552	563	574	584	595	606	617	627
404	638	649	660	670	681	692	703	713	724	735
405	60 746	60 756	60 767	60 778	60 788	60 799	60 810	60 821	60 831	60 842
406	853	863	874	885	895	906	917	927	938	949
407	959	970	981	991	61 002	61 013	61 023	61 034	61 045	61 055
408	61 066	61 077	61 087	61 098	109	119	130	141	151	162
409	172	183	194	204	215	225	236	247	257	268
410	61 278	61 289	61 300	61 310	61 321	61 331	61 342	61 353	61 363	61 374
411	384	395	405	416	426	437	448	458	469	479
412	490	500	511	521	532	542	553	563	574	585
413	595	606	616	627	637	648	658	669	679	690
414	700	711	721	732	742	753	763	773	784	794
415	61 805	61 815	61 826	61 836	61 847	61 857	61 868	61 878	61 888	61 899
416	909	920	930	941	951	962	972	982	993	62 003
417	62 014	62 024	62 034	62 045	62 055	62 066	62 076	62 086	62 097	107
418	118	128	138	149	159	170	180	190	201	211
419	221	232	242	253	263	273	284	294	304	315
420	62 325	62 335	62 346	62 356	62 366	62 377	62 387	62 397	62 408	62 418
421	428	439	449	459	470	480	490	500	511	521
422	531	542	552	562	572	583	593	603	614	624
423	634	644	655	665	675	685	696	706	716	726
424	737	747	757	767	778	788	798	808	819	829
425	62 839	62 849	62 859	62 870	62 880	62 890	62 900	62 910	62 921	62 931
426	941	951	961	972	982	992	63 002	63 012	63 022	63 033
427	63 043	63 053	63 063	63 073	63 084	63 094	104	114	124	134
428	144	155	165	175	185	195	205	215	226	236
429	246	256	266	276	286	296	306	317	327	337
430	63 347	63 357	63 367	63 377	63 387	63 397	63 407	63 418	63 428	63 438
431	448	458	468	478	488	498	508	518	528	538
432	548	558	569	579	589	599	609	619	629	639
433	649	659	669	679	689	699	709	719	729	739
434	749	759	769	779	789	799	809	819	829	839
435	63 849	63 859	63 869	63 879	63 889	63 899	63 909	63 919	63 929	63 939
436	949	959	969	979	989	998	64 008	64 018	64 028	64 038
437	64 048	64 058	64 068	64 078	64 088	64 098	108	118	128	138
438	147	157	167	177	187	197	207	217	227	237
439	247	256	266	276	286	296	306	316	326	335
440	64 345	64 355	64 365	64 375	64 385	64 395	64 404	64 414	64 424	64 434
441	444	454	464	473	483	493	503	513	523	532
442	542	552	562	572	582	591	601	611	621	631
443	640	650	660	670	680	689	699	709	719	729
444	738	748	758	768	777	787	797	807	817	826
445	64 836	64 846	64 856	64 865	64 875	64 885	64 895	64 904	64 914	64 924
446	934	943	953	963	972	982	992	65 002	65 011	65 021
447	65 031	65 041	65 050	65 060	65 070	65 079	65 089	099	108	118
448	128	138	147	157	167	176	186	196	205	215
449	225	234	244	254	263	273	283	292	302	312
450	65 321	65 331	65 341	65 350	65 360	65 370	65 379	65 389	65 398	65 408
N.	0	1	2	3	4	5	6	7	8	9

450 to 500 Logarithms of Numbers

N.	0	1	2	3	4	5	6	7	8	9
450	65 321	65 331	65 341	65 350	65 360	65 370	65 379	65 389	65 398	65 408
451	418	427	437	447	456	466	475	485	495	504
452	514	524	533	543	552	562	572	581	591	600
453	610	619	629	639	648	658	667	677	686	696
454	706	715	725	734	744	753	763	773	782	792
455	65 801	65 811	65 820	65 830	65 839	65 849	65 858	65 868	65 877	65 887
456	897	906	916	925	935	944	954	963	973	982
457	992	66 001	66 011	66 020	76 030	66 039	66 049	66 058	66 068	66 077
458	66 087	096	106	115	125	134	143	153	162	172
459	181	191	200	210	219	229	238	248	257	266
460	66 276	66 285	66 295	66 304	66 314	66 323	66 332	66 342	66 351	66 361
461	370	380	389	398	408	417	427	436	445	455
462	464	474	483	492	502	511	521	530	539	549
463	558	568	577	586	596	605	614	624	633	642
464	652	661	671	680	689	699	708	717	727	736
465	66 745	66 755	66 764	66 773	66 783	66 792	66 801	66 811	66 820	66 829
466	839	848	857	867	876	885	895	904	913	922
467	932	941	950	960	969	978	988	997	67 006	67 015
468	67 025	67 034	67 043	67 052	67 062	67 071	67 080	67 090	099	108
469	117	127	136	145	154	164	173	182	191	201
470	67 210	67 219	67 228	67 238	67 247	67 256	67 265	67 274	67 284	67 293
471	302	311	321	330	339	348	357	367	376	385
472	394	403	413	422	431	440	449	459	468	477
473	486	495	505	514	523	532	541	550	560	569
474	578	587	596	605	615	624	633	642	651	660
475	67 669	67 679	67 688	67 697	67 706	67 715	67 724	67 733	67 742	67 752
476	761	770	779	788	797	806	815	825	834	843
477	852	861	870	879	888	897	906	916	925	934
478	943	952	961	970	979	988	997	68 006	68 015	68 025
479	68 034	68 043	68 052	68 061	68 070	68 079	68 088	097	106	115
480	68 124	68 133	68 142	68 151	68 160	68 169	68 178	68 187	68 196	68 206
481	215	224	233	242	251	260	269	278	287	296
482	305	314	323	332	341	350	359	368	377	386
483	395	404	413	422	431	440	449	458	467	476
484	485	494	503	511	520	529	538	547	556	565
485	68 574	68 583	68 592	68 601	68 610	68 619	68 628	68 637	68 646	68 655
486	664	673	682	690	699	708	717	726	735	744
487	753	762	771	780	789	798	806	815	824	833
488	842	851	860	869	878	887	895	904	913	922
489	931	940	949	958	966	975	984	993	69 002	69 011
490	69 020	69 029	69 037	69 046	69 055	69 064	69 073	69 082	69 091	69 099
491	108	117	126	135	144	152	161	170	179	188
492	197	205	214	223	232	241	249	258	267	276
493	285	294	302	311	320	329	338	346	355	364
494	373	382	390	399	408	417	425	434	443	452
495	69 461	69 469	69 478	69 487	69 496	69 504	69 513	69 522	69 531	69 539
496	548	557	566	574	583	592	601	609	618	627
497	636	644	653	662	671	679	688	697	706	714
498	723	732	740	749	758	767	775	784	793	801
499	810	819	828	836	845	854	862	871	880	888
500	69 897	69 906	69 914	69 923	69 932	69 940	69 949	69 958	69 966	69 975
N.	0	1	2	3	4	5	6	7	8	9

Logarithms of Numbers **500 to 550**

N.	0	1	2	3	4	5	6	7	8	9
500	69 897	69 906	69 914	69 923	69 932	69 940	69 949	69 958	69 966	69 975
501	984	992	70 001	70 010	70 018	70 027	70 036	70 044	70 053	70 062
502	70 070	70 079	088	096	105	114	122	131	140	148
503	157	165	174	183	191	200	209	217	226	234
504	243	252	260	269	278	286	295	303	312	321
505	70 329	70 338	70 346	70 355	70 364	70 372	70 381	70 389	70 398	70 407
506	415	424	432	441	449	458	467	475	484	492
507	501	509	518	527	535	544	552	561	569	578
508	586	595	604	612	621	629	638	646	655	663
509	672	680	689	697	706	714	723	732	740	749
510	70 757	70 766	70 774	70 783	70 791	70 800	70 808	70 817	70 825	70 834
511	842	851	859	868	876	885	893	902	910	919
512	927	936	944	952	961	969	978	986	995	71 003
513	71 012	71 020	71 029	71 037	71 046	71 054	71 063	71 071	71 079	088
514	096	105	113	122	130	139	147	155	164	172
515	71 181	71 189	71 198	71 206	71 214	71 223	71 231	71 240	71 248	71 257
516	265	273	282	290	299	307	315	324	332	341
517	349	358	366	374	383	391	399	408	416	425
518	433	441	450	458	467	475	483	492	500	508
519	517	525	534	542	550	559	567	575	584	592
520	71 600	71 609	71 617	71 625	71 634	71 642	71 650	71 659	71 667	71 675
521	684	692	700	709	717	725	734	742	750	759
522	767	775	784	792	800	809	817	825	834	842
523	850	859	867	875	883	892	900	908	917	925
524	933	941	950	958	966	975	983	991	999	72 008
525	72 016	72 024	72 033	72 041	72 049	72 057	72 066	72 074	72 082	72 090
526	099	107	115	123	132	140	148	156	165	173
527	181	189	198	206	214	222	231	239	247	255
528	263	272	280	288	296	305	313	321	329	337
529	346	354	362	370	378	387	395	403	411	419
530	72 428	72 436	72 444	72 452	72 460	72 469	72 477	72 485	72 493	72 501
531	510	518	526	534	542	550	559	567	575	583
532	591	599	608	616	624	632	640	648	656	665
533	673	681	689	697	705	713	722	730	738	746
534	754	762	770	779	787	795	803	811	819	827
535	72 835	72 844	72 852	72 860	72 868	72 876	72 884	72 892	72 900	72 908
536	917	925	933	941	949	957	965	973	981	989
537	997	73 006	73 014	73 022	73 030	73 038	73 046	73 054	73 062	73 070
538	73 078	086	094	102	111	119	127	135	143	151
539	159	167	175	183	191	199	207	215	223	231
540	73 239	73 247	73 256	73 264	73 272	73 280	73 288	73 296	73 304	73 312
541	320	328	336	344	352	360	368	376	384	392
542	400	408	416	424	432	440	448	456	464	472
543	480	488	496	504	512	520	528	536	544	552
544	560	568	576	584	592	600	608	616	624	632
545	73 640	73 648	73 656	73 664	73 672	73 680	73 687	73 695	73 703	73 711
546	719	727	735	743	751	759	767	775	783	791
547	799	807	815	823	831	838	846	854	862	870
548	878	886	894	902	910	918	926	934	941	949
549	957	965	973	981	989	997	74 005	74 013	74 021	74 028
550	74 036	74 044	74 052	74 060	74 068	74 076	74 084	74 092	74 099	74 107
N.	0	1	2	3	4	5	6	7	8	9

550 to 600 — Logarithms of Numbers

N.	0	1	2	3	4	5	6	7	8	9
550	74 036	74 044	74 052	74 060	74 068	74 076	74 084	74 092	74 099	74 107
551	115	123	131	139	147	155	162	170	178	186
552	194	202	210	218	225	233	241	249	257	265
553	273	280	288	296	304	312	320	328	335	343
554	351	359	367	375	382	390	398	406	414	422
555	74 429	74 437	74 445	74 453	74 461	74 468	74 476	74 484	74 492	74 500
556	508	515	523	531	539	547	554	562	570	578
557	586	593	601	609	617	625	632	640	648	656
558	663	671	679	687	695	702	710	718	726	733
559	741	749	757	765	772	780	788	796	803	811
560	74 819	74 827	74 834	74 842	74 850	74 858	74 865	74 873	74 881	74 889
561	896	904	912	920	927	935	943	950	958	966
562	974	981	989	997	75 005	75 012	75 020	75 028	75 035	75 043
563	75 051	75 059	75 066	75 074	082	089	097	105	113	120
564	128	136	143	151	159	166	174	182	190	197
565	75 205	75 213	75 220	75 228	75 236	75 243	75 251	75 259	75 266	75 274
566	282	289	297	305	312	320	328	335	343	351
567	358	366	374	381	389	397	404	412	420	427
568	435	443	450	458	465	473	481	488	496	504
569	511	519	527	534	542	549	557	565	572	580
570	75 588	75 595	75 603	75 610	75 618	75 626	75 633	75 641	75 648	75 656
571	664	671	679	686	694	702	709	717	724	732
572	740	747	755	762	770	778	785	793	800	808
573	816	823	831	838	846	853	861	869	876	884
574	891	899	906	914	921	929	937	944	952	959
575	75 967	75 974	75 982	75 989	75 997	76 005	76 012	76 020	76 027	76 035
576	76 042	76 050	76 057	76 065	76 072	080	088	095	103	110
577	118	125	133	140	148	155	163	170	178	185
578	193	200	208	215	223	230	238	245	253	260
579	268	275	283	290	298	305	313	320	328	335
580	76 343	76 350	76 358	76 365	76 373	76 380	76 388	76 395	76 403	76 410
581	418	425	433	440	448	455	462	470	477	485
582	492	500	507	515	522	530	537	545	552	559
583	567	574	582	589	597	604	612	619	626	634
584	641	649	656	664	671	679	686	693	701	708
585	76 716	76 723	76 730	76 738	76 745	76 753	76 760	76 768	76 775	76 782
586	790	797	805	812	819	827	834	842	849	856
587	864	871	879	886	893	901	908	916	923	930
588	938	945	953	960	967	975	982	989	997	77 004
589	77 012	77 019	77 026	77 034	77 041	77 048	77 056	77 063	77 071	078
590	77 085	77 093	77 100	77 107	77 115	77 122	77 129	77 137	77 144	77 151
591	159	166	173	181	188	196	203	210	218	225
592	232	240	247	254	262	269	276	284	291	298
593	306	313	320	327	335	342	349	357	364	371
594	379	386	393	401	408	415	423	430	437	444
595	77 452	77 459	77 466	77 474	77 481	77 488	77 496	77 503	77 510	77 517
596	525	532	539	547	554	561	568	576	583	590
597	597	605	612	619	627	634	641	648	656	663
598	670	677	685	692	699	706	714	721	728	735
599	743	750	757	764	772	779	786	793	801	808
600	77 815	77 822	77 830	77 837	77 844	77 851	77 859	77 866	77 873	77 880
N.	0	1	2	3	4	5	6	7	8	9

Logarithms of Numbers **600 to 650**

N.	0	1	2	3	4	5	6	7	8	9
600	77 815	77 822	77 830	77 837	77 844	77 851	77 859	77 866	77 873	77 880
601	887	895	902	909	916	924	931	938	945	952
602	960	967	974	981	989	996	78 003	78 010	78 017	78 025
603	78 032	78 039	78 046	78 053	78 061	78 068	075	082	089	097
604	104	111	118	125	132	140	147	154	161	168
605	78 176	78 183	78 190	78 197	78 204	78 211	78 219	78 226	78 233	78 240
606	247	254	262	269	276	283	290	297	305	312
607	319	326	333	340	348	355	362	369	376	383
608	390	398	405	412	419	426	433	440	448	455
609	462	469	476	483	490	497	505	512	519	526
610	78 533	78 540	78 547	78 554	78 562	78 569	78 576	78 583	78 590	78 597
611	604	611	618	625	633	640	647	654	661	668
612	675	682	689	696	704	711	718	725	732	739
613	746	753	760	767	774	782	789	796	803	810
614	817	824	831	838	845	852	859	866	873	880
615	78 888	78 895	78 902	78 909	78 916	78 923	78 930	78 937	78 944	78 951
616	958	965	972	979	986	993	79 000	79 007	79 014	79 022
617	79 029	79 036	79 043	79 050	79 057	79 064	071	078	085	092
618	099	106	113	120	127	134	141	148	155	162
619	169	176	183	190	197	204	211	218	225	232
620	79 239	79 246	79 253	79 260	79 267	79 274	79 281	79 288	79 295	79 302
621	309	316	323	330	337	344	351	358	365	372
622	379	386	393	400	407	414	421	428	435	442
623	449	456	463	470	477	484	491	498	505	512
624	519	525	532	539	546	553	560	567	574	581
625	79 588	79 595	79 602	79 609	79 616	79 623	79 630	79 637	79 644	79 651
626	657	664	671	678	685	692	699	706	713	720
627	727	734	741	748	755	761	768	775	782	789
628	796	803	810	817	824	831	837	844	851	858
629	865	872	879	886	893	900	907	913	920	927
630	79 934	79 941	79 948	79 955	79 962	79 969	79 975	79 982	79 989	79 996
631	80 003	80 010	80 017	80 024	80 031	80 037	80 044	80 051	80 058	80 065
632	072	079	085	092	099	106	113	120	127	134
633	140	147	154	161	168	175	182	188	195	202
634	209	216	223	230	236	243	250	257	264	271
635	80 277	80 284	80 291	80 298	80 305	80 312	80 318	80 325	80 332	80 339
636	346	353	359	366	373	380	387	394	400	407
637	414	421	428	434	441	448	455	462	469	475
638	482	489	496	503	509	516	523	530	537	543
639	550	557	564	571	577	584	591	598	604	611
640	80 618	80 625	80 632	80 638	80 645	80 652	80 659	80 666	80 672	80 679
641	686	693	699	706	713	720	726	733	740	747
642	754	760	767	774	781	787	794	801	808	814
643	821	828	835	841	848	855	862	868	875	882
644	889	895	902	909	916	922	929	936	943	949
645	80 956	80 963	80 969	80 976	80 983	80 990	80 996	81 003	81 010	81 017
646	81 023	81 030	81 037	81 043	81 050	81 057	81 064	070	077	084
647	090	097	104	111	117	124	131	137	144	151
648	158	164	171	178	184	191	198	204	211	218
649	225	231	238	245	251	258	265	271	278	285
650	81 291	81 298	81 305	81 311	81 318	81 325	81 331	81 338	81 345	81 351
N.	0	1	2	3	4	5	6	7	8	9

650 to 700 — Logarithms of Numbers

N.	0	1	2	3	4	5	6	7	8	9
650	81 291	81 298	81 305	81 311	81 318	81 325	81 331	81 338	81 345	81 351
651	358	365	371	378	385	391	398	405	411	418
652	425	431	438	445	451	458	465	471	478	485
653	491	498	505	511	518	525	531	538	545	551
654	558	564	571	578	584	591	598	604	611	618
655	81 624	81 631	81 637	81 644	81 651	81 657	81 664	81 671	81 677	81 684
656	690	697	704	710	717	724	730	737	743	750
657	757	763	770	776	783	790	796	803	809	816
658	823	829	836	842	849	856	862	869	875	882
659	889	895	902	908	915	922	928	935	941	948
660	81 954	81 961	81 968	81 974	81 981	81 987	81 994	82 000	82 007	82 014
661	82 020	82 027	82 033	82 040	82 046	82 053	82 060	066	073	079
662	086	092	099	106	112	119	125	132	138	145
663	151	158	165	171	178	184	191	197	204	210
664	217	223	230	236	243	250	256	263	269	276
665	82 282	82 289	82 295	82 302	82 308	82 315	82 321	82 328	82 334	82 341
666	347	354	361	367	374	380	387	393	400	406
667	413	419	426	432	439	445	452	458	465	471
668	478	484	491	497	504	510	517	523	530	536
669	543	549	556	562	569	575	582	588	595	601
670	82 608	82 614	82 620	82 627	82 633	82 640	82 647	82 653	82 659	82 666
671	672	679	685	692	698	705	711	718	724	731
672	737	743	750	756	763	769	776	782	789	795
673	802	808	814	821	827	834	840	847	853	860
674	866	872	879	885	892	898	905	911	918	924
675	82 930	82 937	82 943	82 950	82 956	82 963	82 969	82 975	82 989	82 988
676	995	83 001	83 008	83 014	83 020	83 027	83 033	83 040	83 046	83 053
677	83 059	065	072	078	085	091	097	104	110	117
678	123	129	136	142	149	155	161	168	174	181
679	187	193	200	206	213	219	225	232	238	245
680	83 251	83 257	83 264	83 270	83 276	83 283	83 289	83 296	83 302	83 308
681	315	321	328	334	340	347	353	359	366	372
682	378	385	391	398	404	410	417	423	429	436
683	442	448	455	461	468	474	480	487	493	499
684	506	512	518	525	531	537	544	550	556	563
685	83 569	83 575	83 582	83 588	83 594	83 601	83 607	83 613	83 620	83 626
686	632	639	645	651	658	664	670	677	683	689
687	696	702	708	715	721	727	734	740	746	753
688	759	765	772	778	784	790	797	803	809	816
689	822	828	835	841	847	853	860	866	872	879
690	83 885	83 891	83 898	83 904	83 910	83 916	83 923	83 929	83 935	83 942
691	948	954	960	967	973	979	986	992	998	84 004
692	84 011	84 017	84 023	84 029	84 036	84 042	84 048	84 055	84 061	067
693	073	080	086	092	098	105	111	117	123	130
694	136	142	149	155	161	167	174	180	186	192
695	84 199	84 205	84 211	84 217	84 224	84 230	84 236	84 242	84 248	84 255
696	261	267	273	280	286	292	298	305	311	317
697	323	330	336	342	348	354	361	367	373	379
698	386	392	398	404	410	417	423	429	435	442
699	448	454	460	466	473	479	485	491	497	504
700	84 510	84 516	84 522	84 528	84 535	84 541	84 547	84 553	84 559	84 566
N.	0	1	2	3	4	5	6	7	8	9

Logarithms of Numbers **700 to 750**

N.	0	1	2	3	4	5	6	7	8	9
700	84 510	84 516	84 522	84 528	84 535	84 541	84 547	84 553	84 559	84 566
701	572	578	584	590	597	603	609	615	621	628
702	634	640	646	652	659	665	671	677	683	689
703	696	702	708	714	720	726	733	739	745	751
704	757	763	770	776	782	788	794	800	807	813
705	84 819	84 825	84 831	84 837	84 844	84 850	84 856	84 862	84 868	84 874
706	881	887	893	899	905	911	917	924	930	936
707	942	948	954	960	967	973	979	985	991	997
708	85 003	85 010	85 016	85 022	85 028	85 034	85 040	85 046	85 052	85 059
709	065	071	077	083	089	095	101	108	114	120
710	85 126	85 132	85 138	85 144	85 150	85 156	85 163	85 169	85 175	85 181
711	187	193	199	205	211	218	224	230	236	242
712	248	254	260	266	272	279	285	291	297	303
713	309	315	321	327	333	339	346	352	358	364
714	370	376	382	388	394	400	406	412	419	425
715	85 431	85 437	85 443	85 449	85 455	85 461	85 467	85 473	85 479	85 485
716	491	497	503	510	516	522	528	534	540	546
717	552	558	564	570	576	582	588	594	600	606
718	612	619	625	631	637	643	649	655	661	667
719	673	679	685	691	697	703	709	715	721	727
720	85 733	85 739	85 745	85 751	85 757	85 763	85 769	85 776	85 782	85 788
721	794	800	806	812	818	824	830	836	842	848
722	854	860	866	872	878	884	890	896	902	908
723	914	920	926	932	938	944	950	956	962	968
724	974	980	986	992	998	86 004	86 010	86 016	86 022	86 028
725	86 034	86 040	86 046	86 052	86 058	86 064	86 070	86 076	86 082	86 088
726	094	100	106	112	118	124	130	136	142	148
727	153	159	165	171	177	183	189	195	201	207
728	213	219	225	231	237	243	249	255	261	267
729	273	279	285	291	297	303	309	314	320	326
730	86 332	86 338	86 344	86 350	86 356	86 362	86 368	86 374	86 380	86 386
731	392	398	404	410	416	421	427	433	439	445
732	451	457	463	469	475	481	487	493	499	505
733	510	516	522	528	534	540	546	552	558	564
734	570	576	581	587	593	599	605	611	617	623
735	86 629	86 635	86 641	86 647	86 652	86 658	86 664	86 670	86 676	86 682
736	688	694	700	706	711	717	723	729	735	741
737	747	753	759	764	770	776	782	788	794	800
738	806	812	817	823	829	835	841	847	853	859
739	864	870	876	882	888	894	900	906	911	917
740	86 923	86 929	86 935	86 941	86 947	86 953	86 958	86 964	86 970	86 976
741	982	988	994	999	87 005	87 011	87 017	87 023	87 029	87 035
742	87 040	87 046	87 052	87 058	064	070	076	081	087	093
743	099	105	111	116	122	128	134	140	146	152
744	157	163	169	175	181	187	192	198	204	210
745	87 216	87 222	87 227	87 233	87 239	87 245	87 251	87 256	87 262	87 268
746	274	280	286	291	297	303	309	315	320	326
747	332	338	344	350	355	361	367	373	379	384
748	390	396	402	408	413	419	425	431	437	442
749	448	454	460	466	471	477	483	489	495	500
750	87 506	87 512	87 518	87 524	87 529	87 535	87 541	87 547	87 552	87 558
N.	0	1	2	3	4	5	6	7	8	9

750 to 800 Logarithms of Numbers

N.	0	1	2	3	4	5	6	7	8	9
750	87 506	87 512	87 518	87 524	87 529	87 535	87 541	87 547	87 552	87 558
751	564	570	576	581	587	593	599	605	610	616
752	622	628	633	639	645	651	656	662	668	674
753	680	685	691	697	703	708	714	720	726	731
754	737	743	749	754	760	766	772	777	783	789
755	87 795	87 800	87 806	87 812	87 818	87 823	87 829	87 835	87 841	87 846
756	852	858	864	869	875	881	887	892	898	904
757	910	915	921	927	933	938	944	950	956	961
758	967	973	978	984	990	996	88 001	88 007	88 013	88 019
759	88 024	88 030	88 036	88 041	88 047	88 053	059	064	070	076
760	88 081	88 087	88 093	88 099	88 104	88 110	88 116	88 121	88 127	88 133
761	139	144	150	156	161	167	173	178	184	190
762	196	201	207	213	218	224	230	235	241	247
763	253	258	264	270	275	281	287	292	298	304
764	309	315	321	326	332	338	343	349	355	361
765	88 366	88 372	88 378	88 383	88 389	88 395	88 400	88 406	88 412	88 417
766	423	429	434	440	446	451	457	463	468	474
767	480	485	491	497	502	508	514	519	525	531
768	536	542	547	553	559	564	570	576	581	587
769	593	598	604	610	615	621	627	632	638	643
770	88 649	88 655	88 660	88 666	88 672	88 677	88 683	88 689	88 694	88 700
771	705	711	717	722	728	734	739	745	751	756
772	762	767	773	779	784	790	796	801	807	812
773	818	824	829	835	840	846	852	857	863	869
774	874	880	885	891	897	902	908	913	919	925
775	88 930	88 936	88 941	88 947	88 953	88 958	88 964	88 969	88 975	88 981
776	986	992	997	89 003	89 009	89 014	89 020	89 025	89 031	89 037
777	89 042	89 048	89 053	059	065	070	076	081	087	092
778	098	103	109	115	120	126	131	137	143	148
779	154	159	165	171	176	182	187	193	198	204
780	89 210	89 215	89 221	89 226	89 232	89 237	89 243	89 248	89 254	89 260
781	265	271	276	282	287	293	299	304	310	315
782	321	326	332	337	343	348	354	360	365	371
783	376	382	387	393	398	404	409	415	421	426
784	432	437	443	448	454	459	465	470	476	481
785	89 487	89 493	89 498	89 504	89 509	89 515	89 520	89 526	89 531	89 537
786	542	548	553	559	564	570	575	581	586	592
787	598	603	609	614	620	625	631	636	642	647
788	653	658	664	669	675	680	686	691	697	702
789	708	713	719	724	730	735	741	746	752	757
790	89 763	89 768	89 774	89 779	89 785	89 790	89 796	89 801	89 807	89 812
791	818	823	829	834	840	845	851	856	862	867
792	873	878	884	889	894	900	905	911	916	922
793	927	933	938	944	949	955	960	966	971	977
794	982	988	993	999	90 004	90 009	90 015	90 020	90 026	90 031
795	90 037	90 042	90 048	90 053	90 059	90 064	90 070	90 075	90 080	90 086
796	091	097	102	108	113	119	124	130	135	140
797	146	151	157	162	168	173	179	184	189	195
798	200	206	211	217	222	228	233	238	244	249
799	255	260	266	271	276	282	287	293	298	304
800	90 309	90 314	90 320	90 325	90 331	90 336	90 342	90 347	90 352	90 358
N.	0	1	2	3	4	5	6	7	8	9

Logarithms of Numbers — 800 to 850

N.	0	1	2	3	4	5	6	7	8	9
800	90 309	90 314	90 320	90 325	90 331	90 336	90 342	90 347	90 352	90 358
801	363	369	374	380	385	390	396	401	407	412
802	417	423	428	434	439	445	450	455	461	466
803	472	477	482	488	493	499	504	509	515	520
804	526	531	536	542	547	553	558	563	569	574
805	90 580	90 585	90 590	90 596	90 601	90 607	90 612	90 617	90 623	90 628
806	634	639	644	650	655	660	666	671	677	682
807	687	693	698	704	709	714	720	725	730	736
808	741	747	752	757	763	768	773	779	784	790
809	795	800	806	811	816	822	827	832	838	843
810	90 849	90 854	90 859	90 865	90 870	90 875	90 881	90 886	90 891	90 897
811	902	907	913	918	924	929	934	940	945	950
812	956	961	966	972	977	982	988	993	998	91 004
813	91 009	91 014	91 020	91 025	91 030	91 036	91 041	91 046	91 052	057
814	062	068	073	078	084	089	094	100	105	110
815	91 116	91 121	91 126	91 132	91 137	91 142	91 148	91 153	91 158	91 164
816	169	174	180	185	190	196	201	206	212	217
817	222	228	233	238	244	249	254	259	265	270
818	275	281	286	291	297	302	307	313	318	323
819	328	334	339	344	350	355	360	366	371	376
820	91 381	91 387	91 392	91 397	91 403	91 408	91 413	91 418	91 424	91 429
821	434	440	445	450	456	461	466	471	477	482
822	487	493	498	503	508	514	519	524	529	535
823	540	545	551	556	561	566	572	577	582	588
824	593	598	603	609	614	619	624	630	635	640
825	91 645	91 651	91 656	91 661	91 666	91 672	91 677	91 682	91 688	91 693
826	698	703	709	714	719	724	730	735	740	745
827	751	756	761	766	772	777	782	787	793	798
828	803	808	814	819	824	829	835	840	845	850
829	856	861	866	871	876	882	887	892	897	903
830	91 908	91 913	91 918	91 924	91 929	91 934	91 939	91 944	91 950	91 955
831	960	965	971	976	981	986	991	997	92 002	92 007
832	92 012	92 018	92 023	92 028	92 033	92 038	92 044	92 049	054	059
833	065	070	075	080	085	091	096	101	106	111
834	117	122	127	132	137	143	148	153	158	163
835	92 169	92 174	92 179	92 184	92 189	92 195	92 200	92 205	92 210	92 215
836	221	226	231	236	241	247	252	257	262	267
837	273	278	283	288	293	299	304	309	314	319
838	324	330	335	340	345	350	356	361	366	371
839	376	381	387	392	397	402	407	412	418	423
840	92 428	92 433	92 438	92 443	92 449	92 454	92 459	92 464	92 469	92 474
841	480	485	490	495	500	505	511	516	521	526
842	531	536	542	547	552	557	562	567	573	578
843	583	588	593	598	603	609	614	619	624	629
844	634	639	645	650	655	660	665	670	675	681
845	92 686	92 691	92 696	92 701	92 706	92 711	92 717	92 722	92 727	92 732
846	737	742	747	752	758	763	768	773	778	783
847	788	794	799	804	809	814	819	824	829	835
848	840	845	850	855	860	865	870	875	881	886
849	891	896	901	906	911	916	922	927	932	937
850	92 942	92 947	92 952	92 957	92 962	92 967	92 973	92 978	92 983	92 988
N.	0	1	2	3	4	5	6	7	8	9

850 to 900 — Logarithms of Numbers

N.	0	1	2	3	4	5	6	7	8	9
850	92 942	92 947	92 952	92 957	92 962	92 967	92 973	92 978	92 983	92 988
851	993	998	93 003	93 008	93 013	93 019	93 024	93 029	93 034	93 039
852	93 044	93 049	054	059	064	069	075	080	085	090
853	095	100	105	110	115	120	125	131	136	141
854	146	151	156	161	166	171	176	181	187	192
855	93 197	93 202	93 207	93 212	93 217	93 222	93 227	93 232	93 237	93 242
856	247	252	258	263	268	273	278	283	288	293
857	298	303	308	313	318	323	329	334	339	344
858	349	354	359	364	369	374	379	384	389	394
859	399	404	409	415	420	425	430	435	440	445
860	93 450	93 455	93 460	93 465	93 470	93 475	93 480	93 485	93 490	93 495
861	500	505	510	515	521	526	531	536	541	546
862	551	556	561	566	571	576	581	586	591	596
863	601	606	611	616	621	626	631	636	641	646
864	651	656	661	667	672	677	682	687	692	697
865	93 702	93 707	93 712	93 717	93 722	93 727	93 732	93 737	93 742	93 747
866	752	757	762	767	772	777	782	787	792	797
867	802	807	812	817	822	827	832	837	842	847
868	852	857	862	867	872	877	882	887	892	897
869	902	907	912	917	922	927	932	937	942	947
870	93 952	93 957	93 962	93 967	93 972	93 977	93 982	93 987	93 992	93 997
871	94 002	94 007	94 012	94 017	94 022	94 027	94 032	94 037	94 042	94 047
872	052	057	062	067	072	077	082	087	092	096
873	101	106	111	116	121	126	131	136	141	146
874	151	156	161	166	171	176	181	186	191	196
875	94 201	94 206	94 211	94 216	94 221	94 226	94 231	94 236	94 241	94 246
876	250	255	260	265	270	275	280	285	290	295
877	300	305	310	315	320	325	330	335	340	345
878	350	354	359	364	369	374	379	384	389	394
879	399	404	409	414	419	424	429	434	438	443
880	94 448	94 453	94 458	94 463	94 468	94 473	94 478	94 483	94 488	94 493
881	498	503	507	512	517	522	527	532	537	542
882	547	552	557	562	567	572	576	581	586	591
883	596	601	606	611	616	621	626	631	635	640
884	645	650	655	660	665	670	675	680	685	689
885	94 694	94 699	94 704	94 709	94 714	94 719	94 724	94 729	94 734	94 739
886	743	748	753	758	763	768	773	778	783	788
887	792	797	802	807	812	817	822	827	832	836
888	841	846	851	856	861	866	871	876	880	885
889	890	895	900	905	910	915	920	924	929	934
890	94 939	94 944	94 949	94 954	94 959	94 963	94 968	94 973	94 978	94 983
891	988	993	998	95 002	95 007	95 012	95 017	95 022	95 027	95 032
892	95 037	95 041	95 046	051	056	061	066	071	075	080
893	085	090	095	100	105	110	114	119	124	129
894	134	139	144	148	153	158	163	168	173	178
895	95 182	95 187	95 192	95 197	95 202	95 207	95 211	95 216	95 221	95 226
896	231	236	241	245	250	255	260	265	270	274
897	279	284	289	294	299	303	308	313	318	323
898	328	333	337	342	347	352	357	362	366	371
899	376	381	386	391	395	400	405	410	415	419
900	95 424	95 429	95 434	95 439	95 444	95 448	95 453	95 458	95 463	95 468
N.	0	1	2	3	4	5	6	7	8	9

Logarithms of Numbers **900 to 950**

N.	0	1	2	3	4	5	6	7	8	9
900	95 424	95 429	95 434	95 439	95 444	95 448	95 453	95 458	95 463	95 468
901	473	477	482	487	492	497	501	506	511	516
902	521	526	530	535	540	545	550	554	559	564
903	569	574	578	583	588	593	598	602	607	612
904	617	622	627	631	636	641	646	651	655	660
905	95 665	95 670	95 675	95 679	95 684	95 689	95 694	95 698	95 703	95 708
906	713	718	722	727	732	737	742	746	751	756
907	761	766	770	775	780	785	789	794	799	804
908	809	813	818	823	828	833	837	842	847	852
909	856	861	866	871	876	880	885	890	895	899
910	95 904	95 909	95 914	95 919	95 923	95 928	95 933	95 938	95 942	95 947
911	952	957	961	966	971	976	980	985	990	995
912	96 000	96 004	96 009	96 014	96 019	96 023	96 028	96 033	96 038	96 042
913	047	052	057	061	066	071	076	080	085	090
914	095	099	104	109	114	118	123	128	133	137
915	96 142	96 147	96 152	96 156	96 161	96 166	96 171	96 175	96 180	96 185
916	190	194	199	204	209	213	218	223	228	232
917	237	242	246	251	256	261	265	270	275	280
918	284	289	294	299	303	308	313	317	322	327
919	332	336	341	346	350	355	360	365	369	374
920	96 379	96 384	96 388	96 393	96 398	96 402	96 407	96 412	96 417	96 421
921	426	431	435	440	445	450	454	459	464	468
922	473	478	483	487	492	497	501	506	511	516
923	520	525	530	534	539	544	548	553	558	563
924	567	572	577	581	586	591	595	600	605	610
925	96 614	96 619	96 624	96 628	96 633	96 638	96 642	96 647	96 652	96 656
926	661	666	671	675	680	685	689	694	699	703
927	708	713	717	722	727	731	736	741	745	750
928	755	760	764	769	774	778	783	788	792	797
929	802	806	811	816	820	825	830	834	839	844
930	96 848	96 853	96 858	96 862	96 867	96 872	96 876	96 881	96 886	96 890
931	895	900	904	909	914	918	923	928	932	937
932	942	946	951	956	960	965	970	974	979	984
933	988	993	998	97 002	97 007	97 011	97 016	97 021	97 025	97 030
934	97 035	97 039	97 044	049	053	058	063	067	072	077
935	97 081	97 086	97 090	97 095	97 100	97 104	97 109	97 114	97 118	97 123
936	128	132	137	142	146	151	155	160	165	169
937	174	179	183	188	193	197	202	206	211	216
938	220	225	230	234	239	243	248	253	257	262
939	267	271	276	280	285	290	294	299	304	308
940	97 313	97 317	97 322	97 327	97 331	97 336	97 341	97 345	97 350	97 354
941	359	364	368	373	377	382	387	391	396	401
942	405	410	414	419	424	428	433	437	442	447
943	451	456	460	465	470	474	479	483	488	493
944	497	502	506	511	516	520	525	529	534	539
945	97 543	97 548	97 552	97 557	97 562	97 566	97 571	97 575	97 580	97 585
946	589	594	598	603	608	612	617	621	626	630
947	635	640	644	649	653	658	663	667	672	676
948	681	685	690	695	699	704	708	713	718	722
949	727	731	736	740	745	750	754	759	763	768
950	97 772	97 777	97 782	97 786	97 791	97 795	97 800	97 804	97 809	97 814
N.	0	1	2	3	4	5	6	7	8	9

950 to 999 — Logarithms of Numbers

N.	0	1	2	3	4	5	6	7	8	9
950	97 772	97 777	97 782	97 786	97 791	97 795	97 800	97 804	97 809	97 814
951	818	823	827	832	836	841	845	850	855	859
952	864	868	873	877	882	887	891	896	900	905
953	909	914	918	923	928	932	937	941	946	950
954	955	959	964	969	973	978	982	987	991	996
955	98 000	98 005	98 009	98 014	98 019	98 023	98 028	98 032	98 037	98 041
956	046	050	055	059	064	069	073	078	082	087
957	091	096	100	105	109	114	118	123	128	132
958	137	141	146	150	155	159	164	168	173	177
959	182	186	191	195	200	205	209	214	218	223
960	98 227	98 232	98 236	98 241	98 245	98 250	98 254	98 259	98 263	98 268
961	272	277	281	286	290	295	299	304	309	313
962	318	322	327	331	336	340	345	349	354	358
963	363	367	372	376	381	385	390	394	399	403
964	408	412	417	421	426	430	435	439	444	448
965	98 453	98 457	98 462	98 466	98 471	98 475	98 480	98 484	98 489	98 493
966	498	502	507	511	516	520	525	529	534	538
967	543	547	552	556	561	565	570	574	579	583
968	588	592	597	601	606	610	614	619	623	628
969	632	637	641	646	650	655	659	664	668	673
970	98 677	98 682	98 686	98 691	98 695	98 700	98 704	98 709	98 713	98 718
971	722	726	731	735	740	744	749	753	758	762
972	767	771	776	780	785	789	793	798	803	807
973	811	816	820	825	829	834	838	843	847	851
974	856	860	865	869	874	878	883	887	892	896
975	98 901	98 905	98 909	98 914	98 918	98 923	98 927	98 932	98 936	98 941
976	945	949	954	958	963	967	972	976	981	985
977	990	994	998	99 003	99 007	99 012	99 016	99 021	99 025	99 030
978	99 034	99 038	99 043	047	052	056	061	065	069	074
979	078	083	087	092	096	100	105	109	114	118
980	99 123	99 127	99 132	99 136	99 140	99 145	99 149	99 154	99 158	99 163
981	167	171	176	180	185	189	194	198	202	207
982	211	216	220	224	229	233	238	242	247	251
983	255	260	264	269	273	277	282	286	291	295
984	300	304	308	313	317	322	326	330	335	339
985	99 344	99 348	99 352	99 357	99 361	99 366	99 370	99 375	99 379	99 383
986	388	392	397	401	405	410	414	419	423	427
987	432	436	441	445	449	454	458	463	467	471
988	476	480	485	489	493	498	502	507	511	515
989	520	524	528	533	537	542	546	550	555	559
990	99 564	99 568	99 572	99 577	99 581	99 585	99 590	99 594	99 599	99 603
991	607	612	616	621	625	629	634	638	642	647
992	651	656	660	664	669	673	677	682	686	691
993	695	699	704	708	712	717	721	726	730	734
994	739	743	747	752	756	761	765	769	774	778
995	99 782	99 787	99 791	99 795	99 800	99 804	99 809	99 813	99 817	99 822
996	826	830	835	839	843	848	852	856	861	865
997	870	874	878	883	887	891	896	900	904	909
998	913	917	922	926	931	935	939	944	948	952
999	957	961	965	970	974	978	983	987	991	996
N.	0	1	2	3	4	5	6	7	8	9

INDEX

A

A and B scales of slide rule....383–386
Acme thread....................298
Acute angles................144, 234
Addendum of gear tooth...........325
Addition
  of decimal fractions..............30
  elimination of unknown by...104, 105
  of fractions....................7–9
  of mixed numbers.................9
  of monomials....................64
  of polynomials..................65
  of positive and negative algebraic numbers..............62, 63
  and subtraction of algebraic expressions.................64–66
  and subtraction formulas........249
  and subtraction of fractions combined..................10, 11
Adjacent angles...................143
Affected quadratic, complete or 110–114
Air cylinder operations............450
Air operated boosters..............451
Air power cylinders...........449–450
Airplane controls.............245, 246
Algebra.......................58–141
Algebraic equations, solving....106–108
Algebraic expressions............59–71
Algebraic numbers and expressions, positive and negative..62–71
Algebraic principles applied to shop and toolroom problems................114–128
Algebraic symbols..............58, 59
Alpha, definition of...............247
Alternate-exterior angles...........151
Alternate-interior angles...........151
American National screw threads..................293–297
American Standard taper pipe thread...................300, 301
American Standards Associated stub-tooth system 328, 330, 334, 335
Analogue computer...........441–443
Anchor ring......................205
Angles
  acute......................144, 234
  adjacent.......................143
  alternate-exterior...............151
  alternate-interior...............151
  bisecting......................186
  central...................164, 165

Angles—*continued*
  complementary........144, 247, 248
  constructed equal to given angles.189
  constructing 30°, 45° and 60°.188, 189
  corresponding..............151, 152
  definition of................142–144
  designation of..................234
  double.........................250
  equal......................152–154
  exterior........................151
  exterior-interior............151, 152
  half...........................250
  helix..................293, 306, 307
  inscribed.............164, 166, 167
  and inscribed circles, problems involving.......271, 272
  interior........................151
  negative.......................248
  obtuse.........................144
  opposite..............143, 149, 150
  of polygons....................145
  pressure.......................324
  problems involving..........272–274
  right..........................143
  straight.......................143
  supplementary........144, 247, 248
  tangent of.....................233
  thread.........................293
  thread helix...............306, 307
  of triangles................154–157
  trigonometric functions of...232–234, 247–251
  types of...................143, 144
  used in engineering practice......262
  vertex of..................142, 143
  vertical...............143, 149, 150
Angular indexing.................352
Angular measurement.........143, 144
Annulus, definition of.............199
Antecedent, definition of...........92
Antilogarithm of logarithm, finding.....................363
Applying statistical quality control to dimension......466–468
Arc, circular.....163, 164, 185–188, 192 213–218, 272–274
Area graphs.....................434
Areas
  of circles......184, 199, 200, 389–391
  definition of...................201
  of fillet.......................200
  of plane geometric figures....197–200

569

570 MATHEMATICS FOR INDUSTRY

Areas—*continued*
  of solids 201–206
  of torus 205
Arithmetic, review of 1–57
Automatic process inspection ... 477–478
Automatic production system ... 447–448
Automation, computation of
  automatic controls for 438–456
Automobile and airplane
  controls, linkage in 245, 246
Axes 433
Axioms and theorems 80, 81, 147–154

B

Bar
  cylindrical 243, 421, 422
  sine 267–270
Base, definition of 60
Base, circle of gear 323
Beam
  cantilever 413
  H 414
  I 414
  supporting and loading of 413, 414
Belts and power transmitted by
  belts 418–420
Bending moments 412–414
Best wire diameter 303–305
Bevel gears
  explanation of 326
  formulas for 338–341
  planetary 327
  ratios of 348
  skew 327, 339
  spiral 339
  straight-tooth 339–341
Binomials, definition of 60
Bisecting
  of angles 186
  of circular arcs 186
  of straight line 186
Bolts, problems in 405–407
Braces, removing 78, 79
Brackets, removing 78, 79
Brake horsepower 417
Briggs system of logarithms 359
Broaching of gear teeth 346
Brown and Sharpe gear system 328
Buttress thread 299, 300

C

C and D scales of slide rule 373–377
Cancellation method 13
Cantilever beam 413
Casting 345
Center
  of circle 163
  of circular arcs 187, 188
  of moment 410

Center distance of spur gears 322
Central angle 164, 165
CF and DF scales of slide rule 390
Characteristic and mantissa 360, 361
Chart, decimal equivalent 39
Charts
  flow 435
  percentage 434
  pictorial 434
Chords
  of the circle, theorems for 168, 169
  definition of 164
  height of 213, 314
CI scale of slide rule 392
Circles, angles and inscribed ... 166, 167,
                              271, 272
  area of 184, 199, 200
  base 323
  center of 163
  chords of 168, 169
  circle tangent to three 264, 265
  circumference of 163, 183, 184
  circumscribed about squares 195
  circumscribed about
    triangles 170, 171, 194
  concentric 164
  construction problems
    relating to 170–172
  definition of 163, 164
  diameter of ... 163, 183, 184, 212, 213
  in which disks fit 215, 216
  drawn between inclined lines 193
  drawn through points ... 192, 218, 219
  hexagon circumscribed about 189
  hexagon inscribed in 171, 172, 189
  inscribed in right triangles ... 167, 216
  inscribed in squares 195
  inscribed in triangles 164
  lines cutting 169, 170
  lines tangent to 165, 166
  octagons circumscribed about 196
  octagons inscribed in 196
  pentagons inscribed in 195
  pitch 322
  polygons circumscribed about 164
  polygons inscribed in 164
  problems involving angles and
    inscribed 271, 272
  properties of 183–185
  radius of 163
  slide rule used to find
    diameters, circumferences,
    and areas of 389–391
  squares inscribed in 171
  tangent externally 167, 168, 218
  tangents drawn to 193
  theorems or propositions
    relating to 165–170
  triangles inscribed in 169

INDEX 571

Circular arcs
  bisecting....................186
  center of.................187, 188
  definition of................163, 164
  diameter of...................213
  drawn with given radius.........192
  length of..............185, 217, 218
  lines drawn tangent to...........192
  obtaining..................216, 217
  problems involving..........272–274
  radius or diameter of
    internal..................214, 215
Circular pitch of spur gears........322
Circumference of circle.......163, 183,
                    184, 389–391
Clearance of gear tooth............325
Coefficient, definition of............60
Cofunctions, functions and.....234, 235
Cologarithms.................363, 364
Commercial tolerance.............438
Common denominator, least........5–7
Common fractions
  extracting cube root of............48
  extracting square root of..........45
  review of....................1–15
Common logarithms...............359
Complementary angles
  definition of...................144
  functions of................247, 248
Complete or affected quadratic..110–114
Composite gear system............328
Composition of forces.............409
Compound indexing...............352
Compound ratios and
  proportions...............97–100
Compound-geared lathes.......313–315
Computation of automatic
  controls for automation....438–456
Computing
  critical costs...................484
  tooling costs...................485
Computers, electronic
  analogue...................441–447
  digital.....................441–447
Concentric circles.................164
Cone...........................203
Cone gears......................327
Consequent, definition of...........92
Construction, geometric
  problems of..............186–197
Control chart....................484
Control devices
  explanation of.............440–441
  hydraulic..................453–454
  pneumatic.................448–449
Control valves, pneumatic.....452–453
Conversion
  of decimals and fractions......37, 38
  inch-millimeter.................39

Coordinated data processing
  systems..................445–447
Corollary, definition of............148
Corresponding angles.........151, 152
Cosines, law of...............253–257
Costs
  direct........................481
  fixed........................479
  indirect......................482
  manufacturing.................479
  tooling......................485
Cotangent formulas..........259–262
Crest of screw threads.............292
Critical point, definition of.........413
Critical life, computing............484
Cube
  imperfect.....................47
  perfect....................46, 47
  total area of surface of..........201
  volume of....................201
Cube roots, extraction of.45–48, 386–389
Curve, involute......191, 192, 323, 432
Cutters used to machine gears......346
Cutting operations, estimating
  time required for various..421, 422
Cutting screw threads on a
  lathe....................311–335
Cutting speeds and feeds for
  machine tools.............420, 421
Cycloid, constructing a............190
Cylinder
  air power..................449–450
  frustum of....................204
  helix constructed on............191
  hydraulic pressure..............546
  right circular..................204
  volume of.................204–206
Cylindrical bar
  cut into hexagon...............243
  problems in...............421, 422

D

D scale of slide rule..............394
Decagon........................147
Decimal equivalent chart...........39
Decimal fractions
  addition of.....................30
  conversion of................37, 38
  division of..................33–36
  extracting cube root of..........48
  extracting square root of......42–45
  multiplication of.............31–33
  review of...................29–39
  slide rule used to find square
    and cube roots of.........387, 388
  subtraction of...............30, 31
Decimal point located on slide
  rule.........................377
Dedendum of gear tooth...........325

Deformation, definition of..........402
Degrees, definition of..............143
Denominator
  adding fractions with like..........8
  adding fractions with unlike........8
  definition of......................1
  finding factors common to
    numerator and................4, 5
  least common...................5-7
  subtracting fractions with like......9
  subtracting fractions with
    unlike......................9, 10
Depreciation......................480
Depth of thread...................293
Diagonal line.....................147
Dial indicators and gages......458-460
Diameter
  best wire...................303-305
  of circles......163, 183, 184, 212, 213,
                                389-391
  of circular arcs.............213-215
  pitch..293, 304, 309, 310, 322, 337, 338
  of screw threads................292
Diametral pitch of spur gears......322
Differential indexing..........351, 352
Digital computer..............441-447
  basic elements of...........443-444
  uses of.........................445
Dimensional control
  explanation of..................457
  through gaging..................458
Dimensional quality control....460-476
Direct indexing...................349
Direct costs......................481
Direct and inverse proportions....95-97
Disks, precision......122-124, 215, 216,
                          270, 271, 280
Dividend, definition of.............14
Dividing lines....................192
Division
  checking.....................18, 19
  of decimal fractions...........33-36
  of fractions..................14, 15
  of mixed numbers................15
  of monomials and polynomials..69-71
  and multiplication with
    logarithms...............365, 366
  and multiplication with slide
    rule.....................378-383
  of positive and negative
    algebraic numbers and
    expressions................68-71
Divisor, definition of..............14
Double angles, functions of........250
Double-thread screw...............293
Dovetails, problems involving..277-279
Drill, tap....................296, 297
Drill jig plate, locating holes in.120-122
Drill press.......................421

E

Economic selections, equipment....483
Effector mechanisms..........448, 453
Elastic limit, definition of..........403
Elasticity, modulus of.............403
Electronic computors
  analogue...................441-443
  digital.....................441-447
Elimination
  of nines process..............15-19
  of unknown by addition or
    subtraction..............104, 105
  of unknown by comparison...105, 106
  unknown found by process of...82-84
  of unknown by substitution......105
Ellipse
  area of........................200
  constructing an.............190, 191
Energy
  definition of...................417
  potential......................417
Engineering computations.....402-433
Engineering practice, angles used
  in............................262
Engineering problems, use of
  graphs in solution of......432-437
Engines
  horsepower of.............416, 417
  torque of various...............415
Equal angles, theorems for.....152-154
Equal triangles, theorems
  for................155, 156, 158
Equations
  algebraic..................106-108
  definition of....................59
  inconsistent...................104
  indeterminate..................103
  quadratic.........110-114, 126, 127
  simple.......................80-84
  solution of simple
    simultaneous.............103-106
  steps followed to solve.......106-108
Equiangular triangle, definition
  of............................146
Equilateral polygons..............147
Equilateral triangle......145, 146, 197
Evaluation of formulas and
  methods of substitution....106-110
Excess of nines
  determining..................15-19
  used to check division of
    decimal fractions...........34-36
  used to check extraction of the
    cube root..................47, 48
  used to check extraction of
    square root....................42
  used to check multiplication of
    decimal fractions...........31, 32

INDEX

E (continued)

Exponents
  definition of..................60, 61
  laws of....................359, 360
  powers and..................75, 76
  rule of...............66, 69, 75, 76
Expressions, algebraic...........59, 71
Exterior angles, definition of.......151
Exterior-interior angles........151, 152
External thread..................292
Extraction
  of cube root of common fractions..48
  of cube root of decimal
    fractions......................48
  of cube root of imperfect cube.....47
  of cube root without K scale of
    slide rule................388, 389
  of cube root of perfect cube....46, 47
  of cube root with slide rule....386, 387
  of root of an expression........76-78
  of root with logarithms......366, 367
  of square root of common
    fractions......................45
  of square root of a decimal
    fraction....   42-45
  of square root of an imperfect
    square........................41
  of square root of a perfect
    square.....................40, 41
  of square root with slide rule.383-386
Extremes of the proportion.........93

F

Factor of safety..................404
Factoring
  affected quadratic found by......111
  by grouping......................74
  by inspection..................72-74
  special products used in......:..71, 72
Factors, investment...............483
Fellows Gear Shaper Company
  system..............328, 330, 331
Fillet
  area of........................200
  definition of....................325
Fillet radius......................325
Flow chart.......................435
Fixed costs......................479
Folded scales of slide rule..........390
Foot-pound, definition of..........410
Force
  composition of.................409
  definition of................402, 409
  lever arm of....................410
  moment of................410-412
Forming on power presses..........345
Forming tool, problem in..........281
Formulas
  addition and subtraction.........249
  for bevel gears..............338-341

Formulas—*continued*
  cotangent..................259-262
  definition of.............59, 79, 80
  double-angle...................250
  evaluation of..............106-110
  for full-depth tooth gears....328-330
  for gear problems...........331-335
  for half angles..................250
  for helical and spiral gears...341-343
  for internal gears...............338
  product...................250, 251
  projection.................257-259
  quadratic..................112-114
  for screw threads.......294-297, 306
  for spur gears..............327-335
  for straight-tooth bevel
    gears...................339-341
  for stub-tooth gears.........330, 331
  three-wire.............303-305, 310
  for worm gears.............343-345
Fractions
  addition of.....................7-9
  combining addition and
    subtraction of..............10, 11
  common...............1-15, 45, 48
  conversion of decimals and.....37, 38
  decimal............29-39, 42-45, 48
  division of...................14, 15
  improper......................2, 3
  multiplication of..............12, 13
  proper...........................2
  reduction of...................2-7
  subtraction of.................9, 10
Frequency distribution........461-464
Frustum
  of cylinder....................204
  definition of...................202
  of prism......................202
  of pyramid or cone.............203
Full-depth tooth gears, formulas
  for......................328-330
Functions
  and cofunctions............234, 235
  of complementary and supple-
    mentary angles..........247, 248
  of double angles................250
  of half angles...................250
  of negative angles..............248
  relations between
    trigonometric............247-251
  of sum and difference of two
    angles......................249
  and tables, trigonometric....232-240
Fundamental identities...........247

G

Gages
  for checking thickness of tooth...337
  dial indicators and..........458-460

574 MATHEMATICS FOR INDUSTRY

Gages—*continued*
  snap thread 310, 311
  taper 280, 281
  taper plug 268, 269, 274–277
  thread 307–311
  thread plug 307, 308
  thread ring 308, 309
Gear blank, weight of the
  castiron 208–211
Gear Shaper Company system,
  Fellows 328, 330, 331
Gear teeth
  addendum of 325
  broaching of 346
  clearance of 325
  dedendum of 325
  involute shape of 323
Gear tooth vernier 336
Gear trains 348
Gears
  base circle of 323
  bevel 326, 338–341
  Brown and Sharpe system of 328
  composite system of 328
  cone 327
  cut on milling machine 350
  definition of 322
  eccentric 327
  formulas for problems of 331–335
  full-depth tooth 328–330
  helical 326, 341–343
  herringbone 326, 343
  hypoid 339
  internal 327, 338
  machining of 345, 346
  pin 327
  planetary bevel 327
  ratios of 346–348
  shaping 345, 346
  sheet-metal stamped 345
  skew bevel and hypoid 327, 339
  spiral 326, 327, 341–343, 348
  spiral bevel 339
  spur 321, 322, 325–327, 335, 337,
                         338, 346–348
  standard involute 328, 331–334
  straight-tooth bevel 339–341
  stub-tooth 328, 330, 331, 334, 335
  types of 325–327
  worm 326, 343, 345, 350, 351
General theorems or
  propositions 148–154
Generating processes 346
Geometric figures, plane 144–147
                              197–200
Geometric principles applied in
  solution of shop problems .. 212–219
Geometric problems of
  construction 186–197

Geometry
  plane 142–200
  solid 201–212
Glass runner of slide rule 372
Graph
  area 434
  bar 433
  line 432
  nomograph 436
  percentage 434
  pictorial 435
Graphical method
  definition of 432
  trigonometric functions found
    by 239, 240
Grouping
  factoring by 74
  symbols of 78, 79

H

H beam 414
Half angles, function of 250
Head, indexing or dividing 348
Helical gear ratios 348
Helical gears 326, 341–343
Helix constructed on cylinder 191
Helix angle
  explanation of 293
  thread 306, 307
Heptagon 147
Herringbone gears 326, 343
Hexagon
  circumscribed about a circle 189
  cylindrical bar cut into 243
  explanation of 147
  inscribed in circle 171, 172, 189
Holes
  checking reamed 243
  diameter of eight 243
  problem in three 244, 245
Hooke's law 403
Horsepower
  brake 417
  definition of 410
  engines and motors 416, 417
Hydraulic
  control devices 453–454
  pressure cylinders 456
  pumps 454–455
Hypoid gears 327, 339

I

I beam 414
Identities, fundamental 247
Imperfect cube, extracting cube
  root of an 47
Imperfect square
  definition of 40
  extracting square root of an 41

INDEX 575

	PAGE
Improper fractions	2, 3
Inch-millimeter conversion	39
Inch-pound, definition of	410
Incomplete or pure quadratic	110
Inconsistent equations	104
Indeterminate equation	103
Index of the root	40, 61
Indexing	348–352
Indexing plate	351
Indirect costs	482
Indicating gage	461
Inscribed angle	164
Inscribed circles	271, 272
Inspection, factoring by	72–74
Inspection and quality control, problems in	457–478
Insurance	480
Integer	2
Interest	480
Interior angles, definition of	151
Internal gears	327, 338
Internal thread	292
International (Metric) thread	300
Interpolation	237, 238, 362, 363
Inverse proportions, direct and	95–97
Inverse ratios	92, 93
Investment, factors	483
Involute curve	191, 192, 323
Involute gears, standard	328
Involute shape of gear tooth	323
Involute systems, full-depth	328
Involute teeth, checking thickness of	335–337
Irregular plane surface, area of	200
Isosceles triangle	145, 157

K

K scale of slide rule	386–389
Kinetic energy	417

L

L scale of slide rule	394
Lathe	
compound-geared	313–315
cutting screw threads on	311–315
cutting speed of	420
simple-geared	311–313
Law	
of cosines	253–257
of exponents	359, 360
Hooke's	403
of sines	251–257
Lead of screw thread	293
Least common denominator	5–7
Lever arm of force	410
Levers, forms of	411
Life, critical	484
Limit, elastic	403
Line graph	432

	PAGE
Lines	
angle constructed on given	189
bisecting straight	186
circles drawn between inclined	193
cutting circle, theorem for	169, 170
cutting triangle, theorem for	159
definition of	142
diagonal	147
dividing	192
drawn tangent to circular arcs	192
parallel	144, 150–152, 154, 186, 187
perpendicular	144, 148–152, 187
pressure	324
tangent	164–166
Linkage in automobile and airplane controls	245, 246
Literal algebraic expressions	
addition and subtraction of	64, 65
explanation of	59, 60
determining values of	61, 62
Load, definition of	402
Logarithmic scales of slide rule	394
Logarithms	
antilogarithms of	363
Briggs system of	359
common	359
finding	361–363
division of numbers by means of	365
multiplication of numbers by means of	364, 365
natural or Napierian system of	359
root extracted with	366, 367
used for multiplication and division combined	366
used to raise number to any power	367, 368

M

Machine detail, problem in	245
Machine equipment, selection	483
Machine shop and toolroom problems	124–128
Machine tools, cutting speeds and feeds for	420, 421
Machines	
mechanical efficiency of	417, 418
milling	346, 350, 420
reciprocating	346
Machining	
of gears	345, 346
of reamer	348
Mannheim slide rule, 10-inch	372
Mantissa, characteristic and	360, 361
Manufacturing costs	479
Mean proportional	93
Means of the proportion	93
Mechanical advantage	412
Mechanical efficiency of machines	417, 418

Micrometer, thread............301, 302
Milling machines.........346, 350, 420
Minuend, definition of...............9
Minutes, definition of.............143
Mixed numbers
   addition of......................9
   definition of....................2
   dividing........................15
   multiplying....................13
   reduction of...................2, 3
   subtracting....................10
Modulus of elasticity..............403
Molding...........................345
Moment
   bending....................412-414
   center of.......................410
   of force...................410-412
   of resistance...................413
   turning........................415
Monomials
   addition of.....................64
   definition of.................59, 60
   division of...................69, 70
   multiplication of...............67
   subtraction of...............64, 65
Motors
   horsepower of..............416, 417
   torque of various...............415
Multiplicand, definition of..........12
Multiplication
   checking.....................17, 18
   of decimal fractions..........31-33
   and division with logarithms..364-366
   and division with slide rule...378-383
   of fractions..................12, 13
   of mixed numbers...............13
   of positive and negative algebraic numbers and expressions....................66-68
Multiplier, definition of............12

N

Napierian system of logarithms,
   natural or....................359
Negative algebraic numbers and
   expressions, positive and.....62-71
Negative angles, functions of.......248
Nines, excess of...15-19, 31, 32, 34-36, 42, 47, 48
Nomograph.......................436
Nonagon.........................147
Numbers
   logarithms of...............361-368
   mixed.............2, 3, 9, 10, 13, 15
   negative........................62
   positive........................62
   whole......................2, 3, 12

Numerator
   definition of.....................1
   and denominator, finding
     factors common to............4, 5
Numerical algebraic expressions...59-61
Numerical values, rounding off...36, 37

O

Oblique triangles
   with all sides given...........256-259
   with any side and its two
     adjacent angles...........259-261
   with any two angles and any
     one side given.................253
   with any two sides and the
     angle opposite one of sides
     given...................254, 255
   with any two sides and angle
     included given............255, 256
   definition of...................145
   problems involving..........251-262
   projection-type problems in..162, 163
Obtuse angle.....................144
Octagon
   circumscribed about a circle......196
   constructed in a square..........197
   construction of.................196
   definition of...................147
   inscribed in circle..............196
Open loop control............439-440
Opposite angles..........143, 149, 150
Optical method of measuring
   threads.......................305

P

Parabola, area of..................200
Parallel lines
   drawing...................186, 197
   definition of...................144
   theorems for..........150-152, 154
Parallelepiped, definition of........201
Parallelogram
   area of........................198
   construction of.................194
   definition of...................146
Parenthesis, removing...........78, 79
Pentagon
   definition of...................147
   inscribed in circle..............195
Percentages, finding..100-103, 391, 392
Percentage chart..................434
Perfect cube, extracting cube
   root of a....................46, 47
Perfect square
   definition of....................40
   extracting square root of......40, 41
Perimeter, definition of............147

INDEX 577

Perpendicular lines
  definition of................... 144
  erecting...................... 187
  theorems for.............. 148–152
Pictorial chart................... 435
Pie chart........................ 434
Pinion, definition of.............. 321
Pitch
  of spur gears, circular.......... 322
  of thread..................... 293
Pitch diameter
  of a screw................ 304, 305
  of screw threads............... 293
  of spur gears.......... 322, 337, 338
  of three-fluted tap......... 309, 310
Plane geometric figures 144–147, 197–200
Plane geometry, principles of... 142–200
Planer.......................... 420
Plate
  indexing...................... 351
  locating holes in a drill jig... 120–122
Plug gage, thread............ 307, 308
Pneumatic
  control devices............ 448–449
  control valves............. 452–453
Point
  circles drawn through... 192, 218, 219
  critical....................... 413
  decimal....................... 377
  definition of.................. 142
  of tangency................... 164
Polygons
  area of regular................ 198
  circumscribed about circles...... 164
  definition of.................. 144
  equilateral.................... 147
  inscribed in circles............. 164
  quadrilateral.................. 146
  sum of angles of............... 145
  types of.................. 145–147
Polynomials
  addition of..................... 65
  definition of................... 60
  division of.................. 69–71
  multiplication of............. 67, 68
  subtraction of............... 65, 66
Positive and negative algebraic
  numbers and expressions..... 62–71
Postulates...................... 147
Potential energy................. 417
Power
  definition of................ 60, 410
  expression raised to desired.... 75, 76
  logarithms used to raise
    number to any........... 367, 368
  required for cutting given
    amount of metal.......... 423–425
  transmitted by belts........ 418–420
Power presses, forming on......... 345

Powers and exponents......... 75, 76
Precision disks... 122–124, 270, 271, 280
Precision screw, pitch diameter
  of..................... 304, 305
Press
  drill.......................... 421
  power........................ 345
Pressure angle................... 324
Pressure line.................... 324
Prime expression.................. 71
Prism
  definition of.................. 201
  frustum of.................... 202
  rectangular................... 201
  regular....................... 202
  right......................... 202
  volume of..................... 202
Problems
  in bolts................... 405, 407
  of construction, geometric.... 186–197
  in cylindrical bars.......... 421, 422
  gear...................... 331–335
  involving angles and circular
    arcs..................... 272–274
  involving angles and inscribed
    circles.................... 271, 272
  involving circles............ 170–172
  involving dovetails.......... 277–279
  involving forming tool........... 281
  involving oblique triangles... 251–262
  involving percentages....... 101–103
  involving precision disks. 270, 271, 280
  involving right triangles..... 240–246
  involving the sine bar...... 267–270
  involving steel rivets........ 405, 406
  involving steel rod.......... 404, 405
  involving strength of
    materials................ 402–409
  involving stub-tooth gears... 334, 335
  involving taper plug gages... 274–277
  involving tapers.... 114–120, 265–267
  involving three holes...... 244, 245
  shop.......... 114–128, 212–219, 394
  tool design................ 279–281
Product, definition of.............. 12
Product formulas............. 250, 251
Product and process control.... 438–440
Products, special............... 71, 72
Projection formulas.......... 257–259
Projection-type problems of
  triangles.................. 162, 163
Proper fractions.................... 2
Proportions
  compound ratios and......... 97–100
  direct and inverse............ 95–97
  extremes of.................... 93
  found with slide rule........ 391, 392
  means of...................... 93
  principles of................ 93–95

Punched card and tape controls, uses of 447
Pure quadratic, incomplete or .. 110, 111
Pyramid
 definition of 202
 frustum of 203
 regular 202
 right 202
 right regular 203
 volume of 203
Pythagorean theorem 161, 162

Q

Quadratic
 complete or affected 110-114
 incomplete or pure 110, 111
Quadratic equations .. 110-114, 126, 127
Quadratic formula, solution of,
 affected quadratic by 112-114
Quadrilateral polygon, definition of 146
Quality control
 automatic process inspection and 477-478
 and inspection, problems in .. 457-478
 and statistical sampling 476-477
Quality control charts
 construction and plotting of .. 474-475
 formation and use of 468-472
 obtaining control limits for .. 472-474
 plotting of 475-476
 sampling procedure and 468-472
Quantity, imaginary 111
Quotient, definition of 3, 14

R

Rack, definition of 327
Rack cutters 346
Rack ratios, spur-gear and 346-348
Radical sign 40
Radius
 of circle 163
 circular arcs drawn with given 192
 or diameter of internal circular arcs 214, 215
 fillet 325
Range 465
Ratios
 explanation of 92, 93
 found with slide rule 391, 392
 gear 346-348
 inverse 92, 93
 and proportions, compound ... 97-100
 reduced to lowest terms 93
Reamed holes, checking 243
Reamer, machining 348
Reciprocal scales of slide rule 392
Reciprocals, definition of 14
Reciprocating machines 346

Rectangle 146, 198
Rectangular prism 201
Reducing ratios to lowest terms 93
Reduction
 of fractions 2-7
 of whole and mixed numbers 2, 3
Relations
 between functions of one angle ... 247
 between trigonometric functions 247-251
Remainder, definition of 9
Resistance, moment of 413
Rhomboid, definition of 146
Rhombus, definition of 146
Right angles 143
Right circular cylinder 204
Right pyramid 202
Right regular prism 202
Right regular pyramid 203
Right triangles
 acute angles of 234
 circles inscribed in 167, 216
 definition of 145
 problems involving 240-246
 problems not appearing in form of 242-246
 projection-type, problem in .. 162, 163
 theorems for 160-162
 trigonometric functions used to solve 232, 233
Ring
 anchor 205
 weight of aluminum 208
Ring gage, thread 308, 309
Rivet
 weight of hollow copper 207, 208
 weight of steel 207
 weight of wrought-iron 206, 207
Rollers
 distance between 268
 locating holes with 122-124
Rolling 345
Roots
 cube 45-48, 387, 388
 extracted with logarithms 366, 367
 extracting 76-78
 index of 40, 61
 of screw threads 292
 square 39-45, 383-386
Rotary side-cutters 346
Rounding off numerical values ... 36, 37
Rule
 of exponents 66, 69, 75, 76
 of signs 66, 68, 69, 75
 slide 372-401

S

S scale of slide rule 392-394
Safe working unit stress 404

INDEX

	PAGE
Safety, factor of	404
Scales of slide rule	373–377, 383–389, 390, 392–394
Screw	
double-thread	293
pitch diameter of	304, 305
single-thread	293
Screw threads	
Acme	298
American National	293–297
American standard taper pipe	300, 301
buttress	299, 300
crest of	292
depth of	293
diameter of	292
external	292
fits for	294
formulas for	294–297, 306
internal	292
International (Metric)	300
on a lathe, cutting	311–315
lead of	293
major diameter of	292
measuring	301–306
minor diameter of	292
optical method of measuring	305
pitch diameter of	293
root of	292
square	298
three-wire method of measuring	302–305
types of	293–301
worm	299
Secants	169
Seconds, definition of	144
Sector	
definition of	164
spherical	205
Segment	
definition of	164
spherical	205
Set points	438
Shaper	420
Shaping gears, processes for	345, 346
Sharp v-thread	297
Shearing, definition of	402
Sheet-metal stamped gears	345
Shop problems	
algebraic principles applied to	114–128
geometric principles applied to	212–219
slide rule applied to	394
Shop trigonometry	262–281
Sign, radical	40
Signs, rule of	66, 68, 69, 75
Similar triangles, theorems for	159–161
Simple indexing	349–351
Simple-geared lathe	311–313
Simultaneous equations, solution of simple	103–106
Sine bar, problems involving	267–270
Sines, law of	251–257
Single-thread screw	293
Skew bevel gears	327, 339
Slide rule	
A and B scales of	383–386
applied to shop problems	394
C and D scales of	373–377
CF and DF scales of	390
CI scale of	392
cubing a number and extracting cube roots with	386, 387
D scale of	394
decimal point located on	377
diameters, circumferences, and areas of circles found with	389–391
division with	381, 382
division and multiplication combined with	382, 383
extracting cube root without K scale of	388, 389
extracting square roots with	383–386
folded scales of	390
K scale of	386–389
L scale of	394
logarithmic scales of	394
Mannheim 10-inch	372
multiplication with	378–381
multiplication and division combined with	382, 383
parts of	372
ratio, proportion, and percentage found with	391, 392
reciprocal scales of	392
S and A scales of	392–394
square and cube roots of decimals found with	387, 388
squaring a number and extracting square roots with	383–386
trigonometric scales of	392–394
Snap, thread gage, checking	310, 311
Solid geometry	201–212
Solids, areas and volumes	201–206
Special products	71, 72
Specific gravities of various materials	211, 212
Sphere, volume and area of	205
Spherical sector, volume of	205
Spherical segment, volume of	205
Spherical zone, volume of	205
Spiral bevel gears	339
Spiral gears	
explanation of	326, 327
formulas for	341–343
ratios of	348

Spur gears
 center distance of............322
 checking pitch diameter of...337, 338
 circular pitch of.............322
 diametral pitch of............322
 explanation of.....321, 322, 325, 326
 formulas for..............327-335
 ratios of..................346-348
 teeth of......................322
Square
 affected quadratic solved by
  completing..................112
 area of........................198
 circle circumscribed about.......195
 circle inscribed in..............195
 definition of...................146
 drawn on given line.............188
 imperfect....................40, 41
 inscribed in circle..............171
 octagon constructed in..........197
 perfect.....................40, 41
 side of........................184
Square root
 checking extraction of........41, 42
 extraction of.........39-45, 383-388
Square screw threads..............298
Squaring a number and extracting square roots with slide
 rule......................383-386
Standard involute gears...328, 331-334
Statistical quality control,
 applying to dimensions....466-468
Statistical sampling, quality
 control and...............476-477
Steel, specific gravity of...........212
Steel rivets, problems in.......405, 406
Steel rod, problems in.........404, 405
Stock of slide rule................372
Straight angle....................143
Straight line, bisecting a..........186
Straight-tooth bevel gears, for-
 mulas for.................339-341
Strain, definition of...............402
Strength
 of materials problems dealing
  with....................402-409
 ultimate.......................404
Stresses
 definition of...............402, 403
 tensile........................402
 torsion........................402
 twisting.......................402
 unit..........................404
Stub tooth.......................324
Stub-tooth gears.328, 330, 331, 334, 335
Substitution
 elimination of unknown by.......105
 evaluation of formulas and
  methods..................106-110

Subtraction
 and addition of algebraic
  expressions................64-66
 and addition formulas...........249
 and addition of fractions
  combined...................10, 11
 of decimal fractions..........30, 31
 elimination of unknown by...104, 105
 of fractions...................9, 10
 of mixed numbers................10
 of monomials................64, 65
 of polynomials...............65, 66
 of positive and negative algebraic numbers..............63, 64
Subtrahend, definition of............9
Supplementary angles.....144, 247, 248
Surface
 of the body....................142
 irregular plane.................200
Symbols
 algebraic...................58, 59
 of grouping..................78, 79

T

Tables, trigonometric functions
 and.....................232-240
Tangency, point of................164
Tangent line.................164-166
Tangents
 of angle.......................233
 of circles..........167, 168, 193, 218
 drawn to circular arcs...........192
Tap
 pitch diameter of three-fluted.309, 310
 v-block used to check.......309, 310
Tap drill...................296, 297
Tape and punched card controls,
 uses of........................447
Taper gage, constructing......280, 281
Taper pipe thread, American
 standard.................300, 301
Taper plug gages
 checking..................268, 269
 problems involving..........274-277
Taper problems, simplified rules
 for solving...............114-120
Tapers, problems involving....265-267
Taxes..........................480
Teeth, gear......322, 325, 335-337, 346
Tensile stress, definition of.........402
Term, definition of.................59
Theorem, pythagorean........161, 162
Theorems
 for angles.........152-154, 165-167
 axioms and................147-154
 for circles.................165-170
 general...................145-154
 for tangent lines............165, 166
 for triangles...............154-162

INDEX

Thickness
  of gear tooth....................325
  of involute teeth, checking...335–337
Thread angle.....................293
Thread gages
  checking..................307–311
  snap......................310, 311
Thread helix angle............306, 307
Thread micrometer...........301, 302
Thread plug gage, checking....307, 308
Thread ring gage, checking....308, 309
Thread series.................293, 294
Thread-cutting tools..........309, 310
Threads (see Screw threads)
Three-wire formula........303–305, 310
Three-wire method of measuring
  threads....................302–305
Tool design problems..........279–281
Tooling costs......................485
Toolmaker's buttons, locating
  holes with................122–124
Tools, thread-cutting..........309, 310
Tooth, stub.......................324
Torque of various engines and
  motors.........................415
Torsion stresses, definition of.......402
Torus, volume and area of.........205
Trains, gear......................348
Trapezium
  area of.........................198
  definition of...................146
Trapezoid
  area of.........................198
  definition of...................146
Triangles
  angles of..............145, 154–158
  area of.........................197
  circles circumscribed about......170, 171, 194
  circles inscribed in......164, 167, 216
  constructed with base and
    sides given...............193, 194
  definition of....................145
  equal, theorems for.....155, 156, 158
  equiangular....................146
  equilateral.............145, 146, 197
  inscribed in circles, theorem for...169
  isosceles..................145, 157
  lines cutting...................159
  oblique................145, 251–262
  projection-type, problems of..162, 163
  right.....145, 160–163, 167, 216, 234, 240–246
  theorems or propositions
    relating to...............154–162
  types of..................145, 146
Trigonometric functions
  found by construction and
    measurement.................235

Trigonometric functions—*continued*
  found by graphical method...239, 240
  found by tables.............235–237
  of more than one angle......248–251
  relations between...........247–251
  and tables.................232–240
  used to solve right triangles..232, 233
Trigonometric scales of slide
  rule.......................392–394
Trigonometry.................232–291
Trigonometry, shop...........262–281
Trinomials........................73
Turning moment..................415
Twisting stresses, definition of......402

U

Ultimate strength.................404
Unit strain, definition of...........403
Unit stress
  definition of...................403
  safe working...................404
Unknown found by process of
  elimination.................82–84

V

v-block
  checking shape of..........270, 271
  used to check tap...........309, 310
v-thread
  measuring.................303, 304
  sharp......................297, 304
Velocity..........................410
Vernier, gear tooth................336
Vertex, definition of...............142
Vertical or opposite angles.........143, 149, 150
Volume of solids, areas and....201–206

W

Weights, methods and short
  cuts in...................206–211
White pine, specific gravity of......212
Whole numbers
  definition of.....................2
  multiplying of a fraction and......12
  reduction of.....................2
Wire diameter, best............303–305
Work.......................409, 410
Worm gears
  explanation of..................326
  formulas for...............343–345
  indexing of................350, 351
Worm thread.....................299

X

x-axis............................433

Y

y-axis............................433

17757
5944
―――――
11793